Workbook for
Organic Synthesis:
The Disconnection Approach

Workbook for Organic Synthesis: The Disconnection Approach

Stuart Warren
*Department of Chemistry and Churchill College,
Cambridge University*

JOHN WILEY & SONS
Chichester · New York · Brisbane · Toronto · Singapore

Copyright © 1982 by John Wiley & Sons Ltd.

Reprinted January 1985.
Reprinted September 1987.
Reprinted February 1990.
Reprinted April 1992.
All rights reserved.

No part of this book may be reproduced by any means, nor transmitted, nor translated into a machine language without the written permission of the publisher.

Library of Congress Cataloging in Publication Data:

Warren, Stuart G.
  Workbook for organic synthesis, the disconnection approach.
  Includes index.
  1. Chemistry, Organic—Synthesis.   I. Title
  QD262.W284   547'.2   81-19694

ISBN 0 471 90082 6

British Library Cataloguing in Publication Data:

Warren, Stuart
  Workbook for organic synthesis: the disconnection approach.
  1. Chemistry, Organic—Synthesis
  I. Title
  547'.2'028   QD262

ISBN 0 471 90082 6

Printed in Great Britain

Preface

This workbook accompanies the text *Organic Synthesis: The Disconnection Approach*. It contains further examples, problems, and solutions under the same chapter headings as those of the main text. I assume that you have read the corresponding chapter in the text before tackling the chapter in the workbook. You should then find examples and problems to help you understand each part of the chapter in the main text.

Within each chapter, the problems and examples are arranged either to follow the same organisation as that of the main text or to present a series graded in difficulty. Particularly easy or difficult problems are so labelled that you can avoid them if you want to. My programmed text *Designing Organic Syntheses*, Wiley, Chichester, 1978, provides another graded series of problems and solutions within a similar framework.

My chief thanks go to Denis Marrian who gave me vigorous encouragement as well as checking all the text, diagrams and references and providing index and formula index. His hard work and that of Marilyn Buck, who typed the printed words, enabled this workbook to be produced at the same time as the main text. I am deeply grateful to them both.

Cambridge 1982   *Stuart Warren*

Author's Note

Cross References: Cross references marked T, e.g. 'chapter T 20' or 'page T 145' refer to the main text. Other cross-references refer to this workbook.

Index: The text and the workbook each have their own index. At the end of this workbook there is also a formula index of all target molecules in both the text and the workbook.

Contents

Chapter 1	The Disconnection Approach	1
Chapter 2	Basic Principles: Synthesis of Aromatic Compounds	5
Chapter 3	Strategy I: The Order of Events	15
Chapter 4	One-Group C—X Disconnections	29
Chapter 5	Strategy II: Chemoselectivity	38
Chapter 6	Two-Group C—X Disconnections	45
Chapter 7	Strategy III: Reversal of Polarity, Cyclisation Reactions, Summary of Strategy	62
Chapter 8	Amine Synthesis	70
Chapter 9	Strategy IV: Protecting Groups	80
Chapter 10	One-Group C—C Disconnections I: Alcohols	89
Chapter 11	General Strategy A: Choosing a Disconnection	96
Chapter 12	Strategy V: Stereoselectivity	107
Chapter 13	One-Group C—C Disconnections II: Carbonyl Compounds	123
Chapter 14	Strategy VI: Regioselectivity	134
Chapter 15	Alkene Synthesis	145
Chapter 16	Strategy VII: Use of Acetylenes	160
Chapter 17	Two-Group Disconnections I: Diels–Alder Reactions	175
Chapter 18	Revision Examples and Problems	188
Chapter 19	Two-Group Disconnections II: 1,3-Difunctionalised Compounds and α,β-Unsaturated Carbonyl Compounds	200
Chapter 20	Strategy IX: Control in Carbonyl Condensations	209
Chapter 21	Two-Group Disconnections III: 1,5-Difunctionalised Compounds: Michael Addition and Robinson Annelation	229
Chapter 22	Strategy X: Use of Aliphatic Nitro Compounds in Synthesis	241
Chapter 23	Two-Group Disconnections IV: 1,2-Difunctionalised Compounds	252
Chapter 24	Strategy XI: Radical Reactions in Synthesis: FGA and its Reverse	267
Chapter 25	Two-Group Disconnections V: 1,4-Difunctionalised Compounds	284

Chapter 26	Strategy XII: Reconnections: Synthesis of 1,2- and 1,4-Difunctionalised Compounds by C=C Cleavage	299
Chapter 27	Two-Group Disconnections VI: 1,6-Difunctionalised Compounds	311
Chapter 28	General Strategy B: Strategy of Carbonyl Disconnections	323
Chapter 29	Strategy XIII: Introduction to Ring Synthesis: Saturated Heterocycles	335
Chapter 30	Three-Membered Rings	353
Chapter 31	Strategy XIV: Rearrangements in Synthesis	368
Chapter 32	Four-Membered Rings: Photochemistry in Synthesis	376
Chapter 33	Strategy XV: Use of Ketenes in Synthesis	389
Chapter 34	Five-Membered Rings	397
Chapter 35	Strategy XVI: Pericyclic Rearrangements in Synthesis: Special Methods for Five-Membered Rings	407
Chapter 36	Six-Membered Rings	417
Chapter 37	General Strategy C: Strategy of Ring Synthesis	429
Chapter 38	Strategy XVII: Stereoselectivity B	440
Chapter 39	Aromatic Heterocycles	452
Chapter 40	General Strategy D: Advanced Strategy	470
References		491
Formula Index		511
Index		527

CHAPTER 1

The Disconnection Approach

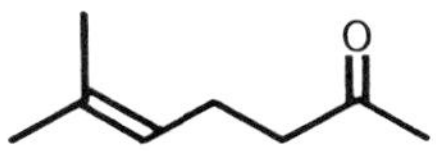

(1)

Two syntheses of ketone (1) are described in Chapter T1. The starting materials for the two syntheses are different: (2), (3), and (4) were used for the industrial synthesis, (5) and (6) for the laboratory synthesis. Analysis shows that the same bond (a) was made in both syntheses.

Industrial Synthesis

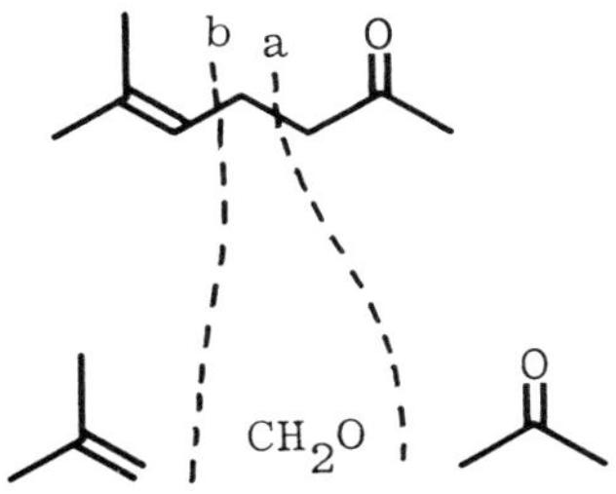

(2)   (3)   (4)

Laboratory Synthesis

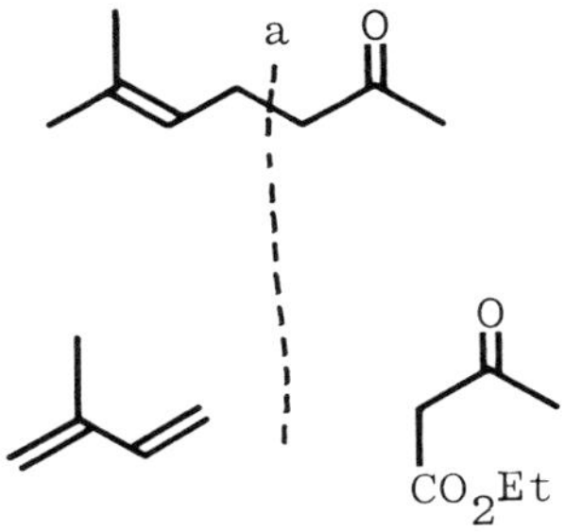

(5)   (6)

If you can analyse published syntheses like this, you will increase your understanding of good places to make good disconnections.

Problem

Two syntheses of multistriatin follow. For these and for the synthesis in the main text (p T 3) draw diagrams like those above to show which parts of multistriatin came from which starting materials in each synthesis. Do not be too concerned about the details of each step in the syntheses.

Synthesis[1] 2

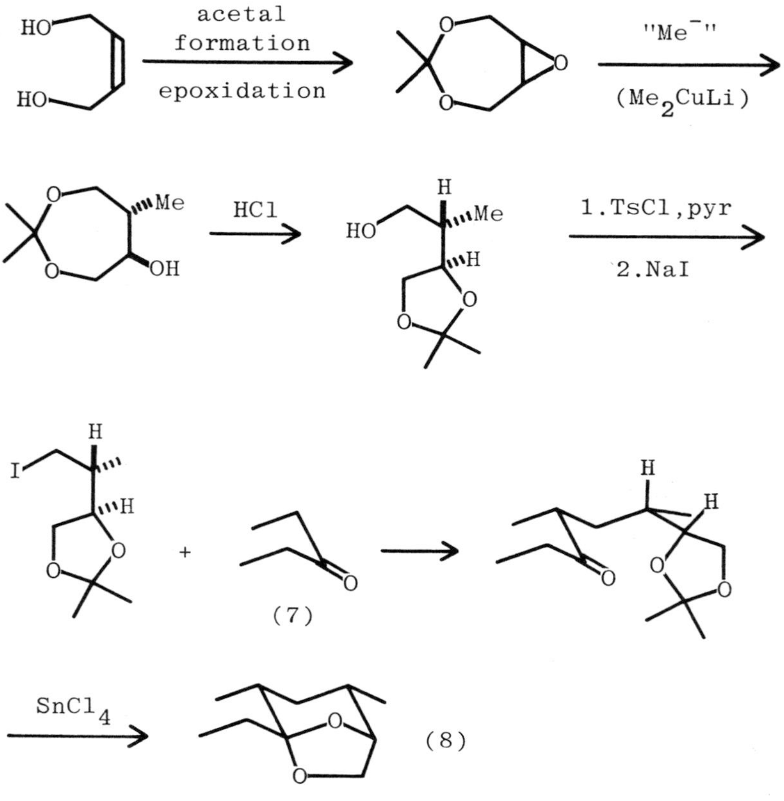

Synthesis[2] 3

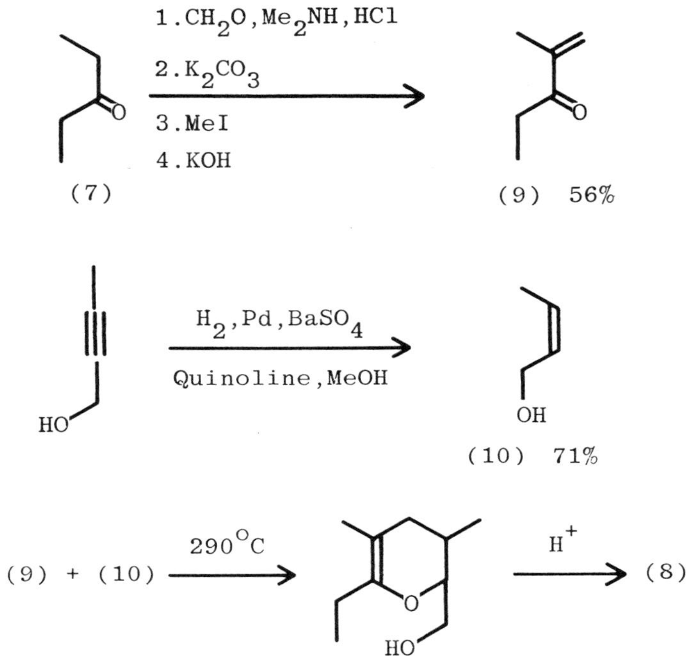

```
                1. CH₂O, Me₂NH, HCl
                2. K₂CO₃
(7)    ———————————————————→    (9)   56%
                3. MeI
                4. KOH
```

```
              H₂, Pd, BaSO₄
              ——————————————→
              Quinoline, MeOH
HO⌇⌇⌇                              (10)   71%
                                   OH
```

(9) + (10) — 290°C → [pyran intermediate] — H⁺ → (8)

Answer In each case, simply draw the structure (8) and trace which atoms came from which original starting materials.

Synthesis 1

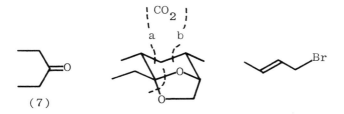

Synthesis 2

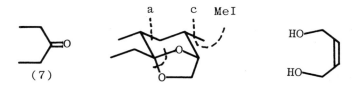

Bond (a) is made in all three syntheses and bond (b) in two of them. We shall see later that the symmetry of ketone (7) is a reason for (a) and the branchpoints in the molecule a reason for (a) and (b).

Synthesis 3

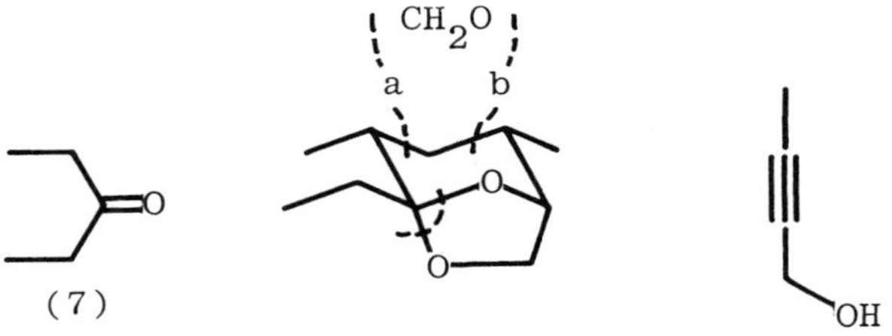

CHAPTER 2

Basic Principles: Synthesis of Aromatic Compounds

Disconnection and FG1

Example : Steps in the synthesis of multistriatin given on page T 3 correspond to disconnections or FG1s as follows:-

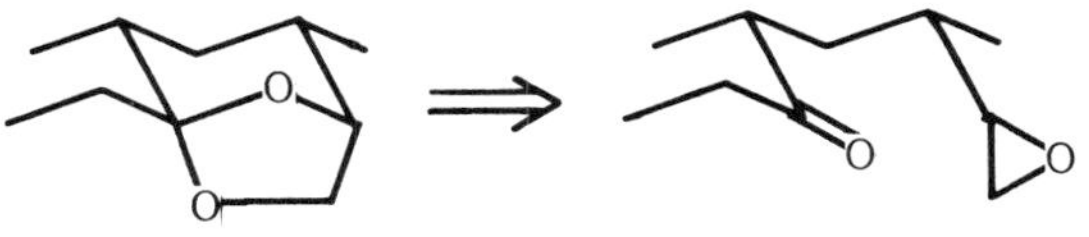

This is a rearrangement and is best described as an FG1 since no new C-C bonds are formed and all carbon atoms retain the same number of bonds to oxygen atoms.

⇒ [structure] + RCO_3H Could be described as two C-O disconnections or as FG1 (olefin to epoxide by oxidation).

⇒ [structure] + [OTs structure] Definitely the major disconnection — a new C-C bond is formed.

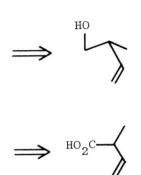

Could be described as a C-O disconnection or as an FG1-conversion of OH to OTs by substitution.

FG1: CO_2H converted to CH_2OH by reduction with $LiAlH_4$.

It may surprise you that there is not always a cut-and-dried answer to whether each step corresponds to FG1 or disconnection, but this is so, particularly when C-X (X = O, N, S etc.) bonds are to be formed. Discussing the type of logic involved in each step helps you to understand the logic of the synthesis : both aspects (FG1 and disconnection) are helpful - choose whichever seems the more helpful in the circumstances.

Problem :

Repeat this exercise for synthesis 2 on page W 2. Give each FG1 a full description (oxidation etc.) and label each disconnection C-O etc. according to the bond formed.

Answer :

As before - a rearrangement best described as FG1.

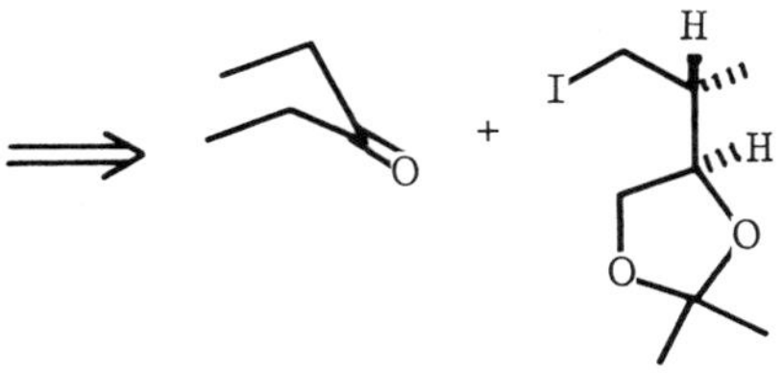

⟹ [ketone] + [iodide with acetonide]   The major C-C disconnection in the synthesis.

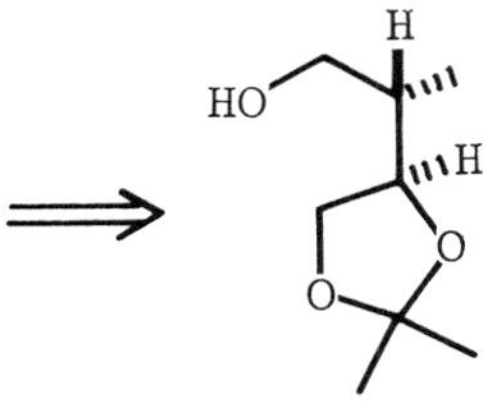

⟹ [primary alcohol with acetonide] (via OTs)   C-I disconnection or two FG1s (alcohol to tosylate to iodide by substitutions).

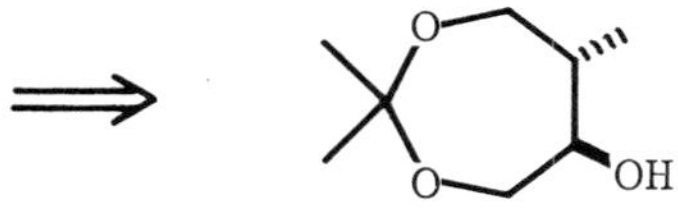

⟹ [7-membered acetal ring with OH]   Rearrangement : helpful to call this FG1.

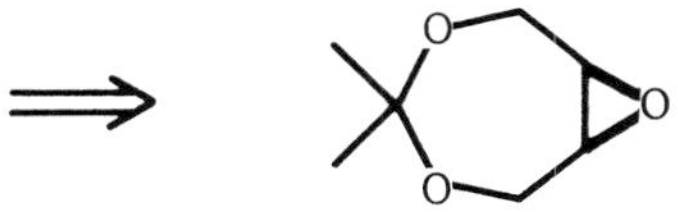

⟹ [acetal-fused epoxide] + Me$_2$CuLi   A C-C disconnection.

⟹ [acetal-containing cycloalkene]   Two C-O disconnections or FG1 (olefin to epoxide by oxidation).

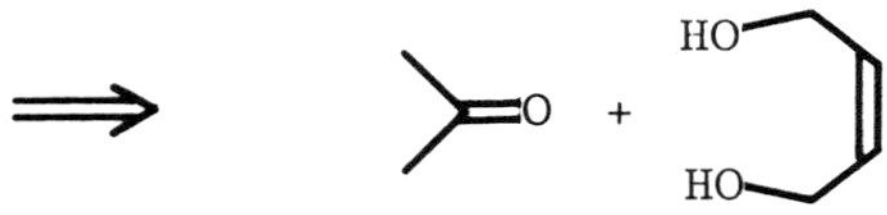

⟹ [acetone] + [cis-2-butene-1,4-diol]   Two C-O disconnections is the most helpful description.

Aromatic Electrophilic Substitution

Examples :

DDT (1) is a very useful insecticide but builds up in the environment with unacceptable effects on wild life. A biodegradable version of DDT such as (2) might solve these problems.[3]

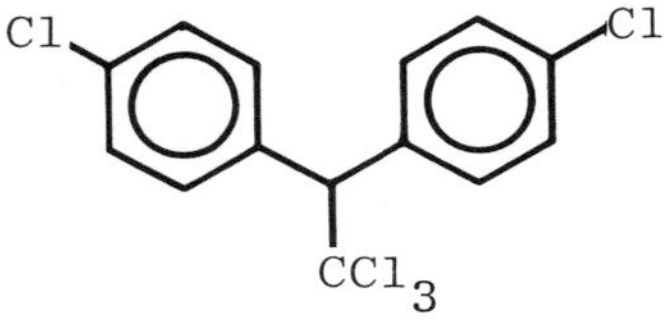

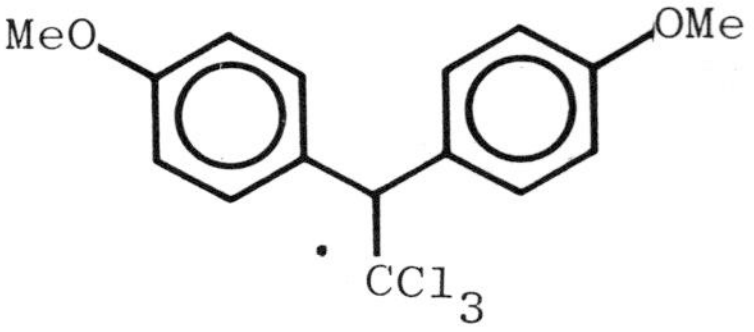

Disconnection of (2) at the bonds joining the aromatic rings to the aliphatic part of the molecule will need a reagent for the doubly charged synthon (3).

Analysis

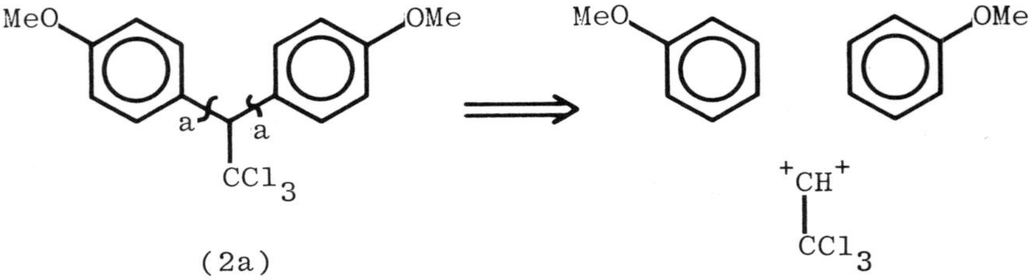

A dihalide will not do as it would be unreactive towards substitution and we already have three halogen atoms in (3). The answer is to use the aldehyde (4) - chloral - in acid solution. One addition gives the alcohol (5) which dehydrates and reacts again *via* a carbonium ion. In practice (5) need not be isolated.

Synthesis[4]

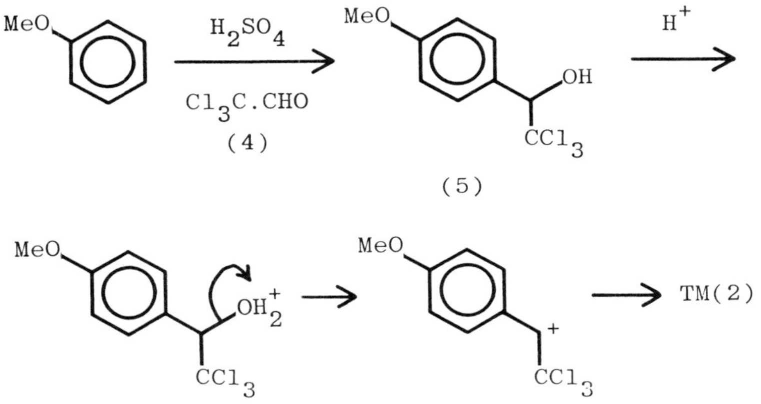

MeO—C₆H₅ $\xrightarrow[\text{Cl}_3\text{C.CHO}]{\text{H}_2\text{SO}_4}$ MeO—C₆H₄—CH(OH)(CCl₃) $\xrightarrow{\text{H}^+}$
(4)   (5)

MeO—C₆H₄—CH(OH₂⁺)(CCl₃) → MeO—C₆H₄—CH⁺(CCl₃) ⟶ TM(2)

As more and more waste products are re-cycled, methods of extracting contaminants from them are needed. Reagent (6) is used to remove boron compounds from alkaline brine wastes.[5]

Analysis

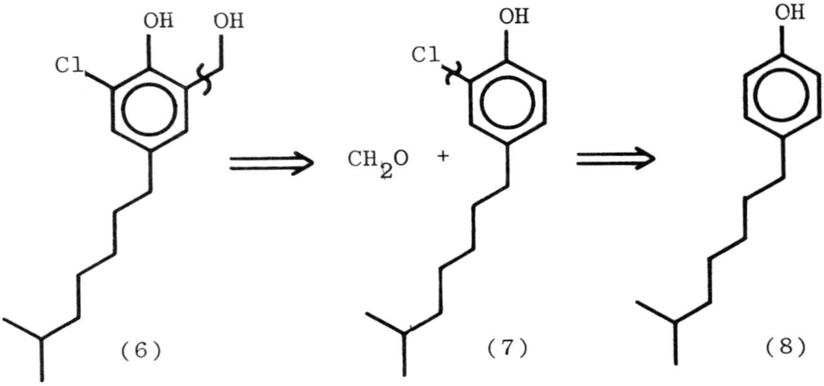

[Structures (6), (7), (8) showing chlorophenol analysis with CH_2O]

Removal of the one-carbon electrophilic formaldehyde (cf p T 10) leaves a chlorophenol (7) made by direct chlorination of (8). The long allyl chain of (8) cannot be put in by Friedel-Crafts alkylation or rearrangement will occur (p T 9) so acylation and reduction are preferred.

Synthesis[5]

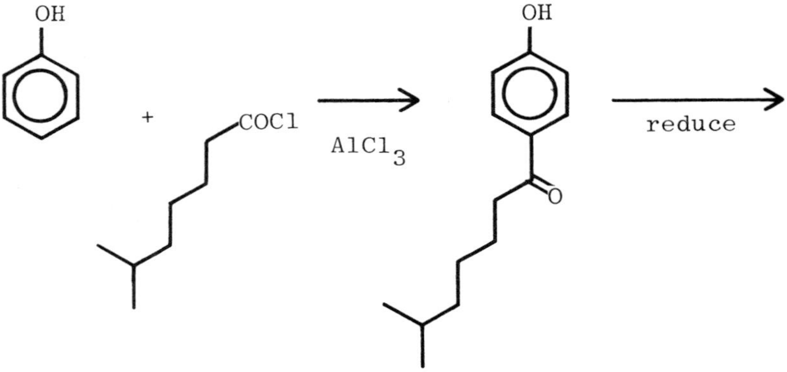

[Phenol + RCOCl, $AlCl_3$ → ketone → reduce]

$$(8) \xrightarrow[90^\circ C]{Cl_2} (7) \xrightarrow[HO^-]{CH_2O} TM(6)$$

Problems :

1. H. C. Brown[6] used ester (9) to test a new reducing agent which did indeed form alcohol (10). How might he have made the ester (9)?

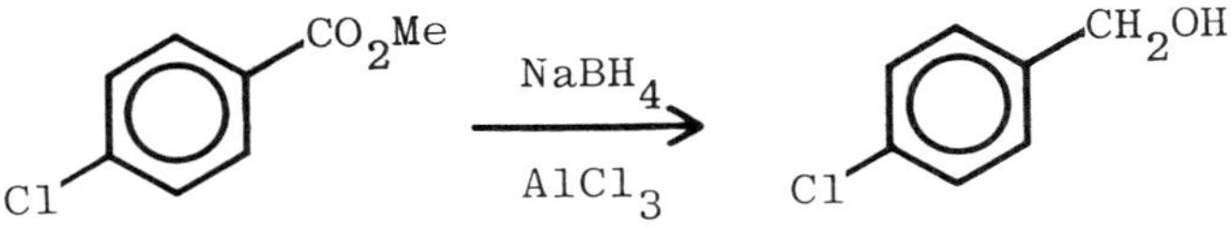

(9) 4-Cl-C$_6$H$_4$-CO$_2$Me $\xrightarrow{\text{NaBH}_4, \text{AlCl}_3}$ (10) 4-Cl-C$_6$H$_4$-CH$_2$OH 84%

Answer :

Ester (9) can easily be made from acid (11). You might consider two approaches to this : a one-carbon electrophile addition via chloromethylation (Table T 2.2) and oxidation or FGI (Table 2.3) back to p-chlorotoluene (12). The latter is easier on a large scale. The p-chlorotoluene (12) can be made either by direct chlorination of toluene or by the diazotisation route (p T 12) again from toluene.

Analysis

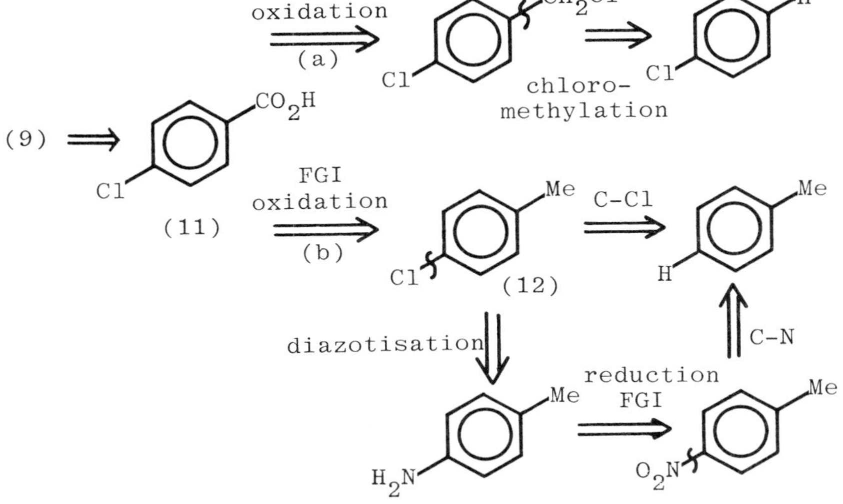

(9) $\Rightarrow$ 4-Cl-C$_6$H$_4$-CO$_2$H (11)

(a) FGI oxidation $\Rightarrow$ 4-Cl-C$_6$H$_4$-CH$_2$Cl $\xrightarrow{\text{chloromethylation}}$ 4-Cl-C$_6$H$_5$

(b) FGI oxidation $\Rightarrow$ 4-Cl-C$_6$H$_4$-Me (12) $\xrightarrow{\text{C-Cl}}$ 4-H-C$_6$H$_4$-Me

(12) $\xrightarrow{\text{diazotisation}}$ 4-H$_2$N-C$_6$H$_4$-Me $\xrightarrow{\text{reduction FGI}}$ 4-O$_2$N-C$_6$H$_4$-Me $\xrightarrow{\text{C-N}}$ 4-H-C$_6$H$_4$-Me

Chlorination with a Lewis acid catalyst[7] at low temperature avoids chlorination in the methyl group but the diazotisation route,[8] though longer, is perhaps easier.*

Synthesis[8]

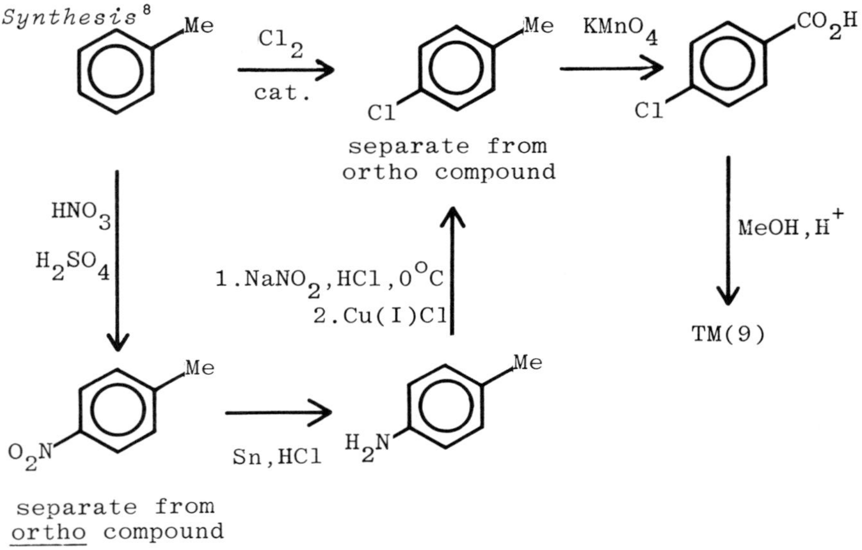

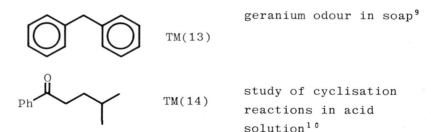

separate from ortho compound

Simple Problems:

For practice, especially for beginners. How would you make TMs (13) - (15), which were needed for the reasons given on the right?

TM(13) geranium odour in soap[9]

TM(14) study of cyclisation reactions in acid solution[10]

* In fact, all these compounds, even TM (9), are available commercially.

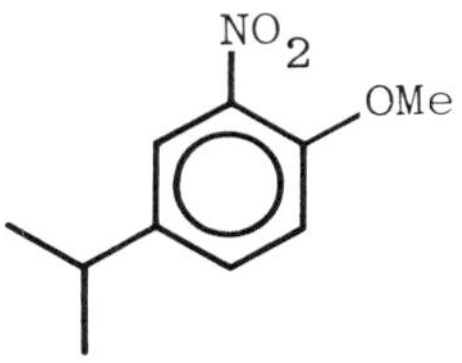

TM(15)

exploration of
industrial uses of HF[11]

Solutions to Simple Problems :

TM (13) : This symmetrical molecule offers only one disconnection (a, below) corresponding to a Friedel-Crafts alkylation (p T 8).

Analysis

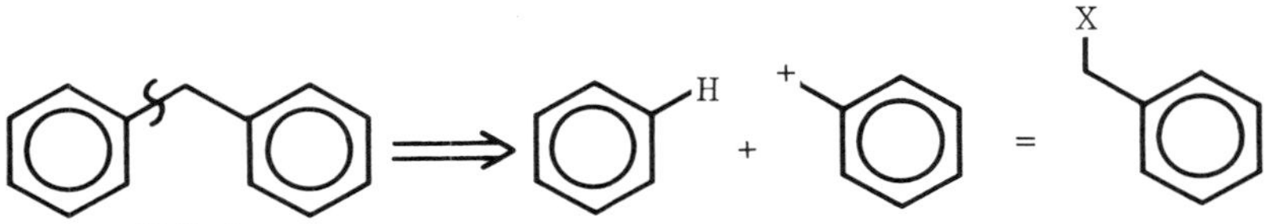

(13a)

Synthesis[12]

$$PhH + PhCH_2Cl \xrightarrow{AlCl_3} TM(13)$$

TM (14) : This time the usual disconnection corresponds to a Friedel-Crafts acylation (p T 7).

Analysis

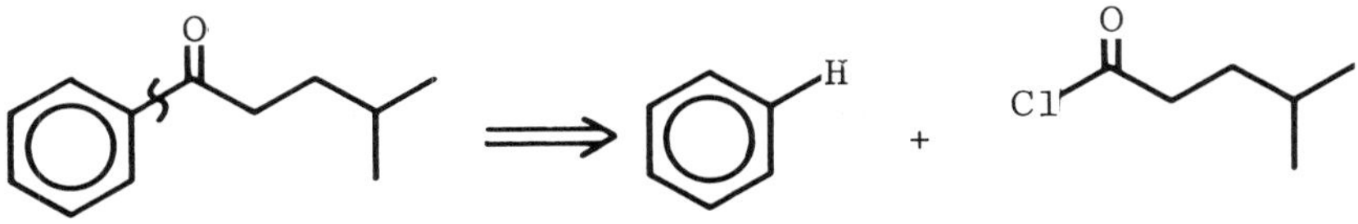

(14a)

Synthesis[13]

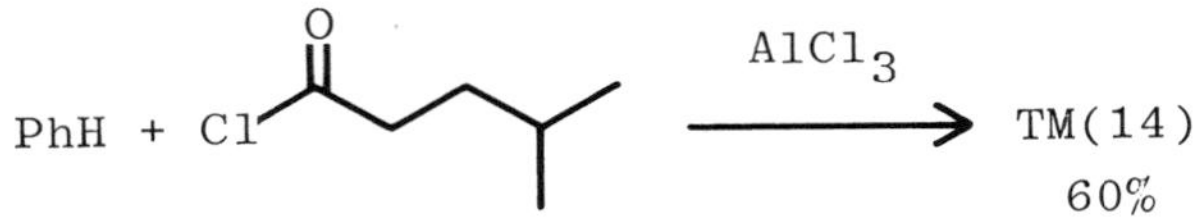

PhH + Cl-C(=O)-CH₂-CH₂-CH(CH₃)₂ →(AlCl₃) TM(14) 60%

TM (15) : Disconnect the nitro group and the alkyl group in either order.

Analysis

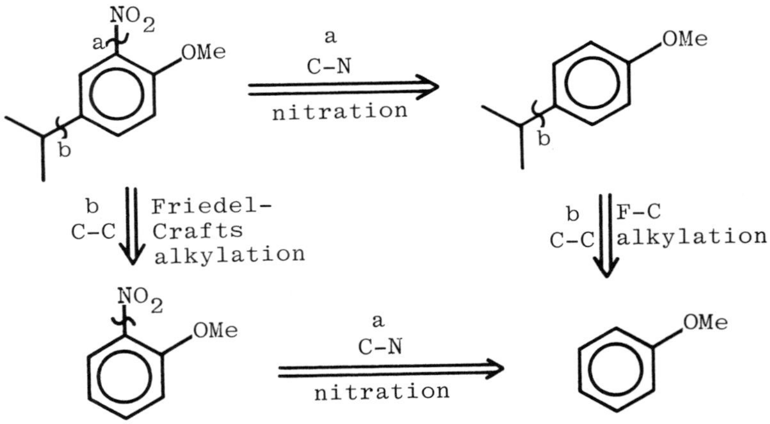

[Scheme showing analysis: 2-nitro-4-isopropylanisole disconnected via route (a) C-N nitration to give 4-isopropylanisole, or via route (b) Friedel-Crafts alkylation to give 2-nitroanisole; 4-isopropylanisole then undergoes F-C alkylation disconnection (b) to anisole; 2-nitroanisole undergoes C-N nitration disconnection (a) to anisole.]

TM (15) has been made by nitration followed by Friedel-Crafts alkylation with HF as catalyst. No doubt the alternative route would be successful too.

Synthesis

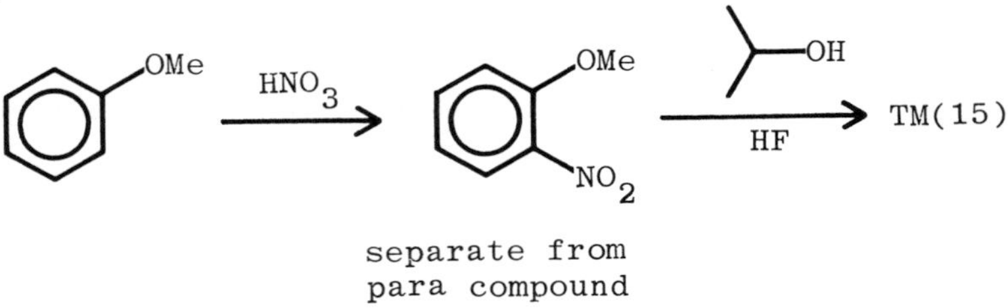

anisole (PhOMe) →(HNO₃) 2-nitroanisole (separate from para compound) →(iPrOH, HF) TM(15)

CHAPTER 3

Strategy I: The Order of Events

General Strategy Examples (see table of guidelines)

Now that you appreciate the reasons why a particular order of events is followed, study the syntheses of benzocaine (p T 6) and piperonal (p T 9). The benzocaine synthesis involves a great deal of FGI : two of the four steps are FGIs. This is because two FGs required for the synthesis (NO_2 and CO_2H) are m-directing and so cannot be used to direct substitution where it is needed (p). (guideline 1 in Chapter T 3). Nitro is the group to put in last (guideline 2) and FGI solves the orientation problem (guideline 3).

The piperonal synthesis (p T 9) is an example where the two *ortho* oxygen substituents would be very difficult to set up. We therefore (guideline 7) start with available catechol (o-dihydroxybenzene).

Table 3.1

Guidelines from Chapter T 3

1. Examine the relationship between the groups, looking for groups which direct to the right position.

2. If there is a choice, disconnect *first* (that is add *last*) the most electron-withdrawing substituent.

3. If FG1 is needed during the synthesis, it may well alter the directing effect of the group and other substituents may be added before or after the FG1.

4. Many groups can be added by *nucleophilic* substitution on a diazonium salt, made from an amine. Adding other groups at the amine stage may be advisable as the amino group is strongly o,p-directing.

5. As a last resort, there is a trick to solve some difficult problems, such as adding two o,p-directing groups, m is one another. A 'dummy' amino group is added, used to set up the required relationship, and then removed by diazotisation and reduction.

6. Look for substituents which are difficult to add. It is often good strategy not to disconnect these at all, but to use a starting material containing the substituents.

7. Look for a *combination* of substituents present in the TM and in a readily available starting material.

8. Avoid sequences which may lead to unwanted reactions at other sites in the molecule.

9. If o,p-substitution is involved, one strategy may avoid separation of isomers in that the other position becomes blocked.

Guidelines may well contradict one another - use your judgement!

General Strategy Problems

1. Analyse the syntheses of (a) BHT (page T 8) and (b) phenol (1) (p T 12) and see which guidelines they follow.

   (1) — phenol with OH, Me, Br substituents

   Answer (a) 7 and 9 in particular
   (b) 3, 4, and 9 in particular.

2. In the trifluralin synthesis (p T 13), what is wrong with this order events, suggested by guidelines 2 and 9?

   (2)

   Answer
   - nucleophilic substitution no longer possible
   - CF_3 m-directing, so difficult to make (2) (guideline 1)
   - nitration might oxidise amine (guideline 8).

Further Worked Examples

1. In 1979, workers investigating[14] the dienone-phenol rearrangement treated (3) with acetic anhydride and got an unknown compound believed to be (4). Treatment with strong acid gave (5), they supposed, but they wanted to synthesise an authentic sample of (5) to check their assignment. This is a tricky problem

as none of the substituents can be easily disconnected.

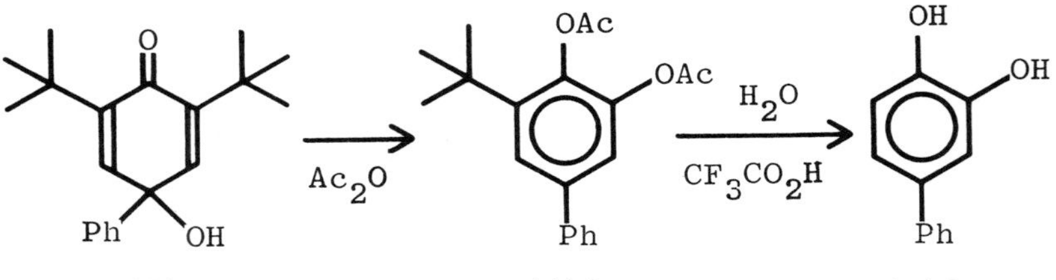

(3) →[Ac$_2$O] (4)? →[H$_2$O, CF$_3$CO$_2$H] (5)?

We really would prefer to keep the two OH groups and start from catechol (guideline 7). This forces us to disconnection (3a). We require a reagent for synthon (6) and the obvious choice is the diazonium salt (7).

Analysis

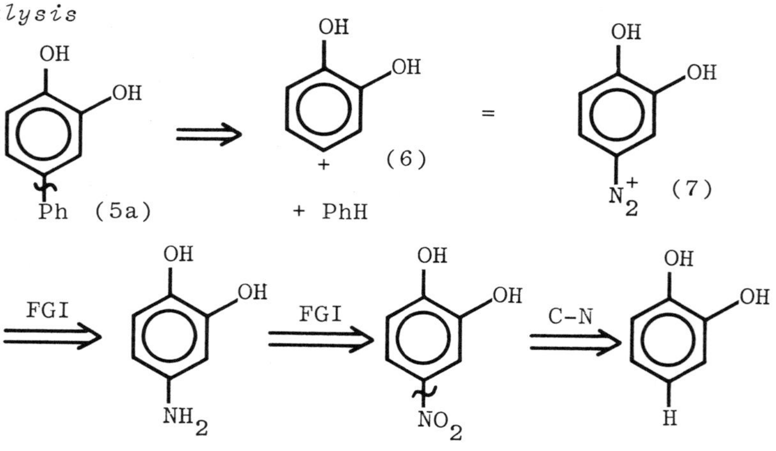

(5a) ⇒ (6) + PhH = (7)

⇒[FGI] (NH$_2$ intermediate) ⇒[FGI] (NO$_2$ intermediate) ⇒[C-N] (H intermediate)

The OH groups must be protected against oxidation during the nitration step and the methyl ether (8) is a convenient starting material. Treatment with HBr at the end of the synthesis, removes the methyl groups.

Synthesis[14]

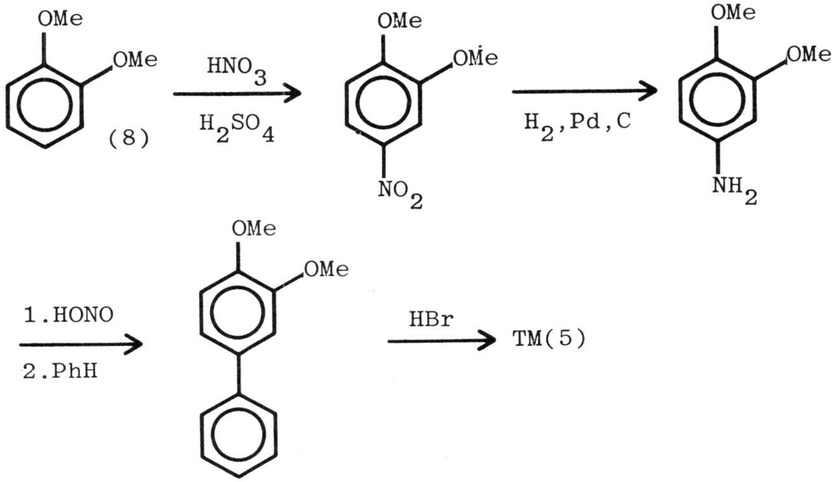

(8) →[HNO₃, H₂SO₄]→ (nitro-veratrole) →[H₂, Pd, C]→ (amino-veratrole)

→[1. HONO; 2. PhH]→ (3,4-dimethoxybiphenyl) →[HBr]→ TM(5)

The product of this synthesis proved to be identical with the unknown compound from the dienone-phenol sequence and the structures are confirmed.

Simple Problems

Suggest syntheses for TMs (9) and (10) needed as intermediates: TM (9) in the synthesis of brominated hydroxybenzoic acids[15] and TM (10) in the synthesis of model compounds for studying biological mechanisms of ester hydrolysis.[16]

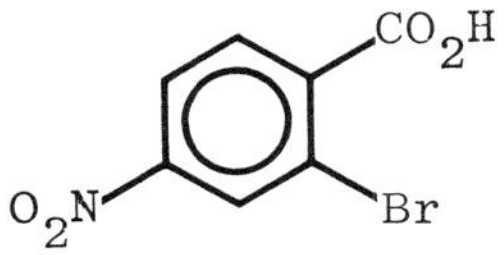

(9) 4-nitro-2-bromobenzoic acid (O_2N, CO_2H, Br)

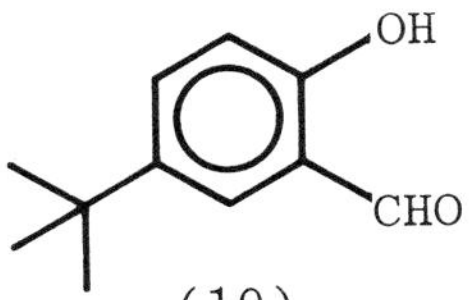

(10) 5-tert-butyl-2-hydroxybenzaldehyde (OH, CHO, t-Bu)

Solutions to Simple Problems

TM (9) With two electron-withdrawing groups, FG1 is

essential to get some control over orientation (guideline 3). The easiest is CO_2H to CH_3. Then disconnection of either Br or NO_2 is possible.

Analysis

[Structure (9): benzene ring with CO_2H, Br (ortho), and O_2N (para to CO_2H)]

$\xrightarrow{\text{FGI oxidation}}$

[Structure (11): benzene ring with CH_3, Br (ortho), and O_2N, with disconnections labeled a (C-N) and b (C-Br)]

a ↙ C-N   b ↓ C-Br

[Structure (12): o-bromotoluene]   [Structure (13): p-nitrotoluene]

The nitration of (12) would probably give (11), but bromination at one of the two identical reactive positions of (13) is the logical route (guideline 9). Nitric acid oxidises the methyl group to CO_2H.

Synthesis[15,17]

[Toluene] $\xrightarrow[H_2SO_4]{HNO_3}$ [(13) p-nitrotoluene] separate from ortho isomer $\xrightarrow[\text{Fe, }100°C]{Br_2}$ (11) 70%

(11) $\xrightarrow[\text{Hg, }H_2O]{HNO_3}$ TM(9) 76%

An alternative synthesis of (11) relies on the partial reduction of m-di-nitro groups (p T 38) but the yield is poor.

Synthesis 2[18]

O_2N-C$_6$H$_3$(Me)(NO$_2$) $\xrightarrow{SnCl_2, HCl}$ O_2N-C$_6$H$_3$(Me)(NH$_2$) (20%) $\xrightarrow{1. NaNO_2, HBr, 0°C \quad 2. Cu(I)Br}$ (11) 86%

TM (10) We would prefer to keep the OH group (guideline 6). Either CHO (Reimer-Tiemann, table T 2.1) or t-Bu (Friedel-Crafts) may then be disconnected as the orientation is correct for either (o, p- to the most electron-donating group -OH) (guideline 1). Guidelines 2 and 9 suggest disconnecting CHO first.

Analysis

t-Bu-C$_6$H$_3$(OH)(CHO) (10a) $\xRightarrow{Reimer-Tiemann}$ t-Bu-C$_6$H$_4$-OH $\xRightarrow{Friedel-Crafts}$ C$_6$H$_5$-OH

Synthesis[16,19]

C$_6$H$_5$OH $\xrightarrow{HF, \; t\text{-BuCl}}$ t-Bu-C$_6$H$_4$-OH (85%) $\xrightarrow{CHCl_3, NaOH}$ TM(10)

More Advanced Example

2. The tetracyclines are important antibiotics often effective against organisms showing penicillin resistance. In 1980, a synthesis of some sulphur containing tetracyclines[20] required thiol (14) as an intermediate. The SH group could be introduced by nucleophilic displacement of a diazonium salt (guideline 4) so amine (15) is an essential intermediate.

Analysis 1

[Structures: (14) 2-chloro-4-methoxy-benzenethiol ⟹ 2-chloro-4-methoxy-benzenediazonium ⟹ (15) 2-chloro-4-methoxy-aniline]

We would rather leave the OMe group alone (guideline 6), the amino group will be added via nitration and reduction which gives us some flexibility of orientation (guideline 3), and the chlorine can be added by direct chlorination or by diazonium displacement. The most obvious disconnection is to remove the chlorine. Unfortunately chlorination of the very electron-rich amine (16) oxidises it to black tars : it would in any case give a mixture of isomers as all positions in the ring in (16) are activated.

Analysis 2

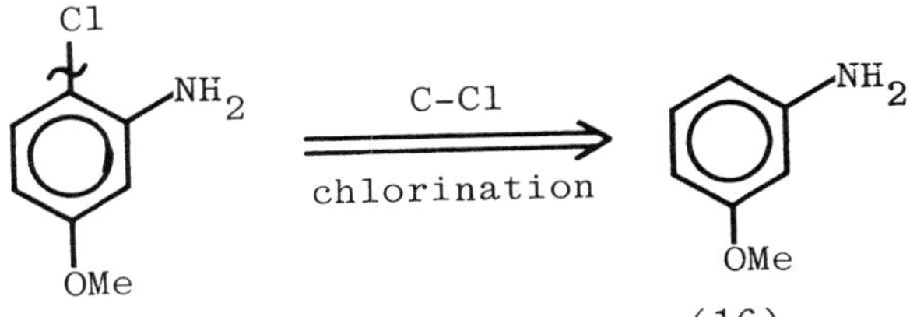

(15a) →[C-Cl chlorination] (16)

One other possible short cut is to disconnect the amino group, hoping to add it by a benzyne mechanism from easily made (17). In fact, this reaction gives a mixture of three isomers of (15) in which (15) does not predominate.[21]

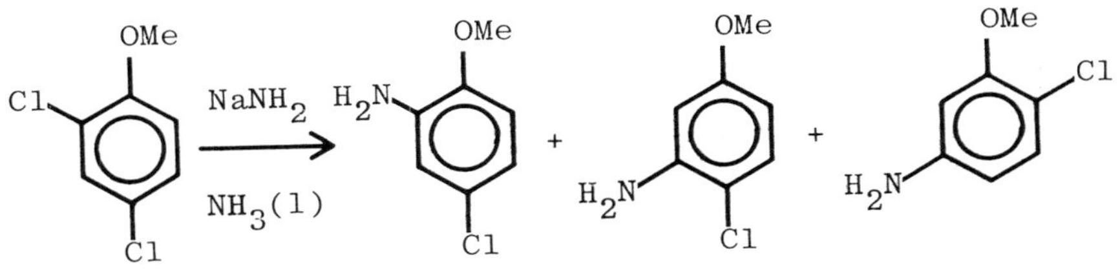

(17) →[NaNH$_2$, NH$_3$(l)] (15) + isomer + isomer

We must therefore invoke FGI and convert NH$_2$ into NO$_2$. This is not immediately of any great help as neither of the obvious disconnections (18a and b) is any good. In both reactions, the strongly electron-donating OMe group will dominate and reaction will occur *ortho* to it[21,22] (b) or give a mixture (a).

Analysis 3

(15) $\xrightarrow[\text{reduction}]{\text{FGI}}$ [structure (18): benzene with Cl (position a), NO$_2$ (position b), and OMe]

(a) ⟹ [benzene with NO$_2$ and OMe]

(b) ⟹ [benzene with Cl and OMe]

The chlorine can be disconnected with the alternative polarity via the diazonium salt to amine (19) (guideline 4). The amino group is *more* powerfully electron-donating than OMe so we can disconnect the NO$_2$ group. The amino group is itself derived from another nitro group.

Analysis 4

(18) [Cl, NO$_2$, OMe benzene] $\xrightarrow[\text{diazonium salt}]{\text{C-Cl}}$ (19) [NH$_2$, NO$_2$, OMe benzene] $\xrightarrow[\text{nitration}]{\text{C-N}}$ (20) [NH$_2$, OMe benzene]

$\xrightarrow[\text{reduction}]{\text{FGI}}$ [NO$_2$, OMe benzene] $\xrightarrow[\text{nitration}]{\text{C-N}}$ [OMe benzene]

It will be wise to protect the amino group of (20) by acetylation to prevent oxidation during the nitration step (cf p T 12).

Synthesis[22,23]

PhOMe →(HNO₃)→ 4-NO₂-C₆H₄-OMe →(H₂, Pd, C)→ 4-NH₂-C₆H₄-OMe (20) →(1. Ac₂O; 2. 11% HNO₃)→

2-NO₂-4-OMe-C₆H₃-NHAc →(10% KOH, EtOH)→ (19) →(1. HNO₂; 2. Cu(I)Cl)→ (18) →(H₂, Ni)→ TM(15) 81%

More Advanced Problems

1. Polyurethanes are often polymerised onto rigid diamines. Compound (21) was needed for a new polyurethane.[24] Suggest a synthesis for it.

3,5-dinitrotoluene (21)

Answer The nitro groups must be put in by nitration but the orientation is wrong. The methyl group cannot be disconnected as a Friedel-Crafts alkylation would never work on the available but extremely unreactive m-dinitro benzene. The solution (guideline 5) is to introduce a dummy amino group in such a position that it can activate the required positions. The best place is *p* to the methyl group.

Analysis

(21) $\xrightarrow{\text{add NH}_2 \text{ group}}$ [2,6-dinitro-4-methylaniline] $\xrightarrow{\text{C-N nitration}}$ [4-methylaniline] $\Rightarrow$ chapter T2

A protective acetyl group is again recommended during the nitration (cf p 25 and p T 12).

Synthesis[24]

chapter T2 $\longrightarrow$ (22) [4-methyl-NHCOMe benzene] $\xrightarrow{90\% \text{ HNO}_3, 20°C}$ [2,6-dinitro-4-methyl-NHAc benzene]

$\xrightarrow{50\% \text{ H}_2\text{SO}_4}$ [2,6-dinitro-4-methylaniline] 80% from (22) $\xrightarrow{\text{NaNO}_2, \text{H}_2\text{SO}_4, \text{EtOH}}$ TM(21)

2. *Bayluscide*

The dreadful tropical disease bilharzia is carried by water snails and one way to tackle the disease is to kill the snails with a moluscicide harmless to mammals and fish. Bayluscide (25) is made by Baeyer[25] for this purpose. It is an amide, made by condensing acid (23) and amine (24). Can you suggest syntheses for these two compounds?

Bayluscide : *Synthesis*

Reagents: PCl_3, xylene, $150°C$

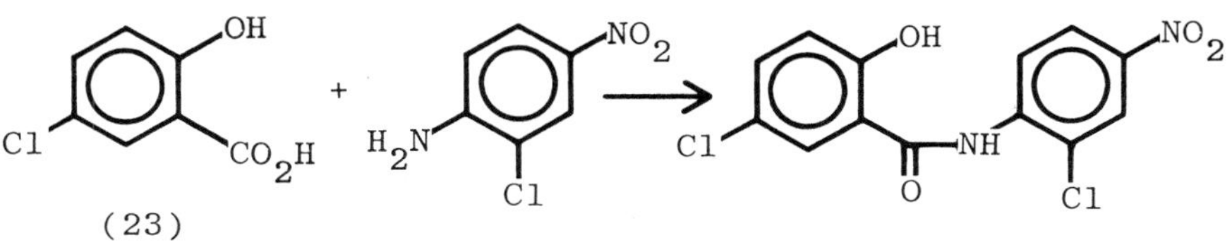

[Structure (23): salicylic acid with Cl; + Structure (24): 4-nitro-2-chloroaniline → Structure (25): the amide product]

Answer
Acid (23)

The combination of OH and CO_2H next to each other suggests salicylic acid (26) as a starting material (guideline 7). Chlorination should occur o, p- to the activating OH group and p- should predominate for steric reasons. This turns out to be correct.

Analysis

[(23) ⇒ (26) salicylic acid]

Synthesis[26]

$$(26) \xrightarrow{Cl_2} TM(23)$$

Amine (24)

Disconnection of chloro or nitro groups is satisfactory as the amino group will direct very strongly o, p. Guideline 2 suggests route (b), guideline 9 route (c). The published synthesis uses route (b).

Analysis

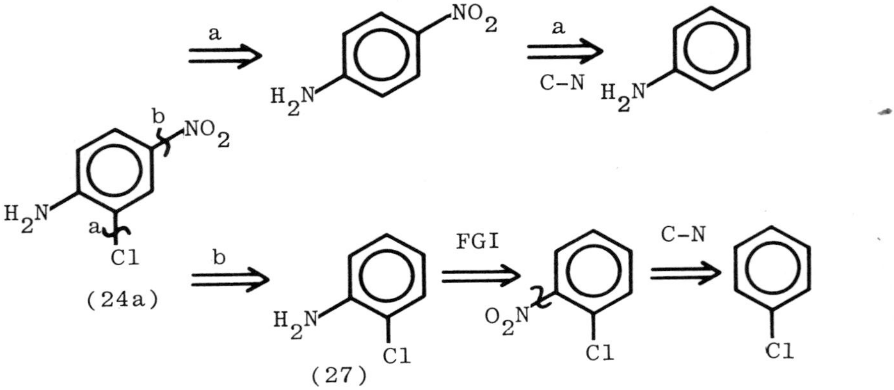

o-Chloroaniline (27) is available nowadays but it can be made from the nitro compound by reduction.[26] Protection during the nitration step was by a 'tosyl' (toluene-p-sulphonamide) derivative.[27]

Synthesis[27,28]

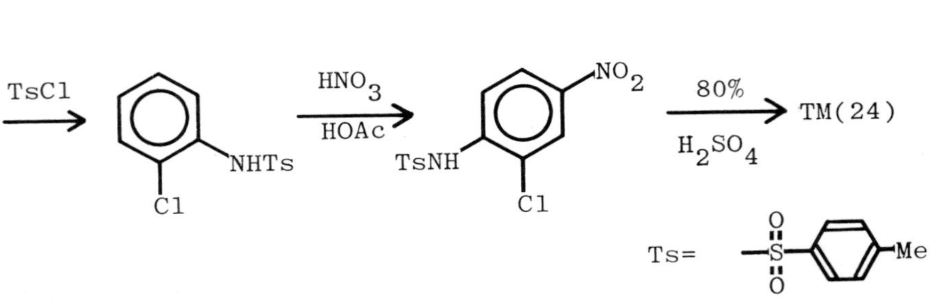

CHAPTER 4

One-Group C—X Disconnections

One-Group C-X Disconnections

Simple Examples

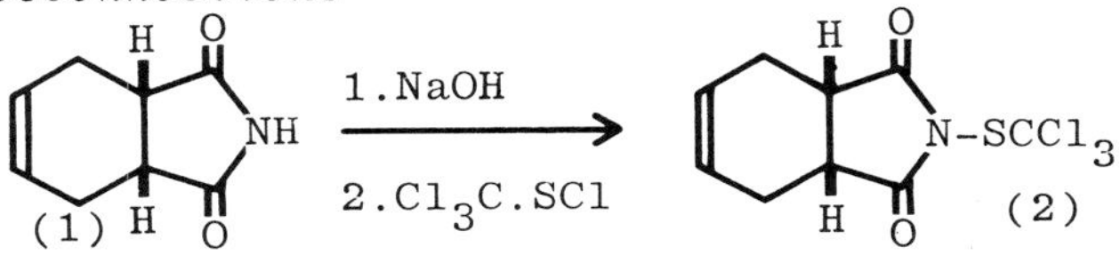

(1) —[1. NaOH; 2. Cl₃C.SCl]→ (2)

The garden fungicide 'Captan' (2) is made from (1). This is an imide - a double amide - and so *both* bonds between carbonyl and nitrogen can be disconnected.

Analysis

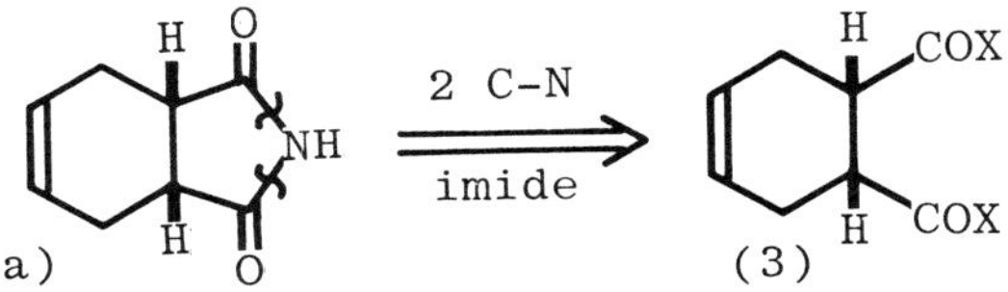

(1a) ⟹[2 C-N imide] (3)

The best reagent for the double acyl derivative (3) is the anhydride (4) whose synthesis is discussed in Chapter T 17. Reaction of (4) with ammonia gives TM (1).

Synthesis[29]

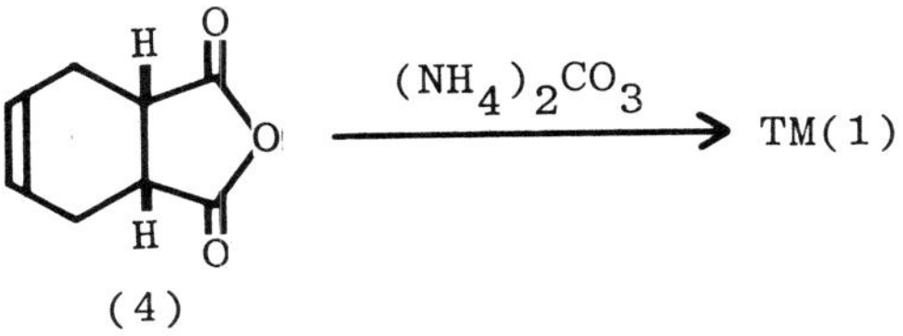

(4) —[(NH₄)₂CO₃]→ TM(1)

Esters

In making esters, either the alcohol or acid may be in short supply. Precious acids are often treated with a large excess of the alcohol (e.g. as solvent) with mineral acid catalysis.

$$RCO_2H \xrightarrow[H^+]{MeOH} RCO_2Me$$

The potential analgesic (5) is an ester of the alcohol (6), which has to be made (see p T 77), and acid $EtCO_2H$. The chosen reagent was the anhydride which wastes half of the cheap acid but ensures efficient conversion of (6) into (5).

Analysis

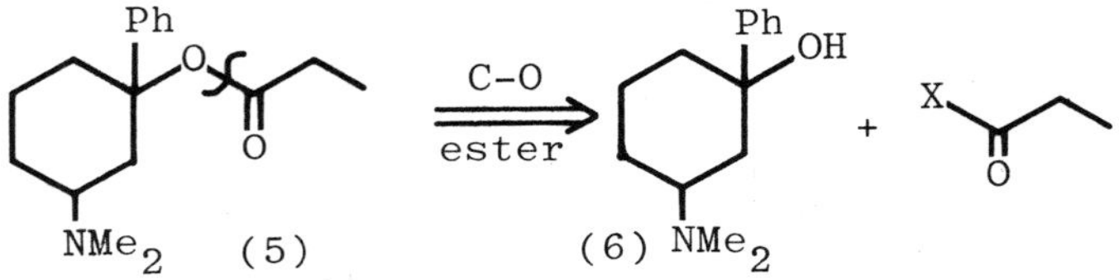

$$\text{(5)} \xRightarrow{\text{C-O ester}} \text{(6)} + X\text{-CO-Et}$$

Synthesis[30(a)]

$$(6) + \text{excess } (EtCO)_2O \xrightarrow{pyr} TM(5)$$

Simple Problems

1. Mark all the C-X bonds you would wish to disconnect in these molecules and say why.

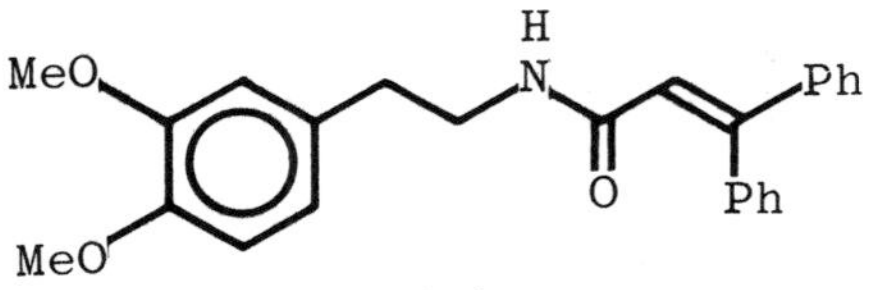

(7)

Intermediate in isoquinoline synthesis[31]

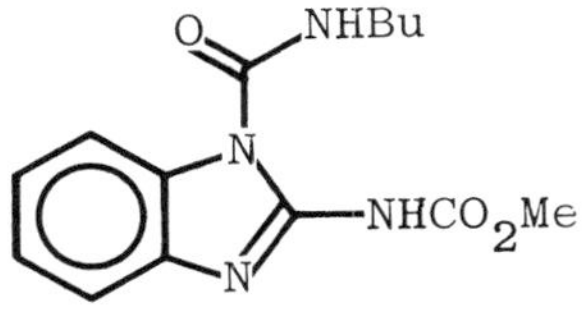

Benomyl[32] fungicide for plants

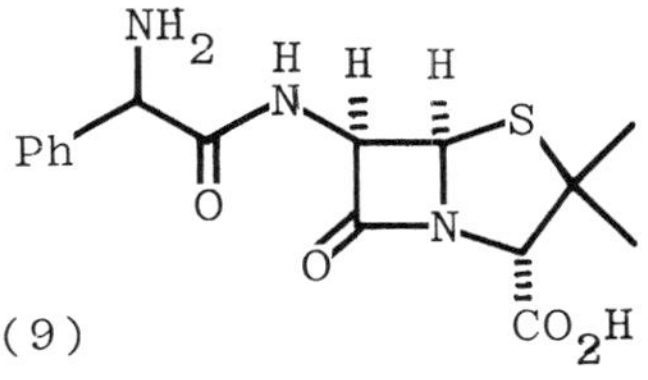

Ampicillin penicillin type of antibiotic

Answer

Choosing all the bonds joining a carbonyl group to a heteroatom (a) or joining a heteroatom (not N, see Chapter T 8) to an alkyl group (b).

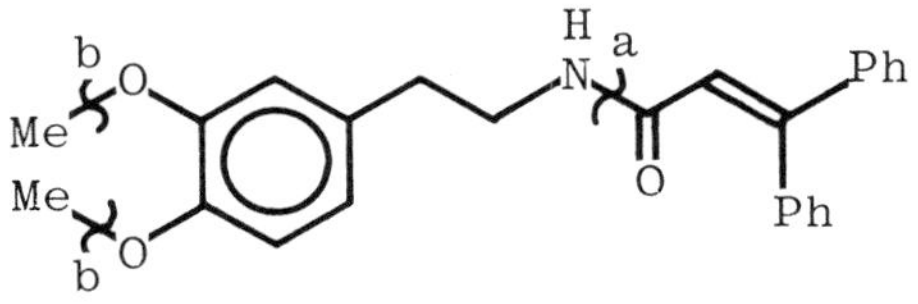

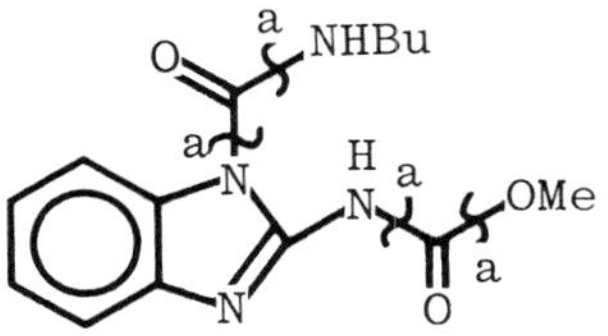

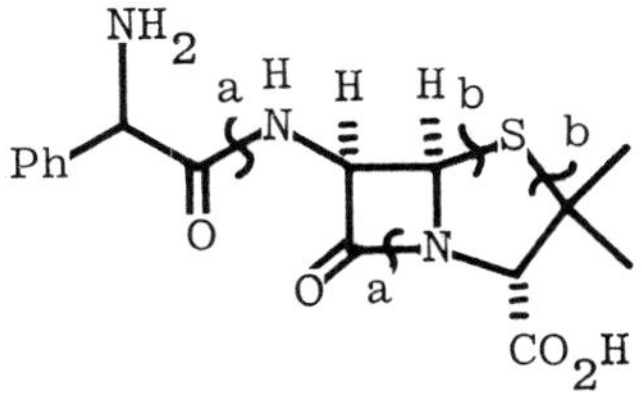

Not all of these bonds would be disconnected in any one synthesis but it is important that you learn to see the possibilities quickly.

2. Choose starting materials for a one-step synthesis of ether (10) saying why you chose these particular compounds.

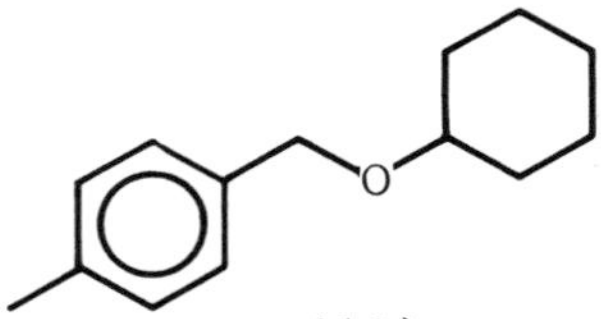

(10)

Answer

An alkyl halide and an alcohol under basic conditions will give the ether.

Analysis

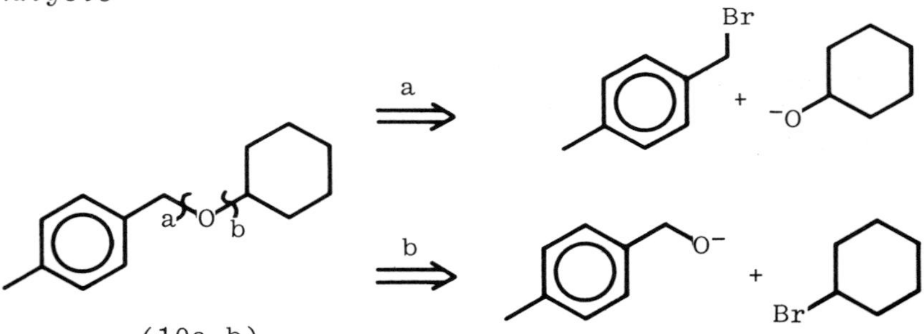

(10a,b)

(11)

Route (a) is preferable since benzylic bromides are reactive towards substitution and cannot eliminate whereas the secondary halide (11) in route (b) is relatively unreactive towards S_N2 and can eliminate.

Synthesis[33]

[Scheme: p-methylbenzyl bromide + cyclohexanol —base→ TM(10)]

More advanced example

Methyl damascanine (12) has a nutmeg-like smell and is used in perfumery. Each group on the aromatic ring has an alkyl group attached - ether, amine, and ester are all present.

[Structure (12): benzene with CO₂Me, NHMe, OMe substituents]

The ester is the easiest to make so we can disconnect that first. The alkyl amine will be made by methylation of the aromatic amine (13) (see Chapter T 8) and we are left with a familiar problem of aromatic synthesis.

Analysis 1

[Scheme: (12a) ⇒ (C-O ester disconnection) ⇒ intermediate with CO₂H, NH—Me, OMe ⇒ (13) with CO₂H, NH₂, OMe]

The amino group must come from a nitro group and direct disconnection of (14) has the correct orientation though whether (15) will nitrate at sites (a) (b) or (c) is doubtful.

Analysis 2

(13) $\xrightarrow[\text{reduction}]{\text{FGI}}$ [structure 14: benzene with CO$_2$H, NO$_2$, OMe] $\xrightarrow[\text{nitration}]{\text{C-N}}$ [structure 15: benzene with CO$_2$H, OMe, positions a and b marked]

(14) (15)

Experiments showed that while nitration of the available hydroxyacid (16) went in position (b) conditions could be found[34] to nitrate (15) at position (a). Disconnection of the methyl group from the ether must therefore be the last step in the analysis.

Analysis 3

(15a) [structure with OMe ether] $\xrightarrow[\text{ether}]{\text{C-O}}$ (16) [structure with OH] $\xrightarrow[\text{H}_2\text{SO}_4]{\text{HNO}_3}$ [structure with OH and NO$_2$]

Dimethyl sulphate (p T 31) is used for the ether synthesis, and esterification was with excess methanol (cf page 30).

Synthesis[34]

[structure with CO$_2$H, OH] $\xrightarrow[\text{NaOH}]{(\text{MeO})_2\text{SO}_2}$ (15) 85% [structure with CO$_2$H, OMe] $\xrightarrow[\text{H}_2\text{SO}_4]{\text{HNO}_3}$

(14) [structure with CO$_2$H, NO$_2$, OMe] $\xrightarrow{\text{Sn,HCl}}$ 60% [structure with CO$_2$H, NH$_2$, OMe] →

[Structure: benzene ring with CO₂H, NHMe (ortho), OMe (other ortho)] → MeOH / HCl → TM(12)

Problem

Suggest a synthesis of the local anaesthetic ambucaine (17). This time carry the analysis of the aromatic starting material further, for revision.* Alcohol (18) is available (and is discussed in Chapter T 6).

[Structure (17): benzene ring with -C(=O)-O-CH₂CH₂-NEt₂ group, -O-nBu group (meta to ester), and -NH₂ group]

[Structure (18): HO-CH₂CH₂-NEt₂]

Answer

Disconnection of the ester removes alcohol (18) and leaves (19). The amino group will have to be put in by nitration and we already know (p 34) that we need the free acid (16) for this. The ether must therefore be disconnected first.

* oxidation of *o*, *m*, or *p*-hydroxy toluenes is not a good way to make the corresponding benzoic acids as other reactions occur.

Analysis 1

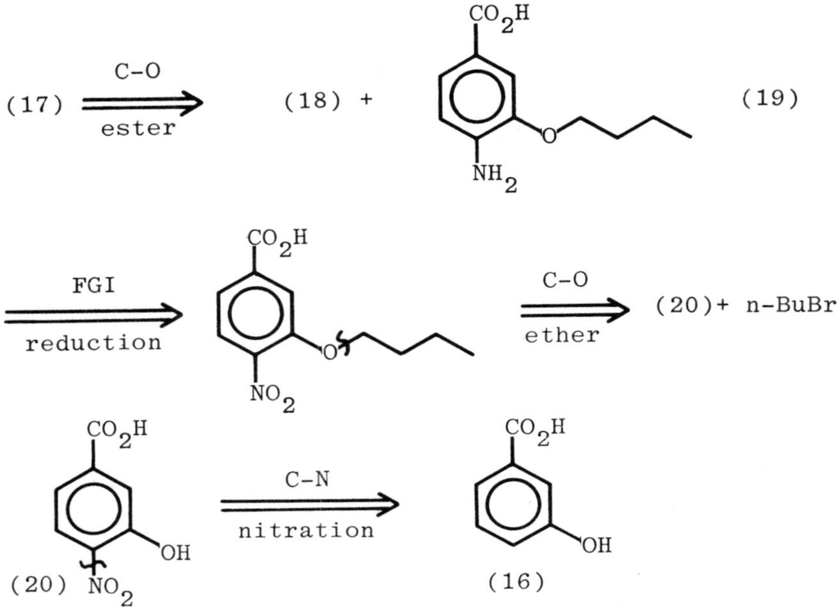

(17) $\xRightarrow[\text{ester}]{\text{C-O}}$ (18) + [4-amino-3-butoxy-benzoic acid] (19)

$\xRightarrow[\text{reduction}]{\text{FGI}}$ [3-butoxy-4-nitro-benzoic acid] $\xRightarrow[\text{ether}]{\text{C-O}}$ (20) + n-BuBr

(20) [3-hydroxy-4-nitro-benzoic acid] $\xRightarrow[\text{nitration}]{\text{C-N}}$ (16) [3-hydroxy-benzoic acid]

The CO_2H group is correctly *meta*-directing in (16) so the best route will be via a diazonium salt back to benzoic acid.

Analysis 2

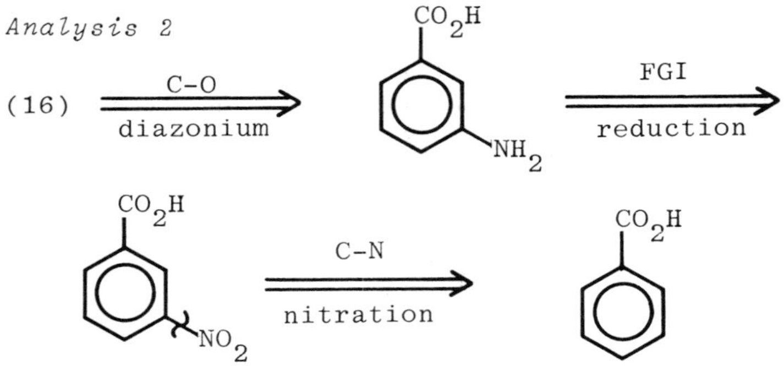

(16) $\xRightarrow[\text{diazonium}]{\text{C-O}}$ [3-amino-benzoic acid] $\xRightarrow[\text{reduction}]{\text{FGI}}$

[3-nitro-benzoic acid] $\xRightarrow[\text{nitration}]{\text{C-N}}$ [benzoic acid]

Experiments[35] showed that nitration of benzoate esters (e.g. 21) is cleaner and the diazotisation[36] is

straightforward. The acid group in (20) was protected
as an ester during ether formation and the acid
chloride method was used for the final esterification.[37]

Synthesis[35-37]

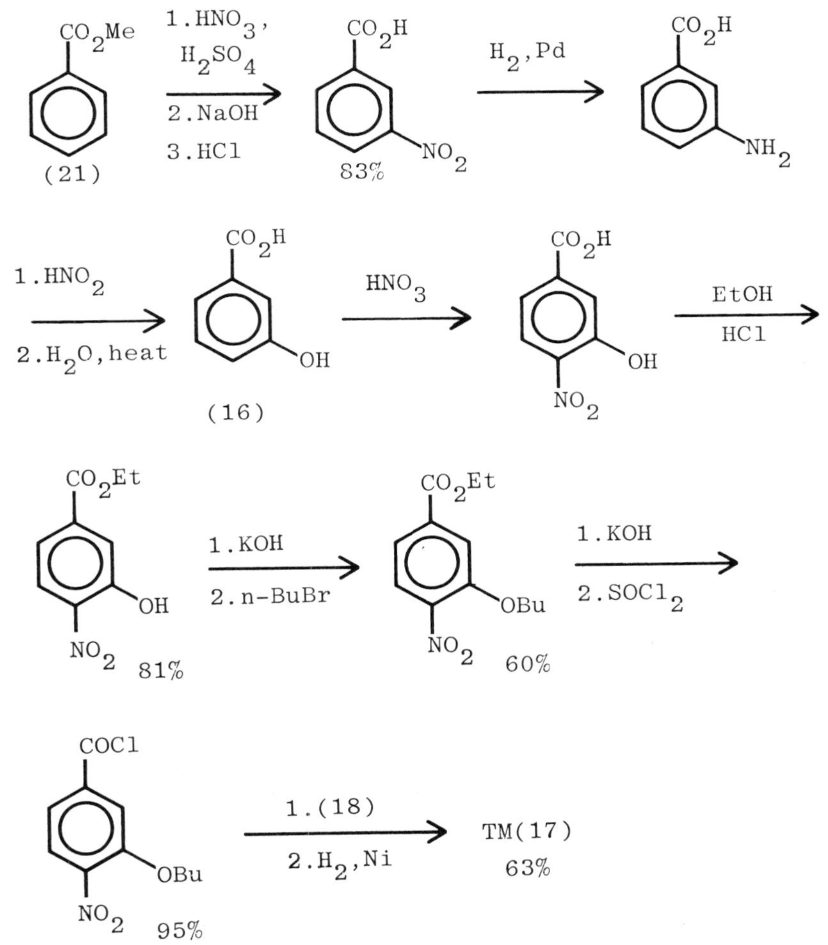

CHAPTER 5

Strategy II: Chemoselectivity

Examples

Ketoacid (1) is an isomer of a compound made in the book (p T 39) and both were needed for the synthesis of potential fungicides.[38]

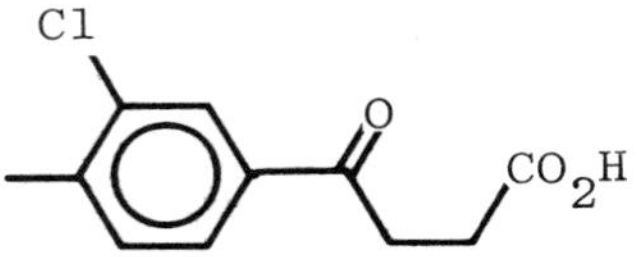

Direct Friedel-Crafts disconnection is no good (p T 39) as it led to the other isomer. Preliminary C-Cl disconnection is the answer, with anhydride (2) providing the chemoselectivity.

Analysis

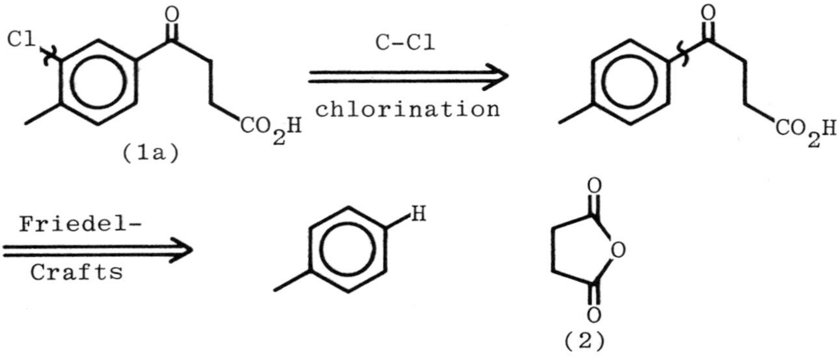

It turned out that the Friedel-Crafts reaction and the chlorination can be done in the same pot. The chlorination needs to be chemoselective as reaction on the methyl group or next to the carbonyl group could occur. Lewis acid catalysis is the answer.

Synthesis[38]

toluene (9.2g) $\xrightarrow[\text{2. Cl}_2, \text{AlCl}_3, 30°C]{\text{1. (2), AlCl}_3, 10°C}$ TM(1) 16.3g

Simple Problem

Why are the chemoselective reactions (a) and (b) successful but (c), used on page 21, unsuccessful.

(a) pyrrolidinone-CO$_2$Et $\xrightarrow{\text{LiBH}_4}$ pyrrolidinone-CH$_2$OH   good yield

The product was used in a synthesis of some specific inhibitors.[39]

(b) 4-O$_2$N-C$_6$H$_4$-NH$_2$ + ClCH$_2$COCl → 4-O$_2$N-C$_6$H$_4$-NHCOCH$_2$Cl   good yield

(c) 2-Me-4,6-(O$_2$N)$_2$-C$_6$H$_3$ $\xrightarrow[\text{HCl}]{\text{SnCl}_2}$ 2-Me-3-NH$_2$-5-O$_2$N-C$_6$H$_3$ about 20%

Answers (The guidelines appear in Chapter T 5)

(a) Esters are more reactive than amides so the right reagent, or in this case LiBH$_4$, does the job.

(b) Amines attack acid chlorides easily as the amide formed is conjugated. The alkyl halide is notably less reactive.

These both fit guideline (1).

(c) The nitro groups are too nearly equal in reactivity for a distinction to be made. Had they been identical, all would have been well : the problem is that they are nearly but not quite identical (guideline 7).

More Advanced Examples

The monoketal of some diketones is formed in reasonable yield by the statistical method. Thus (3) gives a reasonable yield of (4) : after separation from (3) and the bis-ketal, 64% of (4) can be isolated.[40]

However, it does not follow that in any individual case, the same applies. The monoketal[41] (6) of the cyclic diketone (5) is much in demand as a synthetic intermediate and in conformational studies. Mono ketalisation is unsatisfactory here - very little of (6) can be isolated. This may be simply because the right conditions have not yet been found. Various solutions to this problem have been published[41]: I shall describe two, both based on the statistical method.

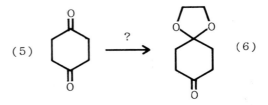

Oxidation[41] of the diol (7) with one equivalent of chromic acid in acetone gives an excellent yield of (8). The rest of the synthesis is trivial, as the two functional groups have been differentiated.

Synthesis 1

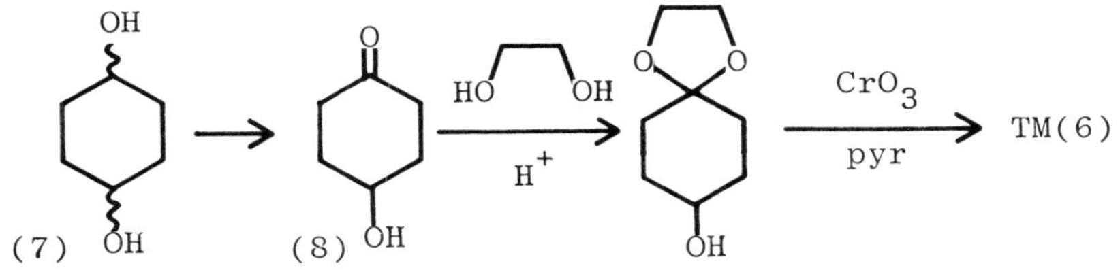

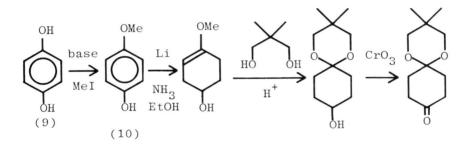

The other approach[42] uses the monoalkylation of quinol (9) as the chemoselective reaction. The rest of the sequence is more advanced but the principle should be obvious.

Synthesis 2

An advantage of this method is that separating acidic (10) from the neutral dimethylated by-product is very easy.

Problems

1. How would you make the half protected aldehyde (12) given that (11) is available from the reaction shown.

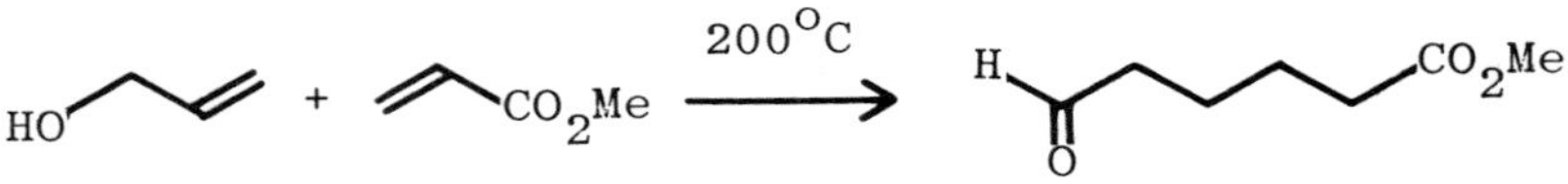

$$HO\diagup\!\!\!\diagdown + \diagup\!\!\!\diagdown CO_2Me \xrightarrow{200°C} \underset{(11)}{H\text{-}CO\text{-}(CH_2)_3\text{-}CO_2Me}$$

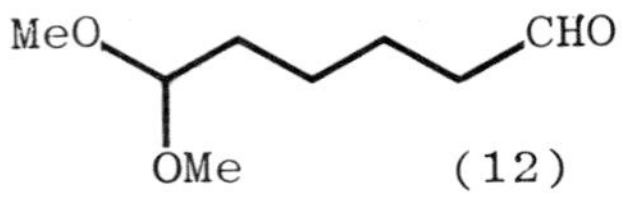

$$\underset{(12)}{MeO\text{-}CH(OMe)\text{-}(CH_2)_4\text{-}CHO}$$

Answer : The two functional groups are already differentiated, it is merely a matter of FG1. One solution is:-

Synthesis[43]

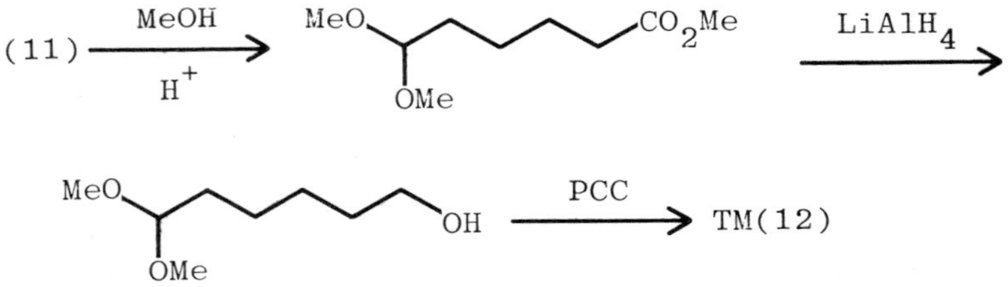

$$(11) \xrightarrow[H^+]{MeOH} MeO\text{-}CH(OMe)\text{-}(CH_2)_4\text{-}CO_2Me \xrightarrow{LiAlH_4}$$

$$MeO\text{-}CH(OMe)\text{-}(CH_2)_5\text{-}OH \xrightarrow{PCC} TM(12)$$

The acetal is used as a protecting group during the reduction of the less reactive ester (guideline 3). The product (12) was used[44] in a synthesis of gascardic acid.

2. Suggest conditions for this synthesis of the analgesic alclofenac.

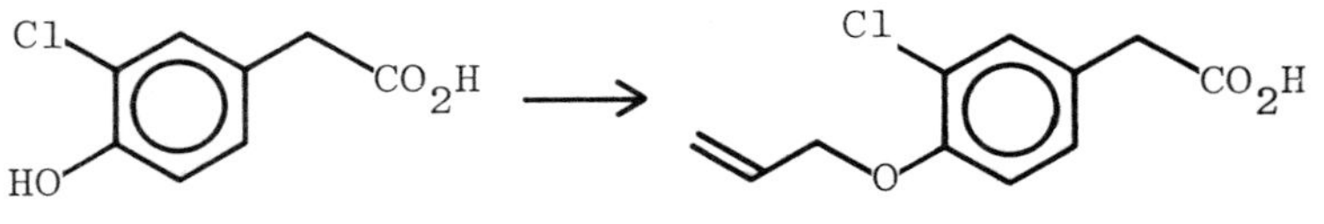

(13) → (14)

Answer : We need to remove both acidic protons from CO_2H (pKa ~ 5) and OH (pKa ~ 10) then ArO^- will be the stronger nucleophile (cf p T 34).

Synthesis[45]

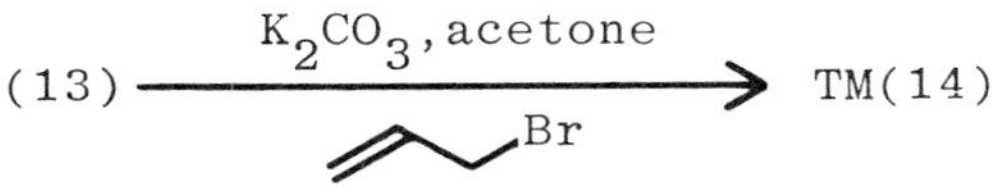

$$(13) \xrightarrow[\text{\ \ \ \ \ }\diagup\!\!\!\diagdown\text{Br}]{K_2CO_3, \text{acetone}} TM(14)$$

3. Suggest a synthesis for (15), needed for studies on oxidative decarboxylation.[46]

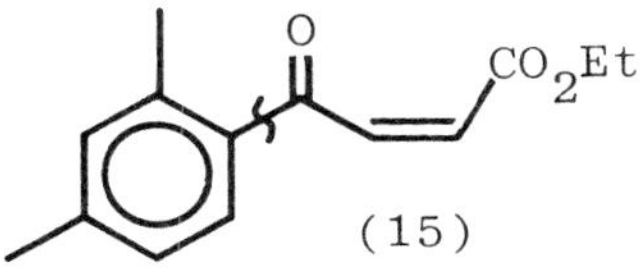

(15)

Answer : Friedel-Crafts disconnection requires the half ester, half acid-chloride (16) which we could make from maleic anhydride (17) as on p T 39. An easier solution is to use the anhydride in the Friedel-Crafts reaction. The two methyl groups cooperate to activate the right position and the yield is excellent.

Analysis

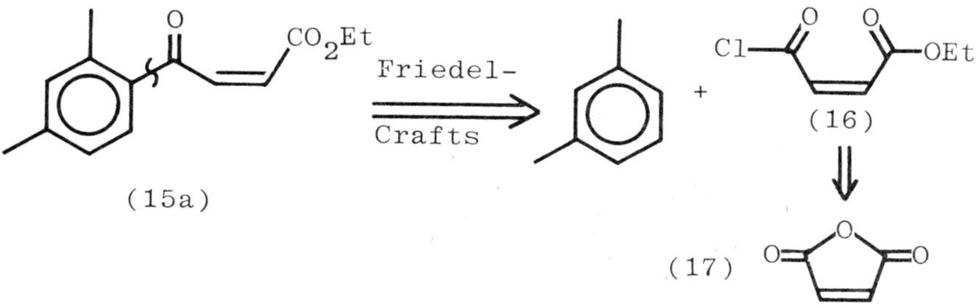

(15a) $\xRightarrow{\text{Friedel-Crafts}}$ + (16)

⇓

(17)

Synthesis[47]

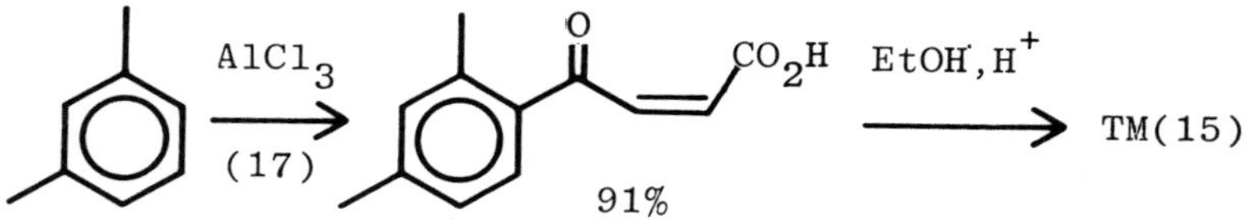

m-xylene + maleic anhydride (17) →(AlCl₃) 1-(2,4-dimethylphenyl)-4-oxo-2-butenoic acid (91%) →(EtOH, H⁺) TM(15)

CHAPTER 6

Two-Group C—X Disconnections

(a) *1,1-Difunctionalised Compounds*

Recognising an acetal is a problem for many chemists in training. Molecule (1) is obviously an acetal, and the first disconnection is easy to write:-

Analysis

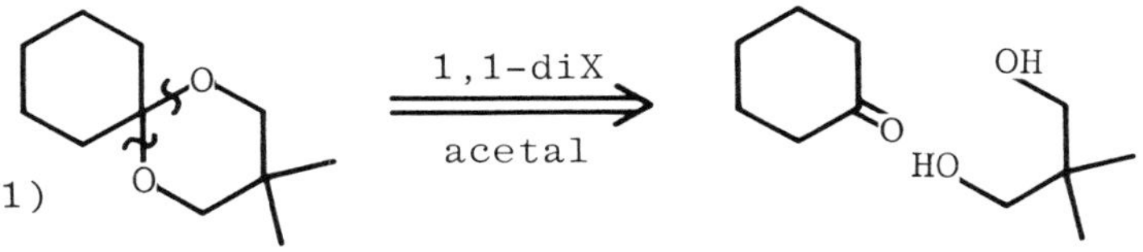

Both starting materials are available, the glycol (2) being a cheap and efficient acetal forming reagent.

Synthesis

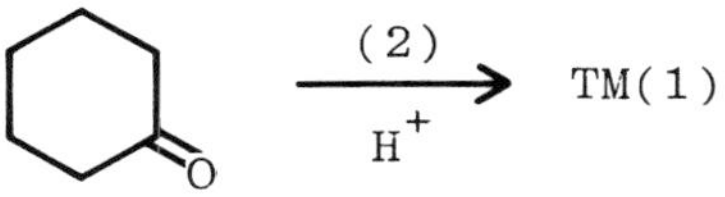

Recognising the acetal in frontalin (3), a pheromone of the western pine beetle,[48] is not so easy. Nevertheless, it is important to look for the two oxygen atoms joined to the same carbon atom (● in 3a) and disconnect the acetal before considering any other steps.

Analysis

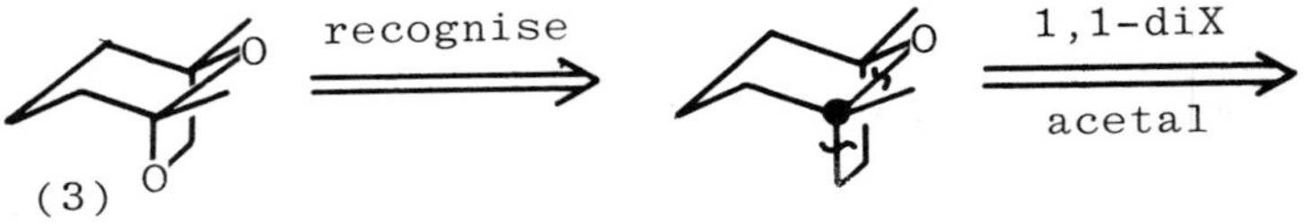

(3)

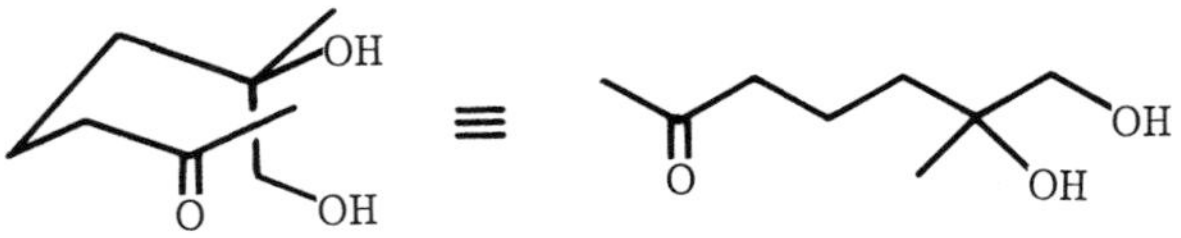

(4)

Intramolecular acetal formation is so favourable that any attempt to make (4) leads at once to frontalin.

Problem :

Recognise the acetal in the following TMs and carry out the first disconnection.

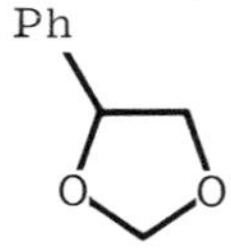

(5) jasmine odour compound[49]

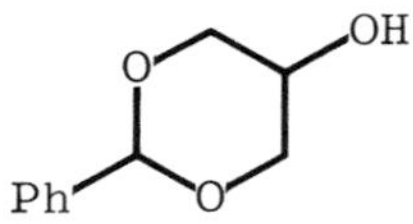

(6) synthetic intermediate[50]

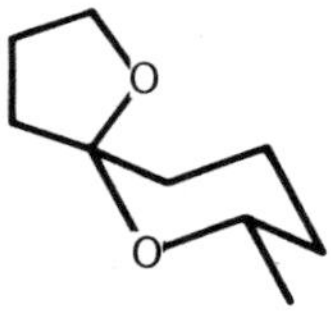

(7) pheromone of the wasp[51]

Answer : The acetal carbon atom is marked ●

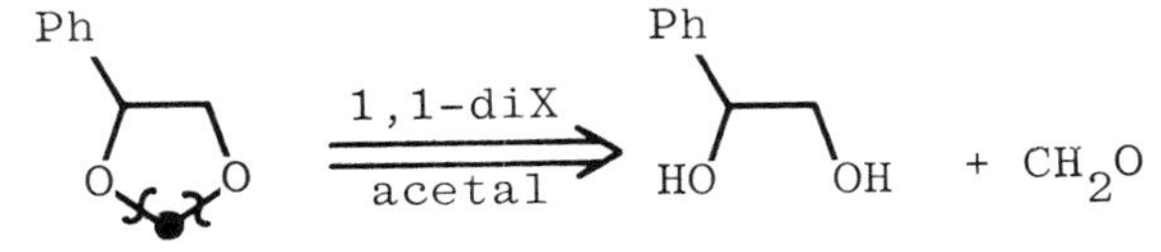

(5a) → [1,1-diX acetal] → PhCH(OH)CH₂OH + CH₂O

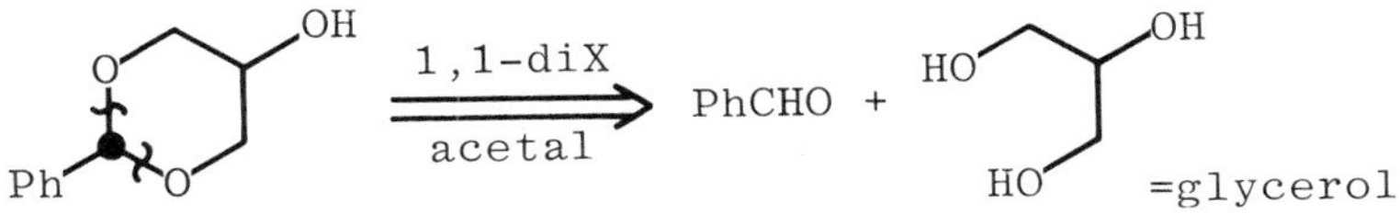

(6a) → [1,1-diX acetal] → PhCHO + glycerol (HOCH₂-CH(OH)-CH₂OH)

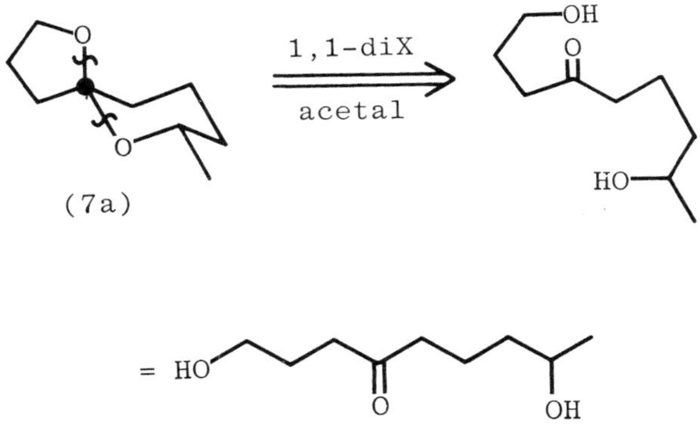

(7a) → [1,1-diX acetal] →

= HO~~~C(=O)~~~CH(OH)CH₃ (5-hydroxy... keto-diol)

Problem :

Suggest a synthesis for the perfume constituent (8).

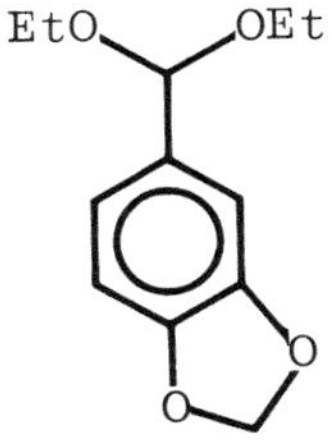

(8) = 3,4-methylenedioxybenzaldehyde diethyl acetal [ArCH(OEt)₂ where Ar = 3,4-(OCH₂O)C₆H₃]

Answer : There are two acetals here! Either could be disconnected first, though not both, since problems of chemoselectivity would arise. Since we have already made piperonal (9) (page T 9) we shall use that as an intermediate.

Analysis

(8) $\xrightarrow[\text{acetals}]{\text{recognise}}$ [aryl diethyl acetal on methylenedioxybenzene] $\xrightarrow[\text{acetal}]{1,1\text{-diX}}$ piperonal (9)

$\xrightarrow[\text{T2}]{\text{chapter}}$ (10) methylenedioxybenzene $\xrightarrow[\text{acetal}]{1,1\text{-diX}}$ available catechol

Acetal (10) is available, though expensive, and is made from catechol and CH_2X_2 (X = Cl, Br, I) in base[52] instead of the usual method.

The final acetal is made by the usual acid-catalysed method, using the ortho-formate $HC(OEt)_3$ as a dehydrating agent.[53]

Synthesis[52,53]

catechol $\xrightarrow[CH_2Cl_2]{\text{base}}$ (10) → (9) $\xrightarrow[HC(OEt)_3]{H^+ \ EtOH}$ TM(8)

(b) *1,2-Difunctionalised Compounds*

Example : Many antiseptics are quaternary ammonium salts with detergent-like action and Benzethonium chloride (11) is particularly efficient. The benzyl group can be added by alkylation of (12) (Chapter T 8) which contains two sets of 1,2-diX relationships.

Analysis 1

Cl⁻ Ph–CH₂–N⁺Me₂–CH₂CH₂–O–CH₂CH₂–O–C₆H₄–C(Me)₂–CH₂–C(Me)₃   (11)

$\xrightarrow{\text{C-N}}$ Me₂N–$\overset{2}{\frown}$–$\overset{1}{\frown}$O$\overset{1}{\frown}$–$\overset{2}{\frown}$O–Ar   (12)

Disconnection at any of these positions is possible (12a). We might prefer a central disconnection (b or c) first (see Chapter T 11) but we shall end up with Me₂NH, two molecules of ethylene oxide, and the phenol (13) as starting materials.

Analysis 2

Me₂N–{a}–{b}–O–{c}–{d}–O–Ar $\xrightarrow{b}$ Me₂N–CH₂CH₂–OH   +   Me₂NH + ethylene oxide (chapter T6)

(12a)

+

HO–CH₂CH₂–OAr   ⇐

ethylene oxide + HO–C₆H₄–C(Me)₂–CH₂–C(Me)₃

(13)

The phenol is clearly a Friedel-Crafts product and the required cation (14) is most conveniently obtained from 'di-isobutylene' - a mixture of (15) and (16) available industrially from the dimerisation of $Me_2 C = CH_2$.

Analysis 3

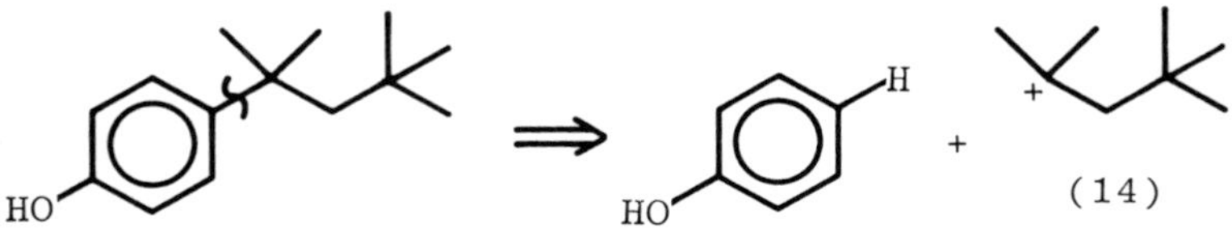

The patented synthesis[54] combines the reagents in a different order and relies on statistical control of chemoselectivity in the reaction of (17) with (13). The advantage is that symmetrical (17) may be made from ethylene chlorohydrin.

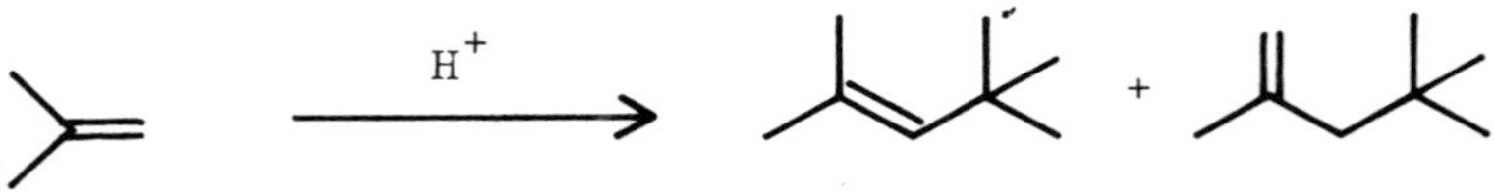

Synthesis[54,55]

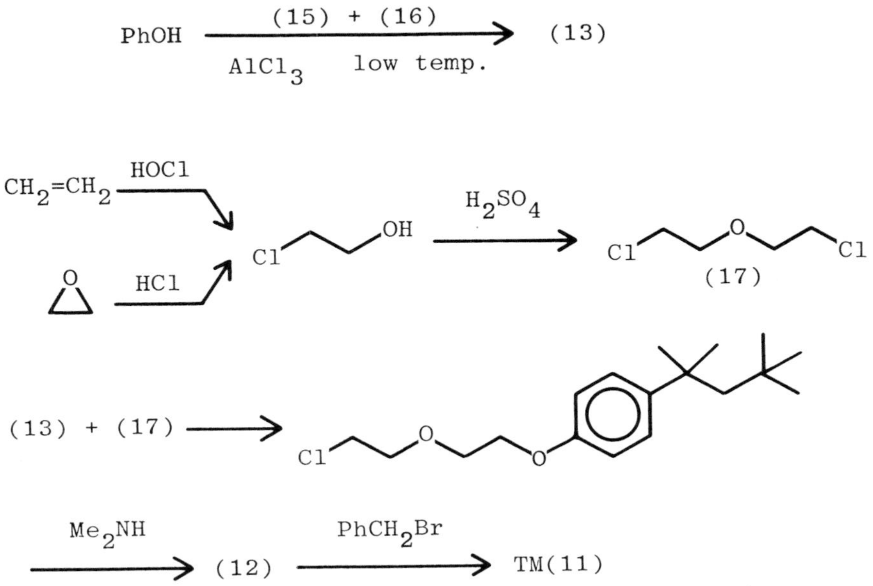

PhOH $\xrightarrow[\text{AlCl}_3 \quad \text{low temp.}]{(15) + (16)}$ (13)

$CH_2=CH_2 \xrightarrow{HOCl}$
$\underset{\triangle}{O} \xrightarrow{HCl}$ Cl~~~OH $\xrightarrow{H_2SO_4}$ Cl~~~O~~~Cl (17)

(13) + (17) ⟶ Cl~~~O~~~O—C₆H₄—C(CH₃)₂CH₂C(CH₃)₃

$\xrightarrow{Me_2NH}$ (12) $\xrightarrow{PhCH_2Br}$ TM(11)

Problems:

Two problems in this section use sulphur nucleophiles as they are not so well represented in the main book.

1. Alcohol (18) is used in the manufacture of demeton, a family of systemic insecticides.[56] How can it be made?

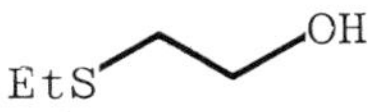

EtS~~~OH
(18)

Answer
Analysis

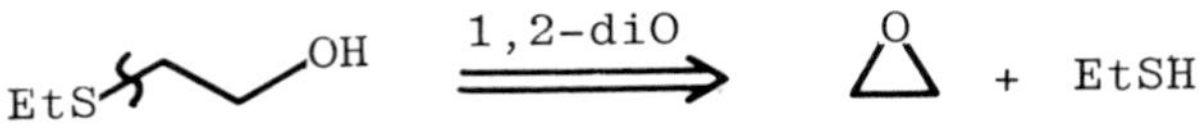

EtS∽∽OH ⇒(1,2-diO) △O + EtSH

(18a)

Ethylene chlorohydrin in base can be used instead of ethylene oxide.

Synthesis

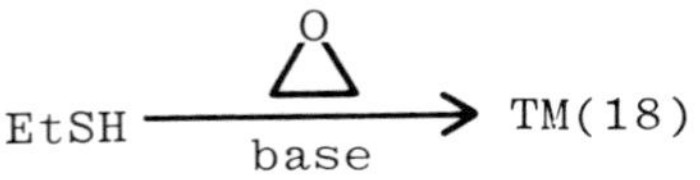

EtSH —△O→(base) TM(18)

The related cockroach repellent (18A) is made from a thiol with ethylene oxide.[56]

n-OctSH + △O ⟶ n-OctS∽∽OH

(18A)

2. Intermediate (19) was needed for a synthesis of the thianaphthalene[57] (20). How would you make (19)?

(19)   (20)

Answer : Friedel-Crafts disconnection reveals 1,2-diX compound (21) easily made from thiol (22) and chloroacetic acid. Thiol (22) can be made by the thiourea route (p T 37).

Analysis

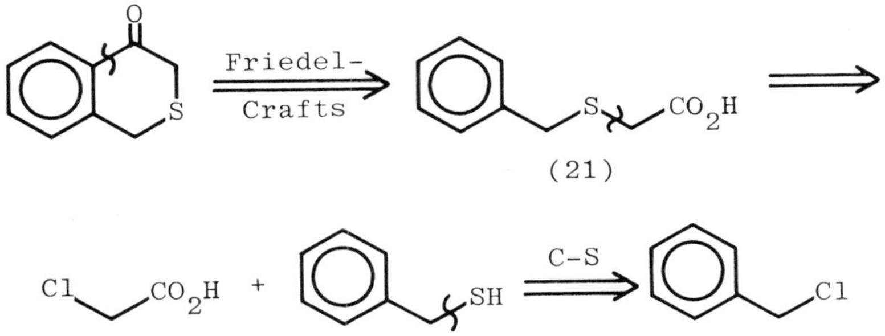

[Structure: isothiochroman-4-one] ⇒ Friedel-Crafts ⇒ PhCH₂-S-CH₂-CO₂H (21) ⇒

Cl-CH₂-CO₂H + PhCH₂-SH (22) ⇒ C-S ⇒ PhCH₂-Cl

Synthesis[58]

$$PhCH_2Cl \xrightarrow[\text{2. HO}^-, H_2O]{\text{1. thiourea}} (22) \xrightarrow[Na_2CO_3]{Cl-CH_2-CO_2H} (21)$$

$$(21) \xrightarrow[\text{benzene}]{P_2O_5} TM(19)$$
100% → 60%

3. Suggest a synthesis for the anti-Parkinson drug orphenadrine (23).

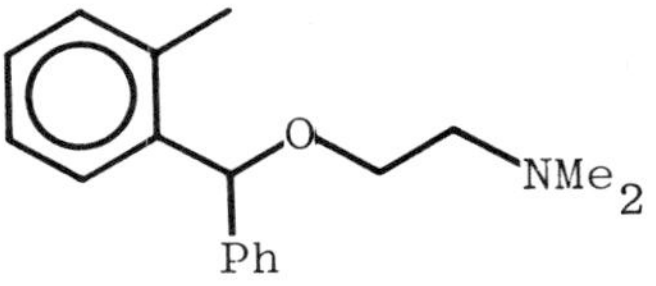

[Structure (23): o-tolyl-CH(Ph)-O-CH₂CH₂-NMe₂]

Answer : Disconnection of the ether reveals a 1,2-diX compound (24) and an alcohol (25) easily made by reduction from a Friedel-Crafts adduct (26).

Analysis

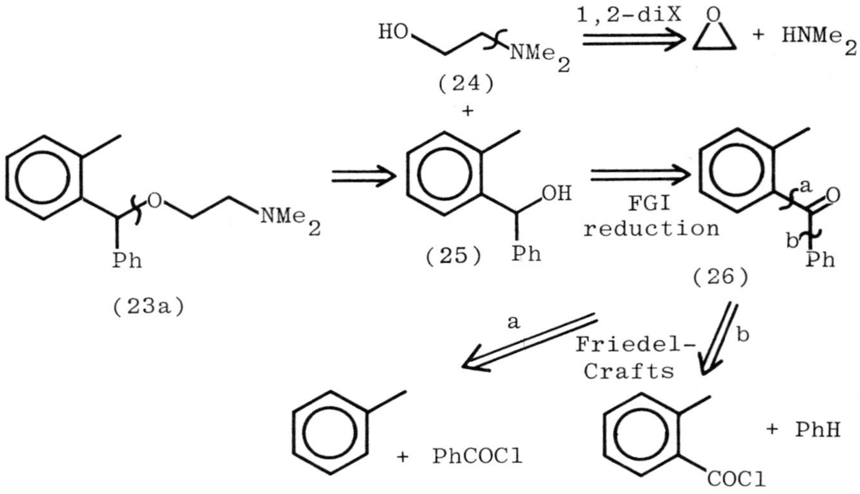

[Retrosynthetic analysis scheme showing:]

HO~~~NMe₂ (24) ⟹ epoxide (1,2-diX) + HNMe₂

(23a) [o-MeC₆H₄CH(Ph)OCH₂CH₂NMe₂] ⟹ (25) [o-MeC₆H₄CH(OH)Ph] ⟹ (FGI reduction) (26) [o-MeC₆H₄COPh]

(26) disconnects via Friedel-Crafts:
- route a: toluene + PhCOCl
- route b: o-MeC₆H₄COCl + PhH

Either Friedel-Crafts disconnection will do, but o-toluic acid (27) is available⁵⁹ from the oxidation of o-xylene so that route is preferred as no separation of o, p-isomers is involved.

Synthesis[59,60]

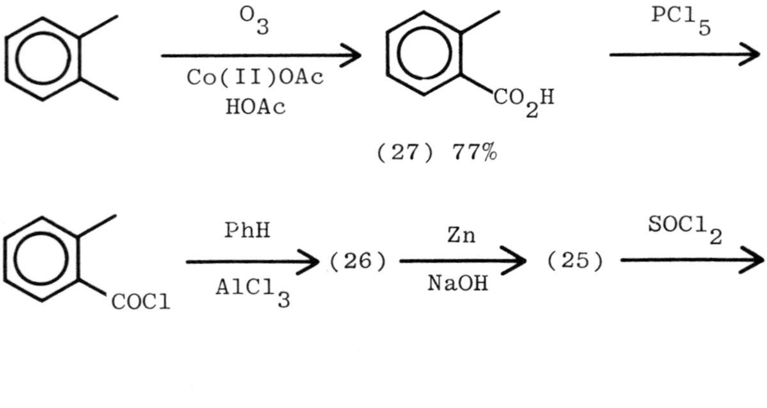

o-xylene $\xrightarrow[\text{HOAc}]{\text{O}_3, \text{Co(II)OAc}}$ o-MeC₆H₄CO₂H (27) 77% $\xrightarrow{\text{PCl}_5}$

o-MeC₆H₄COCl $\xrightarrow[\text{AlCl}_3]{\text{PhH}}$ (26) $\xrightarrow[\text{NaOH}]{\text{Zn}}$ (25) $\xrightarrow{\text{SOCl}_2}$

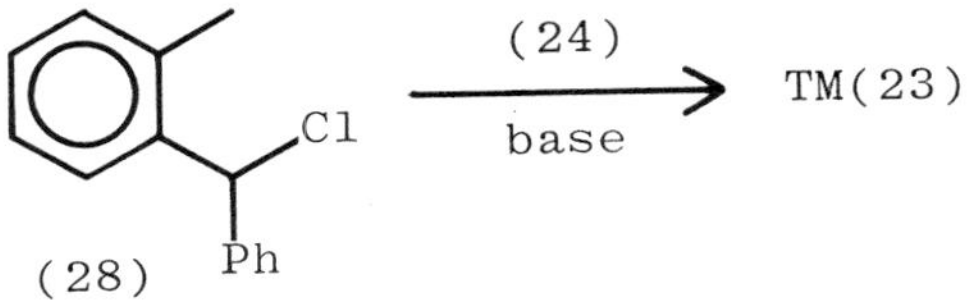

(28) [structure: 2-methylbenzyl chloride with Ph substituent] → (24)/base → TM(23)

The reactive benzylic chloride (28) is used for the final ether synthesis.

(c) *1,3-difunctionalised Compounds*

Example

Compound (29) was needed[61] for photochemical cyclisation to (30). The obvious ether disconnection looks promising but experiments had shown[62] that it was difficult to alkylate (31) without polymerisation.

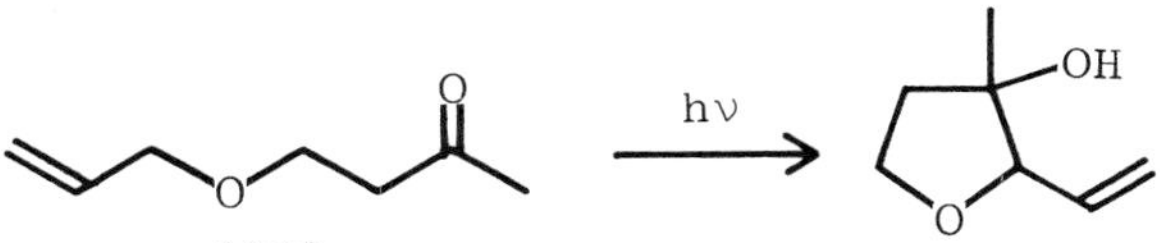

(29) [allyl 3-oxobutyl ether] → hν → (30) 94% [tetrahydrofuran with OH and vinyl substituents]

Analysis 1

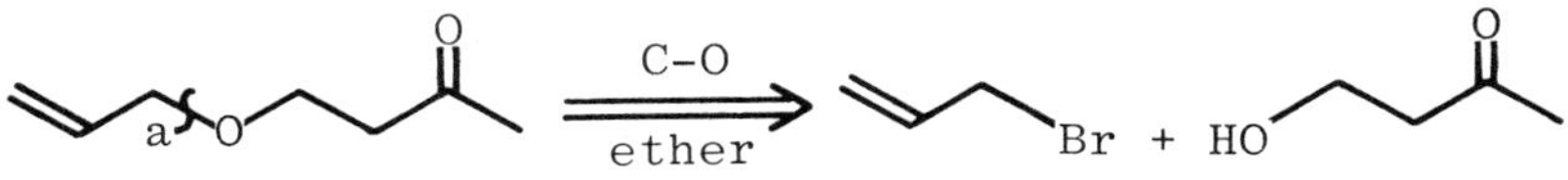

(29a) ⇒ C-O ether ⇒ allyl-Br + HO-CH$_2$CH$_2$-C(O)-CH$_3$ (31)

The alternative 1,3-diX disconnection (29b) requires the addition of allyl alcohol to enone (32) and this worked well with BF$_3$ catalysis.

Analysis 2

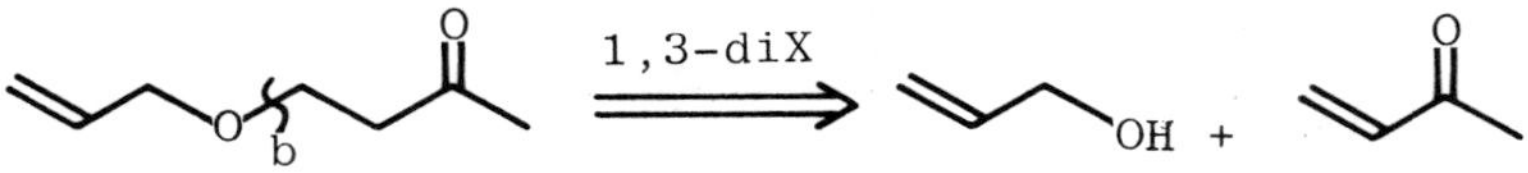

[Structure (29b): allyl-O-CH2-CH2-C(=O)-CH3, with bond 'b' marked] ⟹ 1,3-diX ⟹ allyl-OH + CH2=CH-C(=O)-CH3 (32)

Synthesis[61]

allyl-OH $\xrightarrow{\text{(32)}, \text{BF}_3}$ TM(29)

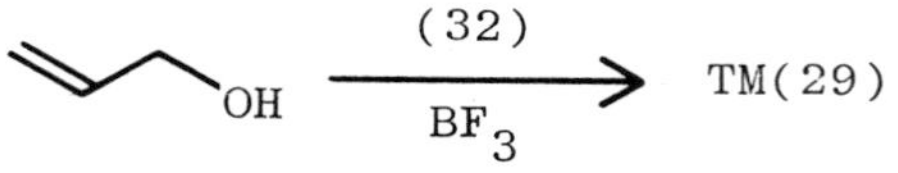

Problems

1. Find the 1,3-diX disconnection in TMs (33 - 35).

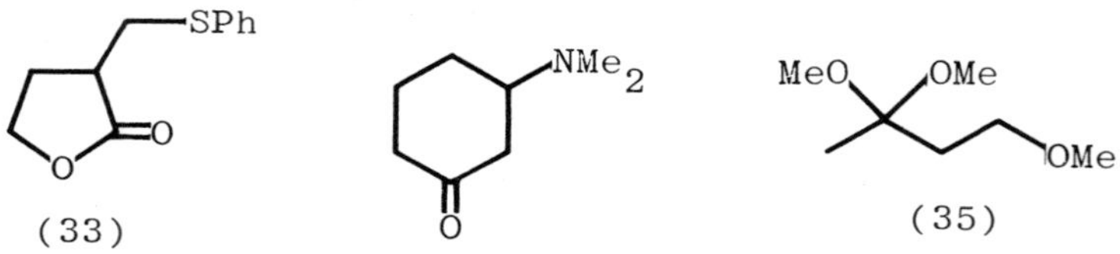

(33) γ-butyrolactone with -CH2-SPh at α-position

(34) 3-(dimethylamino)cyclohexanone

(35) (MeO)2C(Me)-CH2-CH2-OMe

Answer

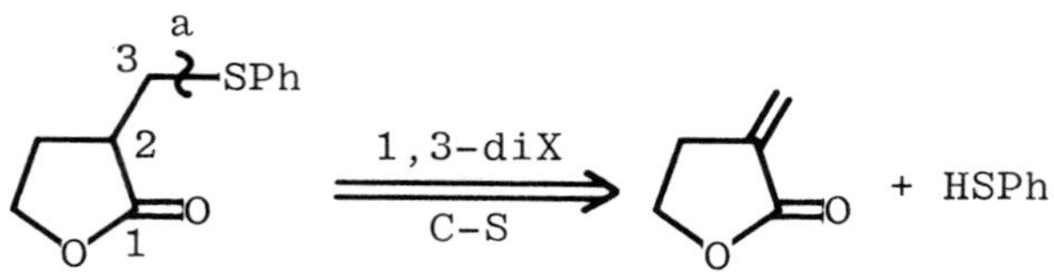

(33a) lactone with positions 1(O), 2(C=O), 3(CH-CH2-SPh); disconnection at 'a' ⟹ 1,3-diX, C-S ⟹ α-methylene-γ-butyrolactone + HSPh

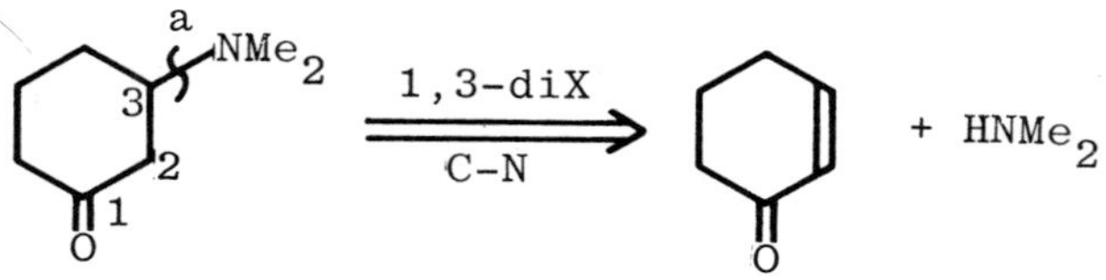

(34a) cyclohexanone with NMe2 at C3; disconnection at 'a' ⟹ 1,3-diX, C-N ⟹ cyclohex-2-enone + HNMe2

The last one requires a 1,1-diX disconnection of the acetal before the 1,3-diX disconnection can be made.

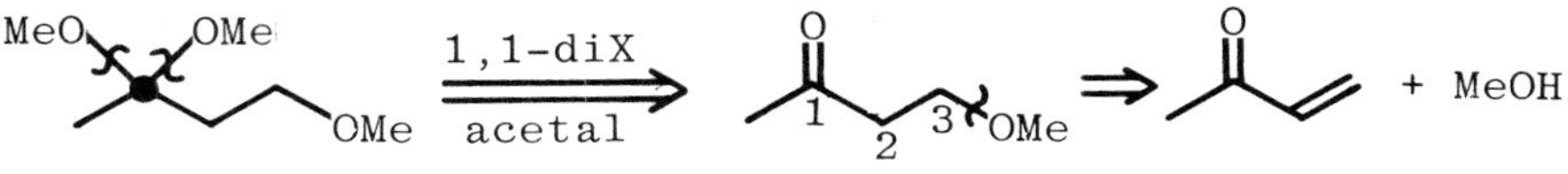

MeO, OMe
    \ /
     C—CH₂CH₂—OMe $\xrightarrow{\text{1,1-diX acetal}}$ CH₃—CO—CH₂—CH₂—OMe $\Rightarrow$ CH₃—CO—CH=CH₂ + MeOH

(35a)

2. What compound is needed to form cyclomethycaine (37) from the intermediate (36) we synthesised on page T 35, and how would you make it?

(36) 4-(cyclohexyloxy)benzoic acid, CO_2H

$\xrightarrow{?}$

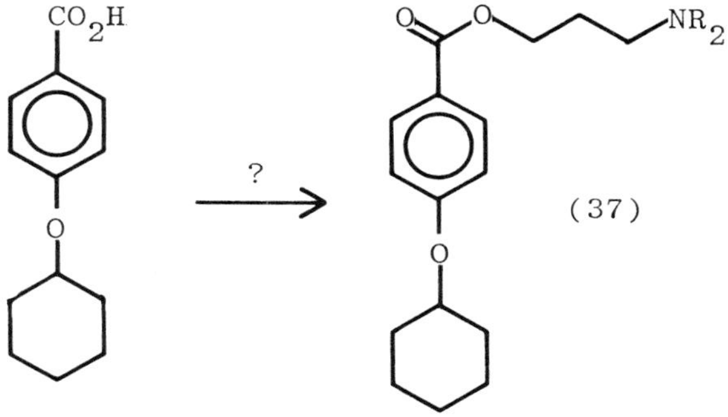

(37) 3-(NR$_2$)propyl 4-(cyclohexyloxy)benzoate

Answer

Alcohol (38) is needed, and it can be made by the logic of 1,3-diX after FGI.

Analysis

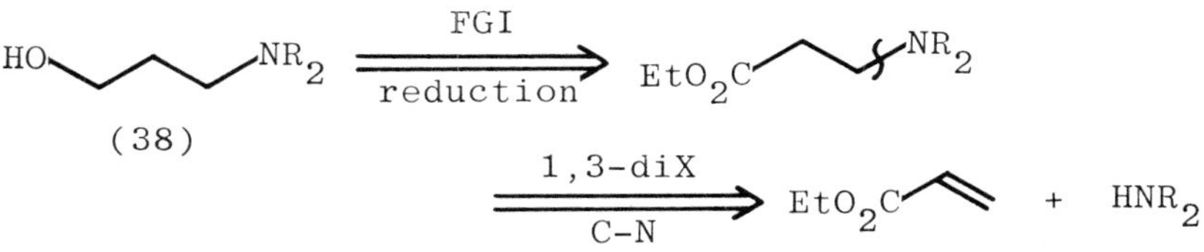

HO—CH₂CH₂CH₂—NR$_2$ $\xrightarrow[\text{reduction}]{\text{FGI}}$ EtO$_2$C—CH₂CH₂—NR$_2$
(38)

$\xrightarrow[\text{C-N}]{\text{1,3-diX}}$ EtO$_2$C—CH=CH₂ + HNR$_2$

Synthesis[63]

$$R^1O_2C{-}CH{=}CH_2 + R_2NH \longrightarrow R^1O_2C{\sim}{\sim}NR_2 \xrightarrow{LiAlH_4} TM(38)$$

This is not the only route[64] to compounds such as (38). The Mannich reaction (p T 158) provides aldehydes and ketones (39) which can be reduced to (38) and its analogues or R_2NH can be alkylated by the chloroalcohol (41), available by the statistical method[65] from symmetrical diol (40).

Synthesis[64]

(a) $R^1{-}CO{-}CH_3 \xrightarrow{CH_2O,\ HNR_2} R^1{-}CO{-}CH_2CH_2{-}NR_2 \xrightarrow{NaBH_4} R^1{-}CH(OH){-}CH_2CH_2{-}NR_2$
                                (39)   (R^1=H, Ar, Alk)

(b) $HO{\sim}{\sim}OH \xrightarrow{HCl} Cl{\sim}{\sim}OH \xrightarrow{HNR_2} (38)$
        (40)   (41)

(d) *General Problems*

1. Amino diester (42) was needed[66] for cyclisation and decarboxylation to (43), (method of Chapter 19) a useful synthetic intermediate. How would you synthesise (42)?

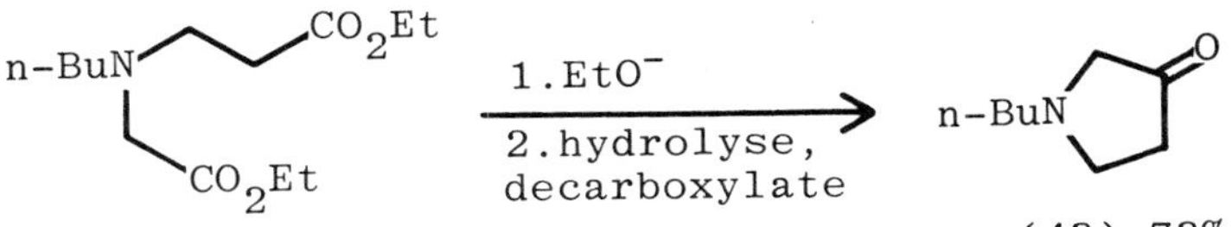

n-BuN(CH₂CH₂CO₂Et)(CH₂CO₂Et) (42) → [1. EtO⁻; 2. hydrolyse, decarboxylate] → n-BuN-pyrrolidinone (43) 73%

Answer : There are 1,2-diX and 1,3-diX disconnections to be made. The order is not very important as the starting materials are the same and the order in the synthesis will be determined by experiment.

Analysis

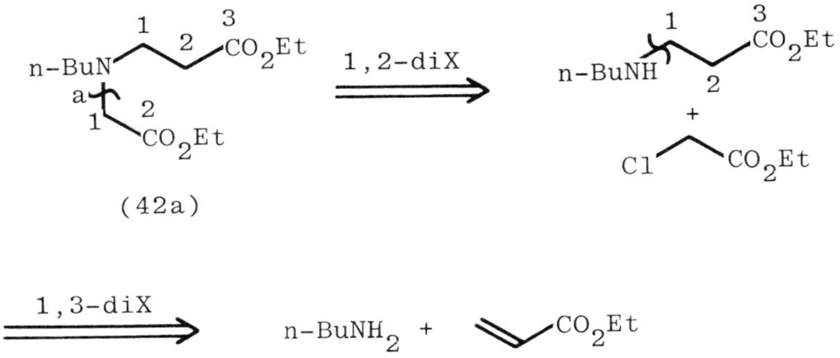

(42a) ⇒ [1,2-diX] n-BuNH–CH₂CH₂CO₂Et + Cl–CH₂CO₂Et

⇒ [1,3-diX] n-BuNH₂ + CH₂=CH–CO₂Et

This was the order followed for TM (42), but in other cases, the Michael reaction came last.

Synthesis[66]

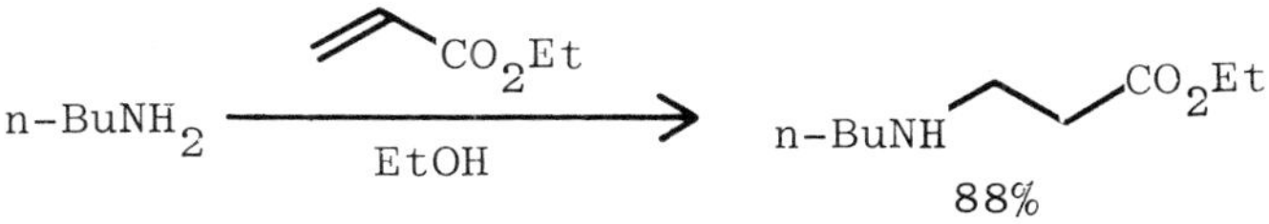

n-BuNH₂ + CH₂=CHCO₂Et —[EtOH]→ n-BuNH–CH₂CH₂CO₂Et   88%

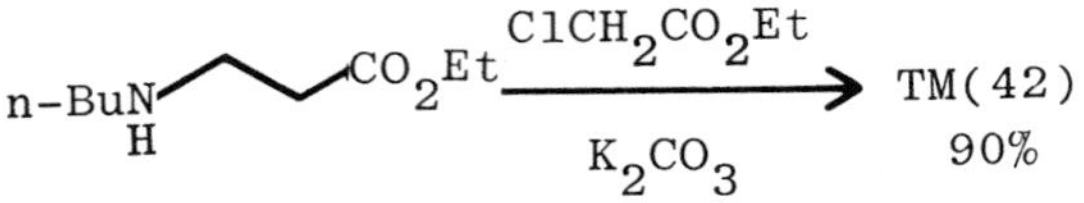

n-BuNH−CH₂CH₂−CO₂Et $\xrightarrow{\text{ClCH}_2\text{CO}_2\text{Et}, \text{ K}_2\text{CO}_3}$ TM(42) 90%

2. Propose a synthesis for the naturally-occurring amino acid methionine (44) using the logic of this chapter.

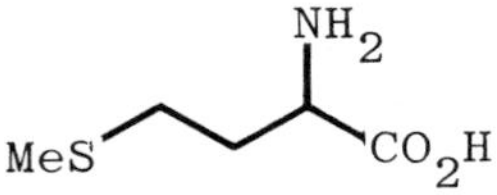

MeS−CH₂CH₂−CH(NH₂)−CO₂H
(44)

Answer: The Strecker amino acid synthesis (p T 43) requires aldehyde (45) for which a 1,3-diX disconnection is appropriate.

Analysis

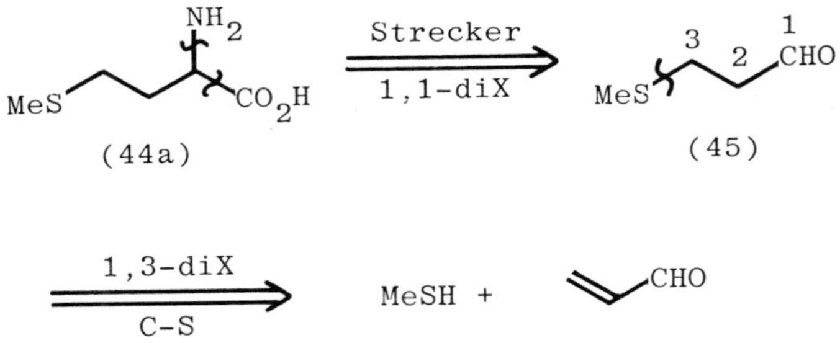

MeS−CH₂CH₂−CH(NH₂)−CO₂H $\underset{1,1\text{-diX}}{\xrightarrow{\text{Strecker}}}$ MeS−CH₂CH₂−CHO (positions 3,2,1)
(44a)   (45)

$\underset{\text{C-S}}{\xrightarrow{1,3\text{-diX}}}$ MeSH + CH₂=CH−CHO

This synthesis is very successful and has been used as the basis for commercial production.

Synthesis[6,7]

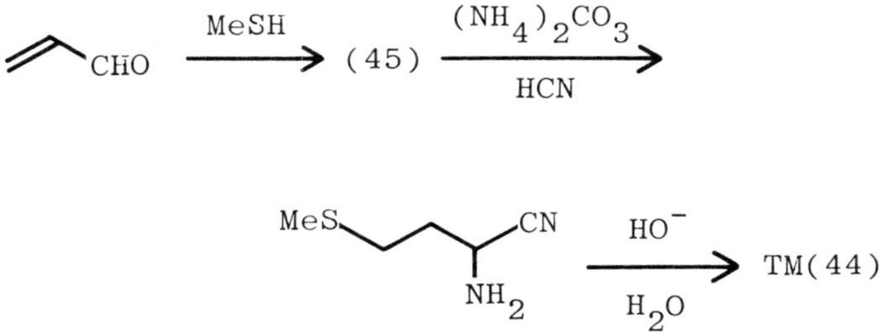

$$\underset{CHO}{\diagup\!\!\!=} \xrightarrow{MeSH} (45) \xrightarrow[HCN]{(NH_4)_2CO_3}$$

$$\underset{(46)*}{MeS\diagdown\!\!\diagdown\!\!\underset{NH_2}{\overset{}{C}}\diagup CN} \xrightarrow[H_2O]{HO^-} TM(44)$$

* (46) was actually formed as [hydantoin with MeS-CH₂CH₂- substituent]

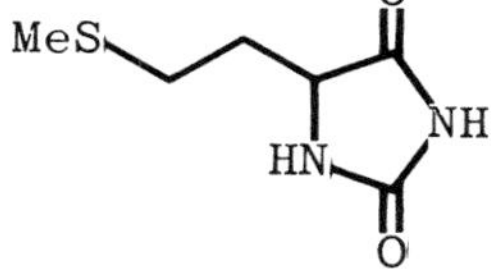

CHAPTER 7

Strategy III: Reversal of Polarity, Cyclisation Reactions, Summary of Strategy

Simple Problems

1. The last stage in most syntheses of the juvenile hormone (2) of *cecropia* is the epoxidation of (1). For example, MCPBA (p T 52) gives[68] 40% of (2) together with 10% epoxide at double bond (b) and 10% diepoxide (a + b). No epoxidation occurs at (c). Epoxidation via the bromohydrin[69] is even more chemoselective, giving 52% of (2).

Suggest some reasons for the chemoselectivity.

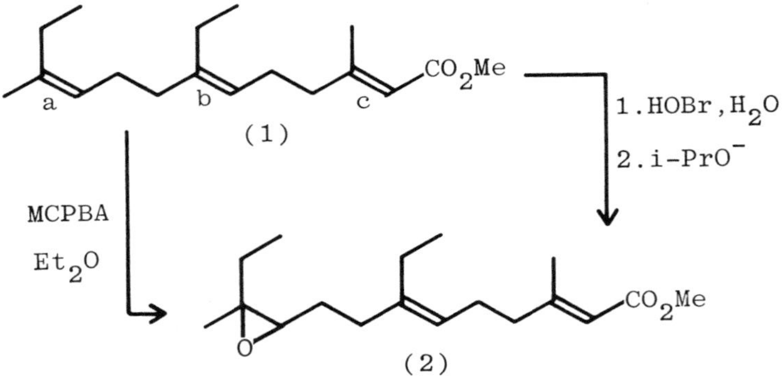

Answer : Both epoxidations involve *electrophilic* attack on the double bond so attack at the electron-deficient bond (c) is unlikely. Preferential attack at (a) can only be steric as (a) and (b) are identical electronically. In polar solvents, (1) probably coils up so that bond (b) is embedded in the hydrocarbon chain and not accessible to attack.[70]

2. Oxidation of (3) with Pb(OAc)$_4$ was thought[71] to give (4). Suggest a synthesis of this compound so that (4) may be compared with an authentic sample.

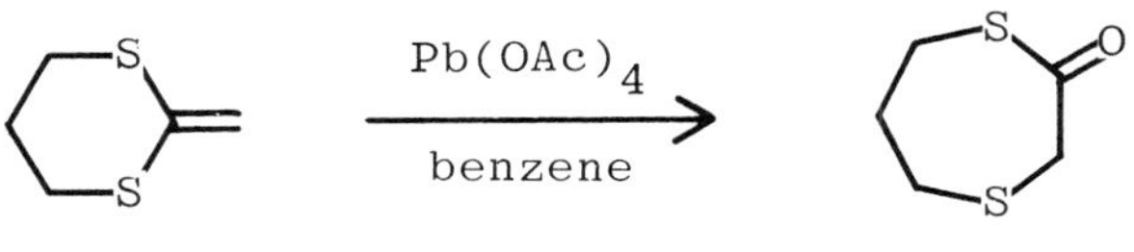

$$(3) \xrightarrow[\text{benzene}]{\text{Pb(OAc)}_4} (4) \ 74\%$$

Answer: Various sequences of C-S disconnections would no doubt all lead to good syntheses. Disconnection (4a) has the advantage of giving a symmetrical dithiol (5) and available (p T 53) chloroacetylchloride.

Analysis

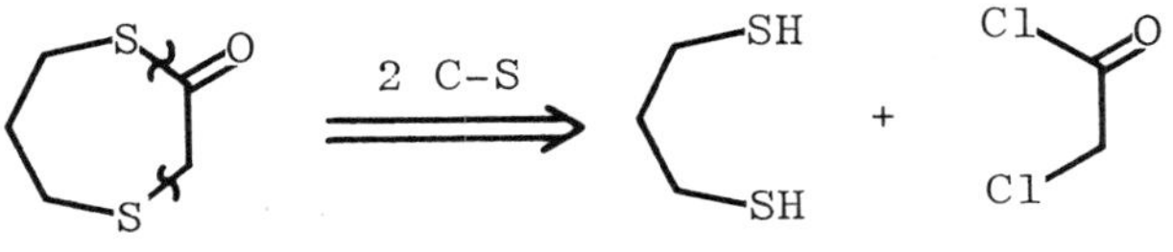

$$(4a) \xRightarrow{2 \ \text{C-S}} (5) + \text{ClCH}_2\text{COCl}$$

Synthesis[71]

$$(5) \xrightarrow{\text{ClCH}_2\text{COCl}} \text{TM}(4)$$

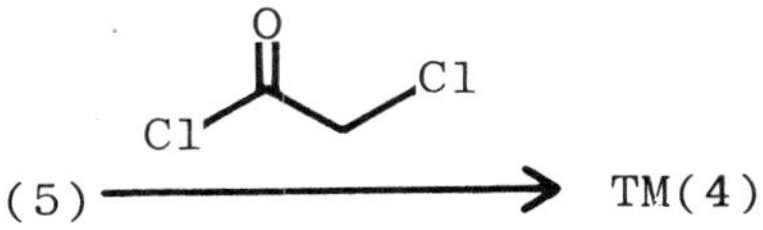

This was indeed how (4) was made though no experimental details are given and the reaction went in 'very low yield'.

More Advanced Problems

1. Compounds such as (6) cannot be made by the opening of epoxides. Why not? Suggest a way round this difficulty.

Analysis (no good)

$$\text{Nu} \underset{R}{\overset{O-R^1}{\diagup\!\!\!\diagdown}} \xrightarrow{\text{C-O ether}} \text{Nu} \underset{R}{\overset{OH}{\diagup\!\!\!\diagdown}} \xrightarrow{1,2\text{-diX}} \text{Nu}^- + \underset{R}{\overset{O}{\triangle}}$$

(6)   (7)

Nu⁻ = nucleophile, such as RO⁻, RS⁻, RNH$_2$

Answer : Attack* would occur at the wrong atom of the epoxide.

$$\underset{R}{\overset{O}{\triangle}} \xleftarrow{} {}^-\text{Nu} \longrightarrow \underset{R}{\overset{HO}{\diagup\!\!\!\diagdown}} \text{Nu}$$

Possible solutions include changing the oxidation level by FGI to (8) or reversing the order of operations so that R'O⁻ acts as the nucleophile.

---

* Considered more fully in Chapters T 6, 10, 12, and 23.

Analysis 1

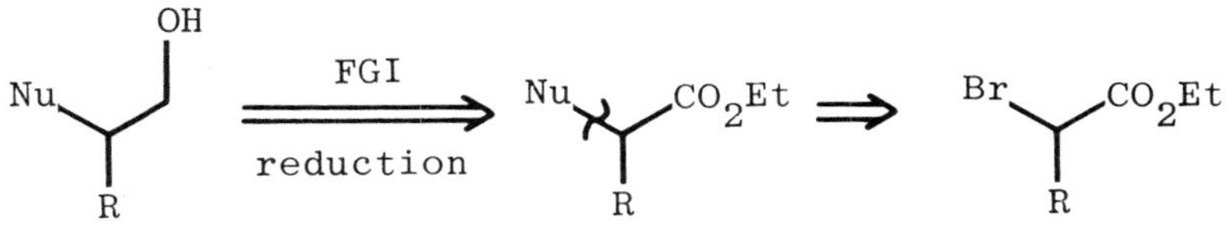

Nu-CH(R)-CH2OH (7a) $\xrightarrow{\text{FGI reduction}}$ Nu-CR-CO2Et (8) $\Rightarrow$ Br-CHR-CO2Et

Synthesis 1

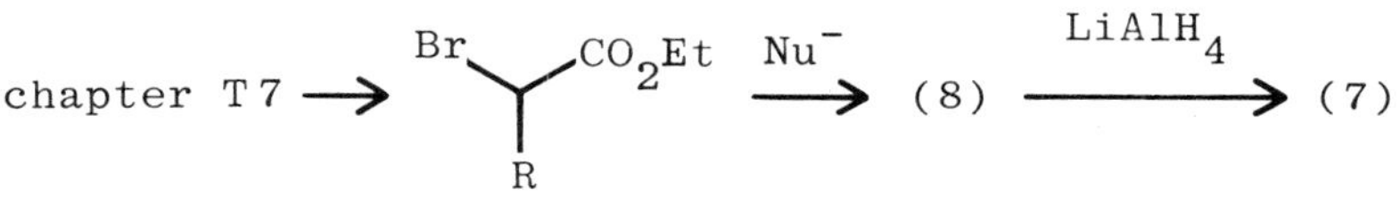

chapter T7 → Br-CHR-CO2Et $\xrightarrow{Nu^-}$ (8) $\xrightarrow{LiAlH_4}$ (7)

Analysis 2

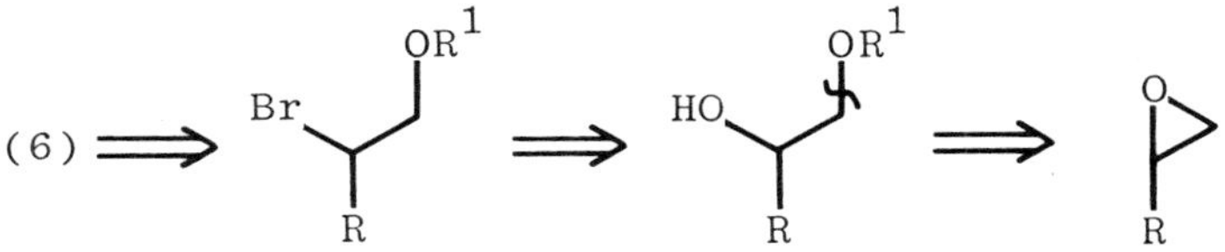

(6) $\Rightarrow$ Br-CH(R)-CH2OR^1 $\Rightarrow$ HO-CH(R)-CH2OR^1 $\Rightarrow$ epoxide (9)

Synthesis 2

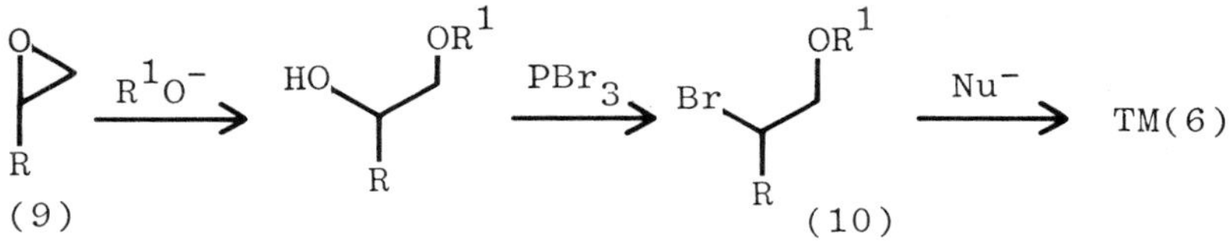

(9) $\xrightarrow{R^1O^-}$ HO-CH(R)-CH2OR^1 $\xrightarrow{PBr_3}$ Br-CH(R)-CH2OR^1 (10) $\xrightarrow{Nu^-}$ TM(6)

Note that it is necessary to open the epoxide (9) and substitute to get (10) - direct attack on a double bond would give[72] the wrong isomer (11). Inversion of

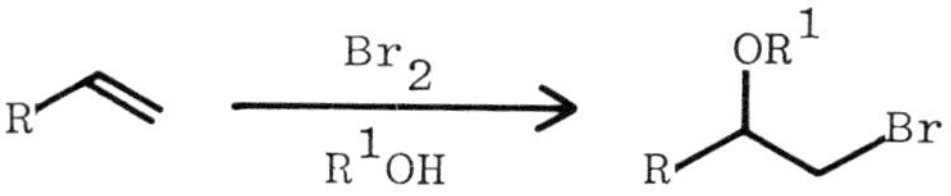

R-CH=CH2 $\xrightarrow[R^1OH]{Br_2}$ R-CH(OR^1)-CH2Br (11)

polarity (p T 52) is vital.

2. Considering these strategies, how might you synthesise the compounds (12) which are used to inhibit the germination of weeds.[73]

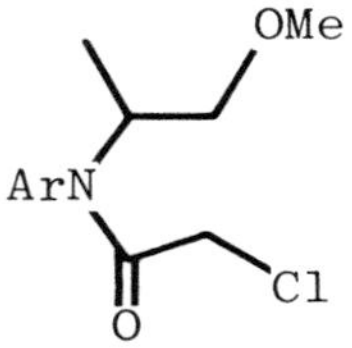

(12)

Answer : The chloroacetyl group can be disconnected first to leave (13) - just the kind of compound we have been discussing. The strategy used in the published synthesis[74] was the second of the two we considered.

Analysis

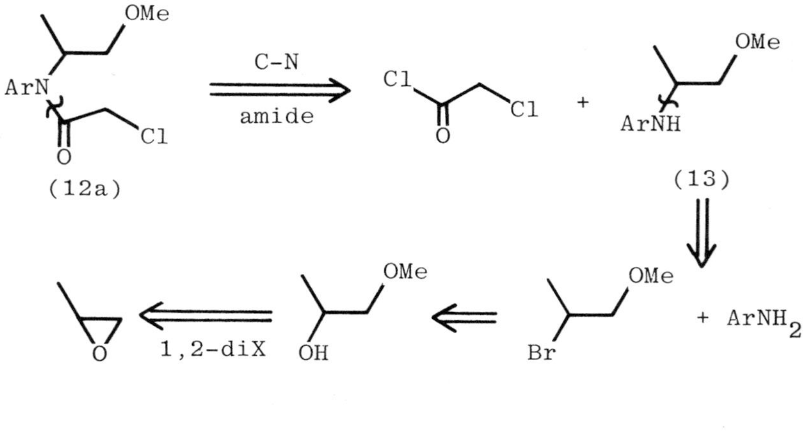

With MeO⁻ as nucleophile, attack occurs cleanly at the right atom in the epoxide,[75] the bromide is made[76] with PBr_3 and the synthesis finished according to plan.

Synthesis[74-76]

$$\text{epoxide} \xrightarrow[\text{MeOH}]{\text{MeO}^-} \text{HOCH}_2\text{CH(Me)OMe (63\%)} \xrightarrow{PBr_3} \text{BrCH}_2\text{CH(Me)OMe (58\%)}$$

$$\xrightarrow{ArNH_2} (13) \xrightarrow{ClCOCH_2Cl} TM(12)$$

Cyclisation
Example

On page T 54 we use disconnection (a) to devise a synthesis of morpholines (e.g. 14). Since cyclisation is so easy, we might consider the alternative double C-N disconnection (b) and use the bis-chloro ether (15), made on page

Analysis

(14) morpholine with N-Ph, disconnections a and b

$\xrightarrow{a}$ HOCH₂CH₂–N(Ph)–CH₂CH₂OH $\xrightarrow{1,2\text{-diX}}$ $PhNH_2$ + epoxide

$\xrightarrow{b}$ $PhNH_2$ + ClCH₂CH₂–O–CH₂CH₂Cl (15)

The cyclisation does indeed turn out to be easy : no intermediates can be isolated.

Synthesis[55]

$$(15) \xrightarrow{PhNH_2} TM(14)$$

Problem

Carry out similar disconnections on TM (16) and suggest how the required reagent might be made.

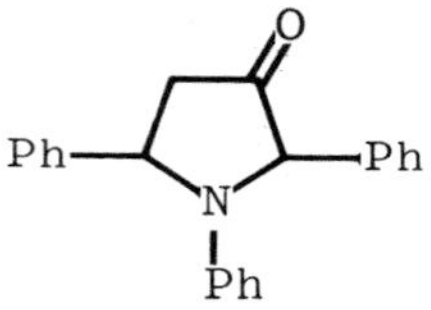

(16) — 1-phenyl-2,5-diphenyl-pyrrolidin-3-one

Answer : There are 1,2-diX and 1,3-diX disconnections to be made, giving (17) as the starting material which could be made by the bromination of enone (18).

Analysis

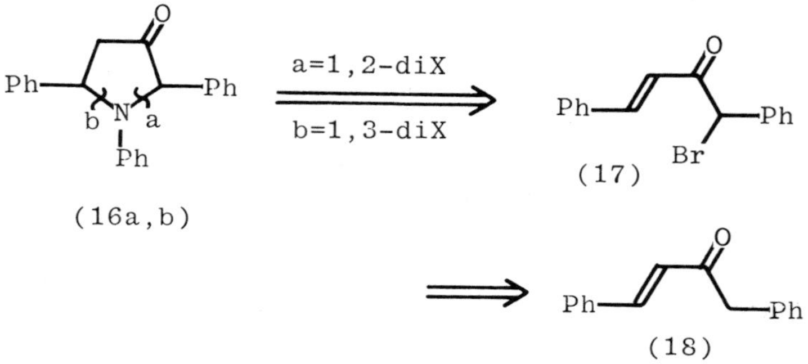

(16a,b) $\xRightarrow{\substack{a=1,2-diX \\ b=1,3-diX}}$ (17)

$\Rightarrow$ (18)

The bromination was carried out with N-bromosuccinimide (NBS, 19), a source of molecular bromine, and the cyclisation proved as easy as expected.
Synthesis[77]

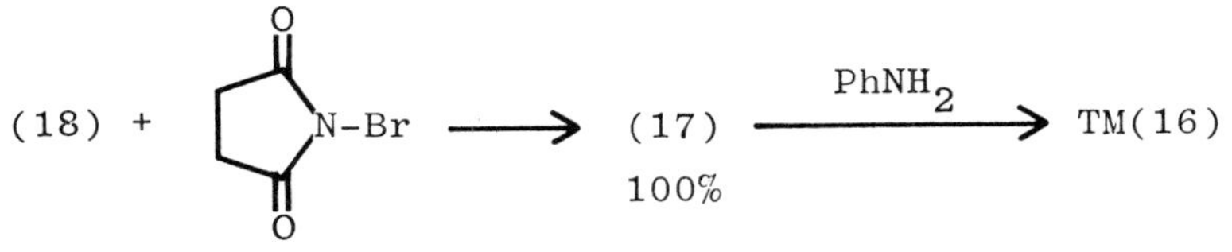

(18) + [succinimide N-Br structure] (19) ⟶ (17) 100% $\xrightarrow{PhNH_2}$ TM(16)

CHAPTER 8

Amine Synthesis

Simple Example

Amine (1) was needed to study the stereochemistry of alkylation reactions. The primary alkyl group had best come from an amide or an imine while the secondary alkyl group must come from an imine. The disconnections may be carried out in any order.

Analysis

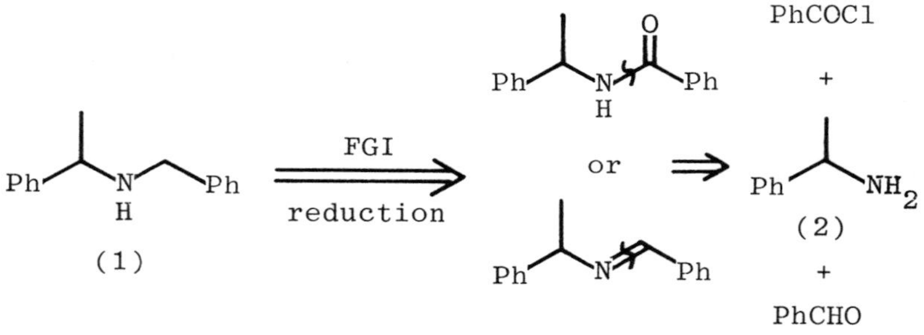

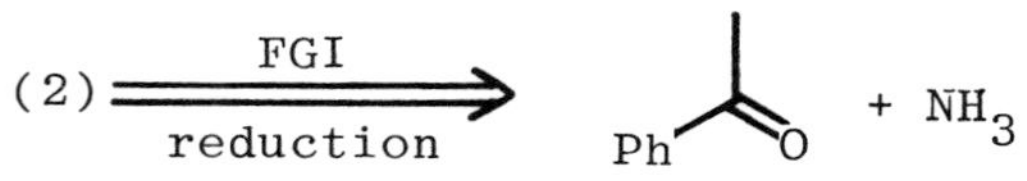

$$Ph\underset{H}{\overset{|}{\text{N}}}\text{-CH}_2\text{-Ph} \;(1) \xrightarrow{\text{FGI reduction}} Ph\underset{H}{\overset{|}{\text{N}}}\text{-COPh} \;\text{or}\; Ph\underset{}{\overset{|}{\text{N}}}\text{=CHPh} \Longrightarrow Ph\overset{|}{\text{CH-NH}_2} \;(2) + PhCOCl + PhCHO$$

$$(2) \xrightarrow{\text{FGI reduction}} PhCOCH_3 + NH_3$$

One synthesis[78] uses the imine method for the second stage.

Synthesis[78,79]

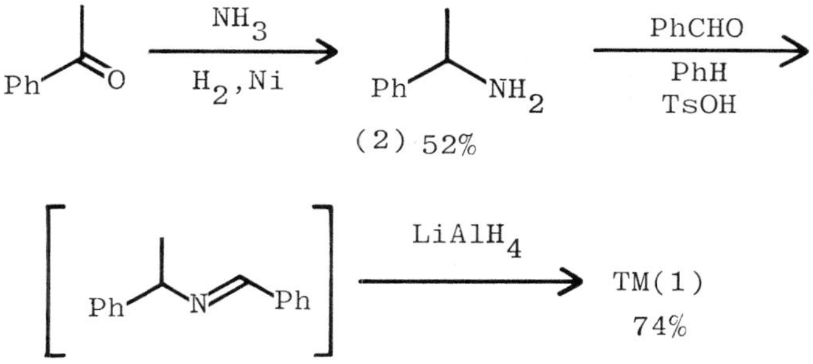

Ph-CO-O →(NH₃, H₂, Ni)→ Ph-CH(CH₃)-NH₂ →(PhCHO, PhH, TsOH)→
(2) 52%

[Ph-CH(CH₃)-N=CH-Ph] →(LiAlH₄)→ TM(1)
74%

Simple Problems

1. Suggest a synthesis for butam (3), a pre-emergent herbicide.[80]

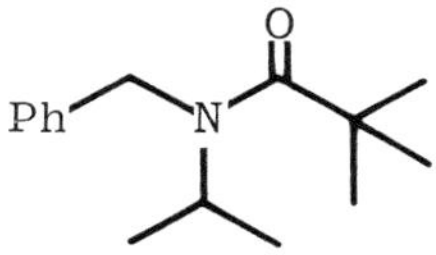

(3) Ph-CH₂-N(iPr)-C(=O)-C(CH₃)₃

Answer : Disconnection of the amide reveals secondary amine (4). This can be disconnected in various ways using reduction of amides (benzyl side only) or imides (either side). Again the order doesn't matter.

Analysis

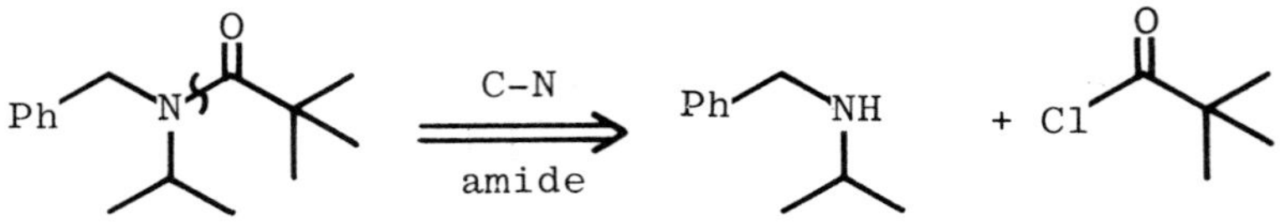

(3a) ⇒[C-N amide] (4) + Cl-CO-C(CH₃)₃

where (3a) is PhCH₂-N(iPr)-CO-C(CH₃)₃ and (4) is PhCH₂-NH-iPr

(4) ⇓ FGI reduction

PhCH=N-iPr   or   PhCO-NH-iPr   or   PhCH₂-N=C(CH₃)₂

⇓   ⇓   ⇓

PhCHO + H₂N-iPr   PhCOCl + H₂N-iPr   PhCH₂NH₂ + (CH₃)₂C=O

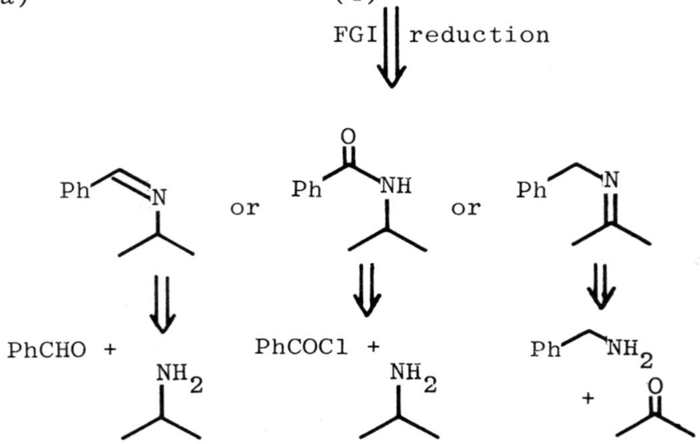

One published synthesis[81] uses the first route with catalytic reduction of the imine.

Synthesis[80,81]

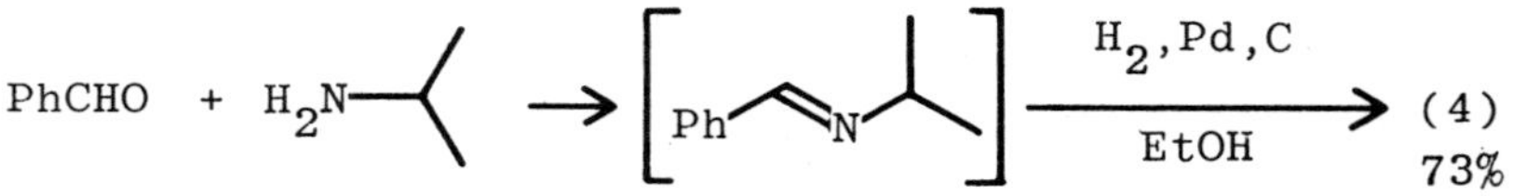

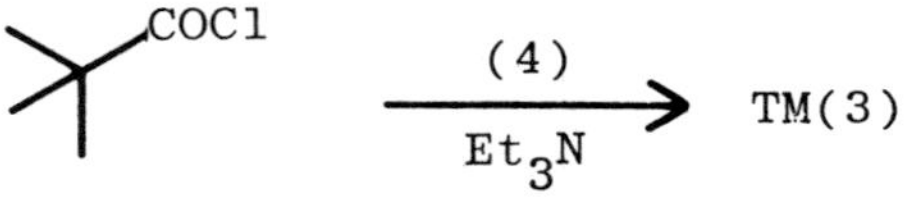

PhCHO + H₂N-iPr → [Ph-CH=N-iPr] —[H₂, Pd, C / EtOH]→ (4) 73%

(CH₃)₃C-COCl —[(4) / Et₃N]→ TM(3)

2. Suggest a synthesis for amine (5), needed to make the potent neuronal excitant α-kainic acid.[82]

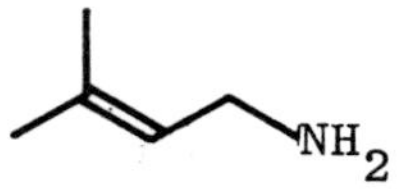

(CH₃)₂C=CH-CH₂-NH₂   (5)

If you cannot see how to do this, look at Chapter T1 and page T 65.

Answer: Reduction routes are risky e.g. from (6) or (7), as we might reduce the double bond too.

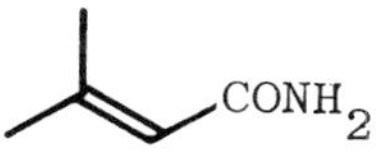

(6) — CH$_3$C(CH$_3$)=CH-CONH$_2$

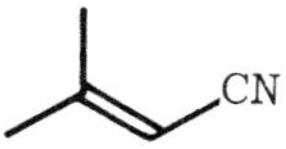

(7) — CH$_3$C(CH$_3$)=CH-CN

The best bet is to use a reagent for the synthon NH_2^- and the available chloride (8) (p T1).

Analysis

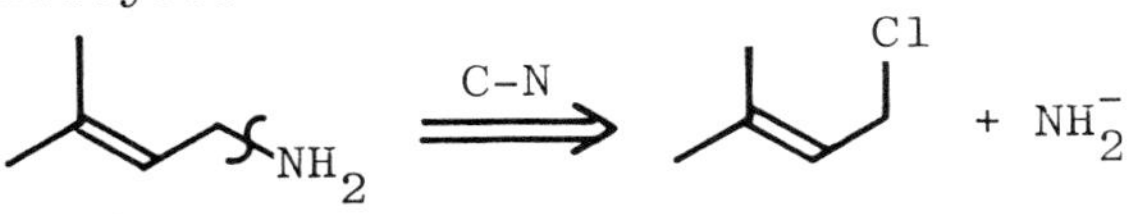

(5a) ⟹ (8) + NH_2^-
   C-N

The published synthesis uses the phthalimide method on the corresponding bromide.

Synthesis[82]

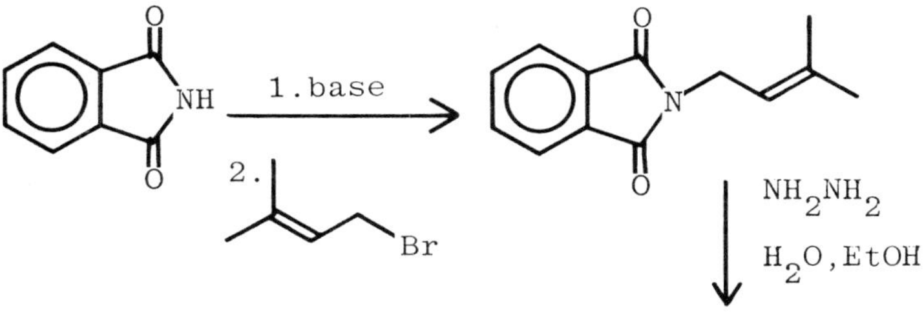

phthalimide-NH → 1. base; 2. (CH$_3$)$_2$C=CH-CH$_2$Br → N-alkylated phthalimide → NH$_2$NH$_2$, H$_2$O, EtOH → TM(5)

The next stage in the route to α-kainic acid was the alkylation of (5) with allylic halide (9). Even in this favourable case (the halide is allylic) only 36% of (10) was isolated and "the low yield was caused by uncontrollable production of large amounts of dialkylation material". Well, you were warned!

$$\text{Me}_2\text{C=CHCH}_2\text{NH}_2 + \text{BrCH}_2\text{CH=CHCO}_2\text{Et} \xrightarrow{\text{K}_2\text{CO}_3, \text{H}_2\text{O}} \text{Me}_2\text{C=CHCH}_2\text{NHCH}_2\text{CH=CHCO}_2\text{Et}$$
(9) → (10)

Example

Phentamine (11), used in the treatment of obesity, is a t-alkyl amine. The Ritter reaction, using HCN as the nitrile (p T63), is ideal for this.

Analysis

Ph–C(Me)$_2$–CH$_2$–NH$_2$ $\xRightarrow{\text{C-N}}$ Ph–C(Me)$_2$–CH$_2^+$ + $^-$NH$_2$
(11)
= Ph–C(Me)$_2$–CH$_2$–OH (HCN)

Synthesis[83]

Ph–C(Me)$_2$–CH$_2$–OH $\xrightarrow[\text{HOAc}]{\text{HCN, H}_2\text{SO}_4}$ Ph–C(Me)$_2$–CH$_2$–NHCHO $\xrightarrow[\text{H}_2\text{O}]{\text{HO}^-}$ TM(11) 92%

Problem

Phentamine analogues, e.g. (12), are also useful drugs. Suggest a synthesis.

(12): 4-MeO-C$_6$H$_4$-CH$_2$-C(Me)$_2$-NH-CH$_2$CH$_2$CH$_3$

Answer : The n-propyl group had best come from an amide (13) which could be the Ritter product if EtCN is used. Making (13) by more conventional methods is fine, too.

Analysis

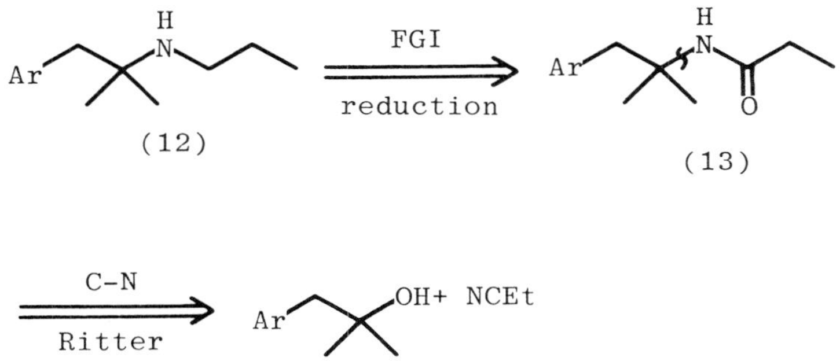

$$Ar\text{-}C(CH_3)_2\text{-}NH\text{-}CH_2CH_2CH_3 \quad (12) \xRightarrow[\text{reduction}]{\text{FGI}} Ar\text{-}C(CH_3)_2\text{-}NH\text{-}CO\text{-}CH_2CH_3 \quad (13)$$

$$\xRightarrow[\text{Ritter}]{\text{C-N}} Ar\text{-}C(CH_3)_2\text{-}OH + NCEt$$

Diborane was used in this reduction, but LiAlH$_4$ should do just as well.

Synthesis[84]

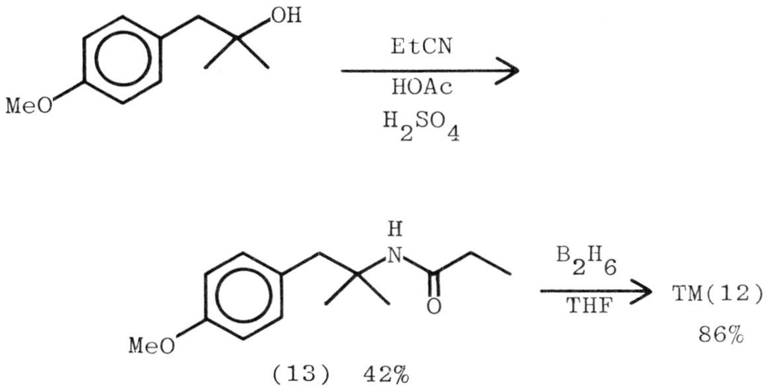

MeO-C$_6$H$_4$-CH$_2$-C(CH$_3$)$_2$-OH $\xrightarrow[\text{H}_2\text{SO}_4]{\text{EtCN, HOAc}}$

MeO-C$_6$H$_4$-CH$_2$-C(CH$_3$)$_2$-NH-CO-CH$_2$CH$_3$ $\xrightarrow[\text{THF}]{\text{B}_2\text{H}_6}$ TM(12) 86%

(13) 42%

More Advanced Example

Diamines find considerable use as pharmaceuticals : here is one example followed by two diamine problems. Histapyrrodine (14) is used as an antihistamine.

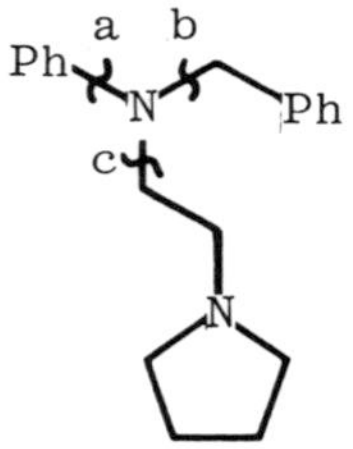

(14)

Disconnecting one of the C-N bonds around the tertiary amine will help : (a) is not possible, (b) would be possible after FGI to the amide but (c) separates the two nitrogens and requires the easily synthesised (16) as intermediate. We shall follow (c).

Analysis

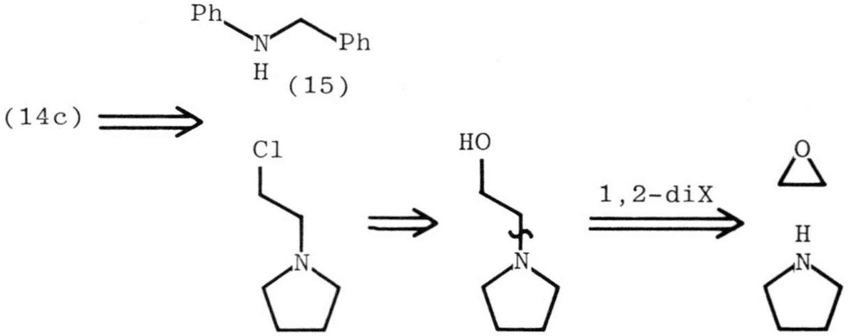

$(14c) \Longrightarrow$ (15) $\Rightarrow$ $\xrightarrow{1,2-diX}$

Some reductive route (via amide or imine) will give (15) and the rest of the synthesis is simple. The final alkylation is good as (14) is crowded and unlikely to form a quaternary salt.

Synthesis[85]

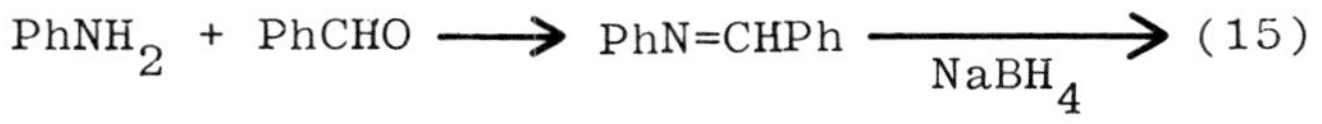

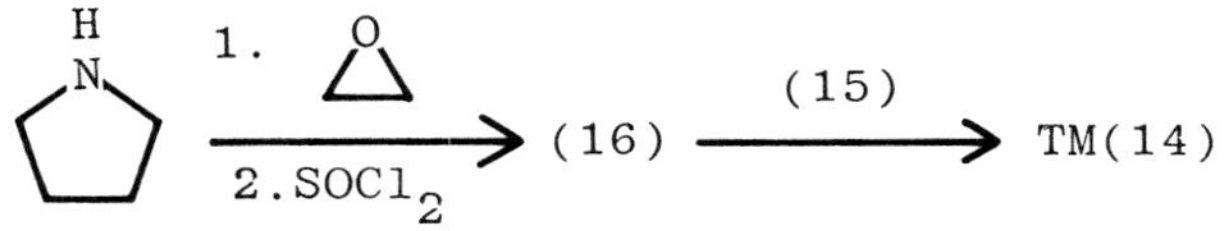

PhNH$_2$ + PhCHO $\longrightarrow$ PhN=CHPh $\xrightarrow{\text{NaBH}_4}$ (15)

[pyrrolidine N-H] $\xrightarrow[\text{2.SOCl}_2]{\text{1. }\triangle\text{O (epoxide)}}$ (16) $\xrightarrow{(15)}$ TM(14)

Problems

1. Diamines (17) are used in the manufacture of pharmaceuticals and agrochemicals. How might they be made?

R$_2$N⁀⁀⁀NH$_2$

(17)

Answer : The primary amine might come from an amide or a cyanide, e.g. (18) : a 1,3-diX disconnection (reverse Michael) is then excellent.

Analysis

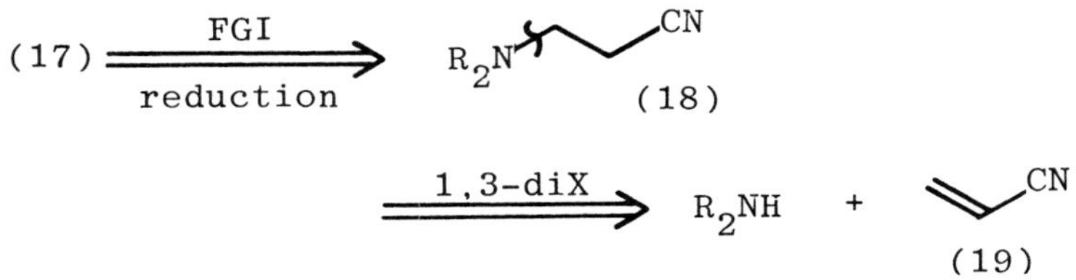

(17) $\xRightarrow[\text{reduction}]{\text{FGI}}$ R$_2$N⁀⁀CN

(18)

$\xRightarrow{\text{1,3-diX}}$ R$_2$NH + ⁀⁀CN

(19)

Amines[86] add very cleanly to acrylonitrile (19) and catalytic reduction of (18) gives TM (17).

Synthesis[86,87]

$$R_2NH \xrightarrow{(19)} (18) \xrightarrow[\text{or Na/EtOH}]{H_2 \text{ cat}} TM(17)$$

2. Gabamide (20) is a psychotropic agent. Suggest a synthesis?

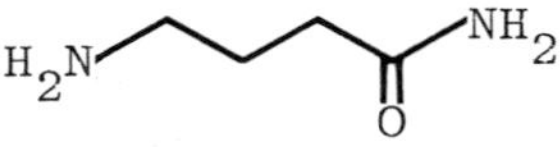

$$H_2N\frown\frown\underset{O}{\overset{\|}{C}}NH_2 \quad (20)$$

Answer : The primary amino group must come from a nitrile (21) and this time the Michael reaction is reversed :

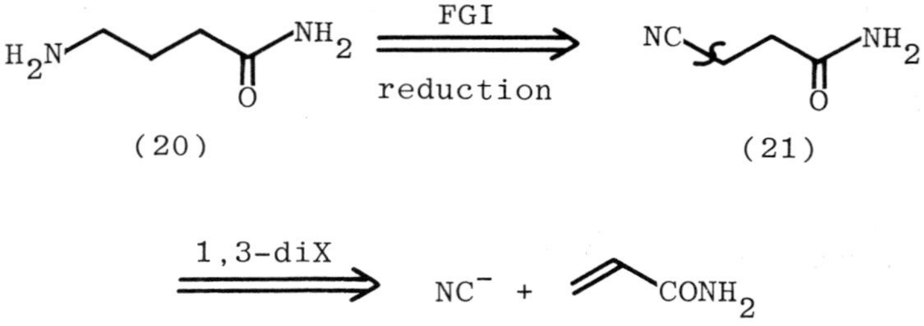

$$H_2N\frown\frown\underset{O}{\overset{\|}{C}}NH_2 \underset{\text{reduction}}{\overset{FGI}{\Longrightarrow}} NC\frown\frown\underset{O}{\overset{\|}{C}}NH_2$$

(20)   (21)

$$\xRightarrow{1,3\text{-diX}} NC^- + \diagup\!\!\!\diagdown CONH_2$$

Catalytic reduction is best for (21) → (20). Chemoselectivity in catalytic reductions depends solely on thermodynamic stability and the amide is much more

stable (conjugation) than the nitrile hence the excellent chemoselectivity. The acid is to ensure the free amine (20) is not formed or it might poison the catalyst.
Synthesis[88]

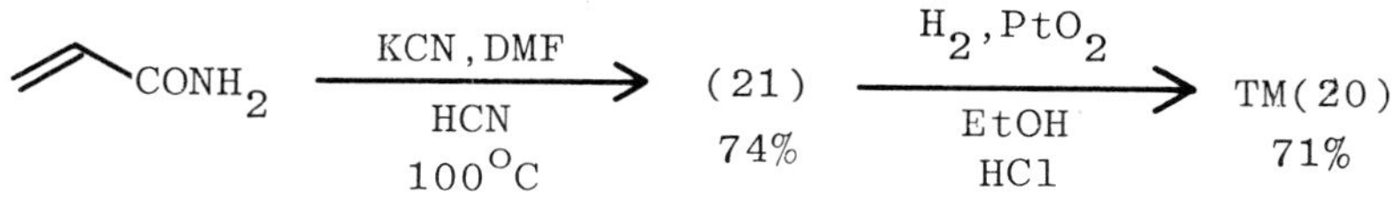

$$\text{CH}_2=\text{CH-CONH}_2 \xrightarrow[\substack{\text{HCN} \\ 100°\text{C}}]{\text{KCN, DMF}} (21) \xrightarrow[\substack{\text{EtOH} \\ \text{HCl}}]{\text{H}_2, \text{PtO}_2} \text{TM(20)}$$

74%　　　71%

CHAPTER 9

Strategy IV: Protecting Groups

New Protecting Groups

New protecting groups are always being invented : the improvements are usually ease of addition, selective stability, and most of all, ease and selectivity of removal. In 1969 a new carboxylic acid protecting group, the ester of alcohol (1), was introduced.[89]

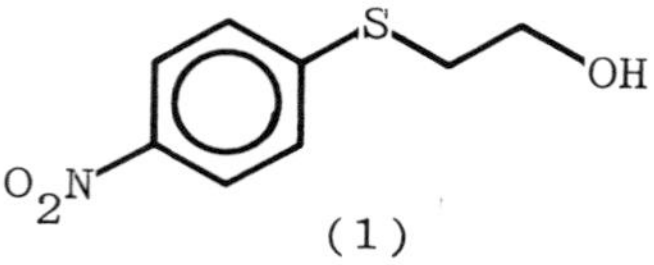

(1)

The reagent (1) is easily made from thiol (3) and ethylene oxide or 2-chloroethanol (2). The acid RCO_2H is then protected by acid-catalysed ester (4) formation.

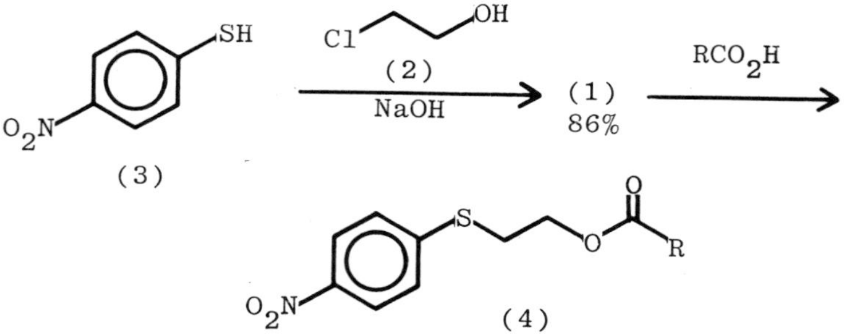

The reaction for which protection was needed, say $R \rightarrow R^1$, is now carried out and now the protecting group must be removed. This may be done by an elimination reaction (6) on the oxidised compound (5) under much milder conditions than normal ester hydrolysis, as the p-nitrophenyl sulphonyl group is strongly anion-stabilising.

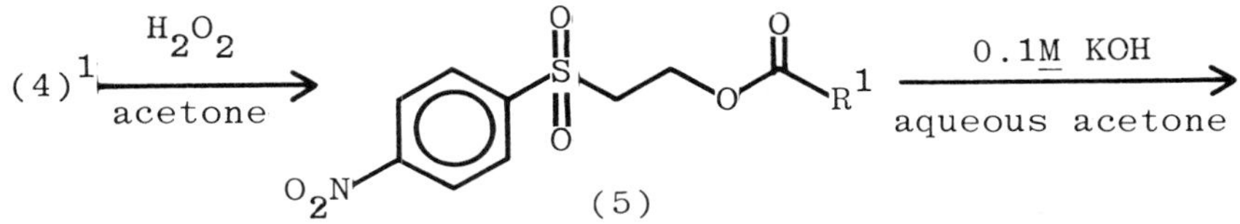

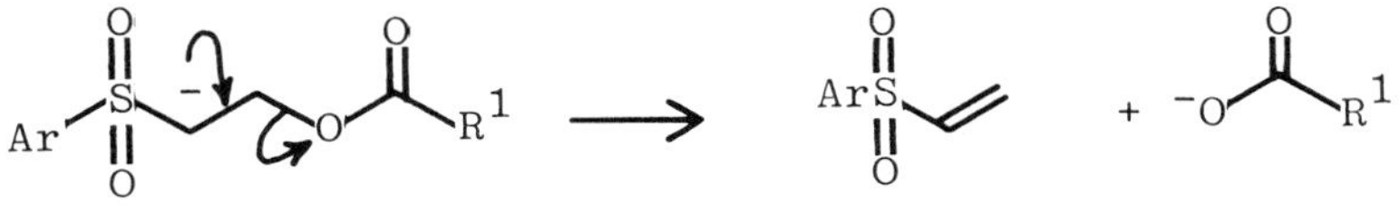

Problem

Amines might be protected as the amides (7). What reagent would you need to put in this protecting group, and how would you make it? The product amine (8) might be released by treatment with zinc. Suggest a mechanism for this reaction.

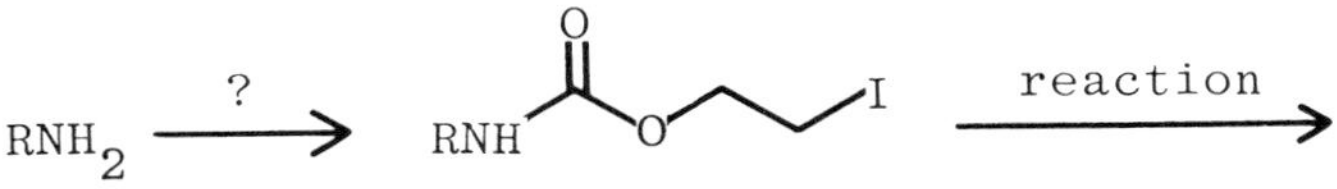

$$\underset{(8)}{\text{AcO-CH}_2\text{CH}_2\text{-I}} \xrightarrow[\text{MeOH}]{\overset{?}{\text{Zn}}} R^1NH_2 + CO_2 + CH_2=CH_2$$

Answer : The chloroformate (9) would be needed. Ester disconnection reveals available 2-iodoethanol and $COCl_2$.

Analysis

$$\underset{(7a)}{RNH\text{-CO-O-CH}_2\text{CH}_2\text{-I}} \xrightarrow[\text{amide}]{C-N} \underset{(9)}{Cl\text{-CO-O-CH}_2\text{CH}_2\text{-I}}$$

$$\xrightarrow[\text{ester}]{C-O} Cl\text{-CO-Cl} + HO\text{-CH}_2\text{CH}_2\text{-I}$$

Chemoselectivity is all right in the synthesis of (9) (p T 36), and the protected amines can be made from (9) with base catalysis. This reaction is also chemoselective as amines prefer to attack carbonyl groups to make stable, conjugated amides rather than attach Sp^3 carbon atoms.

Synthesis[90]

$$HO\text{-CH}_2\text{CH}_2\text{-I} \xrightarrow{COCl_2} \underset{(9)}{Cl\text{-CO-O-CH}_2\text{CH}_2\text{-I}} \xrightarrow[\substack{\text{PhH}\\\text{NaOH}}]{RNH_2} (7)$$

Zinc is a two-electron donating metal so it attacks the iodine atom in an E2 reaction.

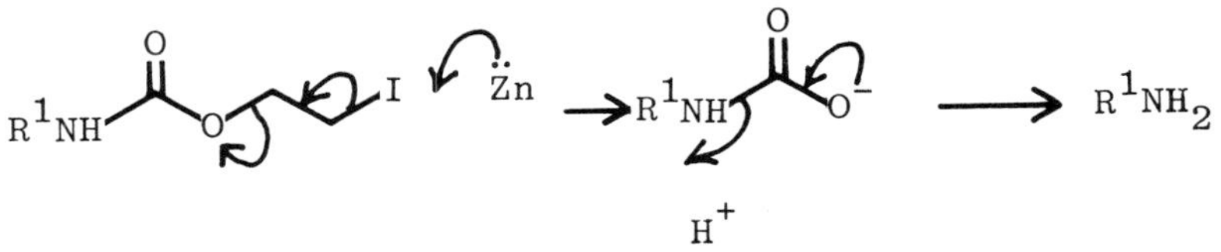

$R^1NH-CO-O-CH_2-I$ + Zn → $R^1NH-CO-O^-$ + H$^+$ → R^1NH_2

Use of Protecting Groups
Simple Problems

1. What unwanted reactions would happen during attempted oxidation of (10) to (11) and reduction of (12) to (13). How might they be prevented?

(a) (10) —KMnO$_4$→ (11) dihydroxycitronellal used in perfumes

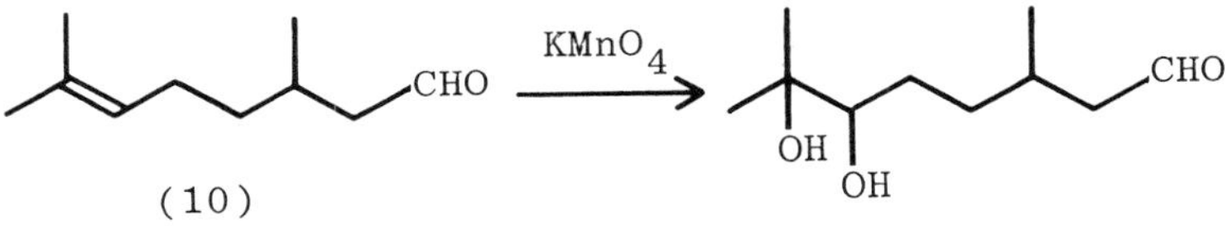

(b) (12) —LiAlH$_4$→ (13) used in a tetracycline antibiotic synthesis

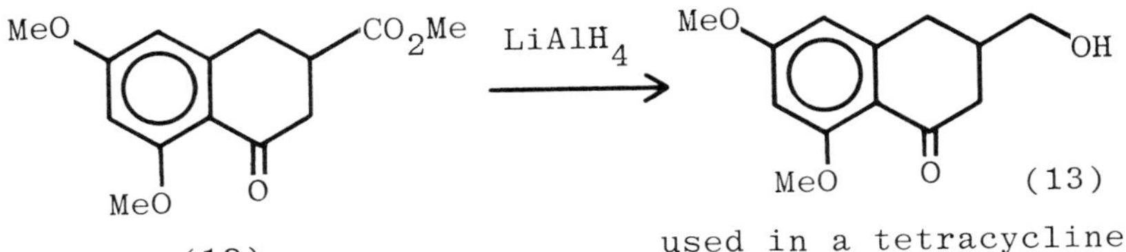

Answer: (a) The aldehyde in (10) would be oxidised to CO$_2$H. Protection by acetal (14) formation is the answer.[91]

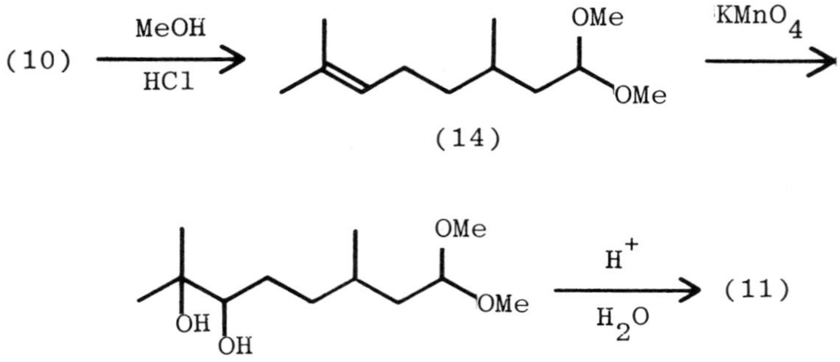

(10) $\xrightarrow[\text{HCl}]{\text{MeOH}}$ [structure (14): CH₂=C(Me)-CH₂-CH₂-CH(Me)-CH₂-CH(OMe)₂] $\xrightarrow{\text{KMnO}_4}$

[structure: (HO)(Me)₂C-CH(OH)-CH₂-CH₂-CH(Me)-CH₂-CH(OMe)₂] $\xrightarrow[\text{H}_2\text{O}]{\text{H}^+}$ (11)

(b) The ketone will be reduced : again acetal protection is the answer.[92] Note the use of the cyclic ketal (15) and the dehydrating agent $HC(OEt)_3$, ethyl orthoformate, both giving higher yields.

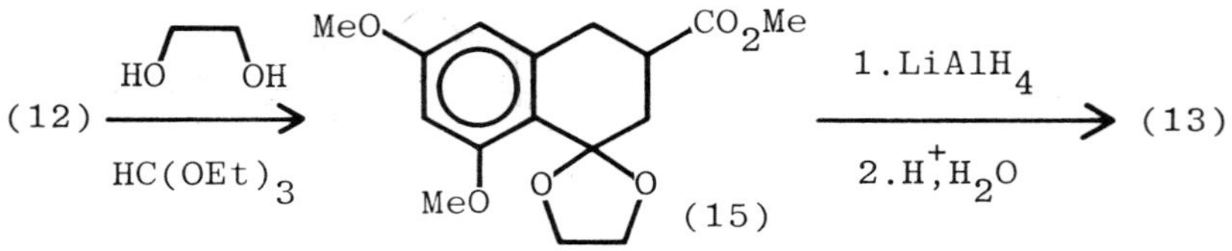

(12) $\xrightarrow[\text{HC(OEt)}_3]{\text{HO}\frown\text{OH}}$ (15) $\xrightarrow[\text{2. H}^+,\text{H}_2\text{O}]{\text{1. LiAlH}_4}$ (13)

2. How would you carry out the conversion (16) → (17), the first steps of Marshall's bulnesol[93] synthesis?

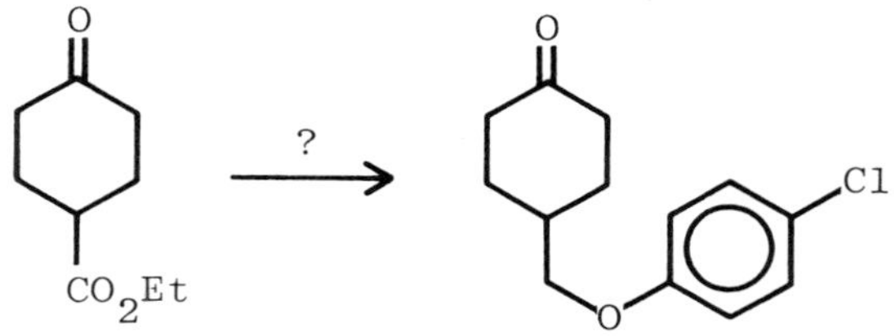

(16) 4-(ethoxycarbonyl)cyclohexanone → (17) 4-[(4-chlorophenoxy)methyl]cyclohexanone

Answer : Ether disconnection (17a) must be on the side away from the aromatic ring. Alcohol (18) is made by reduction of the ester group in (16) and this will require protection of the ketone.

Analysis

[Scheme: cyclohexanone with CH2-O-C6H4-Cl (17a) ⇒ (C-O ether) cyclohexanone with CH2-X ⇒ cyclohexanone with CH2-OH (18)]

The cyclic ketal (19) was again used as protection and the methane sulphonyl (MeSO$_2$ or Ms, mesylate) group was used as the leaving group X (p T 28) (20). The protecting group was left in until the end to avoid any side reactions.

Synthesis[93]

[Scheme: (16) + HO-OH, H$^+$ → ketal (19) with CO$_2$Et → LiAlH$_4$ → ketal with CH$_2$OH → MeSO$_2$Cl, Et$_3$N → (20) with OMs → ArOH, base → ketal with OAr → H$^+$, H$_2$O → TM(17)]

The p-chlorophenyl group is in fact itself a protecting group which was not removed until a late stage in the synthesis.

More Advanced Example

In the synthesis[94] of the A-chain of insulin, 21 amino acids must be combined in a known order. The first steps use a protected dipeptide (22) and protected cysteine (21)

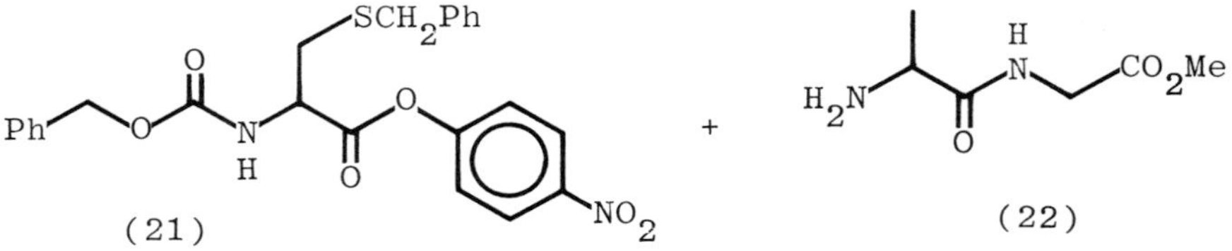

(21) + (22)

The only nucleophile is the free amine in (22) and the most reactive electrophile is the p-nitro-phenyl ester in (21). These combine to give the tripeptide (23).

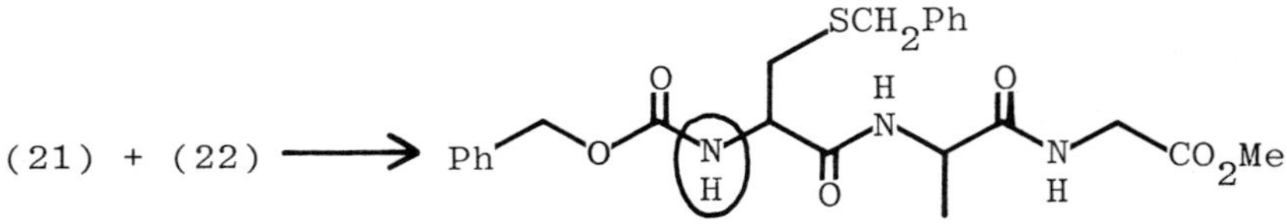

(21) + (22) ⟶ (23)

The next peptide bond must be made to the ringed nitrogen atom, so treatment with HBr frees this amine without affecting the other protecting groups (table T 9.1).

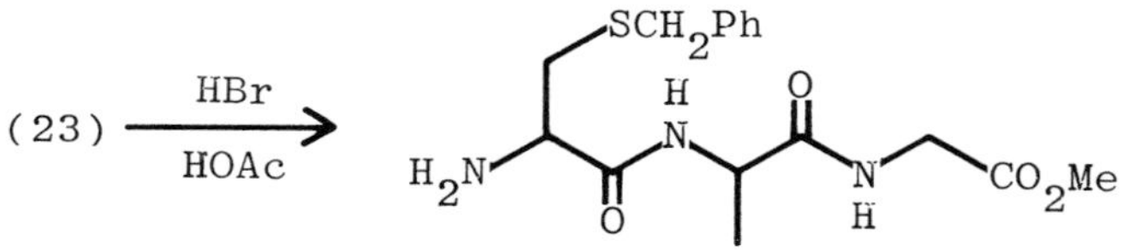

(23) $\xrightarrow[\text{HOAc}]{\text{HBr}}$ (24)

Reaction with another suitably protected p-nitro-phenyl ester now adds the next amino acid (another cysteine).

Problem

Another part of the same insulin synthesis involves this sequence.

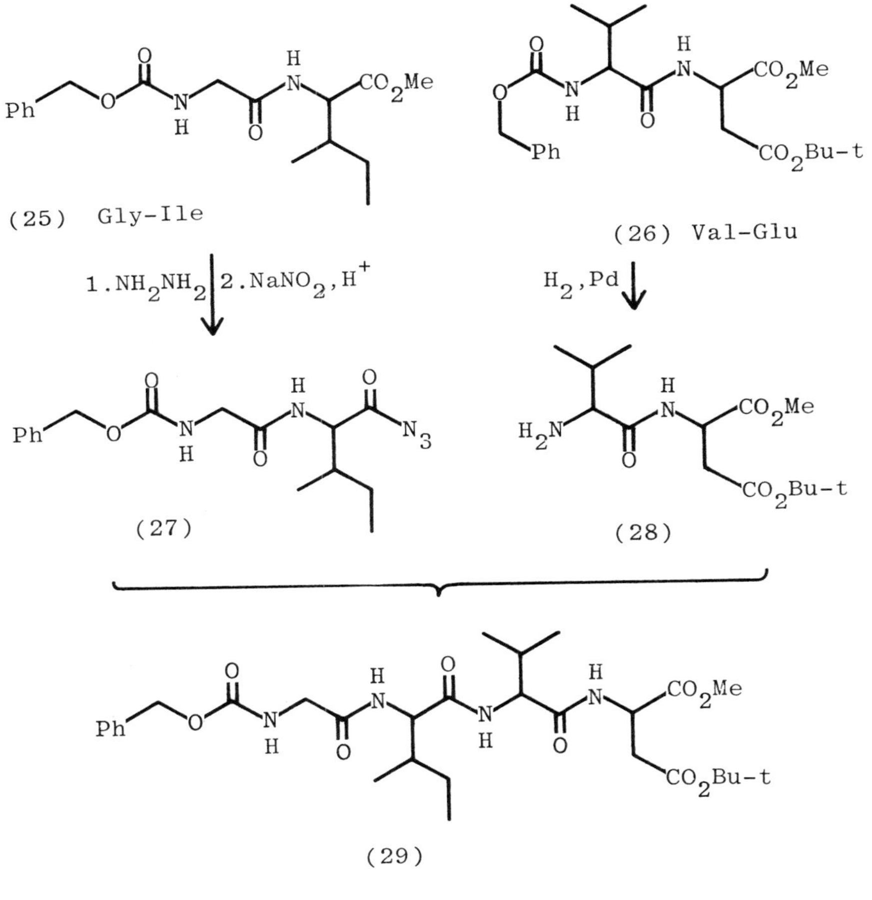

(25) Gly-Ile

(26) Val-Glu

1. NH_2NH_2 | 2. $NaNO_2$, H^+

H_2, Pd

(27)

(28)

(29)

(a) Explain the chemoselectivity of (25) → (27)?

(b) Why was catalytic hydrogenation used for (26) → (28) while HBr was used for the same reaction on (23)?

(c) How would you free the terminal NH_2 group of (29) to react with another activated ester (e.g. 20), or activate the CO_2H end to react with another free amine (e.g. 24)?

Answers

(a) Hydrazine, NH_2NH_2, is a good basic nucleophile and will attack the most electrophilic carbonyl group available. In (25) the methyl ester is more electrophilic than either of the amides. Diazotisation then gives the azide :

$$R-\underset{\underset{}{}}{C(=O)}-NH-NH_2 \xrightarrow{\text{'NO}^+\text{'}} R-\underset{\underset{}{}}{C(=O)}-\overset{+}{N}=N=\overset{-}{N}$$

(27)

(b) The t-butyl ester protecting the side-chain of Glu in (26) would be hydrolysed by acid (p T 71). In (23) the S-benzyl group on the cysteine side chain might be cleaved by hydrogenolysis.

(c) Use H_2-Pd again for the N-terminus as the t-butyl ester is still present, and hydrazine and diazotisation again for the carboxyl end.

Selective protection and activation of amino acids and peptides has now reached a highly sophisticated level and requires specialised knowledge for the most efficient use.

CHAPTER 10

One-Group C—C Disconnection—I: Alcohols

Examples :

Few carbanions are stable enough to be formed in solution as genuine intermediates by removal of a C-H proton by base : $^-CX_3$ (X = Cl, Br, I) are examples. The ester (1) is used as a rose perfume.[95] Disconnection of the ester reveals alcohol (2) which can be disconnected to $^-CCl_3$ and benzaldehyde.

Analysis

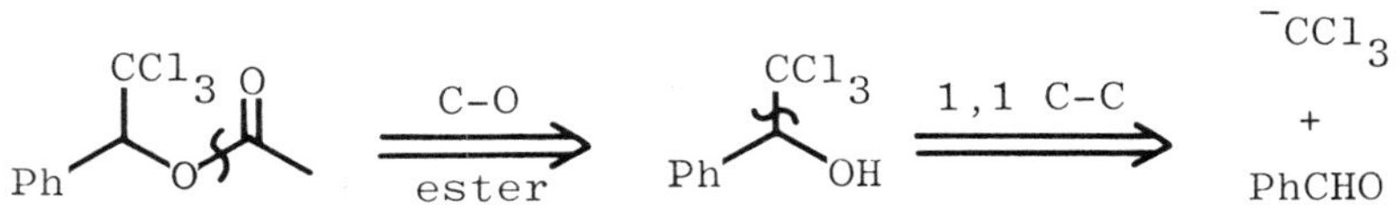

$$Ph\underset{(1)}{\overset{CCl_3}{\diagup}}O\diagdown\overset{O}{\diagup}\quad\underset{ester}{\overset{C-O}{\Longrightarrow}}\quad Ph\underset{(2)}{\overset{CCl_3}{\diagup}}OH\quad\overset{1,1\ C-C}{\Longrightarrow}\quad {}^-CCl_3 + PhCHO$$

The carbanion is made from $CHCl_3$ with KOH in the presence of benzaldehyde.[96]

Synthesis[95,96]

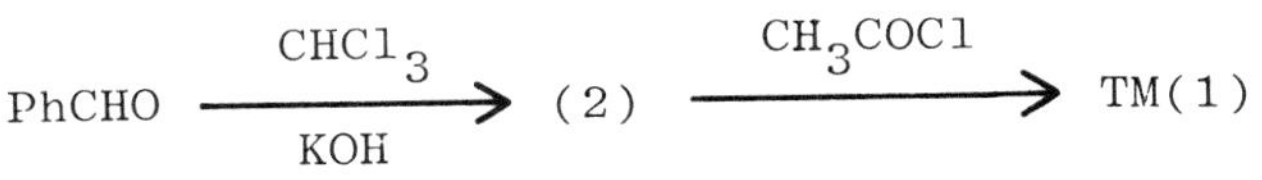

$$PhCHO\ \xrightarrow[KOH]{CHCl_3}\ (2)\ \xrightarrow{CH_3COCl}\ TM(1)$$

2. More typical examples are the amino ethers (3) used as anti-histamine or anti-Parkinson drugs according to the substituents.[97] These are obviously derived from the alcohols (4) which are made from an aryl Grignard reagent and a benzaldehyde. Either starting material may bear the substituent X : the choice can be made according to availability and so that side reactions are avoided.

Analysis

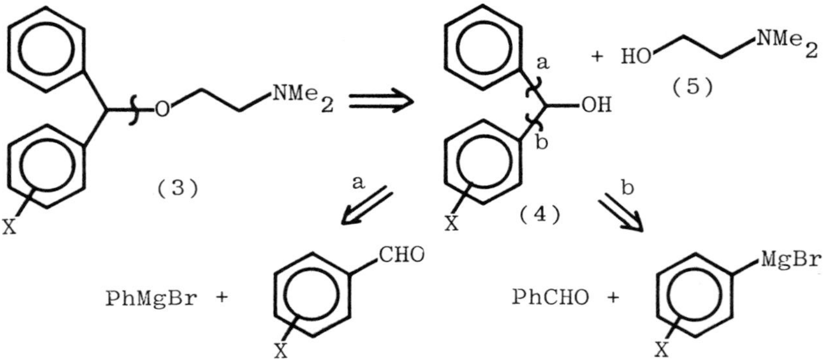

The ether bond is best made from alcohol (5) and the reactive benzylic halide (6).

Synthesis[97]

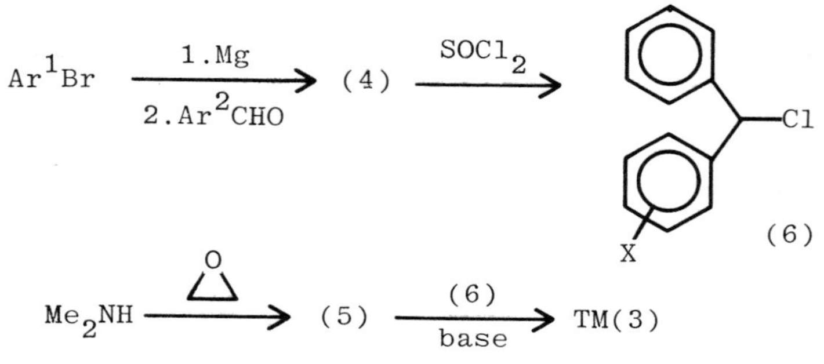

Problems
1. The alcohol (7) was used on page T 30 in a Friedel-Crafts sequence. How would you make it?

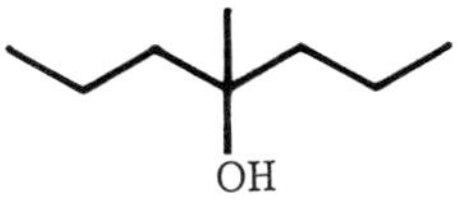

(7)

Answer

Two of the substituents are the same (n-Pr) so they can be added as n-PrMgBr in one step :

Analysis

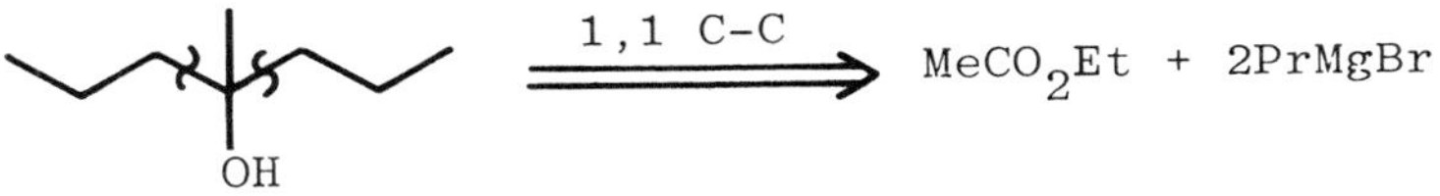

$$\text{(7a)} \xrightarrow{1,1 \text{ C-C}} \text{MeCO}_2\text{Et} + 2\text{PrMgBr}$$

Synthesis[98]

$$\text{n-PrBr} \xrightarrow[\text{2.MeCO}_2\text{Et}]{\text{1.Mg}} \text{TM(7)}$$

2. Aldehyde (8) was needed for a butenolide synthesis. How would (8) be made?

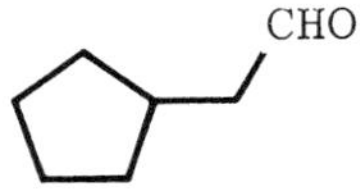

(8)

Answer

Return to the alcohol (9) and disconnect to the simple cyclopentyl compound (10) and ethylene oxide.

Analysis

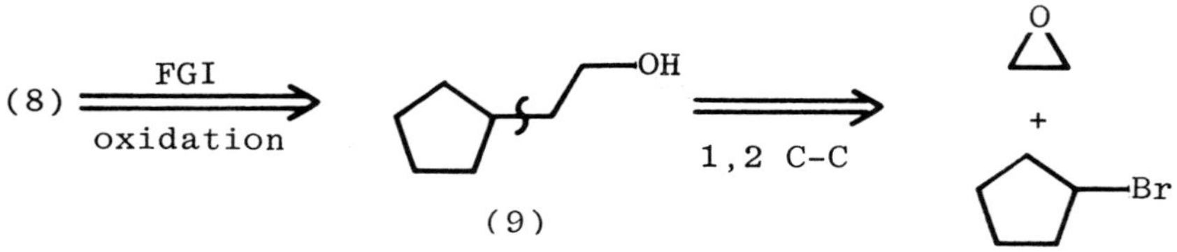

(8) $\xrightarrow[\text{oxidation}]{\text{FGI}}$ [cyclopentyl-CH$_2$CH$_2$-OH] (9) $\xrightarrow{\text{1,2 C-C}}$ [ethylene oxide] + [cyclopentyl-Br] (10)

PCC (table T10.2) is the preferred oxidant here.

Synthesis[99]

(10) $\xrightarrow[\text{2. }\triangle\text{O}]{\text{1. Mg}}$ (9) $\xrightarrow{\text{PCC}}$ TM(8) 72% from (10)

Acids
Problems

1. Acids (11) and (12) were both made by Grignard addition to CO_2 rather than by cyanide displacement (p T 80). Why?

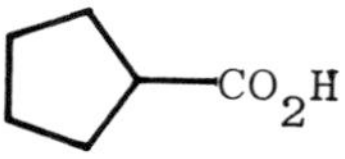

[cyclopentyl-CO$_2$H] (11)   MeO-C$_6$H$_4$-CO$_2$H (12)

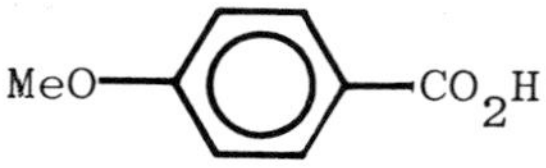

Answer

Cyanide displacement is impossible on halide (14) and gives much elimination on secondary halide (13). The Grignard addition avoids these problems and gives high yields.[100]

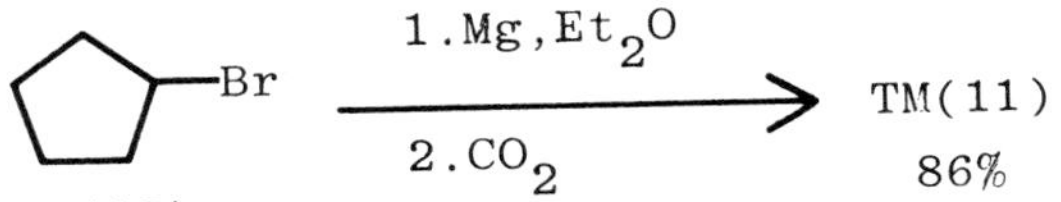

(13) cyclopentyl-Br → 1. Mg, Et$_2$O 2. CO$_2$ → TM(11) 86%

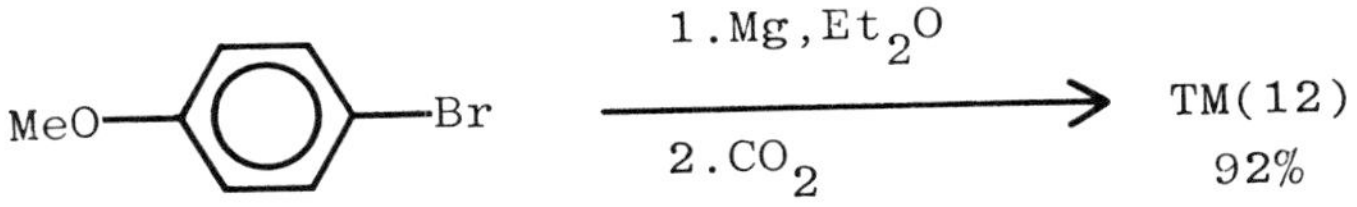

(14) MeO-C$_6$H$_4$-Br → 1. Mg, Et$_2$O 2. CO$_2$ → TM(12) 92%

2. Suggest a synthesis of acid (15) used in an investigation of oxidation reactions.

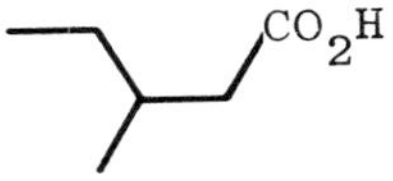

(15)

Answer

The cyanide route via bromide (16) back to bromide (17) would be one obvious method, but the alternative disconnection of two carbon atoms from (18) is shorter.

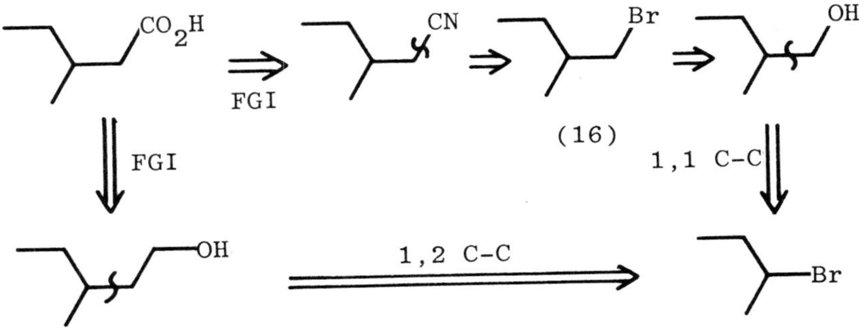

CO_2H ⇒ FGI ⇒ CN ⇒ Br ⇒ OH
                                (16)   1,1 C-C

↓ FGI

OH (18) ⇒ 1,2 C-C ⇒ Br (17)

The shorter route gives good yields : AgO was used as the oxidising agent.[101]

Synthesis[101,102]

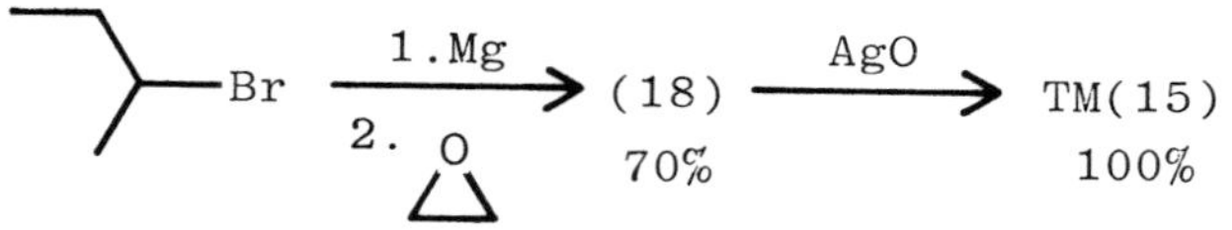

$$\text{\textgreater}\!\!-\!\text{Br} \xrightarrow[\text{2. } \triangle\!\text{O}]{\text{1. Mg}} (18) \xrightarrow{\text{AgO}} \text{TM}(15)$$
$$\qquad\qquad\qquad\qquad 70\% \qquad\qquad 100\%$$

3. Suggest a synthesis of acid (19) used in an investigation of steric hindrance.[103]

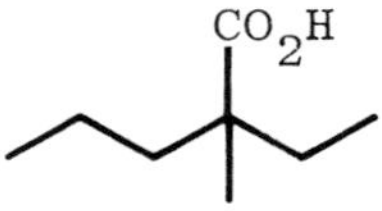

(19)

Answer

Grignard addition to CO_2 must be the answer with this crowded acid and this leads us back to alcohol (20). Disconnection of any group now gives a simple ketone and an available alkyl halide. As we shall see in Chapter 11, it is best to divide the molecule into two nearly equal parts, so we shall use disconnection (a).

Analysis

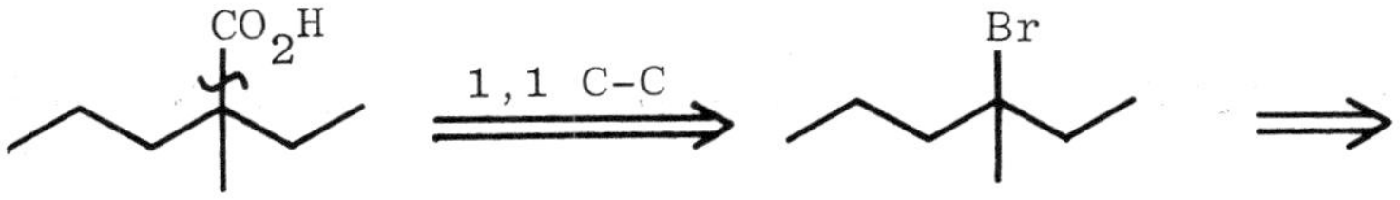

[Scheme: (19a) 3-ethyl-3-methylhexanoic acid type structure with CO₂H ⟹ (1,1 C-C) tertiary bromide ⟹]

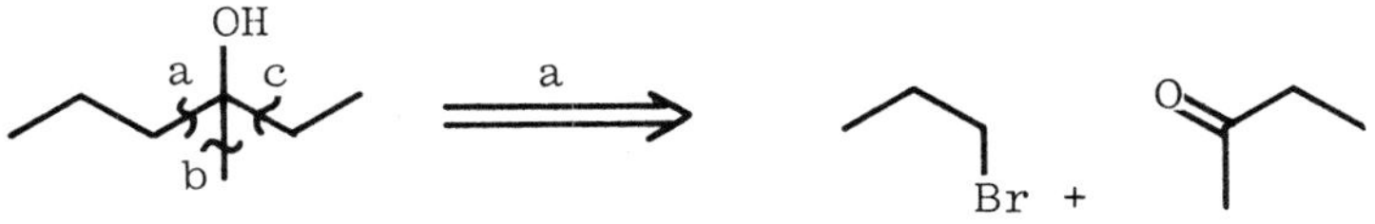

[Scheme: (20) tertiary alcohol with OH, disconnections a, b, c ⟹ (a) propyl bromide + butan-2-one]

This route has been followed with reasonable success, though the yield in the carboxylation step was poor.

Synthesis[103,104]

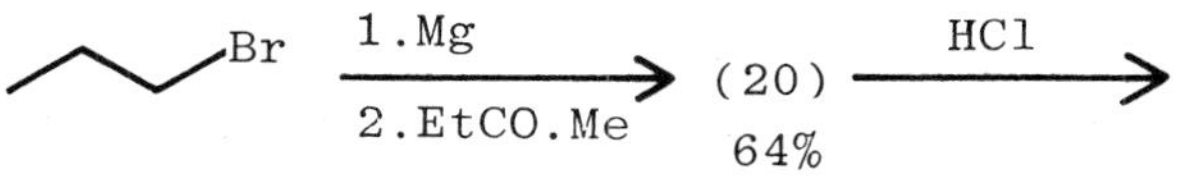

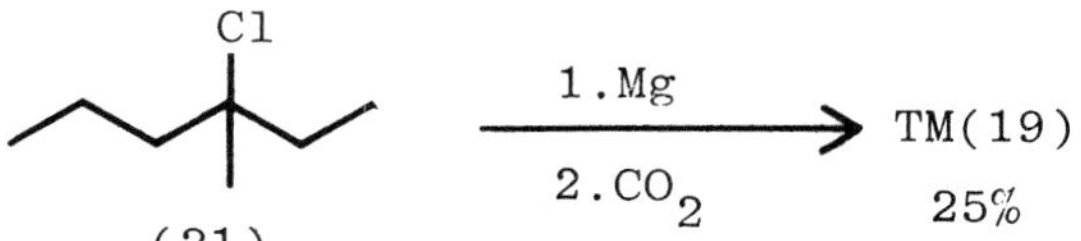

$$\text{PrBr} \xrightarrow[\text{2. EtCO.Me}]{\text{1. Mg}} (20) \xrightarrow{\text{HCl}} $$

$$(21) \xrightarrow[\text{2. CO}_2]{\text{1. Mg}} \text{TM}(19) \quad 25\%$$

(21) tertiary chloride

64%

CHAPTER 11

General Strategy A: Choosing a Disconnection

The guidelines for good disconnections suggested in Chapter T11 are:
1. Greatest simplification (in middle of molecule, at branchpoint, rings from chains.
2. Symmetry.
3. High-yielding steps.
4. Recognisable starting materials.

Example :

The synthesis of the natural product citronellol (1) (used in perfumery) shows guidelines 1 and 4 in action. Disconnection at the branchpoint is possible (1a) and the required alcohol (2) comes from available ketone (3) (p T 1) by reduction.

Analysis

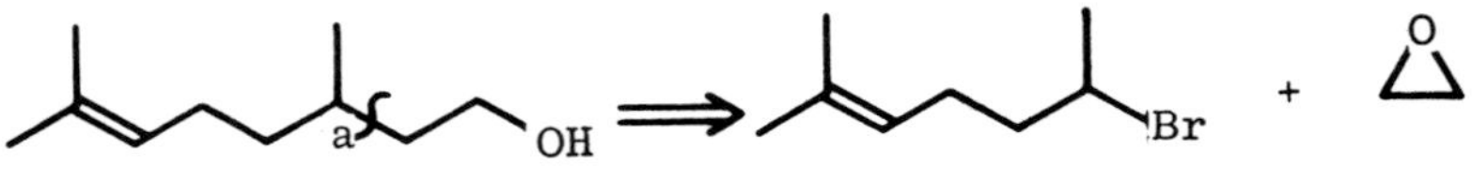

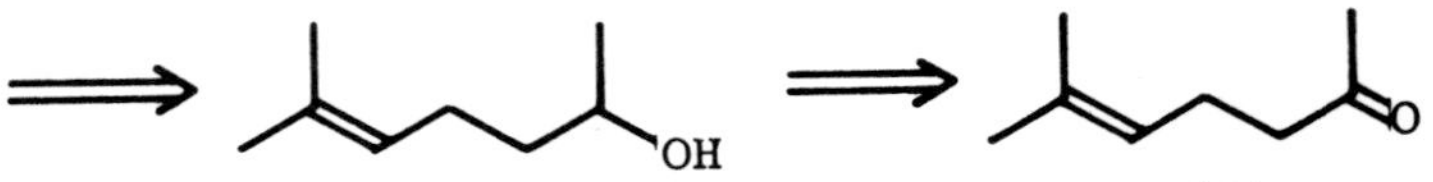

Synthesis[105]

$$(3) \xrightarrow{\text{NaBH}_4} (2) \xrightarrow{\text{PBr}_3}$$

[structure: (CH₃)₂C=CH-CH₂-CH₂-CH(CH₃)-Br] $\xrightarrow[\text{2. } \triangle\text{O}]{\text{1. Mg}}$ TM(1)

Problem : Does the synthesis of multistriatin on page T3 follow these guidelines?

Answer : The main disconnection (corresponding to step 10 → 12) is chosen because it is at a branchpoint and because it uses the symmetry of Et_2CO. In addition all steps were high yielding and compound (11) could easily be made, but you could only guess at this from the chart.

Problem : From the syntheses in Chapter 10 pick examples following each guideline.

Answer : Every synthesis uses guideline 1. We have branchpoint disconnections (all TMs), ring-from-chain disconnections, (TMs 8, 11, 12), disconnections towards the middle of the molecule (TM 19).

Guideline 2 is naturally less general but TMs (7), (8), (11), and to some extent (3) all use it.

Every synthesis naturally uses guideline(3), but this is obvious only in the route to (11) and (12), the choice of conditions for e.g. the synthesis of intermediates (2) and (21), and in the reagents chosen for the synthesis of (3), (8), and (15).

Every synthesis starts from available materials. Note particularly that alcohol (5) appears often in these pages, and that the simple ring compound, (10) [= (13)] appears twice. The details of the synthesis of (4) would be decided largely on this guideline.

Example : The substituted 1,4-diol (4) must surely be made by a method which uses the symmetry of the structure. Disconnection (a) would require Grignard reagent (5) whose OH group would have to be protected. Disconnection (b) gives hydroxy ester (6) and here no protection is necessary as the internal ester (lactone) (7) serves the purpose.

Analysis

HO~~~C(Ph)(Ph)~~~OH (4) $\xrightarrow{a}$ HO~~~MgBr (5) + Ph-C(=O)-Ph

(4) $\xrightarrow{b}$ HO~~~CO$_2$R (6) + 2PhMgBr and lactone (7)

Synthesis[106]

PhBr $\xrightarrow[2.(7)]{1.\text{Mg},\text{Et}_2\text{O}}$ TM(4)

Problem : Suggest syntheses of TMs (8) and (9), using the symmetry to guide you.

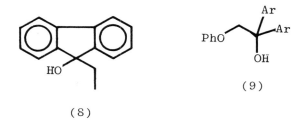

Answer : Disconnection (a) on TM (8) gives symmetrical ketone (10). The alternative disconnection (b) would require bis-Grignard (11) - a doubtful species.

Analysis

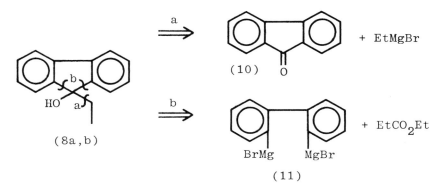

Ketone (10) (fluorenone) is available and route (a) has been used successfully.

Synthesis[107]

$$(10) \xrightarrow{\text{EtMgBr}} \text{TM}(8) \quad 73\%$$

The same two disconnections on TM (9) suggests (b) as the better route. Grignard reagent (12) might be difficult to make, but ester (13) is a simple 1,2-diX problem.

Analysis

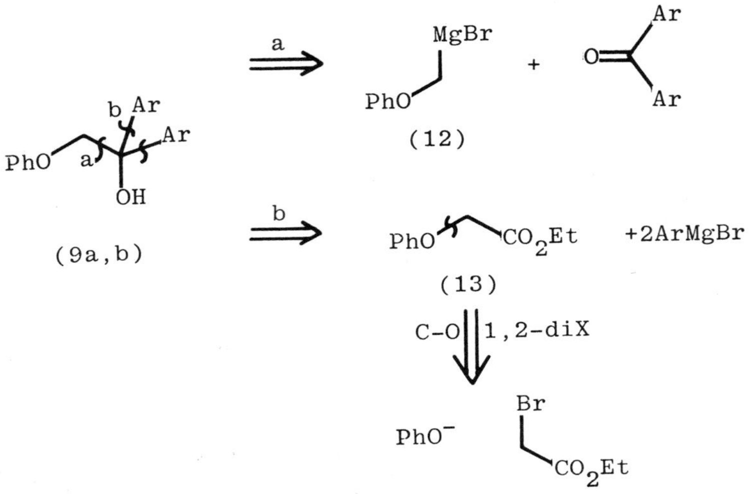

PhO–C(Ar)(Ar)(OH)–CH₂–OPh (9a,b)

a ⇒ PhO–CH(MgBr) (12) + O=C(Ar)(Ar)

b ⇒ PhO–CH(–)–CO₂Et (13) + 2ArMgBr

C–O ⇓ 1,2-diX

PhO⁻ + Br–CH(–)–CO₂Et

Synthesis[107]

Cl–CH₂–CO₂H $\xrightarrow{\text{1. base PhOH}; \text{2. EtOH, H}^+}$ (13) $\xrightarrow{\text{ArMgBr}}$ TM(9) e.g. 65%, Ar=p-Tol

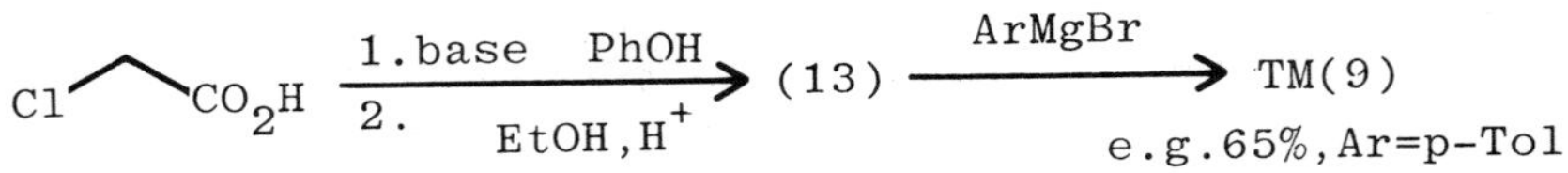

Problem : Alcohol (14) was needed to make the corresponding Grignard reagent.[108] Suggest a synthesis guided by branch-point disconnections.

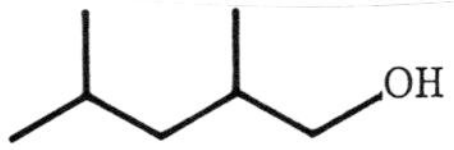

(CH₃)₂CH–CH₂–CH(CH₃)–CH₂OH (14)

Answer : Disconnection back to the first branchpoint (● in 14a) requires formaldehyde and halide (15) which comes from alcohol (16). Disconnection back to the second branchpoint (● in 16) gives two available starting materials.

Analysis

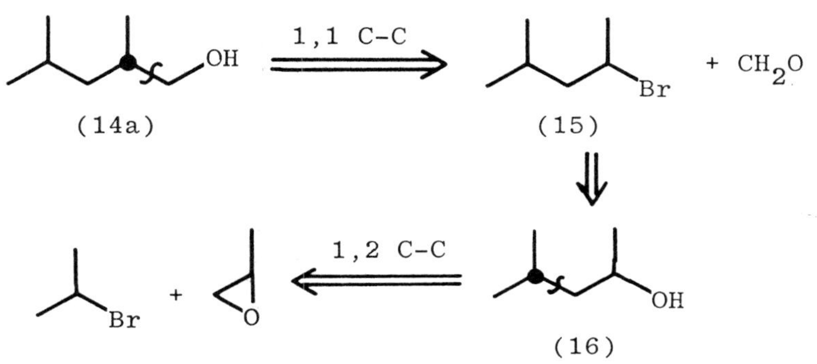

(14a) $\xrightarrow{1,1 \text{ C-C}}$ (15) + CH_2O

⇓

$\underset{Br}{\bigwedge}$ + $\underset{O}{\triangle}$ $\xleftarrow{1,2 \text{ C-C}}$ (16)

The synthesis has been carried out by this route;[109] there is an alternative for the first step which gives a higher yield.[108]

Synthesis[108,109]

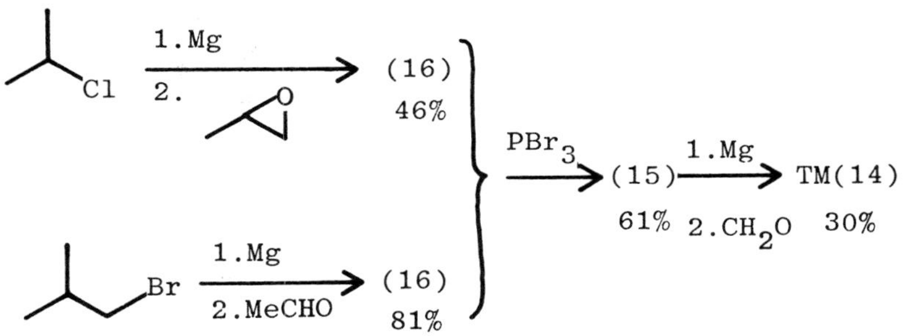

iPr-Cl $\xrightarrow[2. \text{ epoxide}]{1. \text{Mg}}$ (16) 46%

iBu-Br $\xrightarrow[2. \text{MeCHO}]{1. \text{Mg}}$ (16) 81%

$\xrightarrow{PBr_3}$ (15) $\xrightarrow[2. CH_2O]{1. \text{Mg}}$ TM(14)
61%   30%

Example : As part of a programme[110] to screen for new perfumery compounds, alcohol (17) was wanted. This has *two* branchpoints (● in 17) and disconnection between them simplifies the problem a great deal. Ketone (18) is an available natural product (thujaketone) and halide (19) can be made via a Grignard reaction.

Analysis

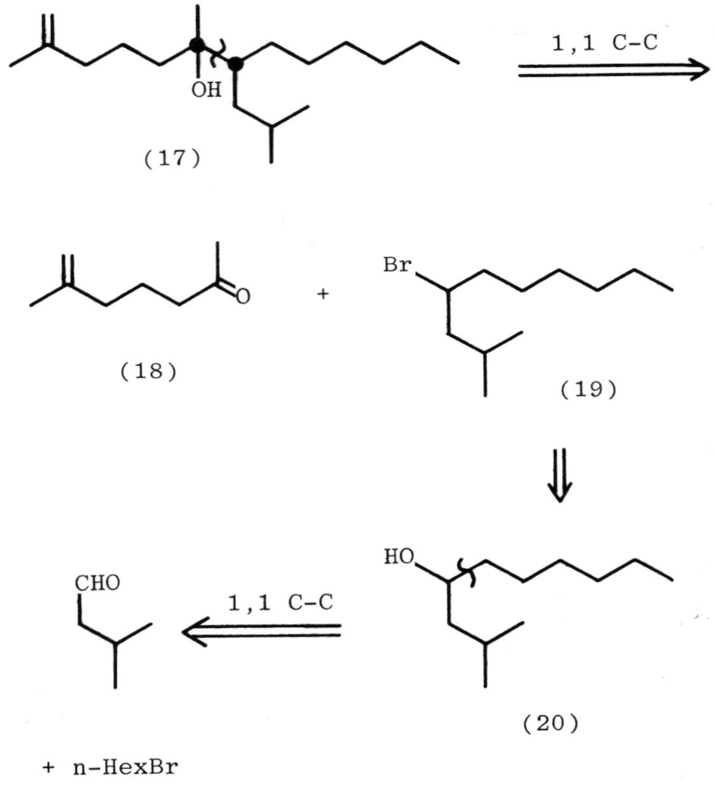

(17) ⟹ (1,1 C-C)

(18) + (19)

⟹

(20) ⟸ (1,1 C-C) CHO (isobutyraldehyde)

+ n-HexBr

Synthesis[110]

n-HexBr $\xrightarrow[\text{2. i-PrCH}_2\text{CHO}]{\text{1. Mg}}$ (20) $\xrightarrow{\text{HBr}}$ (19)
                                      73%   60%

$\xrightarrow[\text{2. (18)}]{\text{1. Mg}}$ TM(17)
                45%

The product (17) had only a mildly pleasant smell and was not adopted by the perfume industry.

Example : By contrast, the synthesis of the potential anti-malarial (21) uses no disconnections at branchpoints at all!

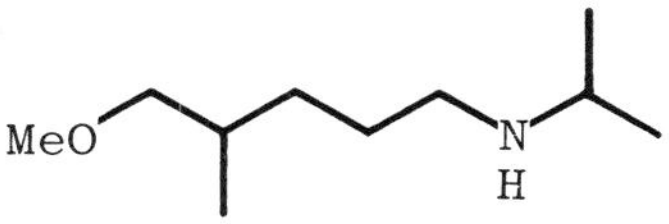

(21)

The amine could be put in by reductive amination on the secondary side or via an amide on the primary side (Chapter T8) but this crowded compound can be made by simple alkylation of (22) with available isopropylamine. Steric hindrance prevents a second alkylation.

Analysis 1

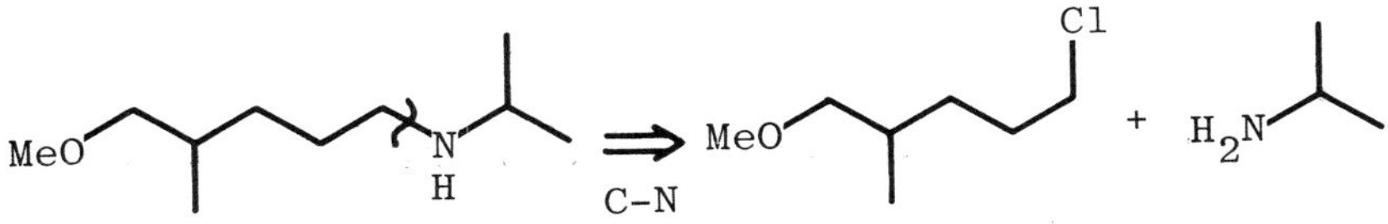

(21a) $\xRightarrow{\text{C-N}}$ (22) + H$_2$N-iPr

The OH group in the required alcohol (23) is too far (*three* atoms) from the branchpoint so the best we can do is use an epoxide to give half ether (24) as the starting material.

Analysis 2

(22) ⟹ MeO-CH(CH₃)-CH₂-CH₂-OH (23) ⟹ epoxide + MeO-CH₂-CH(CH₃)-CH₂-OH (24) ⇐ MeO-CH₂-CH(CH₃)-CH₂-Br

This looks at first sight like a formidable problem of chemoselectivity, but changing the oxidation level to an aldehyde (25) gives a simple 1,3-diX disconnection to available aldehyde (26). The ester would have done as well.

Analysis 3

(24) $\xrightarrow[\text{reduction}]{\text{FGI}}$ MeO-CH₂-CH(CH₃)-CHO (25) $\xrightarrow{\text{1,3-diX}}$ MeOH + CH₂=C(CH₃)-CHO (26)

Synthesis[111]

(26) $\xrightarrow[\text{MeOH}]{\text{MeO}^-}$ (25) 51% $\xrightarrow{\text{H}_2 \text{ Ni}}$ (24) 94% $\xrightarrow{\text{SOCl}_2}$ MeO-CH₂-CH(CH₃)-CH₂-Cl 92%

$\xrightarrow[\text{2. epoxide}]{\text{1. Mg}}$ (23) 67% $\xrightarrow{\text{SOCl}_2}$ (22) 90% $\xrightarrow{\text{iPrNH}_2}$ TM(21) 88%

Problem : Suggest a synthesis of acid (27) needed to confirm the structure of an unknown compound.[112]

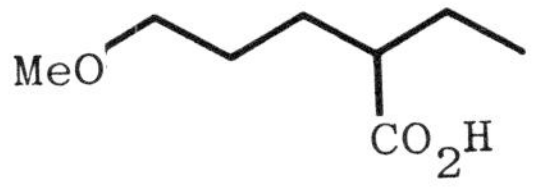

(27) [structure: MeO-CH₂CH₂CH₂CH₂-CH(CO₂H)-CH₂CH₃]

Answer : The Grignard method will be best for making (27) from secondary halide (28) as it avoids elimination reactions likely with cyanide ion. Disconnection (29a) now leads back to an alcohol (30) which can be made in a similar manner as (24).

Analysis

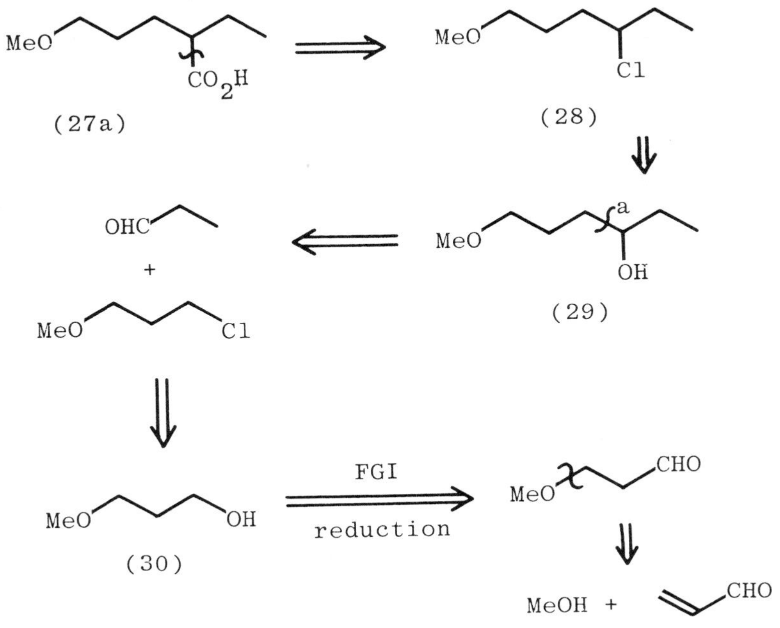

(27a) ⟹ (28)

⇓

(29)

OHC-CH₂CH₃ + MeO-CH₂CH₂CH₂-Cl ⟸ MeO-CH₂CH₂CH₂-CH(OH)-CH₂CH₃

⇓

MeO-CH₂CH₂CH₂-OH (30) —FGI reduction⟶ MeO-CH(...)-CH₂-CHO

⇓

MeOH + CH₂=CH-CHO

Synthesis[112]

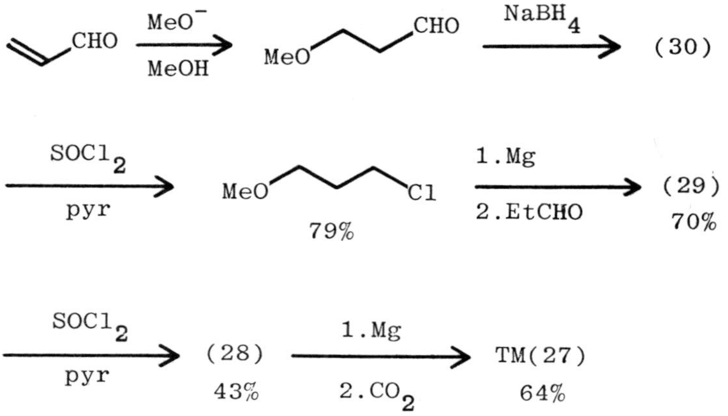

$$\text{CH}_2=\text{CHCHO} \xrightarrow[\text{MeOH}]{\text{MeO}^-} \text{MeOCH}_2\text{CH}_2\text{CHO} \xrightarrow{\text{NaBH}_4} (30)$$

$$\xrightarrow[\text{pyr}]{\text{SOCl}_2} \text{MeOCH}_2\text{CH}_2\text{CH}_2\text{Cl} \xrightarrow[\text{2.EtCHO}]{\text{1.Mg}} (29)$$
79%　　　　　　70%

$$\xrightarrow[\text{pyr}]{\text{SOCl}_2} (28) \xrightarrow[\text{2.CO}_2]{\text{1.Mg}} \text{TM}(27)$$
　　　43%　　　　　64%

CHAPTER 12

Strategy V: Stereoselectivity A

Optically Active Compounds

Problem : The sex-pheromone of the rove-beetle (1) can be made by the following sequence of reactions. Optically active (1) is required for biological activity. How would you obtain it?

Synthesis[113]

n-OctBr $\xrightarrow[\text{2. } \underset{H}{\overset{O}{\underset{\|}{C}}}\sim\sim\text{CN}]{\text{1. Mg}}$ n-Oct–CH(OH)–CH$_2$–CH$_2$–CN

$\xrightarrow{\text{HO}^-/\text{H}_2\text{O}}$ n-Oct–CH(OH)–CH$_2$–CH$_2$–CO$_2$H $\xrightarrow[\text{benzene}]{\text{TsOH}}$ n-Oct–[γ-lactone]

(1)

Answer : A resolution is necessary and the alcohol (2) is the first compound which has the chiral centre (● in 2) so this was resolved by combining it (as a urethane) with an available optically active compound.

n-Oct–C*H(OH)–CH$_2$–CH$_2$–CO$_2$H

(2)

Example : The discarded (-) phenylethylamine (3) from the resolution on page T 95 can be used to make other optically active compounds.

(-)-(3)   (4)   (5)

The strange amino-acid (4) is a 'fat' version of phenylalanine (5) having a side chain which is rigid and inert, but which is also space filling rather than flat.[114] Optically active (4) was needed to study peptide conformation and the biological activity of drugs.

A Strecker (p T 43) synthesis would require aldehyde (6) which can be made from available alcohol (7). We shall now use R for the adamantyl group.

Analysis

$$(4) \xrightarrow[\text{Strecker}]{1,1\text{-diX}} \text{(6)}=R\text{-CHO} \xrightarrow[\text{oxidation}]{\text{FGI}} R\text{-OH} \quad (7)$$

The one-step Strecker synthesis offers no opportunity for resolution and (6) is achiral. However, by using available (-)-(3) in a two-step Strecker synthesis, asymmetric induction can be used to make only the wanted enantiomer of (4). The PhCHMe group in (8) provides a chiral environment so that cyanide adds preferentially to one side of the C=N bond. Crystallisation of the major diastereoisomer of (9) (note *two* chiral centres) ensures the integrity of the new centre (● in 9) and hydrolysis and hydrogenolysis give pure (+)-(4).

The (−)-(3) is recovered and can be used again.
Synthesis[114]

$$R\text{-}CH_2CH_2OH \quad (7) \; (R=\text{adamantyl}) \xrightarrow[\substack{\text{pyr} \\ CF_3CO_2H \\ DCC}]{DMSO} (6) \xrightarrow{(-)-(3)}$$

$$R\text{-}CH_2\text{-}CH=N\text{-}C(Me)(Ph)(H) \;(8) \xrightarrow{HCN} R\text{-}CH_2\text{-}CH(CN)\text{-}NH\text{-}C(Ph)(Me)(H) \;(9) \xrightarrow[\text{conc. HCl}]{EtOH}$$

(9) 75% crystallises out

$$R\text{-}CH_2\text{-}CH(CONH_2)\text{-}NH\text{-}C(Ph)(Me)(H) \xrightarrow[\text{2. conc. HCl}]{1.\, H_2 \; Pd\text{-}C} TM \; (S)\text{-}(+)\text{-}(4)$$

Example : Optically active aldehyde (10) was needed for a synthesis of biotin.[115] The compound has a 1,1-diX disconnection (10a) clearly available and a C-N (amide) disconnection leaving (11) which has the same skeleton as the amino acid cysteine (12).

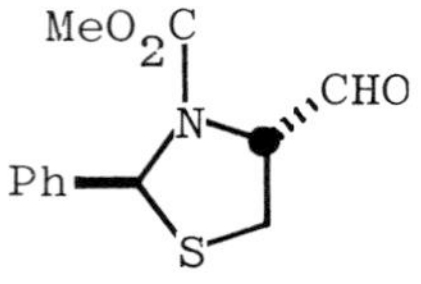

(10)

Analysis

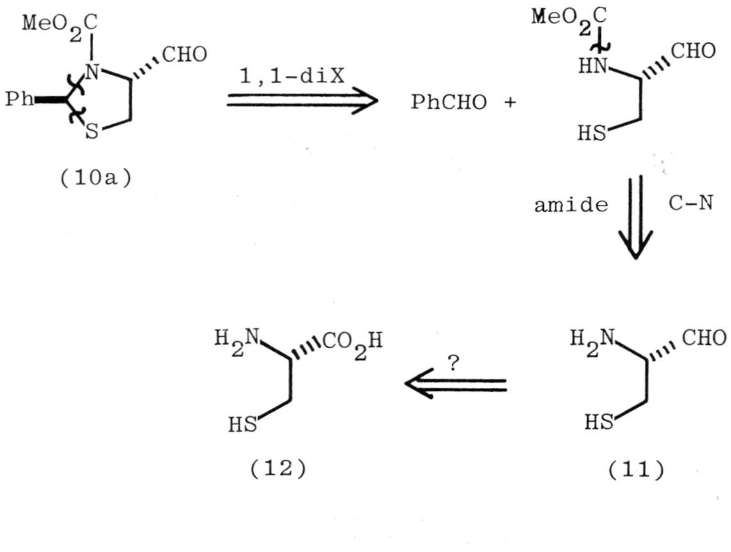

(10a) $\Longrightarrow$ 1,1-diX $\Longrightarrow$ PhCHO + [structure]

amide C-N $\Downarrow$

(12) $\overset{?}{\Longleftarrow}$ (11)

It is better to add the benzaldehyde and the CO_2Me group first so that conversion of CO_2H to CHO does not affect the reactive NH_2 or SH group. Note that diborane, B_2H_6, reduces CO_2H even in the presence of an ester.

Synthesis[115]

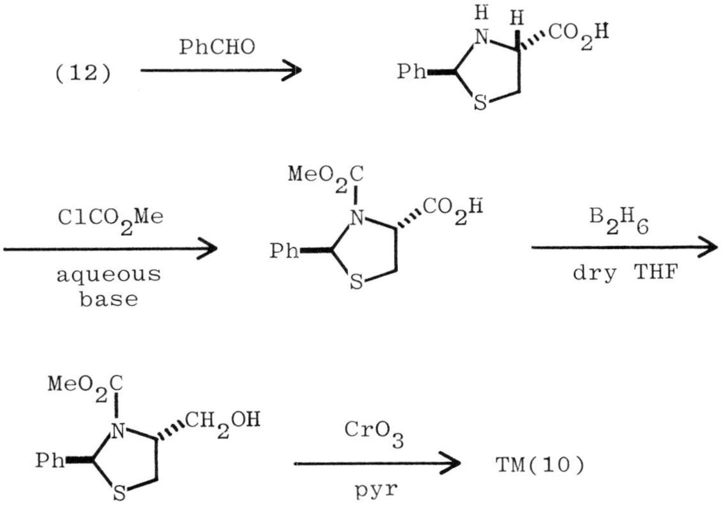
(12)
PhCHO
$ClCO_2Me$
aqueous base
B_2H_6
dry THF
CrO_3
pyr
TM(10)

Problem : Suggest a synthesis of optically active S-(+)-sulcatol (13), the aggregation pheromone of the wood-boring ambrosia beetle, from available ethyl (S)-(-)-lactate (14).

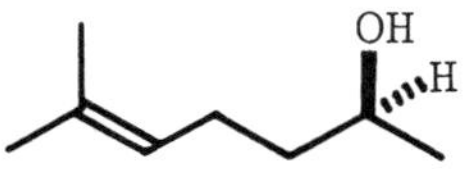

(S)-(+)-(13)

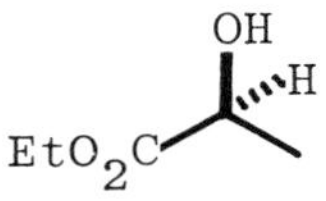

(14)

Answer : The required disconnection is (13a) which clearly needs optically active epoxide (15). This must be made from (14) without inverting the chiral centre so reduction of the CO_2Et group and conversion to a leaving group are needed.

Analysis

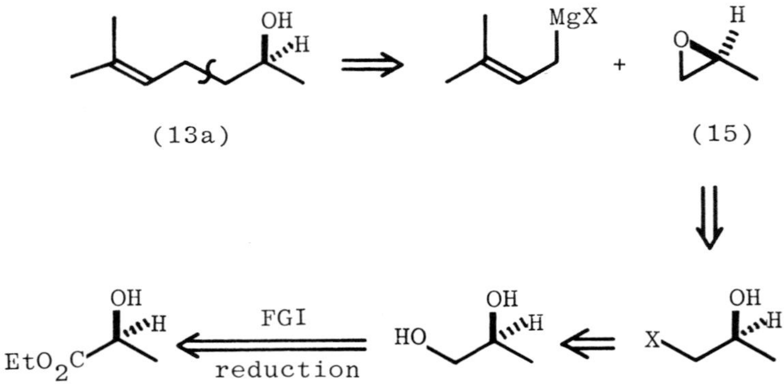

[Scheme: (13a) ⇒ prenyl-MgX + (15) epoxide]

[Scheme: (14) EtO₂C-CH(OH)-CH₃ ⇐ FGI reduction ⇐ HO-CH₂-CH(OH)-CH₃ (16) ⇐ X-CH₂-CH(OH)-CH₃]

The chemoselectivity of converting one OH in (16) into a leaving group will be easier if the other OH is protected, so it is best to introduce protection at the start. The easily removed THP group is ideal (Table T 9.1). The required Grignard reagent is from available halide (17) (p T 1).

Synthesis[116,117]

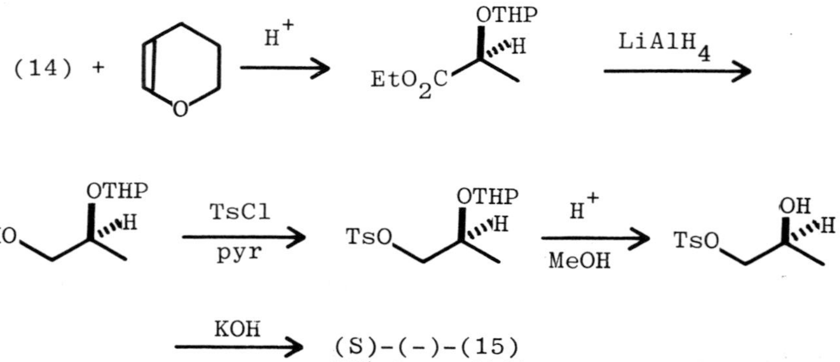

(14) + dihydropyran $\xrightarrow{H^+}$ EtO₂C-CH(OTHP)-CH₃ $\xrightarrow{LiAlH_4}$

HO-CH₂-CH(OTHP)-CH₃ $\xrightarrow[\text{pyr}]{TsCl}$ TsO-CH₂-CH(OTHP)-CH₃ $\xrightarrow[\text{MeOH}]{H^+}$ TsO-CH₂-CH(OH)-CH₃

$\xrightarrow{KOH}$ (S)-(−)-(15)
45% from (14)

$$\text{(prenyl chloride)} \xrightarrow[\text{2.(S)-(-)-(15)}]{\text{1.Mg}} \text{(S)-(+)-(13)} \quad 75\%$$

Epoxide (15) has become an important intermediate in chiral syntheses. How would you make the *other* enantiomer (R)-(+)-(15) from (14)?

Answer : An inversion *is* now necessary, so we must make the tosylate of the secondary OH group before reducing the ester. Diborane will reduce the ester in this case.

Synthesis[116]

$$(14) \xrightarrow[\text{pyr}]{\text{TsCl}} \text{EtO}_2\text{C}\overset{\text{OTs}}{\underset{}{\diagdown}}^{\text{H}} \xrightarrow{\text{B}_2\text{H}_6} \text{HO}\overset{\text{OTs}}{\underset{}{\diagdown}}^{\text{H}} \xrightarrow{\text{base}} \overset{\text{O}}{\triangle}^{\text{H}}$$

(R)-(+)-(15)

Example : There have been various attempts to generate optical activity without the investment of some optically active reagent. The most notorious is the alleged synthesis of optically active santonin (17) from inactive precursors.[118] This was quickly exposed as it transpired the reactions didn't even give the right products, let alone optically active ones.[119]

In more modern times the epoxidation of (18) in a 'confined vortex' (14,000 r.p.m. in a turbine) has been reported[120] to give a product with a tiny optical rotation, but doubt has been cast on this observation too.[121]

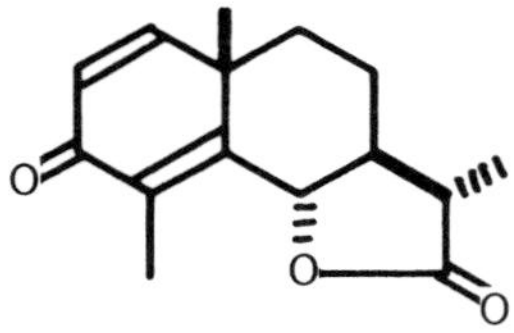

(17)

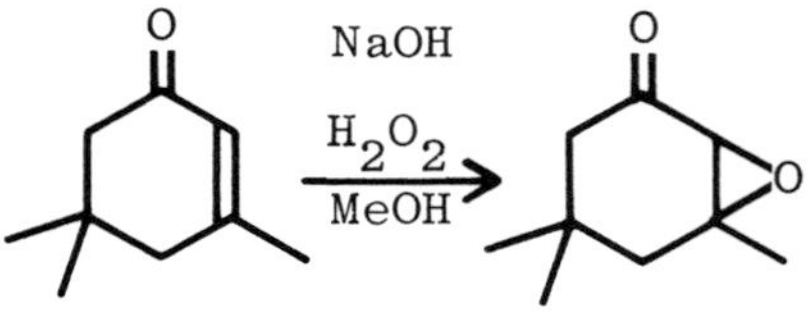

(18)

So far, the methods described here and in the text are the only ones which work : optically active starting materials must be used, or asymmetric induction with an optically active substituent or reagent, or the product or an intermediate must be resolved.

Stereospecific and Stereoselective Reactions

Problem : In the synthesis of a chiral crown ether,[122] compound (19) was needed. Suggest a synthesis for it.

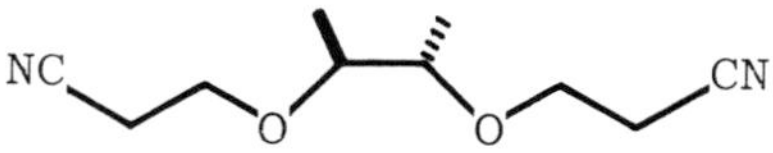

(19)

Answer : Two 1,3-diX (C-O) disconnections (19a) reveal the *trans* diol (20) as the vital intermediate. This can be made from *cis* epoxide (21) by stereospecific inversion.

$$\text{(19a)} \;\Longrightarrow\; 2\;\text{C-O} \;\|\; 1,3\text{-diX}$$

$$\text{NC}\diagup\!\!\diagdown + \underset{\text{HO}\;\;\;\text{OH}}{\diagup\!\diagdown} + \diagdown\!\diagup\text{CN}$$

(20)

$$\diagup\!\!\diagdown \;\Longleftarrow\; \underset{\text{O}}{\triangle} \quad (21)$$

Synthesis[122]

$$\diagup\!\!\diagdown \xrightarrow{\text{MCPBA}} (21) \xrightarrow[\text{H}_2\text{O}]{\text{HO}^-} (20) \xrightarrow[\substack{\text{MeO}^- \\ \text{MeOH}}]{\diagup\!\!\diagdown\text{CN}} \text{TM}(19)$$

Problem : Why do the base catalysed reactions of (35) and (37) in the text (p T 102)   give such different products?

Answer : This *cis* compound (35) cyclises because the anion (22) is held, by the almost rigid five membered ring, in the perfect conformation for stereospecific S_N2 displacement.

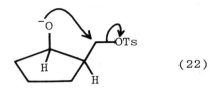

(22)

O^-, CH_2, OTs linear

The *trans* compound (36) fragments because cyclisation is impossible and because it, unlike (35) can adopt the perfect W conformation (all anti-periplanar bonds) ideal for fragmentation (23).

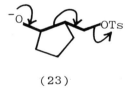

(23)

Problem : Two of the following reactions are stereospecific and one is stereoselective : explain which is which and account for the results.

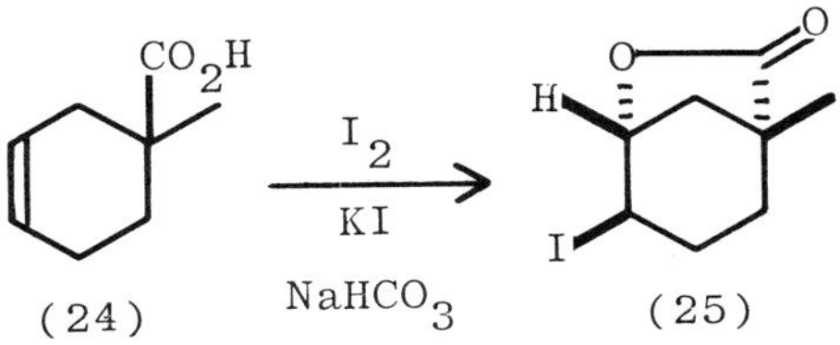

(24) → [I₂, KI, NaHCO₃] → (25)

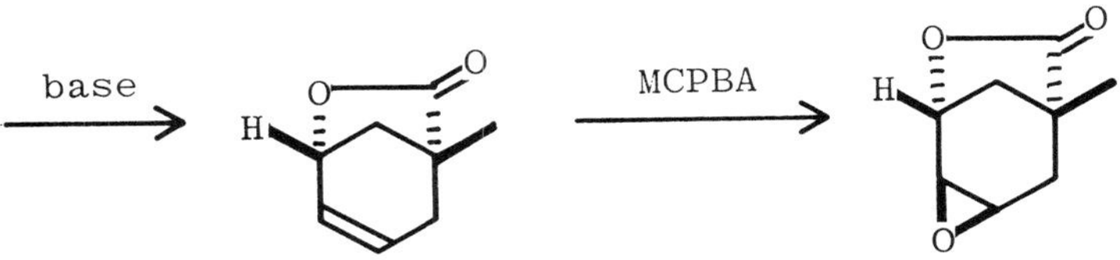

base → (26) → [MCPBA] → (27)

Answer : The first reaction is iodolactonisation and is a stereospecific *trans* addition to the double bond. The iodine adds to form intermediate (28) which opens by S_N2 inversion (arrows) to give (25).

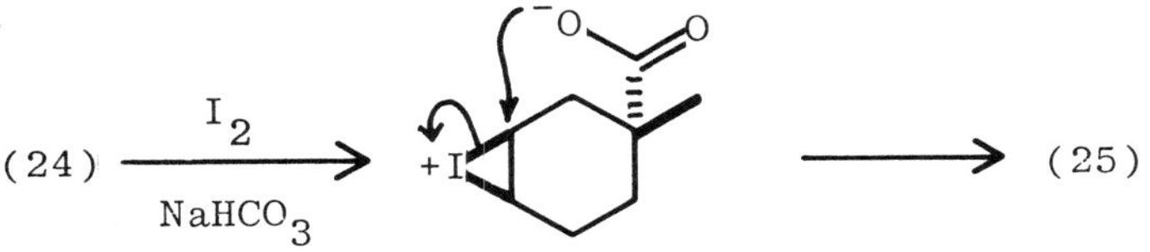

(24) → [I₂, NaHCO₃] → (28) → (25)

The second reaction is a stereospecifically *trans* elimination (E2). The marked proton in (25) cannot be lost as it is *cis* to the leaving group (I). *Trans* elimination (29) is possible only to give (26).

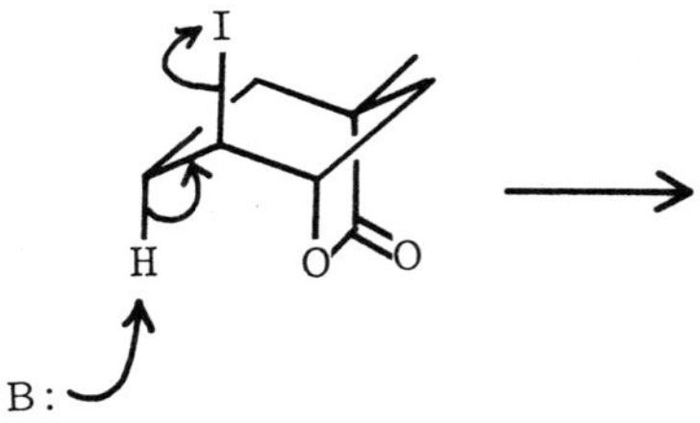

(29) → (26)

The third reaction is the epoxidation of a double bond which can take place from either side. The peracid attacks from the less hindered side away from the lactone bridge (30). This is the stereoselective reaction.

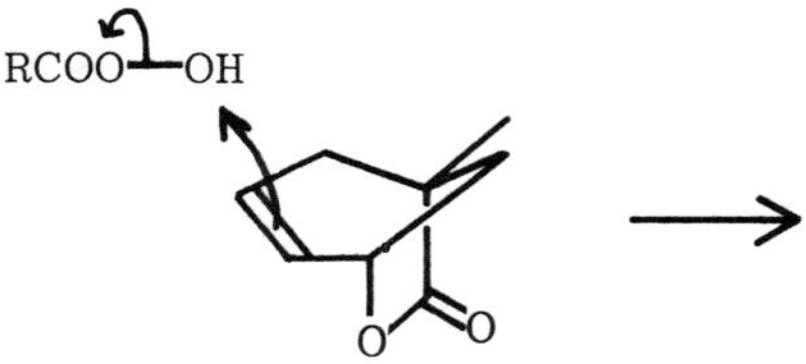

(30) → (27)

Epoxide (27) has been elaborated into useful synthetic intermediates.[123]

Problem : Suggest a synthesis of (31), needed to make prostaglandins.[124]

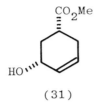

(31)

Answer : This compound (31) is nearly a ring-opened version of (26) so a similar sequence will be satisfactory :
Synthesis[124]

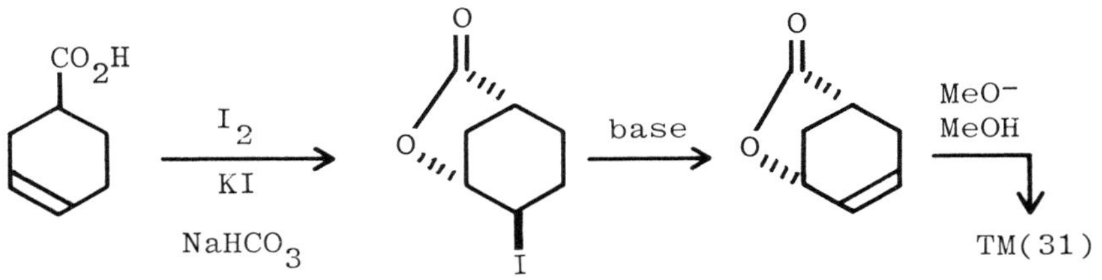

CO$_2$H → (I$_2$, KI, NaHCO$_3$) → iodolactone → (base) → unsaturated lactone → (MeO$^-$, MeOH) → TM(31)

*Conformational Analysis**

Example : Grignard reagents do not normally react cleanly with alkyl halides to give hydrocarbons (32)

$$RMgX + R'Br \rightarrow RR'$$

because of metal exchange, radical reactions and elimination. However, Grignard reagents add cleanly and in high yield to α-haloketones such as (33) to give the coupled product (34).

---

* also involved in the last two problems.

$$\text{(33)} \quad \xrightarrow{\text{PhMgBr}} \quad \text{(34)}$$

This reaction is not what it seems : work[125] at low temperatures reveals that an intermediate (35) is formed by addition to the carbonyl group and that this rearranges to the product.

$$\text{(33)} \xrightarrow[0°C]{\text{PhMgBr}} \text{(35) 83\%} \xrightarrow{\text{heat}} \text{(34) 86\%}$$

Note that (35) must be the stereoisomer with Ph and Cl *trans* for the anti-peri-planar arrangement required for the stereospecific rearrangement, and so the Grignard addition must be highly stereoselective. This all makes sense by conformational analysis.

Chloroketone (33) prefers to adopt the axial conformation (33a) because of dipole repulsion between C-Cl and C=O. The less hindered side of the carbonyl group is therefore *trans* to the chlorine giving (35) in the right conformation (35a) - *trans* diaxial - for rearrangement.

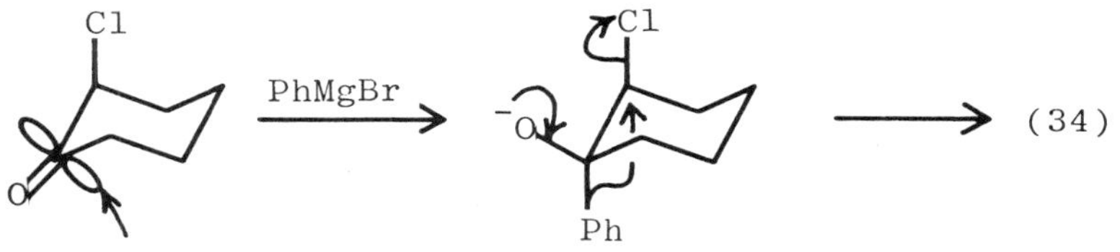

(33a) → (35a) → (34)

Problem : Cyanide ion addition to (36) gives[126] a 2:1 mixture of (37) and (38). Explain. (The reagent and dipolar aprotic solvent simply ensure no confusing hydrogen-bonding.

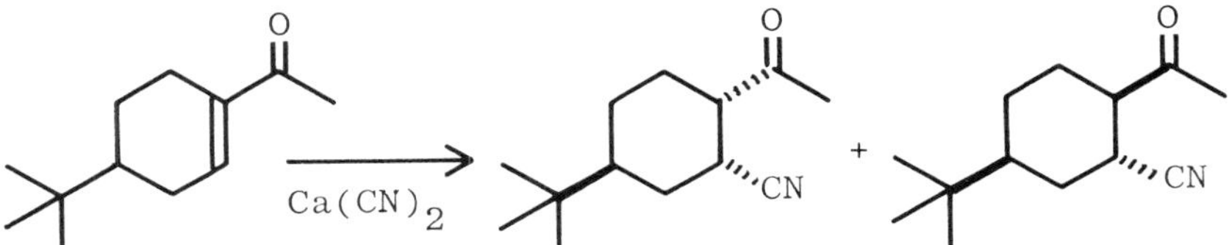

(36) Ca(CN)$_2$, Me$_2$N.CHO → (37) + (38)
                                    2 :   : 1

Answer : Conformational drawings of the products show that the cyanide is *axial* in both cases.

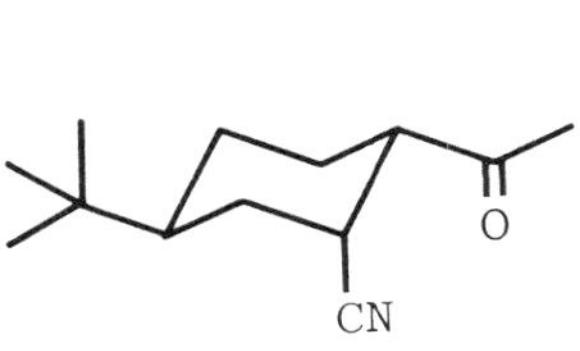

(37a) 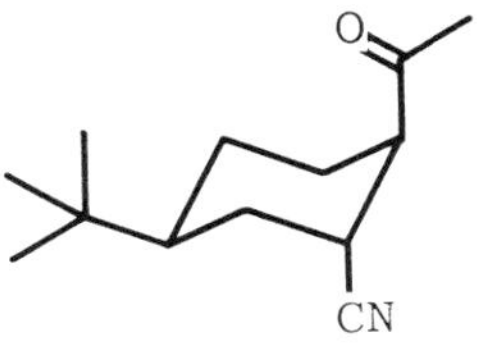   (38a)

The reaction must occur by *axial* addition (39) to give (40) which may be protonated on either side of the double bond, but gives the all equatorial product (37) stereoselectively.

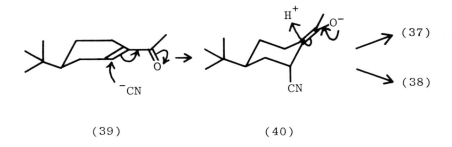

(39) (40)

CHAPTER 13

One-Group C—C Disconnections II: Carbonyl Compounds

(a) Carbonyl compounds by 1,1 C-C Disconnections

Examples :

1. On page 94 appeared a problem on the synthesis of acid (1), needed for conversion to ketone (2). The problem under investigation was steric hindrance in nucleophilic additions to ketones and a crowded ketone with a chiral centre was needed.

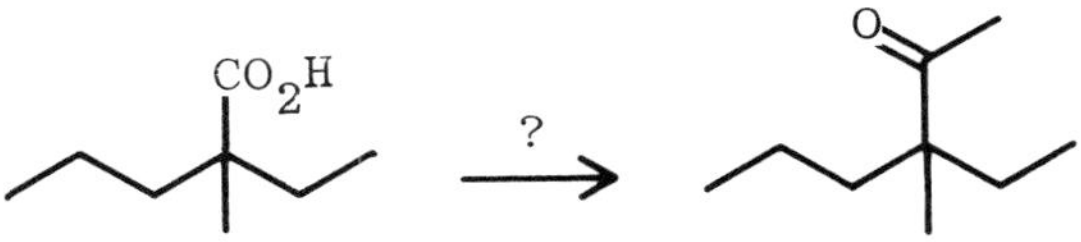

$$(1) \xrightarrow{\text{?}} (2)$$

The conversion (1) → (2) was carried out by the organo-cadmium method (p T 106).

Synthesis[103]

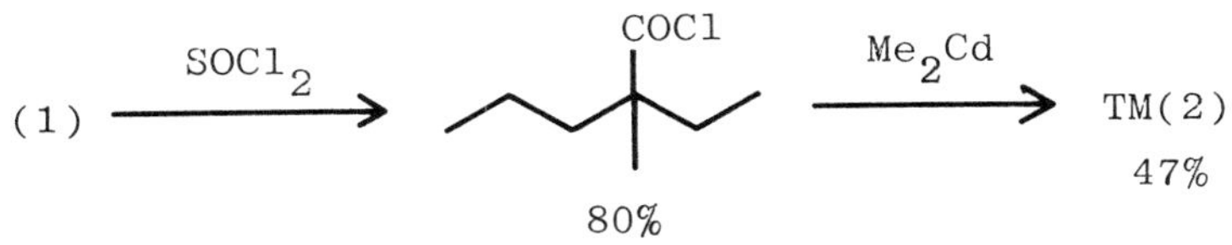

$$(1) \xrightarrow{SOCl_2} \underset{80\%}{[COCl]} \xrightarrow{Me_2Cd} TM(2) \quad 47\%$$

123

2. Among the many anti-malarial drugs tried[111] in the 1940s and 1950s was (3) (cf p 104). The better C-N disconnection is towards the middle of the molecule, revealing ketone (4) as a useful intermediate. C-C disconnection is again preferred towards the middle of the molecule and leads back to available (5) as starting material.

Analysis

MeO~~~(3)~~~HN—iPr ⇒ MeO~~~(4)~~~C(=O)~~~

⇓ 1,1 C-C

Br~~~Cl ⇐ C-O ⇐ MeO~~~Cl + EtCN
(5)

Bromide is displaced more easily from (5) than chloride. The second of our two methods, Grignard attack on a nitrile (p T 107), is used here. The final amination is carried out by catalytic hydrogenation of a mixture of amine and ketone.

Synthesis[111]

(5) $\xrightarrow{\text{MeO}^-/\text{MeOH}}$ MeO~~~Cl $\xrightarrow{\text{1.Mg}, \text{2.EtCN}}$ (4) 54% $\xrightarrow{\text{iPrNH}_2, \text{H}_2, \text{Ni}}$ TM(3) 46%

Problems:
1. Suggest a synthesis of (6) found in lavender oil to which it contributes[127] a 'remarkable freshness'.

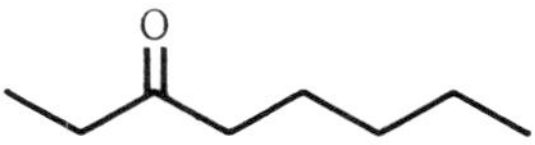

(6) [structure: octan-3-one, CH₃CH₂-CO-C₅H₁₁]

Answer : Disconnection towards the middle of the molecule provides a simple synthesis by either of our methods. Another alternative is preliminary FGI to alcohol (7) followed by a Grignard disconnection.

Analysis

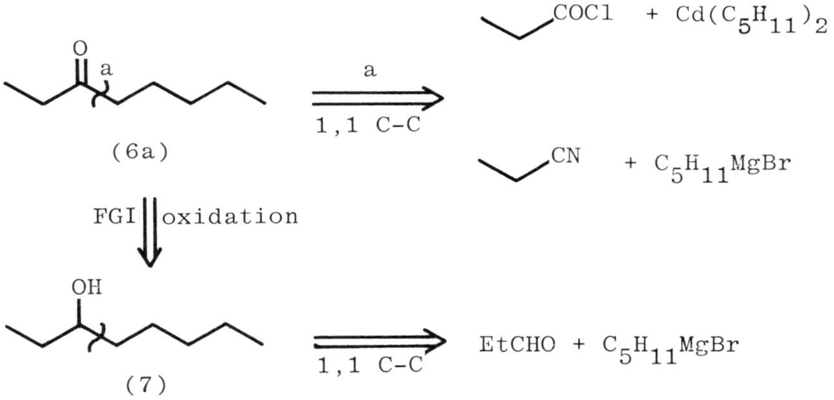

(6a) ⇒ (a, 1,1 C-C) → COCl + Cd(C₅H₁₁)₂

or → CN + C₅H₁₁MgBr

FGI ↓ oxidation

(7) ⇒ (1,1 C-C) → EtCHO + C₅H₁₁MgBr

The synthesis has been carried out by the nitrile route.

Synthesis[128]

n-C₅H₁₁Br $\xrightarrow{\text{1. Mg} \atop \text{2. EtCN}}$ TM(6) 61%

(b) *Carbonyl Compounds by Alkylation of Enols*
Examples : Direct alkylation of nitriles is possible if a strong base is used (often $NaNH_2$) : acid (8) can be made this way.

Analysis

Cy-CH(Ph)-CO_2H $\xrightarrow{\text{FGI}}_{\text{hydrolysis}}$ Cy-CH(Ph)-CN $\Rightarrow$ Cy-X
(8)   (9)   + $PhCH_2CN$

Synthesis[129]

Ph-CH_2-CN $\xrightarrow[\text{2. Cy-Cl}]{1. NaNH_2, NH_3(l)}$ (9) $\xrightarrow[H_2O]{KOH}$ TM(8)
                                                  80%   82%

It is also possible to fill the last place in an otherwise blocked carbonyl compound since further alkylation is impossible and the molecule is too crowded to condense with itself. Again strong bases are needed, such as Ph_3C^- used in the alkylation[130] of ester (10).

Me_2CH-CO_2Et $\xrightarrow[\text{2. EtI}]{1. Ph_3CNa}$ $Me_2C(Et)-CO_2Et$
(10)   58%

Problem : Suggest a synthesis of ketone (11), used in the manufacture of phentermine, a drug used to control obesity.

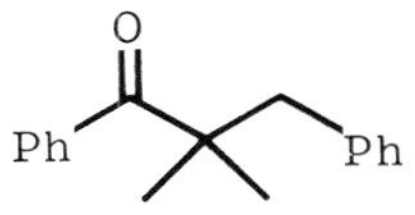

$$\underset{(11)}{\text{Ph}\diagdown\overset{\overset{\text{O}}{\|}}{\text{C}}\diagdown\overset{\text{Me}}{\underset{\text{Me}}{\text{C}}}\diagdown\text{CH}_2\text{Ph}}$$

Answer : This is a fully blocked ketone so we can remove one alkyl group, and we shall get the greatest simplification as well as the best reaction if we remove the benzyl group (11a).

Analysis

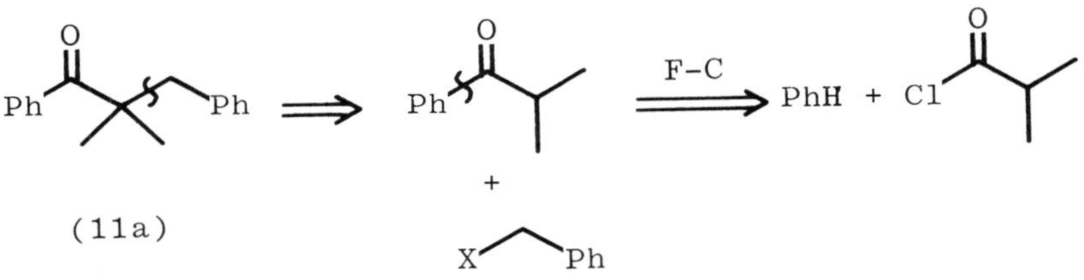

Ph-CO-C(Me)₂-CH₂Ph ⟹ Ph-CO-CHMe₂ $\xrightarrow{\text{F-C}}$ PhH + Cl-CO-CHMe₂

(11a)   +

X-CH₂-Ph

Synthesis[131]

PhH + Cl-CO-CHMe₂ $\xrightarrow{\text{AlCl}_3}$ Ph-CO-CHMe₂ (76%) $\xrightarrow[\text{2. PhCH}_2\text{Cl}]{\text{1. NaNH}_2}$ TM(11)

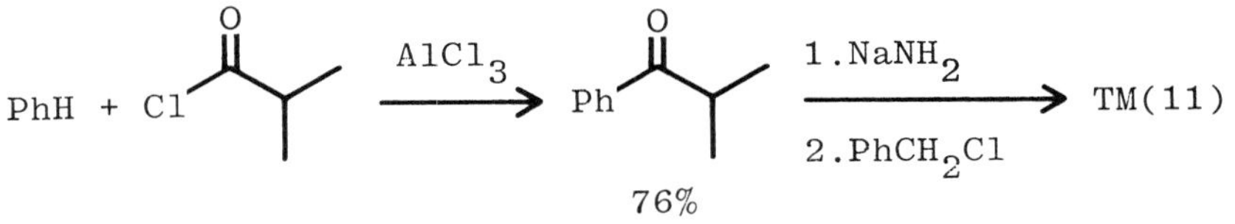

Simple Problem : Suggest syntheses of the perfumery ketones (12) (carnation) and (13) (gardenia).

(12)  CH₃-CO-CH₂CH₂CH₂CH₃   Ph-CH₂CH₂-CO-CH₃ (13)

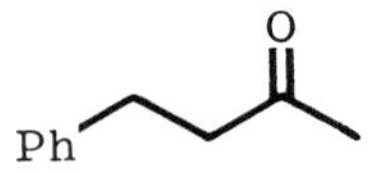

Answer : These are ideal TMs for alkylation of ethyl acetoacetate (15) since 1,2 C-C disconnection gives acetone enolate (14) for which (15) is the synthetic equivalent (p T 108).

Analysis

(12a)

(13a)

1,2 C-C (14) (15)

+ RBr

The alkylation[132] and decarboxylation[133] methods described for the synthesis of (12) are typical.

Synthesis

(15)   1.EtO⁻ / 2.BuBr →   72%   1.NaOH / 2.H⁺,heat → TM(12) 61%

Example : Optically active acid (16) was needed (p T 107) for the synthesis of an ant alarm pheromone. The branch point (● in 16) is also the chiral centre so it is better to avoid disconnections there. The 1,2 C-C disconnection (16a) is ideal as it gives synthon (17), for which we use a malonate ester, and halide (18), available from optically active alcohol (19), a major by-product from fermentation.

Analysis

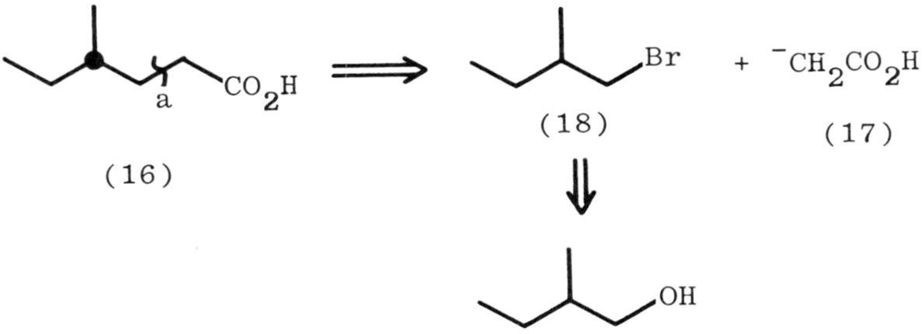

<img of scheme: (16) 3-methylpentanoic acid-like → (18) 1-bromo-2-methylbutane + ⁻CH₂CO₂H (17); (18) ⇒ (19) 2-methylbutan-1-ol>

Synthesis[134]

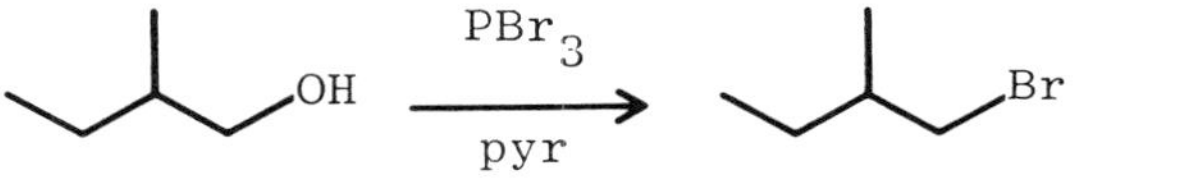

(−)-(S)-(19) —PBr₃, pyr→ (+)-(S)-(18) 60%

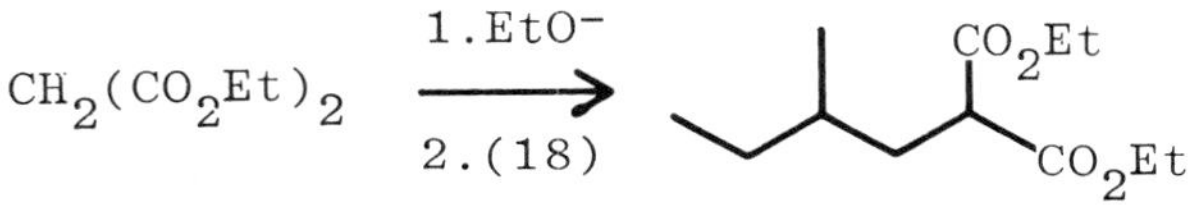

$CH_2(CO_2Et)_2$ —1. EtO⁻; 2. (18)→ [diester intermediate] 75%

—1. HO⁻/H₂O; 2. H⁺, heat→ (+)-(S)-(16) 85%

Problems:

1. Suggest a synthesis of amine (20), needed to study[135] whether cyclisation would occur during bromination of the double bond.

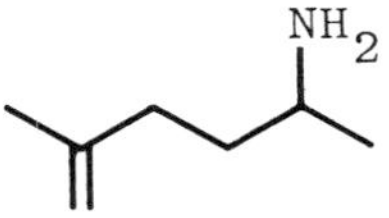

(20)

Answer : This branched primary amine can be made from ketone (21) via the oxime (p T 63). A 1,2 C-C disconnection on (21) is good as it needs the symmetrical allylic halide (22).

Analysis

(20) $\xrightarrow[\text{reduction}]{\text{FGI}}$ [NOH compound] $\xrightarrow[\text{oxime}]{\text{C-N}}$

[ketone 21] $\xrightarrow{\text{1,2 C-C}}$ [allylic chloride 22] + [ethyl acetoacetate]

(21)   (22)   CO₂Et

Synthesis[135,136]

[ethyl acetoacetate] $\xrightarrow[\text{2.(22)}]{\text{1.base}}$ [alkylated β-ketoester] $\xrightarrow[\text{2.H}^+\text{,heat}]{\text{1.KOH}}$

(21) $\xrightarrow[\text{pyr,EtOH}]{\text{NH}_2\text{OH.HCl}}$ [oxime] $\xrightarrow{\text{LiAlH}_4}$ TM(20)

51%   85%   42%

The cyclisation experiments were disappointing[135]: some unsaturated amines did cyclise but many, including (20), were simply brominated on the double bond.

2. A further study on cyclisation reactions[137] needed unsaturated diol (22). Can you suggest a synthesis?

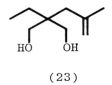

(23)

Answer : This is clearly a reduction product of a malonate ester (24) (cf p T 110) which can be disconnected in the usual way.

Analysis

$$(23) \xRightarrow[\text{reduction}]{\text{FGI}} \underset{(24)}{\underset{\text{EtO}_2\text{C} \quad \text{CO}_2\text{Et}}{\diagdown\kern-0.5em\diagup}} \Longrightarrow \text{EtBr} \quad \underset{(22)}{\text{Cl}\diagdown\kern-0.5em\diagup}$$

$$+ \ CH_2(CO_2Et)_2$$

It will be better to add the more reactive allylic halide (22) in the second more difficult alkylation.

Synthesis[137]

$$CH_2(CO_2Et)_2 \xrightarrow[\text{2.EtBr}]{\text{1.EtO}^-} \xrightarrow[\text{4.(22)}]{\text{3.EtO}^-} (24) \xrightarrow{\text{LiAlH}_4} TM(23)$$

75%

This time the cyclisation was more successful, the five membered cyclic ether (25) being formed in excellent yield in acid.

(23) 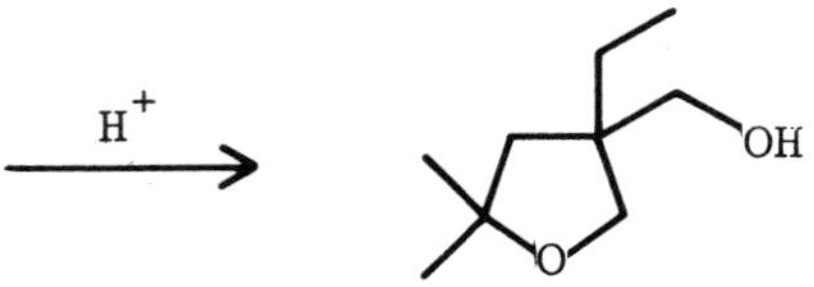 $\xrightarrow{H^+}$ (25)

(c) *Carbonyl Compound Synthesis by Michael Addition*
Example : One advantage of this route is that it allows the addition of an 'angular' alkyl group in compounds like (26). Many natural products, such as steroids, contain an angular group and it can be difficult to set up this quaternary carbon atom.

Analysis

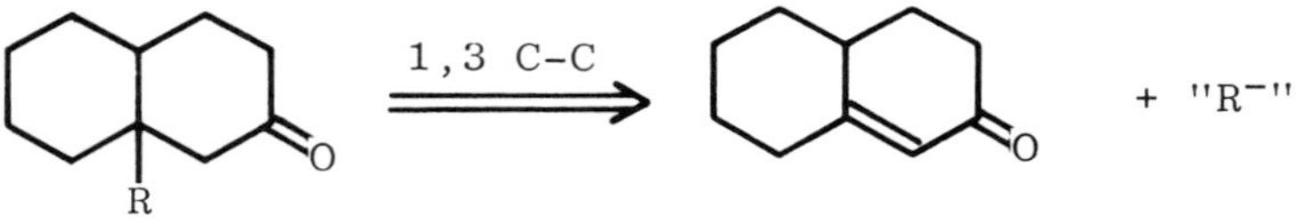

(26) $\xRightarrow{1,3\ C-C}$ (27) + "R⁻"

Synthesis[138]

(27) $\xrightarrow[\text{Cu(OAc)}_2]{\text{RMgBr}}$ TM(26) 71%

Problem : Suggest a synthesis of ketone (28).

(28)

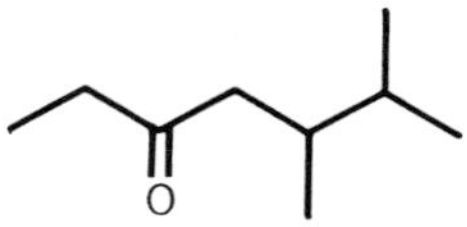

Answer : We obviously want to disconnect on the more substituted side and (a), (b) and (c) are all possible. Disconnection (c) has the advantage that it is between two branchpoints and so gives two simple starting materials.

Analysis

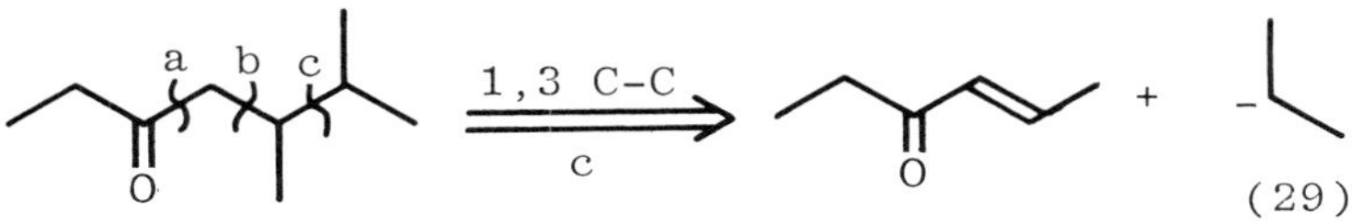

$$\underset{(28a,b,c)}{\text{a b c}} \xrightarrow{\underset{c}{1,3\ C-C}} \quad + \quad (29)$$

A Grignard reagent can be used for (29) without the usual copper catalyst as it happens.

Synthesis[139]

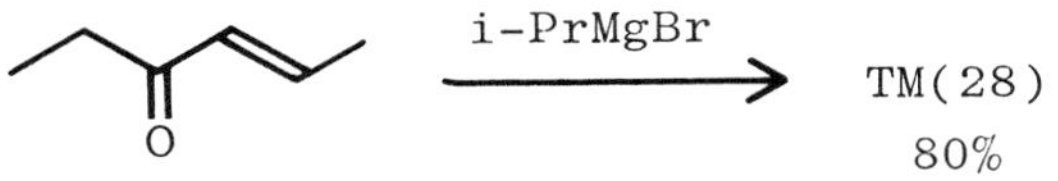

$$\xrightarrow{i\text{-PrMgBr}} \text{TM(28)} \quad 80\%$$

CHAPTER 14

Strategy VI: Regioselectivity

(a) *Regioselective Alkylation of Ketones*
Example : There are alternative strategies to the one outlined in Chapter T 14. Acid derivatives can enolise on one side only, the enols can be alkylated (p. 126) and conversion to the ketone achieved by one of the methods from pages T 106-8. For both steps nitriles are ideal functional groups.

Hence ketone (1) might be made by regioselective alkylation of (2) but this is doubtful. A safer route is to disconnect the ethyl group to leave nitrile (3) which can certainly be made by alkylation of nitrile (4) as there is only one site for enolisation.

Analysis

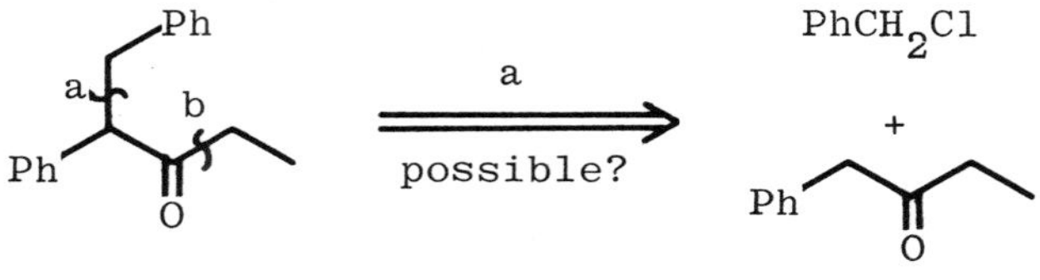

[Structure of (1): PhCH(CH₂Ph)C(O)CH₂CH₃ with disconnection labels a and b] ⟹ (a, possible?) PhCH₂Cl + PhCH₂C(O)CH₂CH₃ (2)

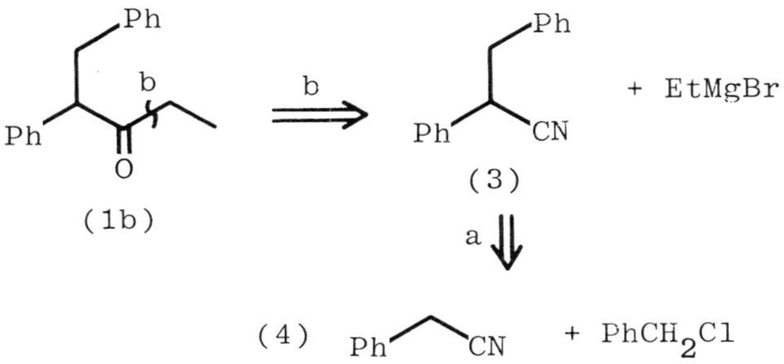

PhCH₂-CH(Ph)-C(=O)-Et (1b) ⟹ PhCH₂-CH(Ph)-CN (3) + EtMgBr

(disconnection b on both sides)

(4) ⇓ a

Ph-CH₂-CN (4) + PhCH₂Cl

Alkylation of nitrile (4) needs only moderately strong base (hydroxide will do[140]) as the benzene ring helps to stabilise the anion (5).

Synthesis[140,141]

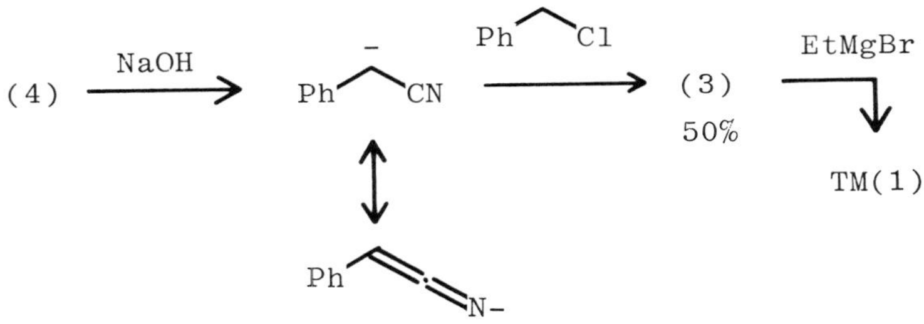

(4) —NaOH→ Ph-CH⁻-CN —PhCH₂Cl→ (3) —EtMgBr→ TM(1)
                                    50%

↕

Ph-CH=C=N⁻

(5)

Optically active ketone (6) was needed for a study of asymmetric induction.[142] It could be made from acid (7) by a Friedel Crafts route or from nitrile (8) by Grignard addition, but neither of these compounds could be made by alkylation as the branchpoint is on the β carbon (● in each). The 1,3 C-C disconnection, e.g. (6b) is not good as it destroys the chiral centre.

Analysis 1

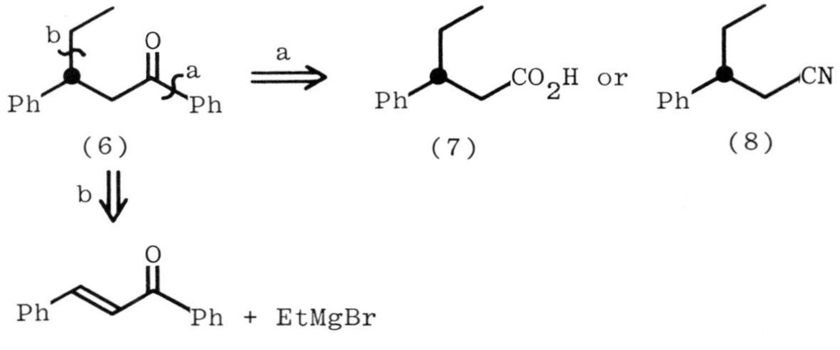

[Structures: (6) Ph-CH(Et)-CH₂-CO-Ph with disconnections a and b; ⇒ (7) Ph-CH(Et)-CH₂-CO₂H or (8) Ph-CH(Et)-CH₂-CN]

[Down arrow b: Ph-CH=CH-CO-Ph + EtMgBr]

The compounds could be made by alkylation if the cyanide (8) were derived from (9) and hence from cyanide (11).

Analysis 2

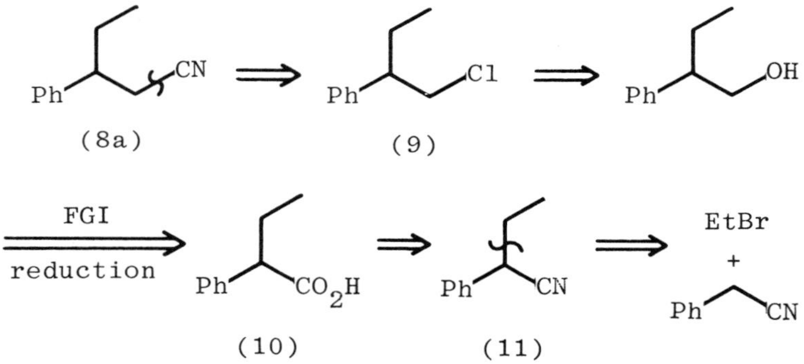

[(8a) Ph-CH(Et)-CH₂-CN ⇒ (9) Ph-CH(Et)-CH₂-Cl ⇒ Ph-CH(Et)-CH₂-OH]

[FGI reduction ⇐ (10) Ph-CH(Et)-CO₂H ⇒ (11) Ph-CH(Et)-CN ⇒ EtBr + Ph-CH₂-CN]

The chiral centre first appears in cyanide (11) but the acid (10) is the ideal compound for resolution as it can form a salt with a naturally-occurring optically active base.

Synthesis[142,143]

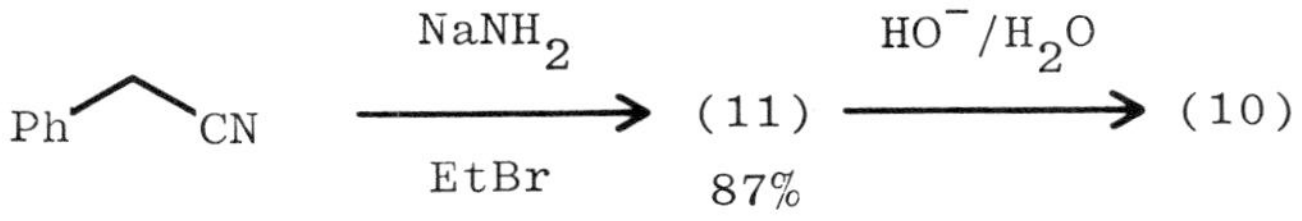

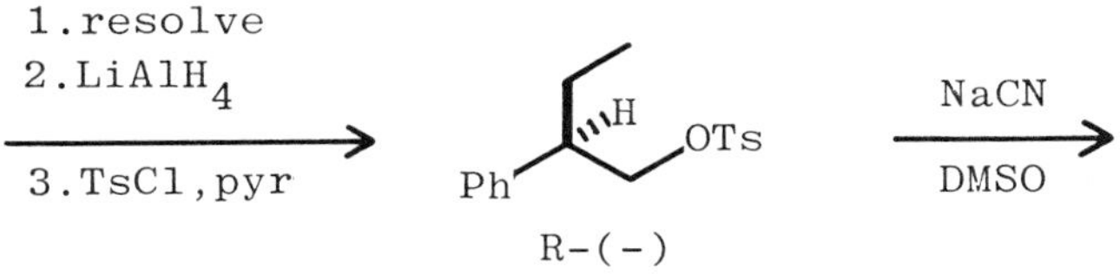

Ph-CH₂-CN →(NaNH₂, EtBr)→ (11) →(HO⁻/H₂O, 87%)→ (10)

1. resolve
2. LiAlH₄
3. TsCl, pyr
→ Ph-C*(H)(Et)-CH₂-OTs   R-(−)   →(NaCN, DMSO)→

R-(+)-(8) →(PhMgBr)→ TM(6) R-(+)

Tosylate was preferred as a leaving group in making cyanide (8) and the Grignard method for the synthesis of (6) to avoid an extra step.

Simple Problem : Suggest a synthesis of ketone (12), analysing the possibility of using alkylation.

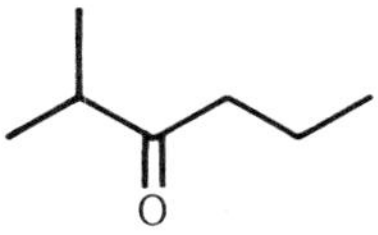

(iPr)C(=O)(CH₂CH₂CH₃)   (12)

Answer : An alkylation disconnection (12a) is indeed possible but regioselective alkylation of (13) is not. We could make (14) by methods to be discussed in Chapters T 19 and 20 or we could revert to nitrile (15), as in the two examples above.

Analysis 1

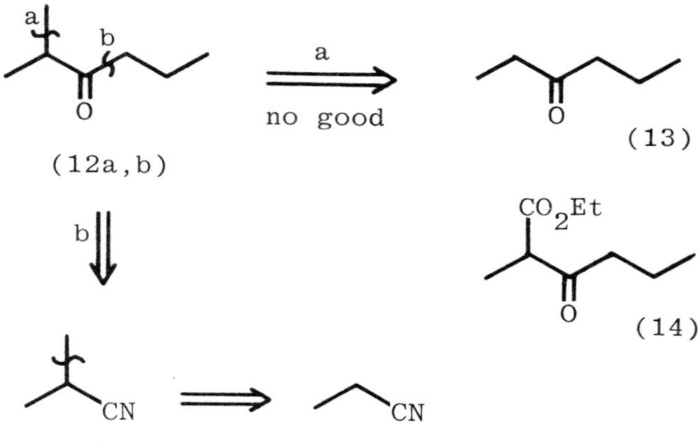

[Scheme showing (12a,b) with disconnections a and b; disconnection a leads to "no good" giving (13) 3-hexanone; disconnection b leads to (15) (Me)₃C-CN + n-PrMgBr, and (15) comes from cyanide displacement; also shown (14) ethyl 2-methyl-3-oxohexanoate]

Easier solutions are to make (15) by cyanide displacement or to use a third disconnection (12c).

Analysis 2

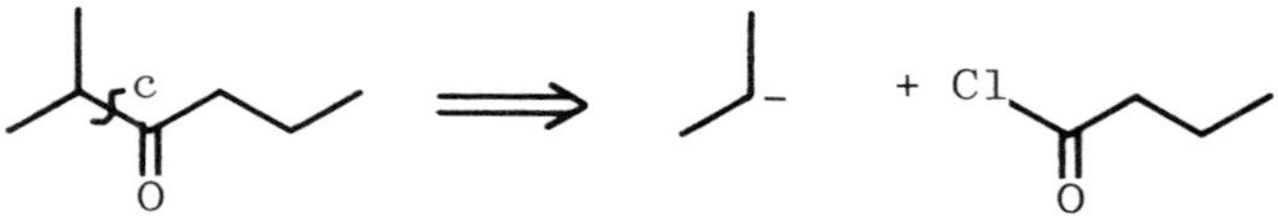

[Scheme: (12c) disconnects at c to give i-Pr anion + Cl-CO-n-Pr (butyryl chloride)]

The synthesis has been achieved by this last method, using an organo-cadmium reagent.

Synthesis[144]

i-PrBr $\xrightarrow{\text{1. Mg, Et}_2\text{O} \atop \text{2. Cd}_2\text{I}}$ i-Pr$_2$Cd $\xrightarrow{\text{n-PrCOCl}}$ TM(12) 60%

Problem 2 : Comment on the feasibility of using disconnection (b) on TM (16) (p 132).

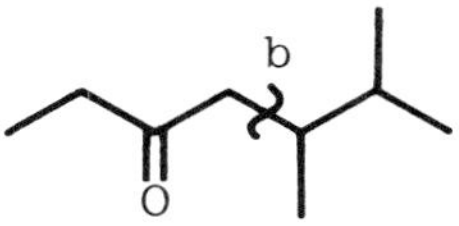

(16)

Answer : We should need keto-ester (17) for this, and even then alkylation with secondary halide (18) is likely to be poor. Alternatively we could use nitrile (19) but this requires the same alkyl halide (18). The Michael synthesis on page 133 is best.

Analysis

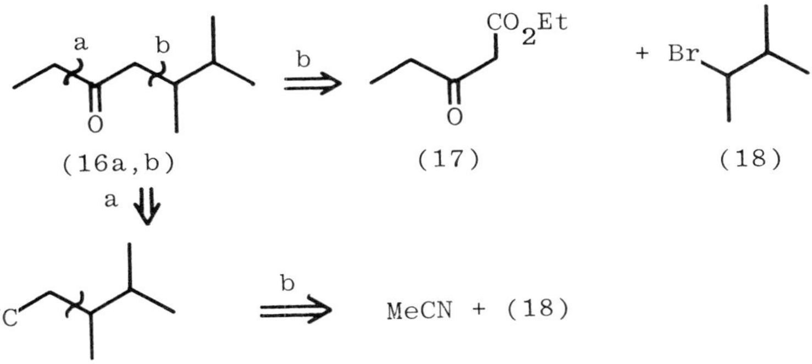

(16a,b) ⇒ (17) + Br (18)

a ⇓

NC... ⇒ MeCN + (18)

Regioselectivity in Michael Reactions

Problem : Identify the factors responsible for regioselectivity in these reactions:

(a) 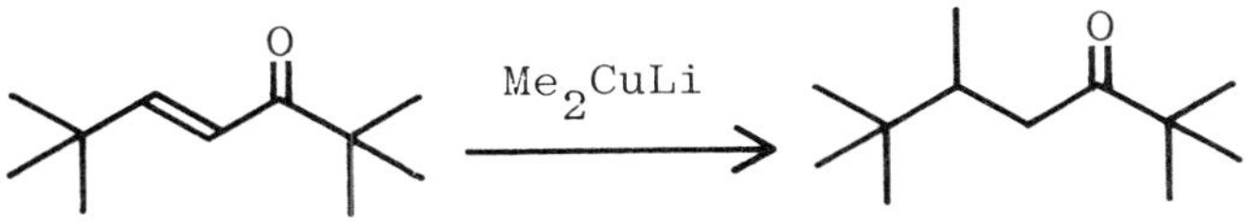 [enone] $\xrightarrow{Me_2CuLi}$ [ketone] 79%

(b)

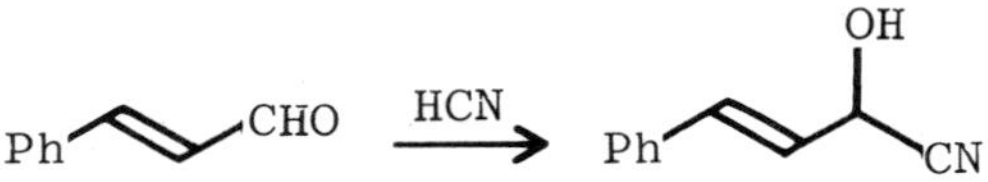

PhCH=CHCHO $\xrightarrow{HCN}$ PhCH=CH-CH(OH)CN

(c)

CH$_2$=C(CH$_3$)CO$_2$Et $\xrightarrow{EtMgI}$ CH$_2$=C(CH$_3$)C(Et)$_2$OH

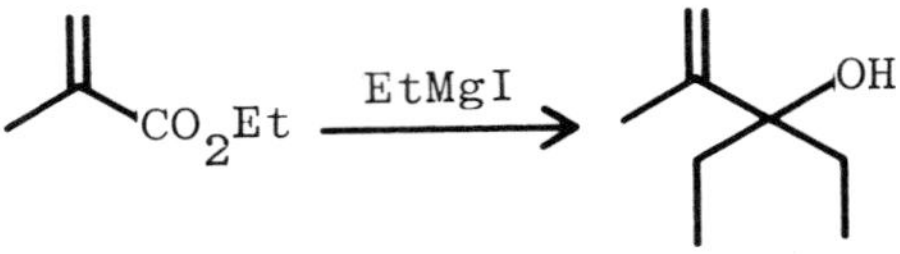

Answer : (a) Steric hindrance equal at either site : copper is the metal to ensure conjugate addition.[145]

(b) Kinetic attack by basic nucleophile on less crowded site.[146]

(c) No copper; Grignard attacks ester directly and displacement of EtO$^-$ is irreversible.[147]

Problem : Make the first disconnection to show how these molecules might be made by organo-copper addition to suitable carbonyl compounds.

(a) 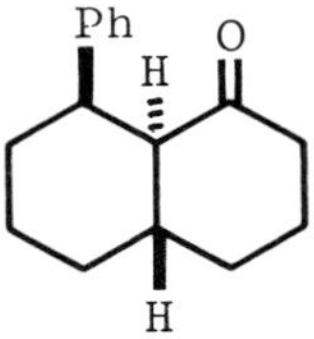 (19) (b) 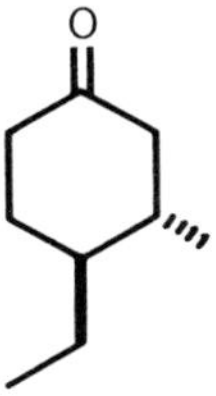 (20)

(c) 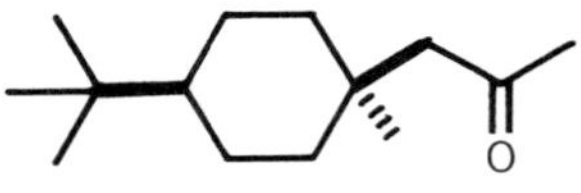 (21) (d) 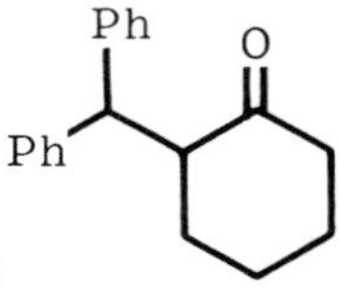 (22)

Answer : The disconnection must be of a bond β,γ to the carbonyl group. It is best to leave the ring intact and disconnect at a ring-chain junction (a-c) or at a branchpoint (d).

Analysis

(a)[148] (19a) ⇒ (23) B + Ph⁻

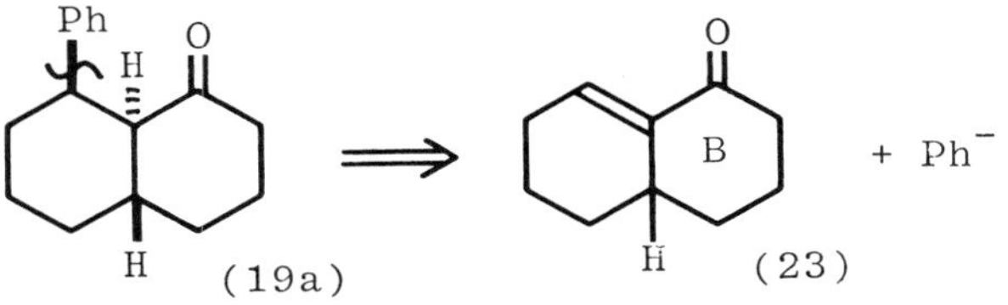

(b)[149] (20a) ⇒ (24) + Me⁻

(c)[150] (21a) ⇒ (25) + Me⁻

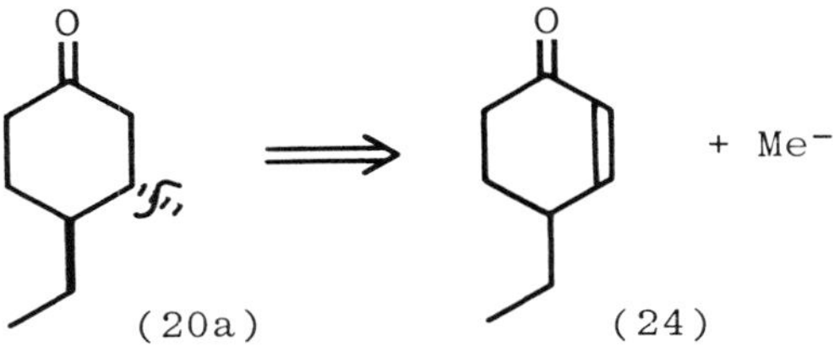

(d)[151] (22a) ⇒ (26) + Ph⁻

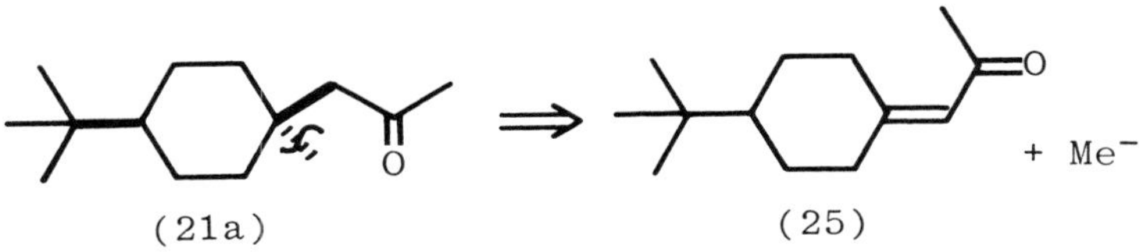

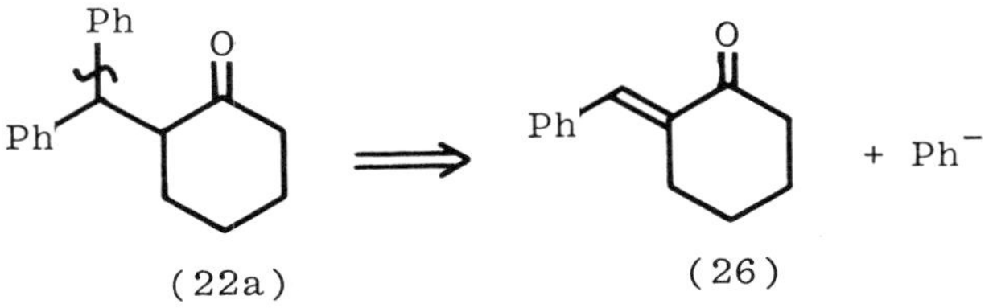

Synthesis : All by RMgBr + Cu(1) + enone.

Compound (22) was used to synthesise some central nervous system stimulants,[151] the others mostly in investigations of the stereochemistry of the reaction. What generalisation can you make on the stereochemistry of the organo-copper additions?

Answer : The reagent approaches axially in a *trans* manner to the largest substituent ring B in (23), Et in (24), and t-Bu in (25). In (19), the last proton is added to enolate (27) to give the more stable *trans* ring junction.

(23) $\xrightarrow{\text{PhMgBr}}_{\text{Cu(I)}}$ (27) $\longrightarrow$ (19)

Example : Sometimes copper solves other regioselectivity problems. Addition of aryl Grignard (28) to enone (29) gives the anomalous product (30) in which the electrophile (29) has been attacked at the right atom but the nucleophile (28, arrows) has attacked with the wrong atom.

Addition of Cu(1) to the reaction mixture prevents this unwanted reaction and the normal 1,4 addition product (31) is formed.[152]

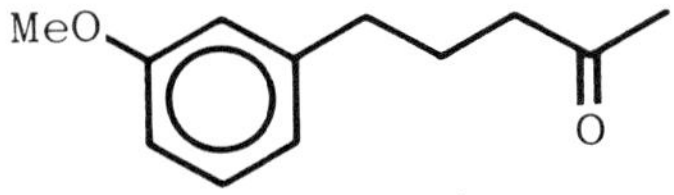

(31)

Further Example :
Regioselectivity in Epoxide Reactions

We have already discussed the regioselectivity of the reactions of epoxide with nucleophiles and devised strategies (p 64-5) to achieve the synthesis of compounds (32).

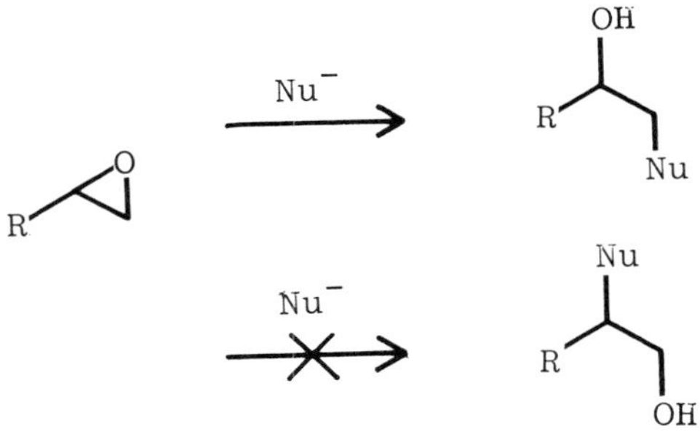

(32)

Another way to make (32) is to carry out the reaction in acid solution when the regioselectivity is reversed.

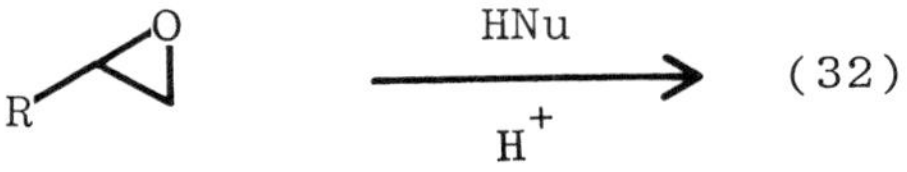

Protonation of the epoxide gives cation (33) which reacts *via* a loose transition state so that a partial positive charge appears on the carbon atom under attack. The *more* stable partial cation is therefore formed.

$$R\text{-epoxide} \xrightarrow{H^+} (33) \xrightarrow{Nu^-} \text{transition state} \rightarrow (32)$$

Strain in the three-membered ring plus the excellence of the leaving group make the C-O bond start to break before the C-Nu bond is fully established.

Not all nucleophiles are compatible with acid conditions, and unfortunately most carbon nucleophiles, especially RMgBr and RLi, definitely cannot be used in this way. The Friedel-Crafts reaction, with an aromatic ring as the carbon nucleophile, is quite satisfactory.[153]

$$\text{PhH} + \text{epoxide} \xrightarrow{AlCl_3} Ph\text{-CH(CH}_3\text{)-CH}_2\text{OH}$$

CHAPTER 15

Alkene Synthesis

By Elimination from Alcohols and Derivatives

Simple Problems:

1. Suggest a synthesis of alkene (1), needed for a morphine synthesis.[154]

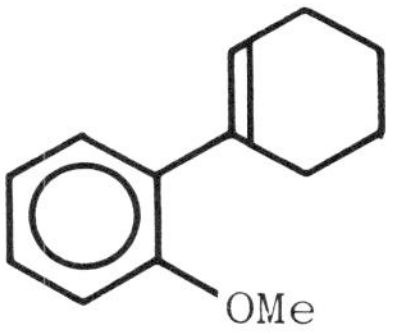

Answer: Alcohol (2) will dehydrate to (1) and the Grignard route gives (2).

Analysis

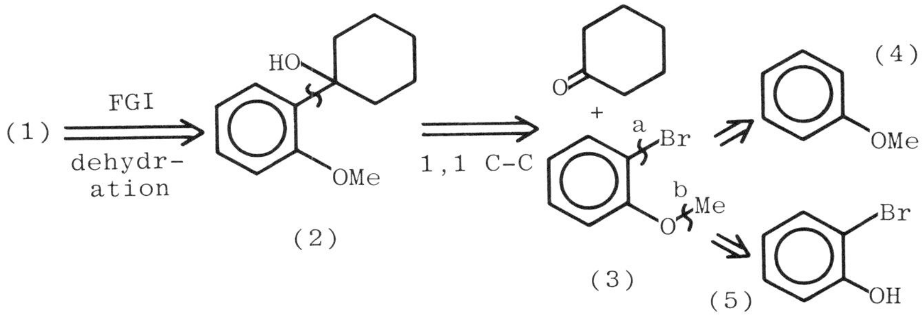

Halide (3) is available, but could be made by bromination of (4) or methylation of (5). In the synthesis, oxalic acid (6) was used as the dehydrating agent. This is a reasonably strong acid (pKa 1.23) and is conveniently solid.

Synthesis[154]

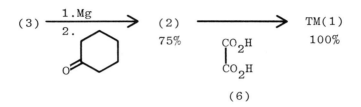

2. Suggest a synthesis of alkene (7).

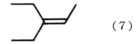

Answer : The OH group is best placed at the branchpoint to give alcohol (8) : any arm of this symmetrical alcohol can now be disconnected.

Analysis

(7) $\xrightarrow[\text{dehydration}]{\text{FGI}}$ [structure with OH] (8) $\xrightarrow{\text{1,1 C-C}}$ [ketone structure] =O + EtMgBr

Synthesis[155]

EtBr $\xrightarrow[\text{2.Et}_2\text{CO}]{\text{1.Mg}}$ (8) 63% $\xrightarrow[\text{reflux}]{\text{(6)}}$ TM(7) 84%

Examples : Unless the alcohol is symmetrical, problems of regioselectivity in elimination occur. The general rule is that elimination in acid gives the more substituted double bond (e.g. 9) while elimination in base gives the less substituted (e.g. 10).[156]

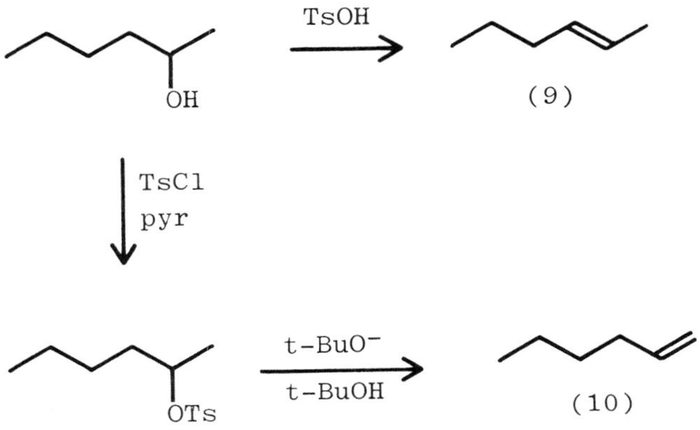

This may change with extra conjugation : the benzylic protons (marked H) in (11) are more acidic than those on the methyl group so elimination in base[157] gives (12) not (13).

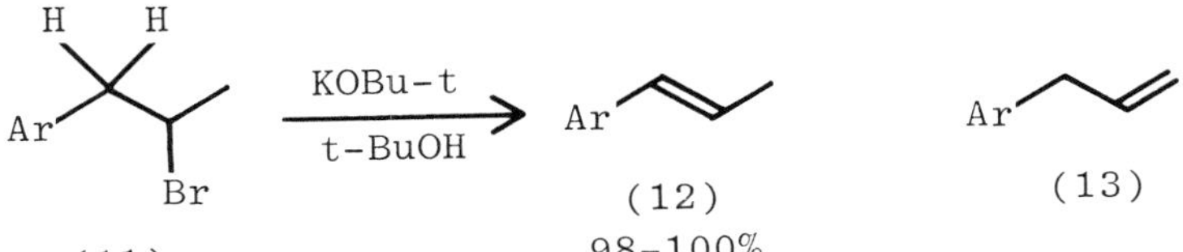

The large base t-BuO⁻ helps to give the less substituted alkene (10) and also helps elimination rather than substitution, as these contrasted results show[158] for (14).

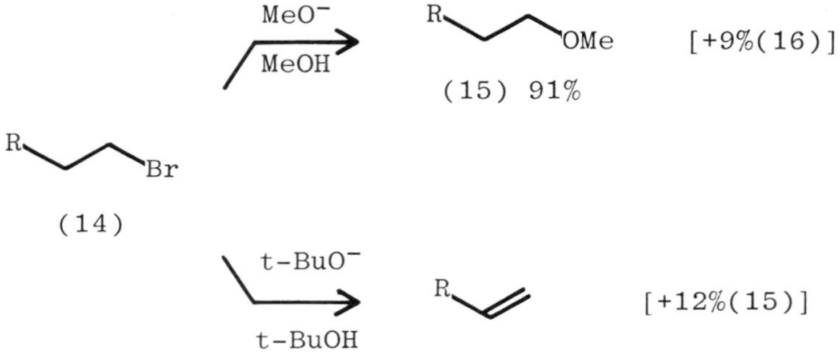

$$\text{(14) R-CH}_2\text{CH}_2\text{Br} \xrightarrow[\text{MeOH}]{\text{MeO}^-} \text{R-CH}_2\text{CH}_2\text{OMe} \quad \text{(15) 91\%} \quad [+9\%(16)]$$

$$\xrightarrow[\text{t-BuOH}]{\text{t-BuO}^-} \text{R-CH=CH}_2 \quad \text{(16) 85\%} \quad [+12\%(15)]$$

Problem : Predict the elimination products from (17) and (18) under the given conditions :

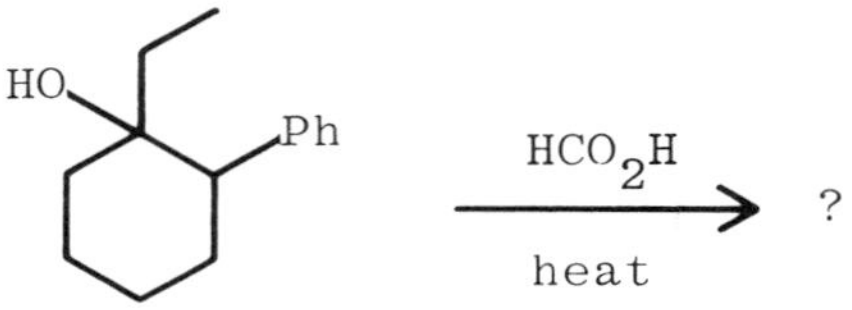

(17) 1-ethyl-2-phenyl-cyclohexan-1-ol $\xrightarrow[\text{heat}]{\text{HCO}_2\text{H}}$?

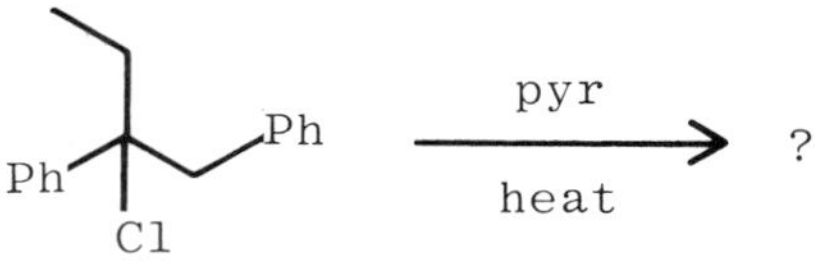

(18) Ph-C(Cl)(Et)-CH$_2$-Ph $\xrightarrow[\text{heat}]{\text{pyr}}$?

Answer : Compound (17) reacts via carbonium ion (20) and gives[158] the most substituted and conjugated double bond, i.e. (21). The product was used to make a biphenyl for studying cyclodehydrogenations and the starting material (17) is available from (19), made on p 119.

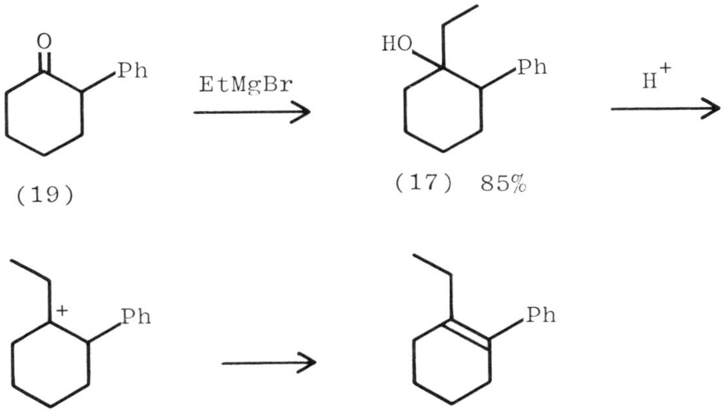

Compound (18) eliminates[159] in base with the loss of the more acidic benzylic proton (22) cf (11) - also giving the most conjugated product.

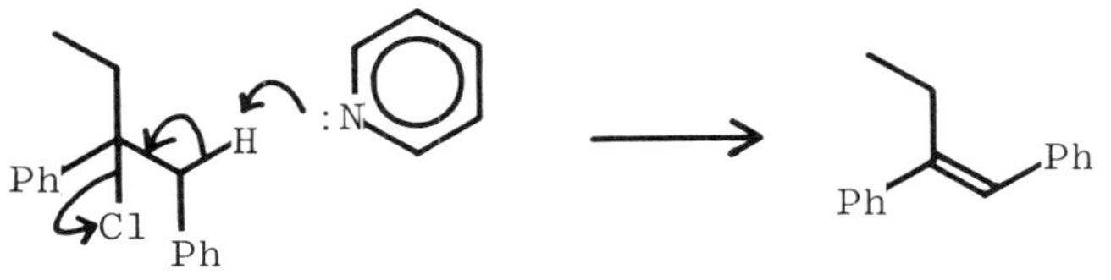

Example : Special bases DBN (23) and DBU (24) are exceptionally reactive in elimination reactions under mild conditions. DBN allowed elimination of HBr from (25) even in the presence of the epoxide so that mono-epoxy-naphthalene (26) could be made for the first time.[160]

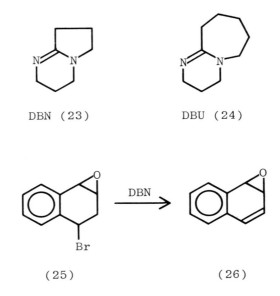

DBN (23)   DBU (24)

(25)   (26)

2. *The Wittig Reaction*

Problem : Suggest a synthesis of the natural product iso-saffrole,[27] used in soap perfumes. Either *cis* or *trans* will do.

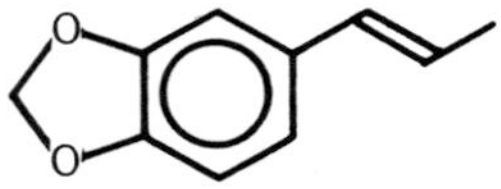

(27)

Answer : The Wittig disconnection suggests two sets of starting materials.

Analysis

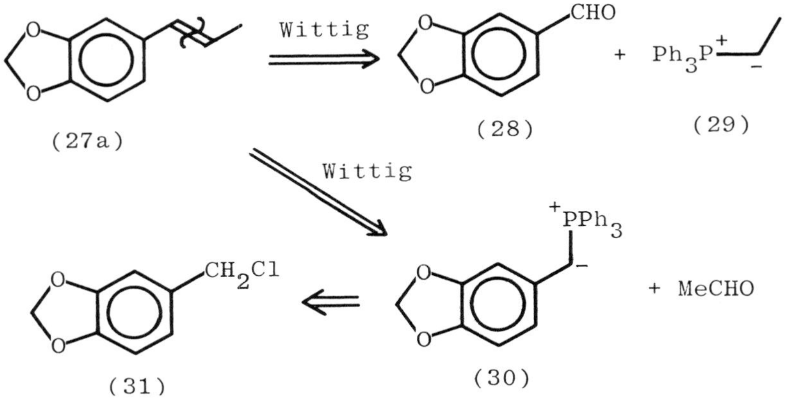

[Wittig disconnection of (27a) giving piperonal (28) + Ph$_3$P$^+$–CH$_2$CH$_3$ ylid (29); alternative Wittig disconnection giving ylid (30) (Ph$_3$P$^+$–CH$_2$–Ar) + MeCHO, with (30) derived from benzyl chloride (31).]

Aldehyde (25) is piperonal (see p T 9) which we made from halide (31), the precursor of ylid (30), so either route is suitable. The route via (28) has been used.

Synthesis[161]

$$Ph_3P \xrightarrow{EtBr} Ph_3P^+Et\ Br^- \xrightarrow{BuLi}$$
$$95\%$$

$$(29) \xrightarrow{(28)} TM(27)$$
$$57\%$$

This route should give more *cis* (27), as the ylid (29) is unstabilised, while the route *via* stabilised (30) should give more *trans* compound.

Example : Ester (32) was needed as a model compound to study mechanisms of biological ester hydrolysis.[16] It is clearly made from acid (33) which could be made from aldehyde (34) by a Wittig reaction. We discussed the synthesis of (35) on page 21.

Analysis

[Scheme: (32) ⇒ (C-O ester) ⇒ (33) + p-nitrophenol; (33) ⇒ Wittig ⇒ acetonide-protected aldehyde + $Ph_3P^+CH_2CO_2H$; that aldehyde ⇐ (C-O ether) ⇐ (35) with OH, CHO, t-Bu substituents; (34) labels the Wittig substrate]

The methyl ether was introduced after the Wittig reaction which used a protected carboxylic acid (ethyl ester). *p*-Nitrophenol is available and is one of the products of the nitration of phenol.

Synthesis[16]

$$Ph_3P + BrCH_2CO_2Et \longrightarrow Ph_3P^+CH_2CO_2Et \xrightarrow{EtO^-} (35)$$

[Reaction scheme: 2-hydroxy-5-tert-butyl cinnamate ester (trans, CO₂Et) → 1. Me₂SO₄, NaOH; 2. H⁺, H₂O → (33) → 1. SOCl₂; 2. (4-nitrophenol, OH) → TM(32)]

The double bond in (32) is *trans* since the Wittig used a stabilised ylid and an aldehyde.

Examples :
1. Unstabilised ylids usually give more *cis* product with aldehydes. Muscalure (36) is a house-fly pheromone used to bait traps and the obvious Wittig synthesis gives 85% *cis* and 15% *trans*.[162]

Synthesis[162]

$$Me(CH_2)_{12}CH_2\overset{+}{-}PPh_3 \xrightarrow[\text{2.Me(CH}_2)_7\text{CHO}]{\text{1.BuLi,DMSO}} Me(CH_2)_7\text{—CH=CH—}(CH_2)_{12}Me$$

(36)   85:15 Z:E

2. Diene alcohol (38) is the pheromone of the codling moth, the creature responsible for the grubs in apples, and of the various possible disconnections (a) is best as it gives most simplification and a stabilised ylid (39) which will produce the required *trans* double bond. Allylic bromide (40) and aldehyde-ester (41) are available.

Analysis

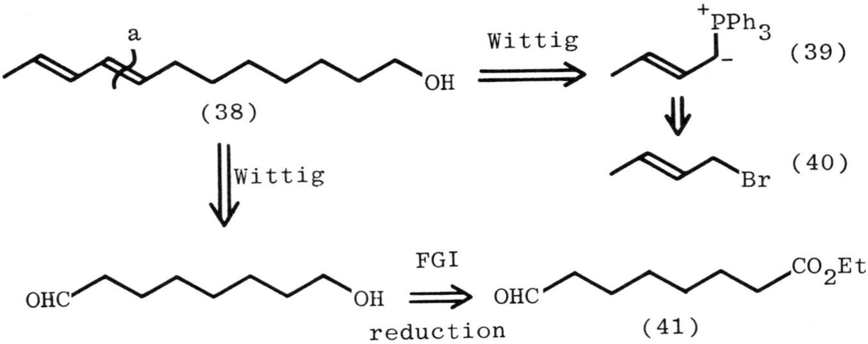

[Scheme: (38) CH₂=CH-CH₂-(CH₂)₇-OH with disconnection 'a' ⟹ Wittig ⟹ allyl-PPh₃⁺ (39) ⟸ allyl bromide (40)]

[(38) ⟹ Wittig ⟹ OHC-(CH₂)₇-OH ⟸ FGI reduction ⟸ OHC-(CH₂)₇-CO₂Et (41)]

If we reduce the ester after completing the Wittig reaction, no protection of the aldehyde will be necessary. In practice, the hindered reducing agent (42) gives good results.

Synthesis[163]

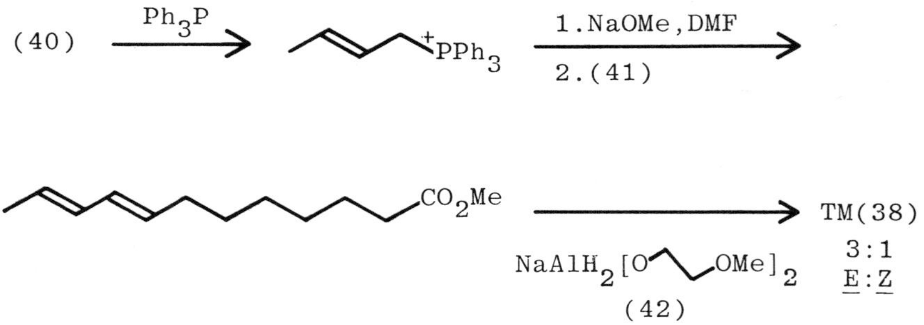

(40) $\xrightarrow{Ph_3P}$ allyl-PPh₃⁺ $\xrightarrow[2.(41)]{1. NaOMe, DMF}$

[structure: CH₃-CH=CH-CH=CH-(CH₂)₆-CO₂Me] $\xrightarrow{NaAlH_2[OCH_2CH_2OMe]_2 \text{ (42)}}$ TM(38)

3:1 E:Z

Problems: Pear ester (43) is an important industrial flavouring compound with a pear-like taste and smell. Consider all possible Wittig disconnections and choose a reaction which should give the right geometrical isomer.

(43) [structure: CH3(CH2)4CH=CH-CH=CH-CO2Me]

Answer : Each disconnection can be carried out in two directions :
Analysis

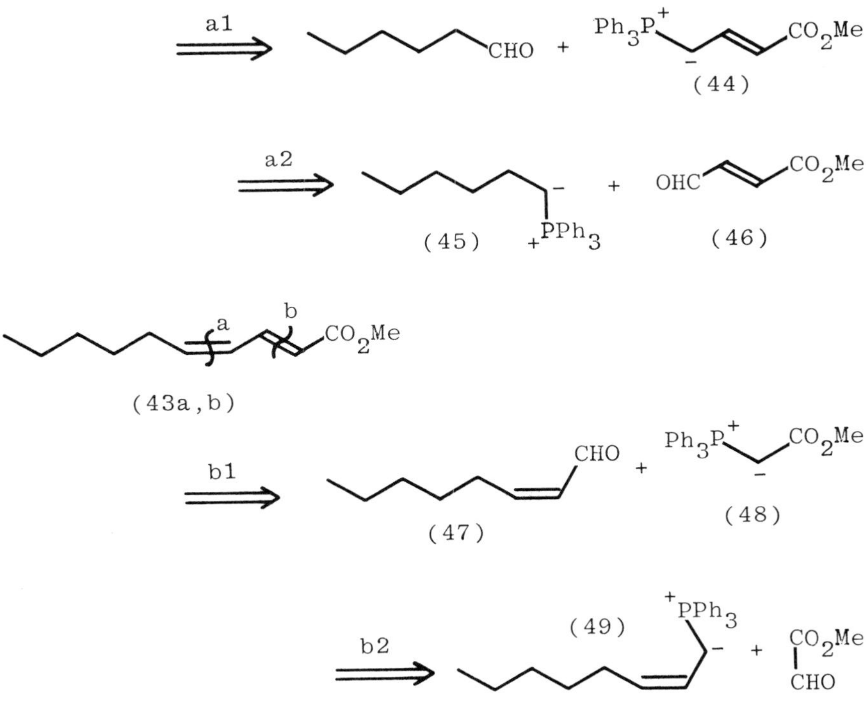

a1 ⟹ ~~~CHO + Ph3P⁺–CH=CH–CO2Me⁻
                              (44)

a2 ⟹ ~~~CH2⁻ PPh3⁺ (45) + OHC–CH=CH–CO2Me (46)

~~~⫩ₐ ⫩ᵦ CO2Me
(43a,b)

b1 ⟹ ~~~CH=CH–CHO (47) + Ph3P⁺–CH2–CO2Me⁻ (48)

b2 ⟹ ~~~CH=CH–CH2 ⁺PPh3⁻ (49) + CO2Me–CHO

The only reaction which should give the wrong isomer is (a1) where stabilised ylid (44) should give an unwanted *trans* double bond. Both (b) disconnections should give the right isomer as stabilised ylids (48) and (49) should give a wanted *trans* double bond. However, the *cis* double bond in (47) and (49) is going to be difficult to make.

Disconnection (a2) leads to the industrial synthesis as the half aldehyde, half ester (46) of fumaric acid (100% *trans*) is available and the Wittig reaction with unstabilised ylid (45) gives 85% *cis* geometry in the new double bond.
Synthesis[164]

$$Ph_3\overset{+}{P}\text{\textasciitilde\textasciitilde\textasciitilde} \xrightarrow[\text{2.(46)}]{\text{1.base}} TM(43) \quad 85\% \underline{E},\underline{Z}$$

2. Bombykol (50), pheromone of the silk moth and the first whose structure was elucidated, can be made by two successive Wittig reactions, each giving the right geometry. Can you find this route?

$$n\text{-Pr}\diagup\diagdown(CH_2)_8CH_2OH \quad (50)$$

Answer : If we disconnect the *cis* double bond first (50a), we must use unstabilised ylid (51) and aldehyde (52) which can be made from stabilised ylid (53) and available half aldehyde, half ester (54).

Analysis

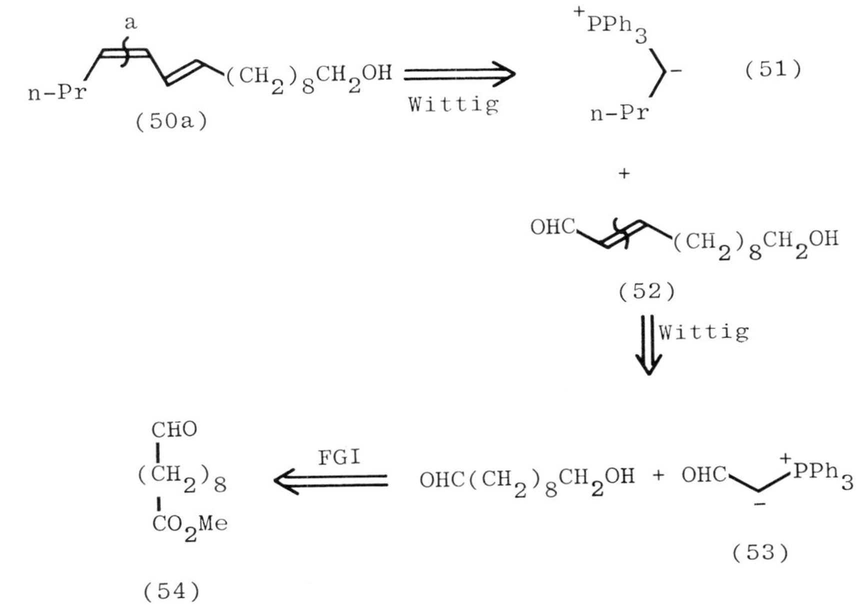

n-Pr–CH=CH–(CH$_2$)$_8$CH$_2$OH (50a) $\xrightarrow{a}_{\text{Wittig}}$ n-Pr–CH(–)–$^+$PPh$_3$ (51)

+

OHC–CH=CH–(CH$_2$)$_8$CH$_2$OH (52) $\Downarrow$ Wittig

CHO–(CH$_2$)$_8$–CO$_2$Me (54) $\xleftarrow{\text{FGI}}$ OHC(CH$_2$)$_8$CH$_2$OH + OHC–CH(–)–$^+$PPh$_3$ (53)

The synthesis has been carried out by this route and by many other routes, mostly involving Wittig reactions.

Synthesis[165]

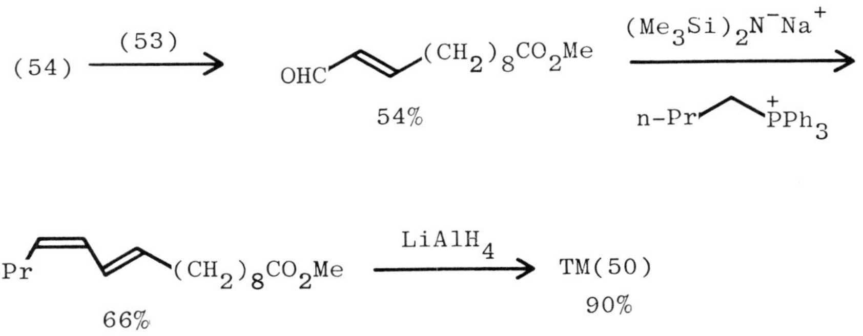

(54) $\xrightarrow{(53)}$ OHC–CH=CH–(CH$_2$)$_8$CO$_2$Me (54%) $\xrightarrow{(Me_3Si)_2N^-Na^+}$ n-Pr–CH$_2$–$^+$PPh$_3$

n-Pr–CH=CH–CH=CH–(CH$_2$)$_8$CO$_2$Me (66%) $\xrightarrow{\text{LiAlH}_4}$ TM(50) (90%)

General Example : Queen substance (55) is a hormone produced by a queen bee to prevent the worker bees' ovaries maturing and to 'train' them to serve her.

The obvious Wittig disconnection gives stabilised ylid (56) and keto-aldehyde (57). We have used many such long-chain dicarbonyl compounds in this Chapter and they are mostly produced from available alkenes by oxidative cleavage (e.g. ozonolysis). In this case, cyclic alkene (58) is the right starting material, and this can be made from alcohol (59) by elimination.

Analysis

[Structure (55): long-chain keto-acid] $\xrightarrow{\text{Wittig}}$ [ylid (56) with CO_2R, $^+PPh_3$]

+

[Structure (57): keto-aldehyde with CHO]

$\Downarrow$

[cycloheptanone] $\xleftarrow[\text{C-C}]{1,1}$ [cycloheptane-OH (59)] $\xleftarrow[\text{dehydration}]{\text{FGI}}$ [cycloheptene (58)]

+ MeMgI

The Wittig reagent (56) is best protected as an ester and reacts chemo-selectively with the aldehyde rather than the less reactive ketone in (57).

Synthesis[166]

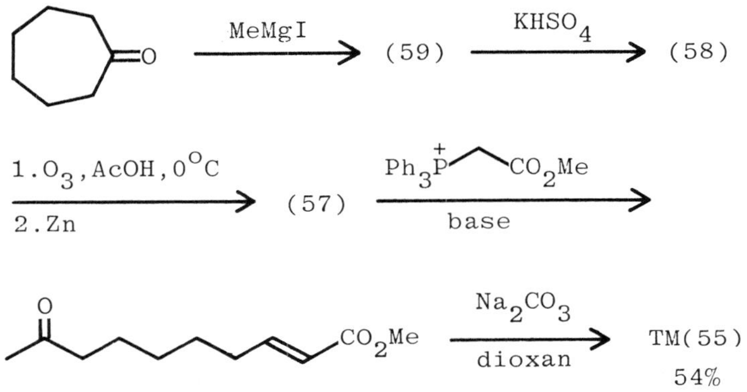

cycloheptanone —MeMgI→ (59) —KHSO$_4$→ (58)

—1. O$_3$, AcOH, 0°C; 2. Zn→ (57) —Ph$_3$P$^+$—CO$_2$Me / base→

MeCO-(CH$_2$)$_5$-CH=CH-CO$_2$Me —Na$_2$CO$_3$ / dioxan→ TM(55) 54%

CHAPTER 16

Strategy VII: Use of Acetylenes

Example : The use of organo copper compounds in Michael additions (p T 118) has a disadvantage in that one of the two starting R groups in (1) is wasted.

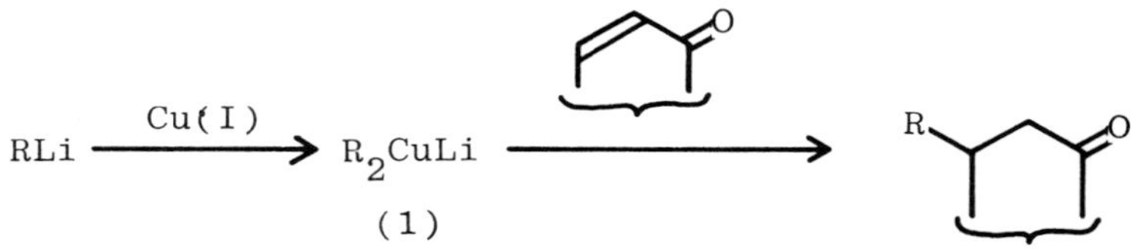

$$RLi \xrightarrow{Cu(I)} R_2CuLi \xrightarrow{\text{[enone]}} R\text{-[product]}$$
(1)

$$RR^1CuLi$$
(2)

One way to avoid this is to have a mixed copper reagent (2) in which R is preferentially transferred to the enone. A good choice[167] for R' is an acetylene and the preferred acetylene is 1-pentyne (3) as it is cheap, easy to make, and reasonably volatile.

Synthesis[167,168]

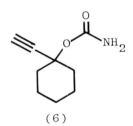

Alkylation of acetylene to (3) goes in high yield, the cuprous derivative (4) is a stable solid, and the required reagent(s) cleanly transfers R to an enone.[167]

Problem: Suggest a synthesis for the sedative meparfynol (6).

Answer: Amide disconnection gives chloroformate (7), made from alcohol (8) with $COCl_2$ (cf p T 37). Alcohol (8) is a simple acetylene adduct.

Analysis

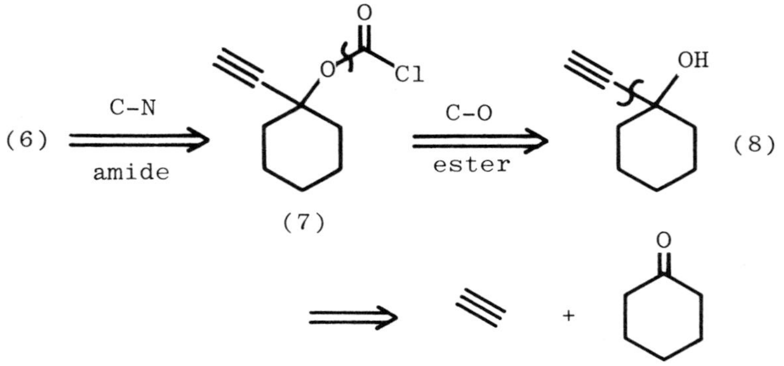

[Structure (7): 1-ethynylcyclohexyl chloroformate-type ester]

(6) ⟹ (C-N, amide) [structure 7] ⟹ (C-O, ester) [structure 8: 1-ethynylcyclohexanol]

⟹ ethyne + cyclohexanone

Synthesis[169]

≡ —(1. Na, NH₃; 2. cyclohexanone)→ (8) —(1. COCl₂; 2. NH₃)→ TM(6)

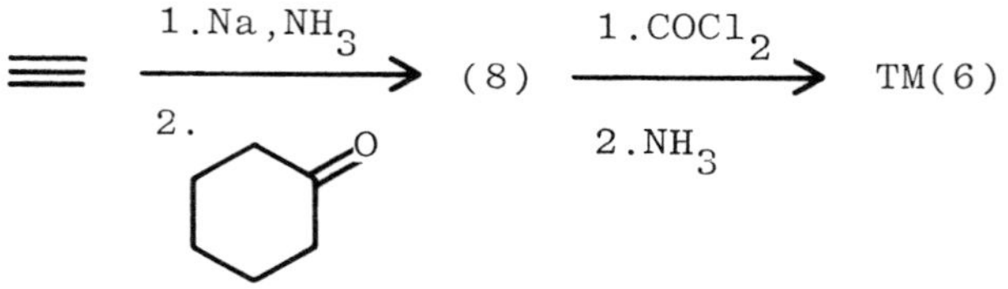

Problem : Explain the chemoselectivity of this synthesis.[170]

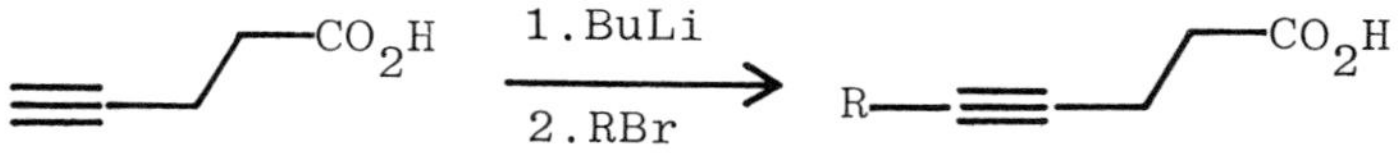

HC≡C–CH₂–CO₂H (9) —(1. BuLi; 2. RBr)→ R–C≡C–CH₂–CO₂H (10)

Answer : Dianion (11) is formed from (9) with BuLi : the acetylene anion (pKa ~ 25) is very much more reactive than the carboxylate ion (pKa ~ 5) so it reacts preferentially.

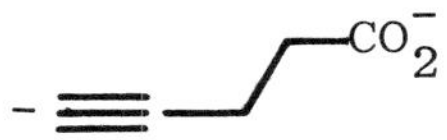

CO2
-

(11)

Problem : Suggest a synthesis for the starting material (9) and a synthesis for (10) based on a different strategy.

Answer :

(a) The simplest synthesis for (9) was alkylation of an enolate with available propargyl bromide (12) : alternatives include Michael addition of acetylene to (13) or oxidation of alcohol (14).

Analysis :

(9)
a
CO2H
b
a
Br
(12)
+ ⁻CH2CO2H
=malonate
b
-
+
CO2R
(13)
FGI
oxidation
OH
(14)

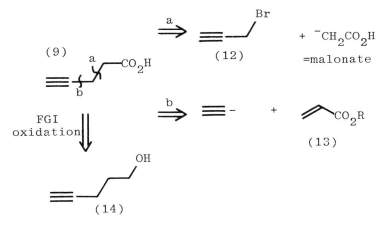

Synthesis (a)[171]

CH2(CO2Et)2
1.EtO⁻
2.(12)
CO2Et
CO2Et
1.KOH
2.H+,heat
TM(9)

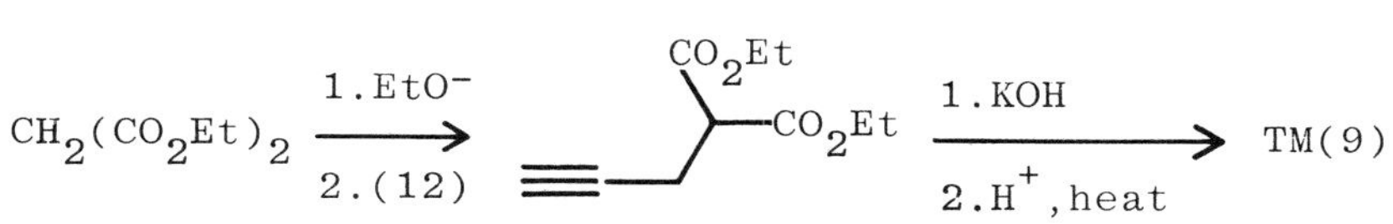

(b) The most obvious alternative strategies for (10) are the enolate alkylation (a) or the other acetylene disconnection (b).

Analysis 1

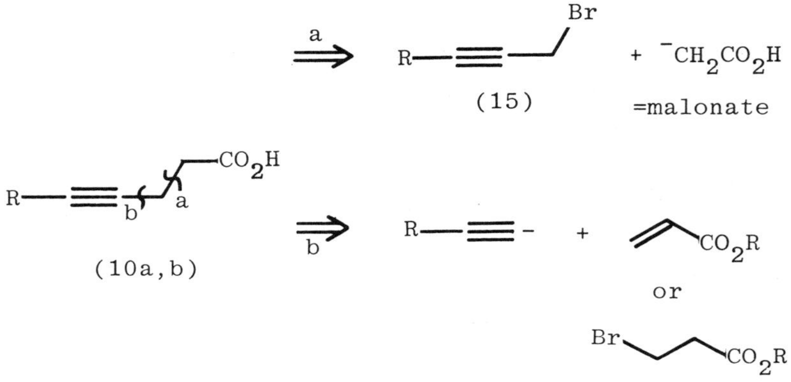

$$R-\equiv-CH_2-CHBr-CO_2H \xrightarrow{a} R-\equiv-CH_2Br \;(15) \;+\; {}^-CH_2CO_2H \;\; (\text{=malonate})$$

$$(10a,b) \xrightarrow{b} R-\equiv^- \;+\; CH_2=CH-CO_2R \;\; \text{or} \;\; Br-CH_2CH_2-CO_2R$$

The propargyl halide (15) can easily be made by acetylene chemistry.

Analysis 2

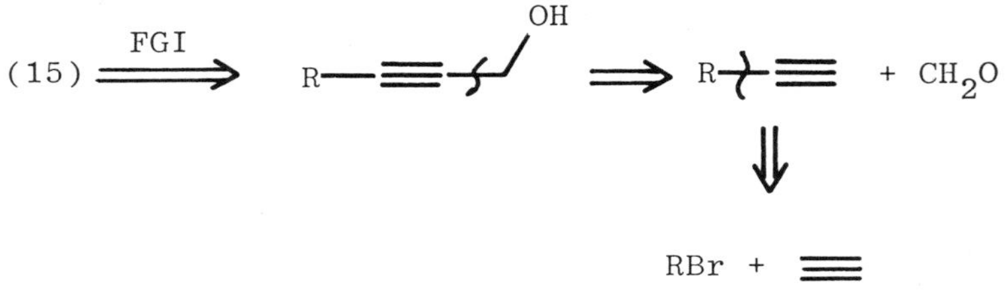

$$(15) \xrightarrow{FGI} R-\equiv-CH_2-OH \Rightarrow R-\equiv-H \;+\; CH_2O$$

$$\Downarrow$$

$$RBr \;+\; \equiv$$

Synthesis

$$\equiv \xrightarrow[2.\,RBr]{1.\,NaNH_2,\,NH_3(l)} R-\equiv \xrightarrow[2.\,CH_2O]{1.\,EtMgBr} R-\equiv-CH_2-OH$$

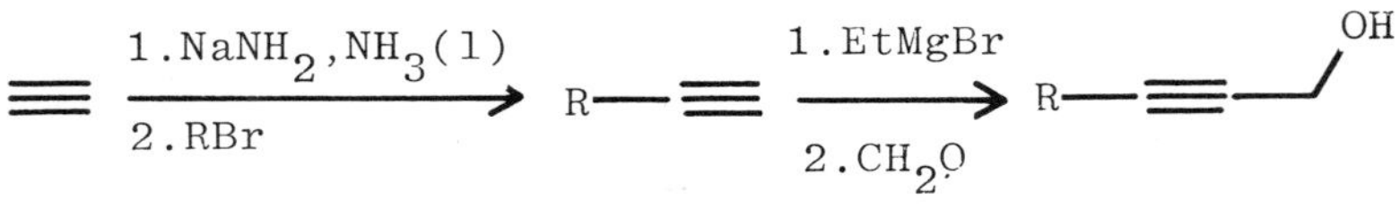

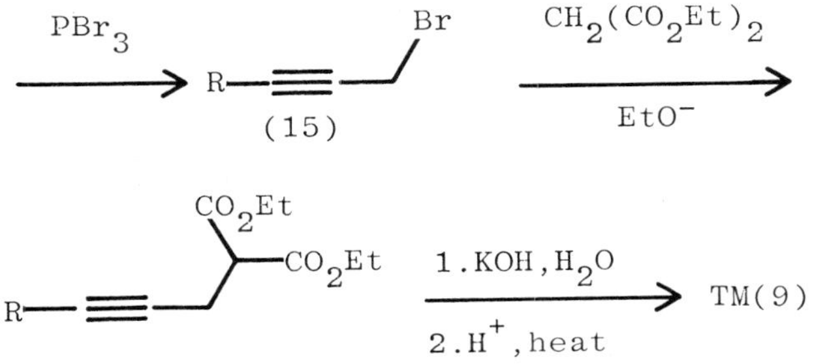

$$\xrightarrow{PBr_3} R\!\!\equiv\!\!\diagup\!\!Br \xrightarrow[EtO^-]{CH_2(CO_2Et)_2}$$

$$R\!\!\equiv\!\!\diagdown\!\!\underset{CO_2Et}{\overset{CO_2Et}{\diagup}} \xrightarrow[2.H^+, heat]{1. KOH, H_2O} TM(9)$$

Alkenes from Acetylenes

Example : Halide (15, R=Pr-n) was used by Butenandt[172] in his original synthesis of bombykol (p 156) (16). Wittig disconnection of the *trans* double bond requires *cis* allylic ylid (17). It is easier to use acetylenic ylid (18) and half aldehyde (19) in this step because of availability.

Bombykol : Analysis

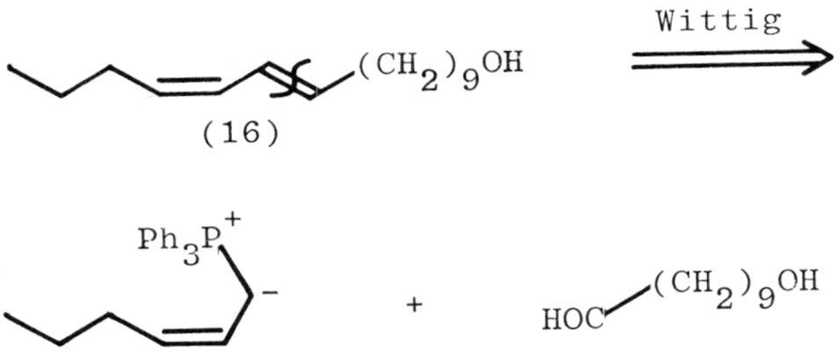

(16) ~~~~=~~~(CH$_2$)$_9$OH $\xRightarrow{\text{Wittig}}$

(17) ~~~~=~Ph$_3$P$^+$ $^-$ + HOC~(CH$_2$)$_9$OH

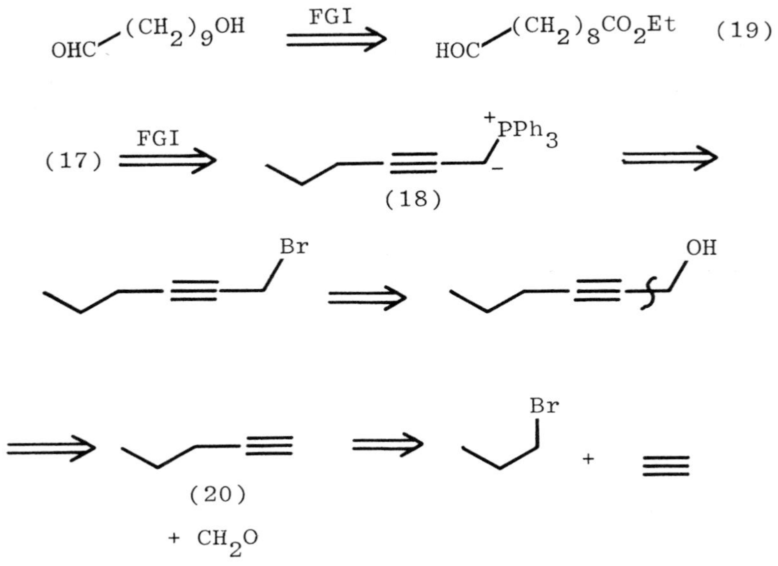

$$\text{OHC}\diagup^{(CH_2)_9 OH} \xRightarrow{FGI} \text{HOC}\diagup^{(CH_2)_8 CO_2 Et} \quad (19)$$

$$(17) \xRightarrow{FGI} \diagup\!\!=\!\!\diagup^{^+PPh_3}_{-} \Rightarrow$$
$$\qquad\qquad\qquad (18)$$

$$\diagup\!\!=\!\!\diagup^{Br} \Rightarrow \diagup\!\!=\!\!\diagup^{OH}$$

$$\Rightarrow \diagup\!\!= \Rightarrow \diagup^{Br} + \;\equiv$$
$$\quad (20)$$
$$+\; CH_2 O$$

Acetylene (20) was the one selected by Corey for use with organo cuprate (p 161).).
Synthesis[172]

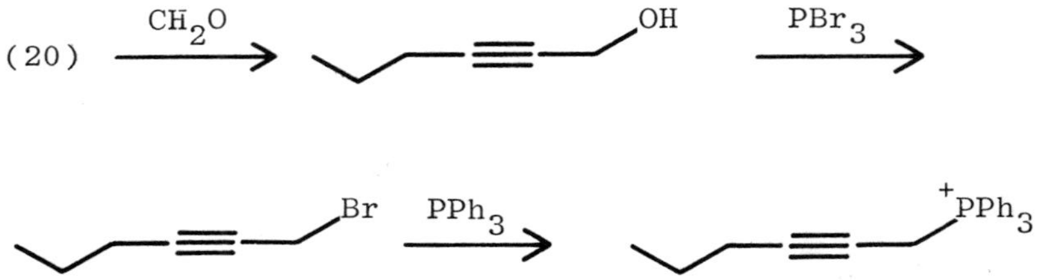

$$(20) \xrightarrow{CH_2 O} \diagup\!\!=\!\!\diagup^{OH} \xrightarrow{PBr_3}$$

$$\diagup\!\!=\!\!\diagup^{Br} \xrightarrow{PPh_3} \diagup\!\!=\!\!\diagup^{^+PPh_3}$$

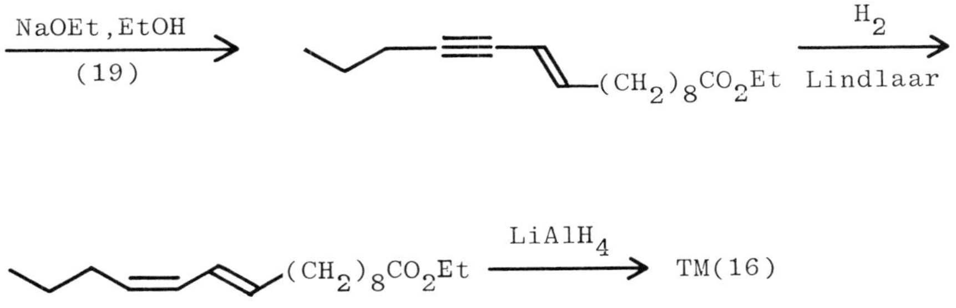

$$\xrightarrow{\text{NaOEt, EtOH}} \text{[CH}_3\text{CH}_2\text{-C}\equiv\text{C-CH=CH-(CH}_2)_8\text{CO}_2\text{Et]} \xrightarrow{\text{H}_2, \text{ Lindlaar}}$$

(19)

$$\text{[cis,cis-diene-(CH}_2)_8\text{CO}_2\text{Et]} \xrightarrow{\text{LiAlH}_4} \text{TM(16)}$$

Problem : The pheromone gossyplure (21) of the pink bollworm moth is a mixture of double bond isomers at bond (a). The *cis* double bond (b) can be made from an acetylene so disconnection (24c) is suggested. How would you make both *cis* and *trans* (23)?

Analysis 1

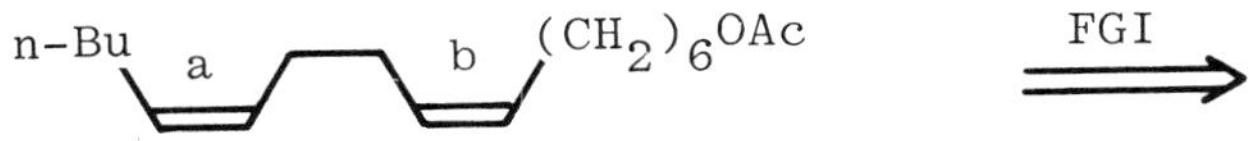

n-Bu—(a)=—(b)=—(CH$_2$)$_6$OAc $\xRightarrow{\text{FGI}}$

E + Z

(21)

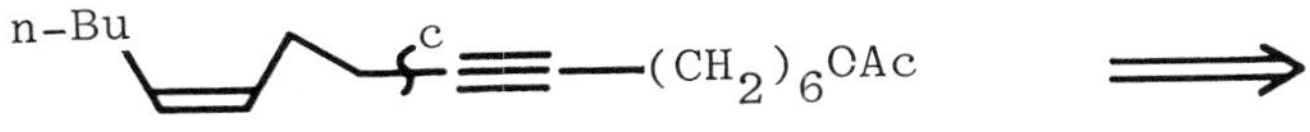

n-Bu—CH=CH—CH$_2$—C$\equiv$C—(CH$_2$)$_6$OAc $\Rightarrow$

(22c)

n-Bu—CH=CH—CH$_2$—Br + HC$\equiv$C—(CH$_2$)$_6$OAc

(23)

Answer : Acetylene (24) will give both by reduction under the right conditions (p T 127). Further disconnections give simple electrophilic starting materials.
Analysis 2

(23) $\xrightarrow{FGI}$ n-Bu—≡—⌒—OH $\Rightarrow$ △O + n-Bu—≡

(24)

$\Downarrow$

n-BuBr + ≡

Reduction of (24) by Lindlaar hydrogenation gives a *cis* double bond and sodium in liquid ammonia gives the *trans* double bond.
Synthesis[173]

≡ $\xrightarrow[\text{2. n-BuBr}]{\text{1. NaNH}_2, \text{NH}_3(l)}$ n-Bu—≡ $\xrightarrow[\text{2. } \triangle O]{\text{1. BuLi}}$ (24)

(24) $\nearrow$ $\xrightarrow[\text{Lindlaar}]{\text{H}_2, \text{Pd-C}}$ n-Bu—⌒=⌒—OH $\xrightarrow[\text{pyr}]{\text{Ph}_3\text{P.Br}_2}$ cis (23)

$\searrow$ $\xrightarrow[\text{NH}_3(l)]{\text{Na}}$ n-Bu—⌒=⌒—OH $\xrightarrow[\text{pyr}]{\text{Ph}_3\text{P.Br}_2}$ trans (23)

The acetate group in (22) is inadequate protection for the terminal OH group and a THP group was used instead[173] (Table T 68).

Synthesis 2

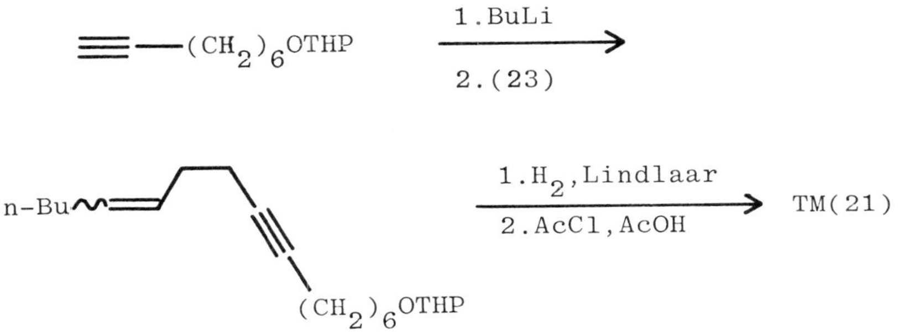

$$\equiv\!\!-\!(CH_2)_6OTHP \xrightarrow[\text{2.(23)}]{\text{1.BuLi}}$$

n-Bu⌒=⌒⌒≡⌒(CH_2)_6OTHP $\xrightarrow[\text{2.AcCl,AcOH}]{\text{1.H}_2\text{,Lindlaar}}$ TM(21)

More Advanced Example : Lactone (25) was needed in the synthesis of pederamide, an inhibitor of protein synthesis found in a beetle. Disconnection of the lactone reveals two stereochemical problems : a *cis* double bond is required and two adjacent chiral centres (● in 26) must be set up correctly.

Analysis 1

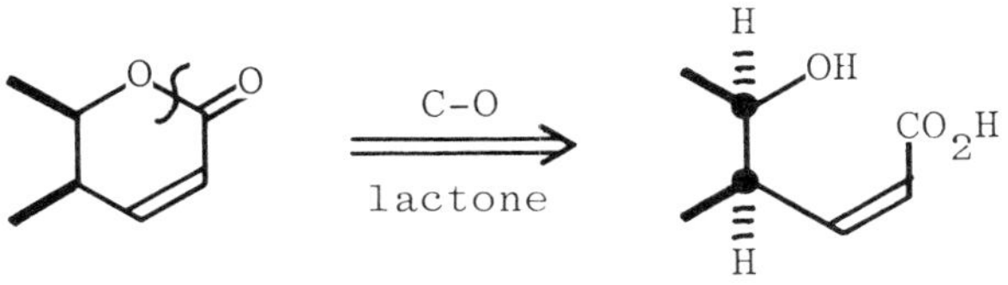

(25) $\xRightarrow{\text{C-O lactone}}$ (26)

If the *cis* double bond were made from an acetylene, disconnection (27a) becomes possible, allowing stereospecific opening of epoxide (28). The *trans* epoxide is needed as the reaction goes with inversion.

Analysis 2

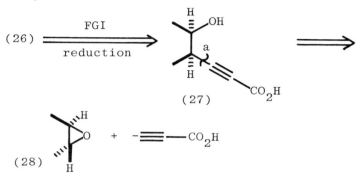

The published synthesis[174] uses (28) with acetylene itself, the acid group being added later. Note that the carboxylation is chemoselective, the dianion of (29) being the reagent (cf p T 129).

Synthesis[174]

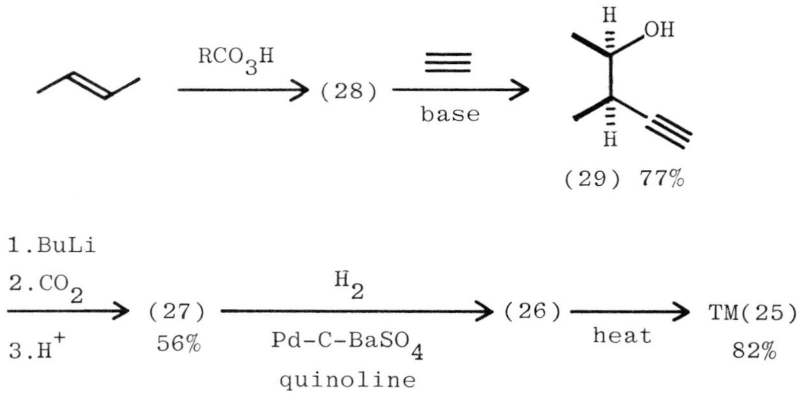

Problem : Suggest a synthesis of diol (30).

(30)

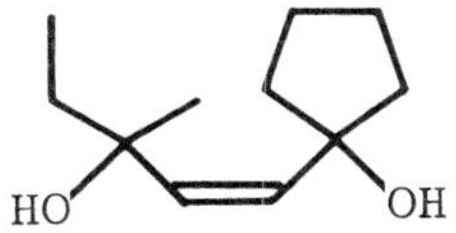

Answer : The *cis* double bond comes from the acetylene (31) made from two ketones.

Analysis

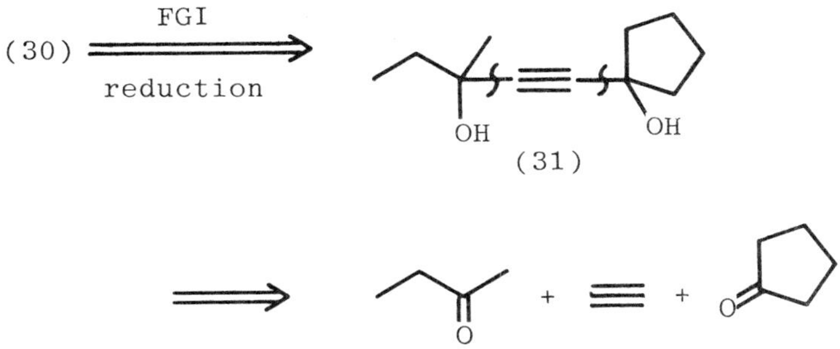

(30) $\xrightarrow[\text{reduction}]{\text{FGI}}$ (31)

$\Longrightarrow$

In the published synthesis, the cyclopentanone is added last, using the dianion of (31). No doubt the other order would be equally effective.

Synthesis[175]

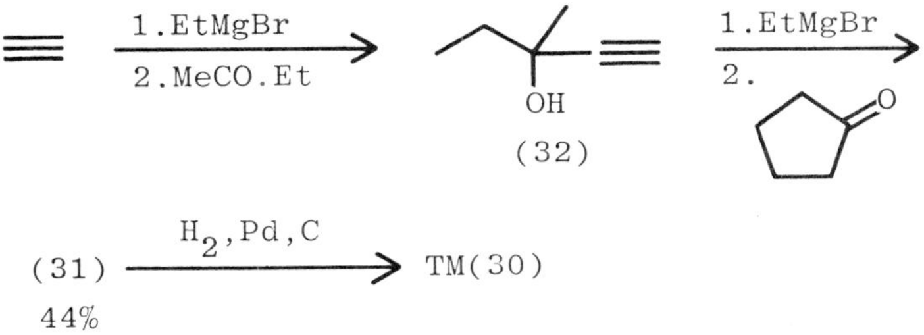

$\equiv$ $\xrightarrow[\text{2.MeCO.Et}]{\text{1.EtMgBr}}$ (32) $\xrightarrow[\text{2.}]{\text{1.EtMgBr}}$

(31) $\xrightarrow{\text{H}_2,\text{Pd},\text{C}}$ TM(30)

44%

Ketones from Acetylenes : Alkylation of carbonyl compounds (p T 108) with propargyl halides gives γ,δ-acetylenic ketones. Hydration then gives a 1,4-diketone of a type we shall use in Chapter T 25 as in the following example.[176]

[Scheme: ethyl 2-oxocyclohexanecarboxylate + propargyl bromide / EtO⁻ → alkylated propargyl intermediate; then MeOH, BF$_3$·Et$_2$O, Cl$_3$C·CO$_2$H → 1,4-diketone, 75%]

Problems :

1. Compare these two routes to the simple ketone (33)[177] and the one used on page T 130. Which disconnection and which synthons are used in each approach?

a) $n\text{-}C_5H_{11}Br$ $\xrightarrow[\text{3. CrO}_3]{\text{1. Mg, Et}_2\text{O} \quad \text{2. MeCHO}}$ (ref. 177) → (33) heptan-2-one

b) [MeCOCH$_2$CO$_2$Et] $\xrightarrow[\text{4. H}^+\text{, heat}]{\text{1. EtO}^-\text{ 2. n-BuBr 3. NaOH, H}_2\text{O}}$ (ref. 132 and 133)

c) acetylene route, see chapter T 16

Answer :

(a) $\xRightarrow{FGI}$ [structure: CH₃CH(OH)CH₂CH₂CH₂CH₂CH₃] ⇒ MeCHO + [n-pentyl]MgBr

[structure (33): methyl ketone with labeled bonds a, b, c] (b)⇒ [acetone] + [n-butyl]Br

(33)

(c)⇒ ≡–CH(structure)–CH₂CH₂CH₂CH₃ ⇒ ≡⁻ + [n-pentyl]Br

(34) [acetate anion structure]

Syntheses (a) and (c) use the same disconnection but with opposite polarity so that the acetylene anion in (c) is a reagent for the synthon (34).

2. In the search for cortisone analogues, compound (35)

[structure (35): cycloheptane with OH and C(O)CH₂OH]   [structure (36): cycloheptane with OH and C(O)CH₂Br]

(35)   (36)

was chosen for biological testing. Suggest a synthesis for (36) the precursor of (35).

Answer : Bromination of (37) should give (36) since only one side of the ketone can enolise (p T 53). This ketone (37) can be made from acetylene (38) and hence from cycloheptanone.

Analysis

(36) 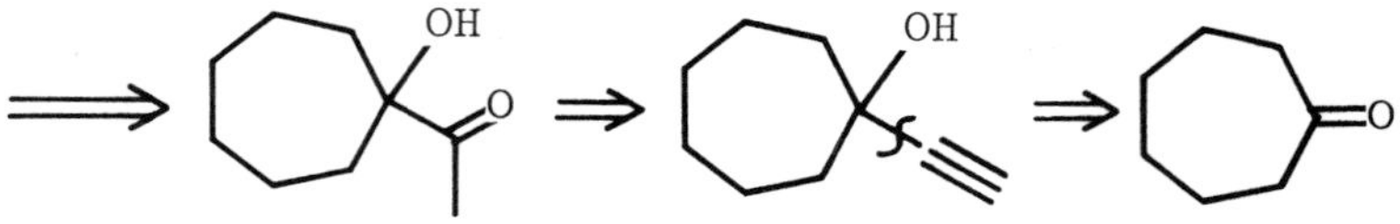 ⇒ [1-hydroxy-1-acetylcycloheptane] (37) ⇒ [1-ethynyl-1-hydroxycycloheptane] (38) ⇒ [cycloheptanone]

The hydration of (38) gave an excellent yield of (37) which was brominated in acid solution.

Synthesis[178]

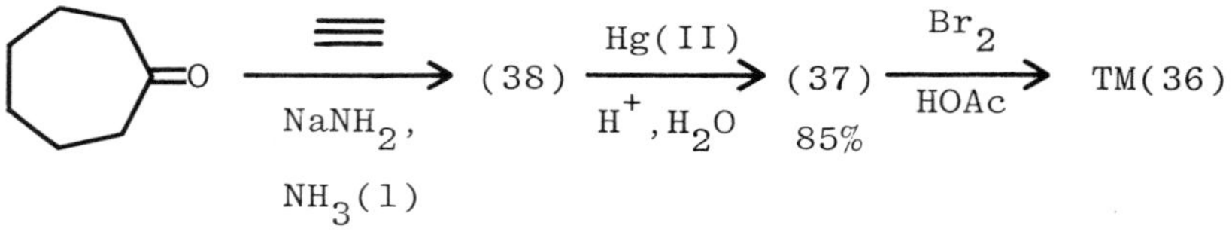

cycloheptanone $\xrightarrow[\text{NH}_3(l)]{\text{HC≡CH, NaNH}_2}$ (38) $\xrightarrow[\text{H}^+, \text{H}_2\text{O}]{\text{Hg(II)}}$ (37) $\xrightarrow[\text{HOAc}]{\text{Br}_2}$ TM(36)
                                                                              85%

Compound (35) was made by this route and did show some cortisone-like activity, but not enough to justify its development.

CHAPTER 17

Two-Group Disconnections I: Diels–Alder Reactions

Simple Problem : Find the Diels-Alder disconnections in the following molecules and draw the starting materials.

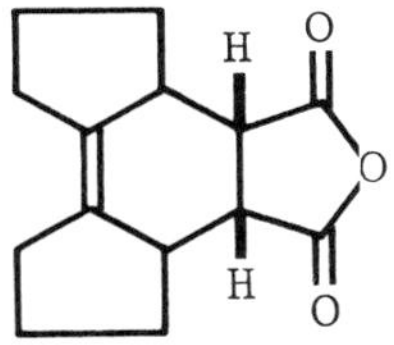

(1)

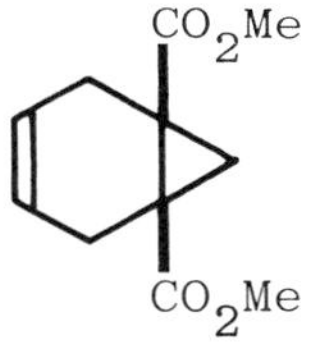

(2)

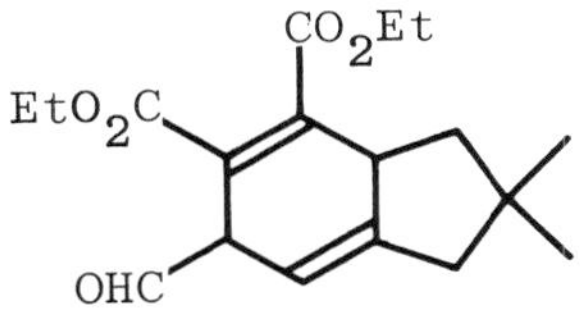

(3)

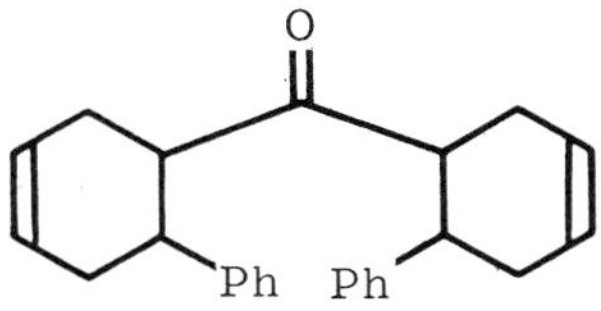

(4)

Answer : Compounds (1) and (2) have only one double bond and one six-membered ring, so only one disconnection is possible.

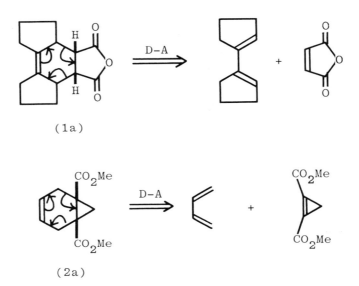

(1a)

(2a)

Anhydride (1), used to make polycyclic ketones,[179] and diester (2), used in a study of oxidative decarboxylation,[180] have each been made in one step by Diels-Alder reactions.

Two disconnections are possible with (3), but we prefer (a) which puts the electron-withdrawing CO_2Et groups on the dienophile and achieves some simplification. This compound was used by Woodward[181] in his synthesis of marasmic acid.

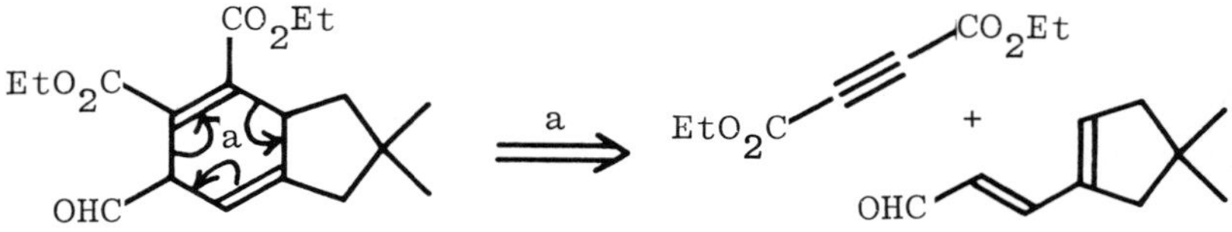

(3a)

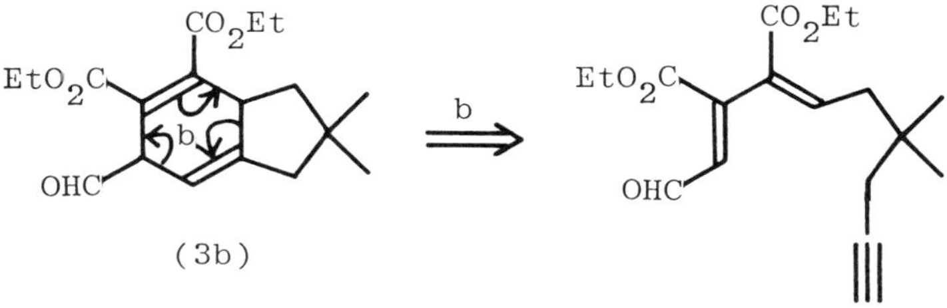

(3b)

Both disconnections on symmetrical (4) should be carried out as very simple starting materials result.[182]

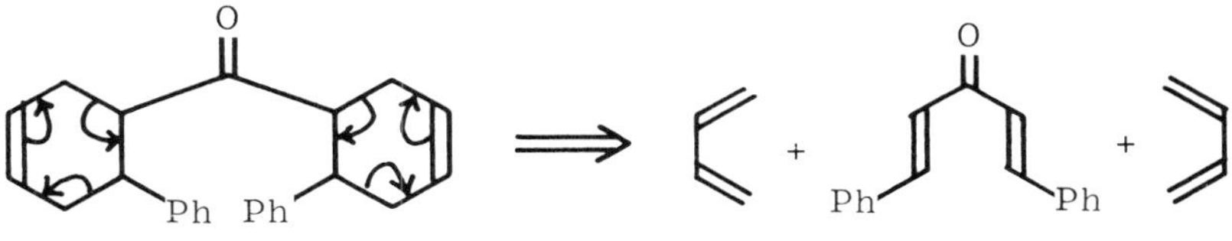

(4a)

Simple Example : Lewis acids, such as $AlCl_3$, catalyse the Diels-Alder reaction. Workers[183] used a three-fold excess of butadiene to react with the $AlCl_3$ complex of 5,6 and 7-membered cyclic enones, e.g. (5), giving excellent yields of *cis* fused bicyclic ketones (6). Me and H must be *cis* in (6) as they were *cis* in (5).

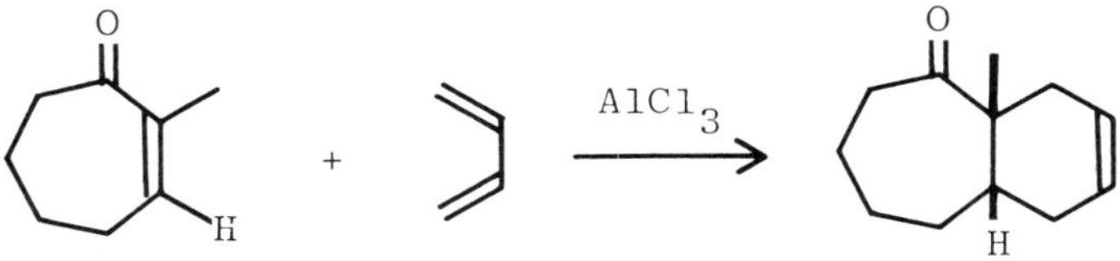

(5)   (6)

Stereospecificity. Problem : What stereochemistry is required in the starting materials for these Diels-Alder reactions:-

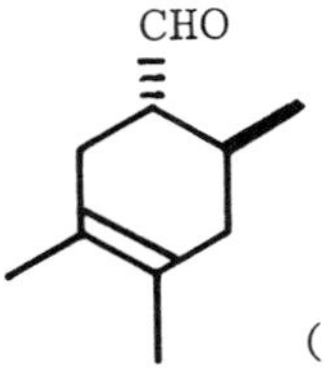

(7)

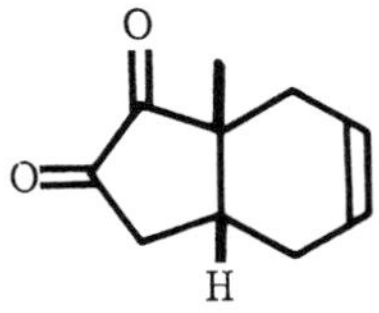

(8)

Answer : Diels-Alder disconnection (7a) reveals a diene (9), with no stereochemistry, and a dienophile (10) which must be *trans* to give *trans* groups in (7). The one-step synthesis is successful.[184]

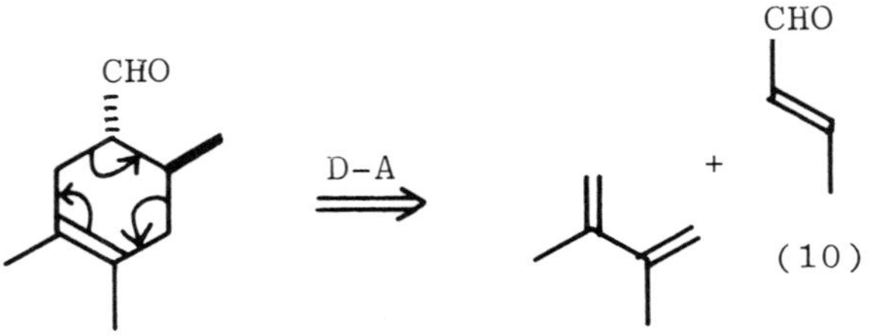

(7a) D-A ⇒ (9) + (10)

Diels-Alder disconnection of (8) reveals a cyclic dienophile (11) in which Me and H must be *cis* since they are also *cis* in (8), and the one step synthesis[185] duly gives *cis* product.

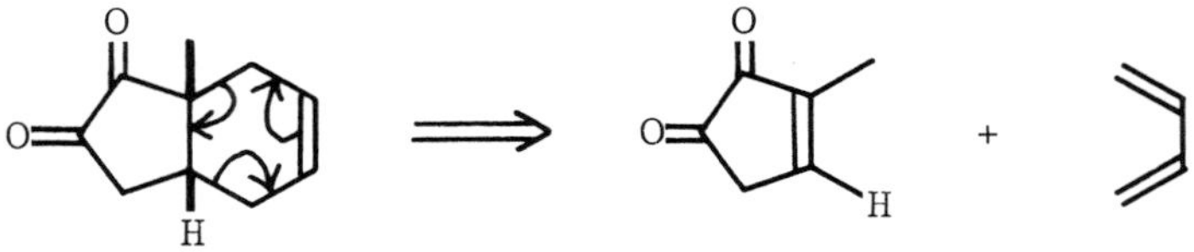

(8a) ⇒ (11) +

Problem : Given the stereochemistry shown on p 176 of the starting materials for the synthesis of Woodward's marasmic acid intermediate (3), what will be the stereochemistry of the product?

Answer : The dienophile has no stereochemistry, the diene (12) has two inside H's - marked H in (12) - which must end up *cis* in (3).

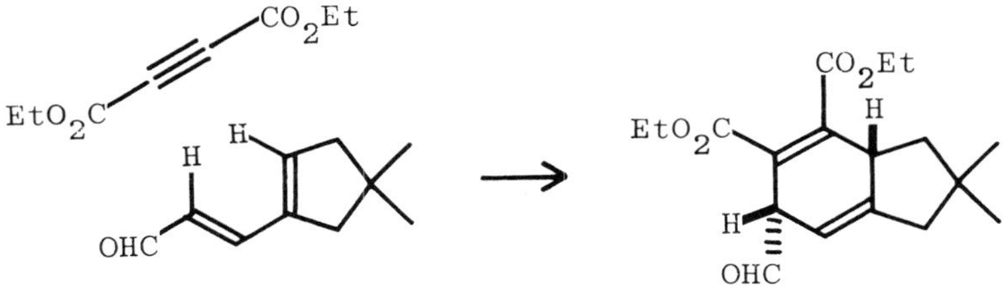

(12) → (3)

More advanced example : The *trans* fused compound (13) was needed[186] for a synthesis of some gelsemium alkaloids. Immediate Diels-Alder disconnection is impossible as it would require an anhydride with a *trans* double bond in a five-membered ring. This would be fumaric anhydride (14) but it does not exist.

Analysis 1

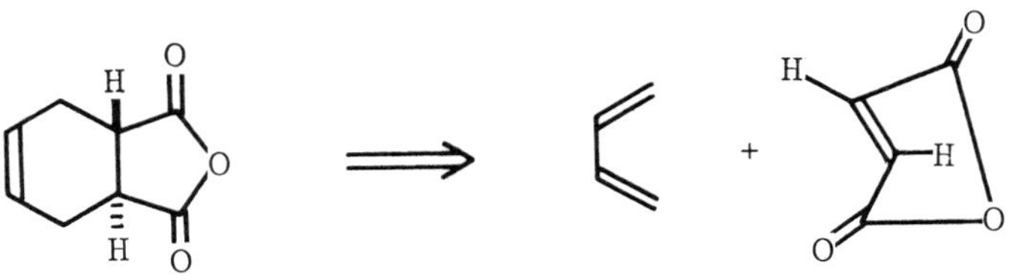

(13) ⟹ + (14)

Fumaric acid (16) does exist, so the best strategy is to go back to the ring opened diacid (15) before Diels-Alder disconnection.

Analysis 2

(13) $\xrightarrow{\text{FGI}}$ (15)
     dehydration

$\xRightarrow{\text{D-A}}$ + (16)

The synthesis turns out to be remarkably easy. The *trans* adduct cyclises to the strained anhydride (13) on heating with acetyl chloride.

Synthesis[187]

(16) $\xrightarrow[\text{sealed tube}]{150°C}$ (15) $\xrightarrow[\text{heat}]{CH_3COCl}$ TM(13)
                                    56%   74%

Endo Selectivity : *Example* : The intra-molecular Diels-Alder reaction was used to make (19) needed for a synthesis of the natural product torreyol.[188] The intermediate (18) could not be isolated as oxidation of alcohol (17) gave (19) directly. The intramolecular Diels-Alder must be very efficient.

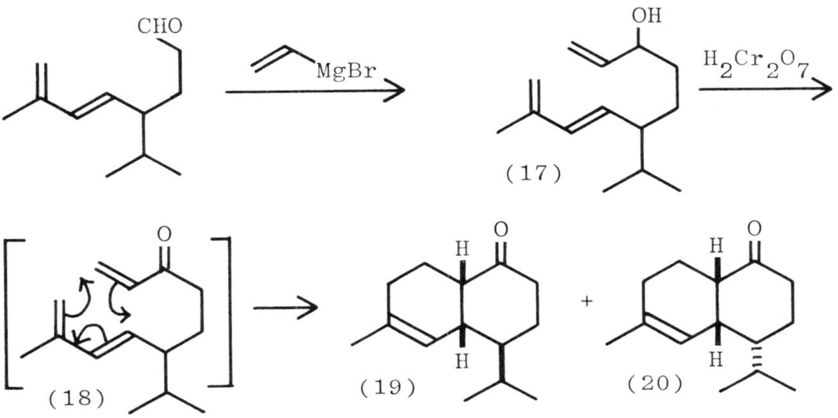

The formation of (19) was very stereoselective, (19) being formed in a 9:1 mixture with (20). Both (19) and (20) are endo adducts, as diagram (21) should make clear: both marked H's have to be *cis* if the CO is to be close to the diene in the transition state.

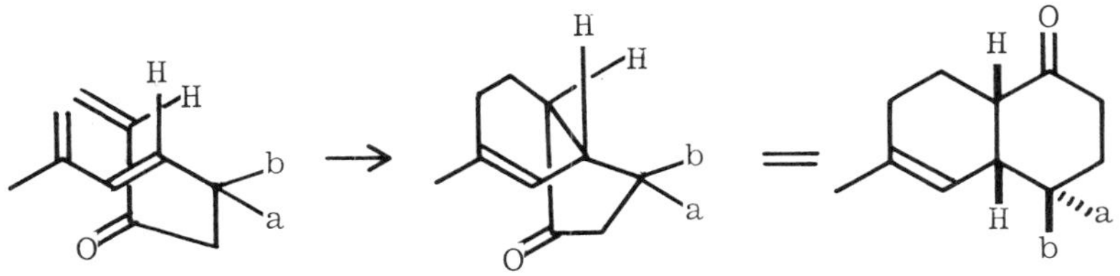

The i-Pr group can be in position a or b : it will obviously prefer to be away from the rest of the molecule at b and when (21) is unrolled as (22) b is indeed the preferred up position.

Problem : The synthesis of bicyclic lactone (25) by the acid-catalysed cyclisation of (24) could be tried if (23) can be made by via Diels-Alder reaction.

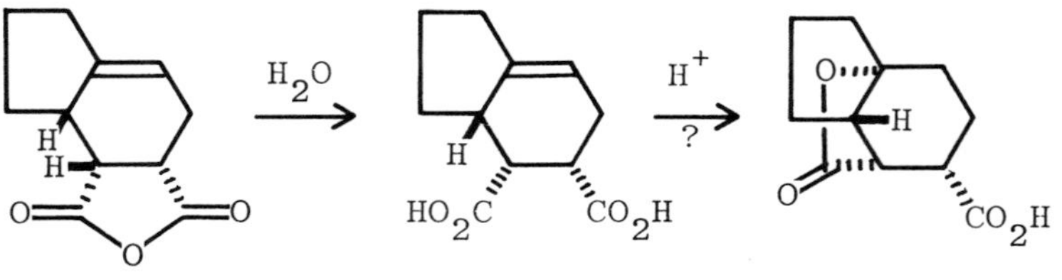

(23) (24) (25)

How can (23) be made? Include a synthesis of the diene. Will the Diels-Alder reaction give the correct stereochemistry?

Answer : Diels-Alder disconnection (23a) reveals maleic anhydride and diene (26) available by a Wittig reaction from (27) or more conveniently from alcohol (28).

Analysis

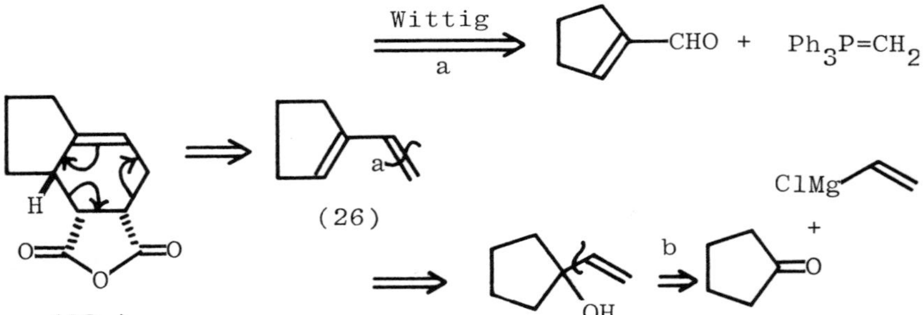

(23a) (26) (28)

Route (b), using acetylide ion, has been used for the synthesis of (26) which did indeed give the right diastereoisomer of (23) with maleic anhydride.[189]

Synthesis

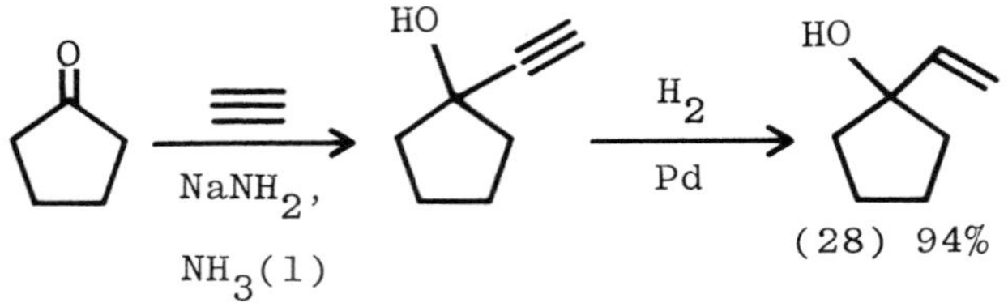

(28) 94%

$$\xrightarrow{KHSO_4} (26) + \text{[maleic anhydride]} \longrightarrow (23)$$
$$48\% \qquad\qquad 77\%$$

The usual diagram (29) shows that (23) is indeed the *endo* adduct : the two marked H's are *cis* in (29) and so must be *cis* in the product (23).

(29)

Regioselectivity : A slightly better way (than the method on p T137) to work out the regio-chemistry of many Diels-Alder reactions is to identify the most nucleophilic atom on the diene and the most electrophilic atom on the dienophile and join them.

Thus the Me_2N group in (30) makes atom a (but not b) more electrophilic. Combining these atoms gives the correct 'ortho' adduct (32) used on page T 137 for the tilidine synthesis.

(30) (31) (32)

Advanced Example : In research on violet perfumes, aldehyde (33) became a key intermediate for a new series of compounds. Diels-Alder disconnection gives acrolein and diene (34) which can be made by the Grignard method from available enone (36).

Analysis

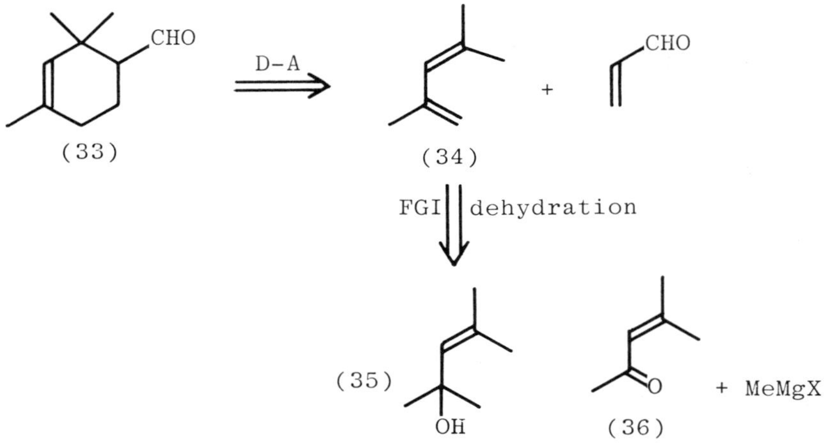

[Scheme: (33) ⇒(D-A) (34) + CH₂=CH-CHO; (34) ⇐(FGI, dehydration) (35) with OH, and (36) ketone + MeMgX]

Note that the Grignard reagent adds regioselectively to the CO group of (36) while the dehydration of (35) is unambiguous: whichever Me group loses the proton the same product is formed.

Synthesis[190]

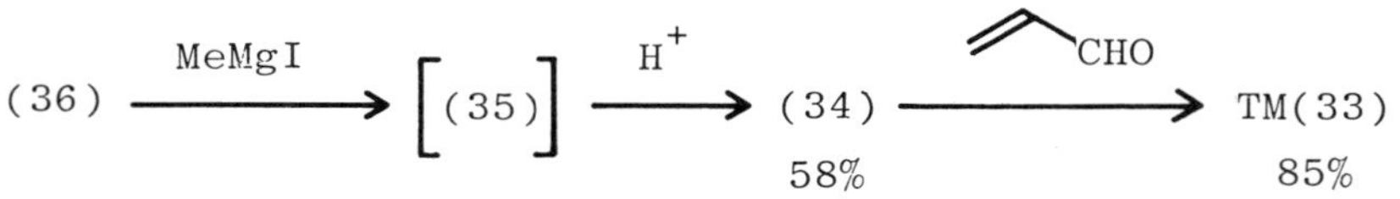

(36) —MeMgI→ [(35)] —H⁺→ (34) 58% —CH₂=CHCHO→ TM(33) 85%

Problem : Comment on the high regioselectivity of this Diels-Alder reaction.

Answer : All three Me groups in diene (34) direct the same way. The product is 'ortho' with respect to CMe_2 and CHO and 'para' with respect to the other Me group. Alternatively we can argue that all three Me groups in-

crease electron density at the same end of the diene
(37) and (38).

(37)   (38)

FG1 on Diels-Alder Adducts

Example : A family of biologically active compounds
isolated from African trees include polygodial (39),
an antifeedant which protects the tree against plagues of
caterpillars.¹⁹¹

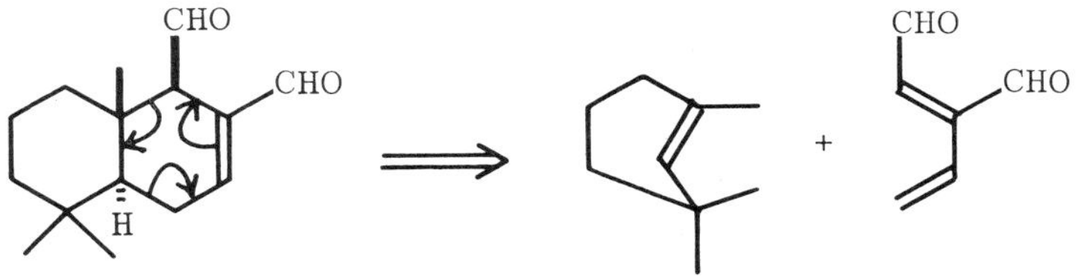

(39)   (40)

Direct Diels-Alder disconnection of (39) requires
the absurd starting materials (40) and has the
activating groups on the diene instead of the dienophile.
The two aldehyde groups are in any case unstable and the
diester (41) would be a more amenable target. If the
double bond were between the ester groups as in (42), a
Diels-Alder disconnection begins to look good, and the
addition of an extra double bond as in (43) provides a
very simple Diels-Alder disconnection.

Analysis

(39) $\underset{\text{reduction}}{\overset{\text{FGI}}{\Longrightarrow}}$ (41) $\overset{?}{\Longrightarrow}$

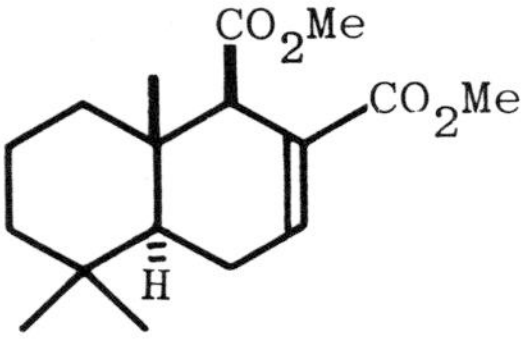

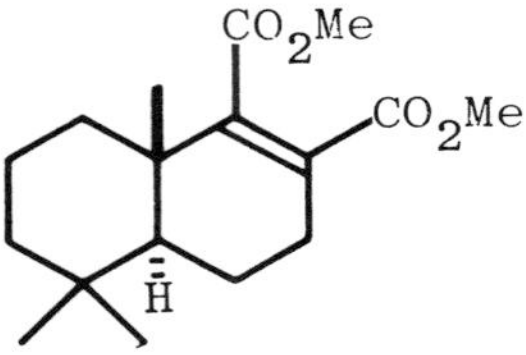

(42) 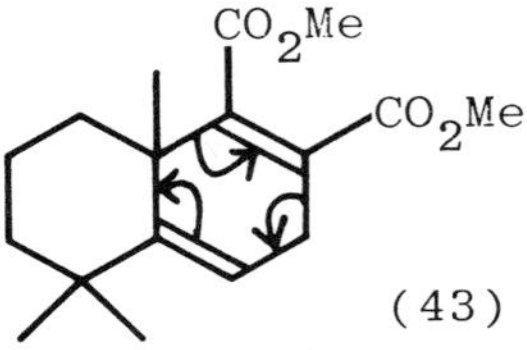 $\overset{?}{\Longrightarrow}$ (43)

$\Downarrow$ D-A

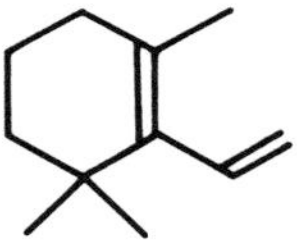

(44) 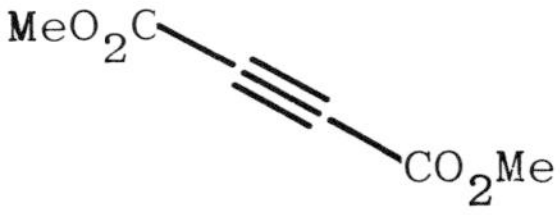 + (45)

Acetylenic diester (45) is available, the synthesis of (44) is known, and the Diels-Alder reaction goes in excellent yield.

Synthesis 1[191]

(44) + (45) $\xrightarrow[\text{16 hours}]{110^\circ C}$ (43)

83%

It now remains to reduce out one of the double bonds. This is straightforward as it is the unconjugated double bond we want to lose and hydrogenation operates on thermodynamic principles. The catalyst approaches (43) from the opposite side to the Me group and therefore adds hydrogen underneath.

Synthesis 2[192]

(43) $\xrightarrow{H_2}$ (46)

The problems of moving the double bond and getting the right stereochemistry at the third chiral centre in (41) turned out to be trivial : isomerisation with HCl achieved both*. The final reduction was done in two stages with special reagents.

Synthesis 3[192]

(46) $\xrightarrow{HCl}$ (41) 80% $\xrightarrow{LiAlH_4}$ 90%

$\xrightarrow[DMSO]{(COCl)_2}$ TM(39) 95%

* *Trans* decalins prefer a double bond in the position of (41) as the other ring has a better chair conformation. The CO_2Me group is equatorial in (41).

CHAPTER 18

Revision Examples and Problems

There is no workbook material corresponding to Chapter T 18 so the opportunity is here to revise material presented so far.

Simple Revision Problems :
1. Suggest a synthesis for the perfumery compound (1).

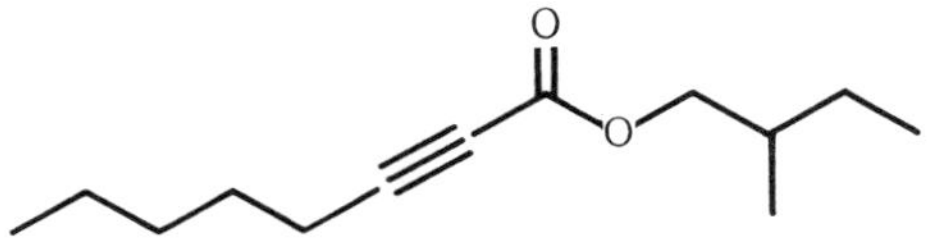

(1)

Answer : Disconnecting the ester first of all gives an acid (2), made by straightforward acetylene chemistry (Chapter 16) and an alcohol (3) which is available, but which could be made by simple Grignard chemistry.

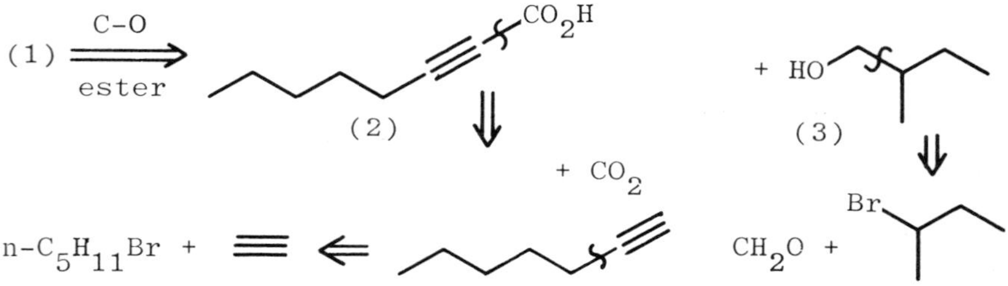

(1) $\underset{\text{ester}}{\overset{\text{C-O}}{\Longrightarrow}}$ (2)   +   (3)

+ CO_2

n-C_5H_{11}Br + ≡ ⇐

CH_2O + Br

The synthesis of the acid[193] and the alcohol[194], and the esterification are straightforward. Commercially, 'amyl alcohol', a mixture of (3) and (4), is used.

Synthesis[193,194]

$$\text{iBuBr} \xrightarrow[\text{2.CH}_2\text{O}]{\text{1.Mg}} (3)\ 67\% \qquad\qquad (4)$$

$$\text{HC≡CH} \xrightarrow[\text{2.C}_5\text{H}_{11}\text{Br}]{\text{1.base}} \text{CH}_3(\text{CH}_2)_4\text{C≡CH}$$

$$\xrightarrow[\text{2.CO}_2]{\text{1.Na}} (2) \xrightarrow{\text{(3)}\ \text{H}^+} \text{TM(1)}$$

2. (a) Suggest a synthesis of the amine (5).

(5) = n-heptyl piperidine

Answer : We wish to disconnect the ring-chain C-N bond but we must first go back to amide (6) to make the synthesis reliable (Chapter 8).

Analysis

$$(5) \underset{\text{reduction}}{\overset{\text{FGI}}{\Longrightarrow}} (6) \Longrightarrow \text{RCOCl} + \text{HN(piperidine)}$$

(6) = hexyl-C(O)-N(piperidine)

The reduction may be carried out with LiAlH$_4$ or, in this instance, by catalytic hydrogenation.

Synthesis[195]

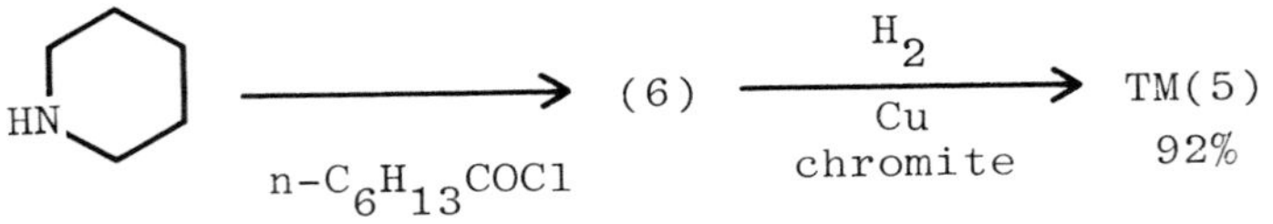

$$HN\text{-piperidine} \xrightarrow{n\text{-}C_6H_{13}COCl} (6) \xrightarrow[\text{chromite}]{H_2 \; Cu} TM(5) \; 92\%$$

(b) But how can you make amine (7), when this approach is impossible?

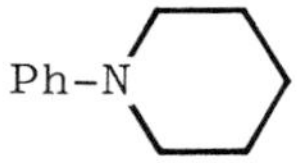

$$Ph-N\text{-piperidine} \quad (7)$$

Answer : The solution must be to disconnect the other two C-N bonds. This can be done directly or after FGI to a double amide (8), when the anhydride (9) is a suitable electrophilic starting material.

Analysis

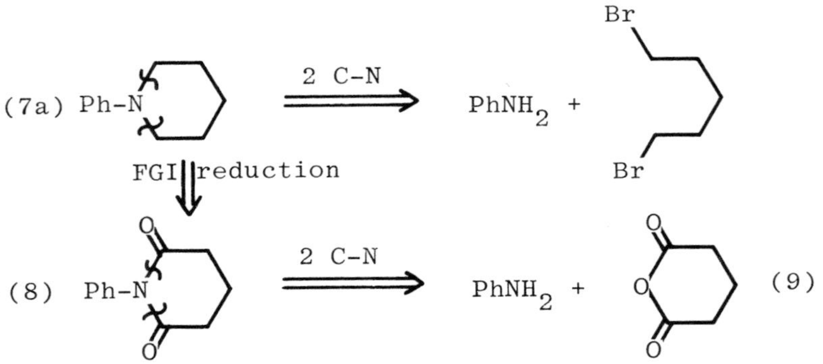

(7a) Ph-N-(piperidine) $\xrightarrow{2 \text{ C-N}}$ PhNH$_2$ + Br-(CH$_2$)$_n$-Br

FGI ↓ reduction

(8) Ph-N-(glutarimide) $\xrightarrow{2 \text{ C-N}}$ PhNH$_2$ + (9) glutaric anhydride

The second of these routes provides a reasonable synthesis.

Synthesis[168]

PhNH$_2$ $\xrightarrow{(9)}$ (8) $\xrightarrow{\text{LiAlH}_4}$ TM(7) 52%

Simple Examples :
1. As part of a programme of synthesising derivatives of branched acids, keto ester (10) was required.

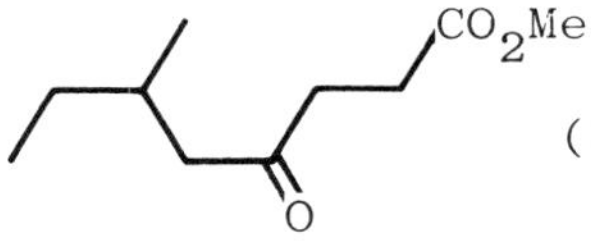

(10)

Disconnection (a) is the best as it splits the molecule into two largish pieces and it allows us to use (13) for the nearly symmetrical synthon (12) (see p T 39).

Analysis 1

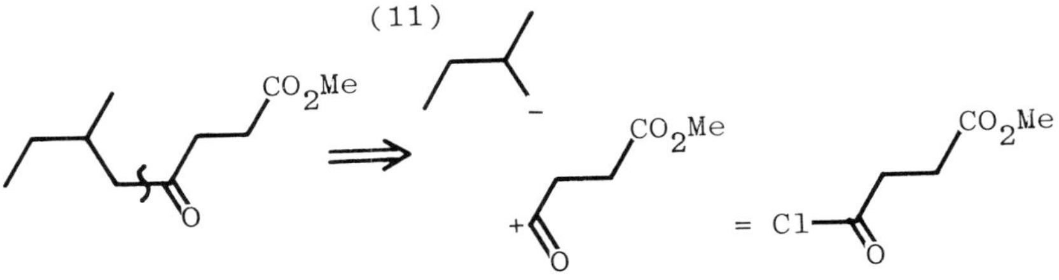

(10a) ⇒ (11) + (12) = Cl— (13)

An organocadmium reagent (14) will react with an acid chloride but not with an ester (p T 106) and so can play the role of synthon (11). Alcohol (3) is discussed above.

Analysis 2

(11) = [iBu]₂Cd ⇒ iBu-Br ⇒ iBu-OH
         (14)   (13)   (3)

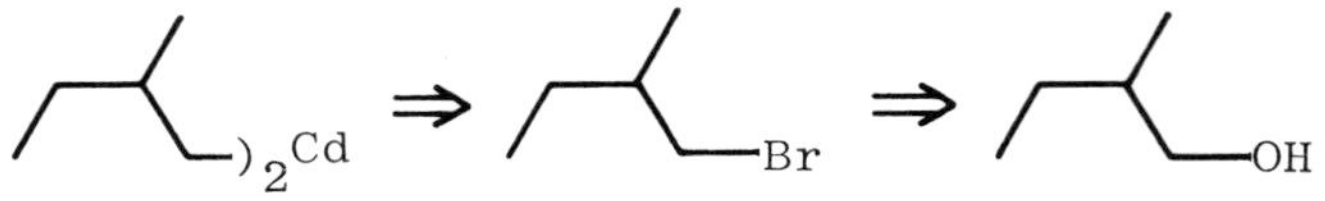

Synthesis[196]

(3) $\xrightarrow{PBr_3}$ iBu-Br $\xrightarrow[2.\,CdCl_2]{1.\,Mg,\,Et_2O}$ (14)

$\xrightarrow{(13)}$ TM(10)
           74%

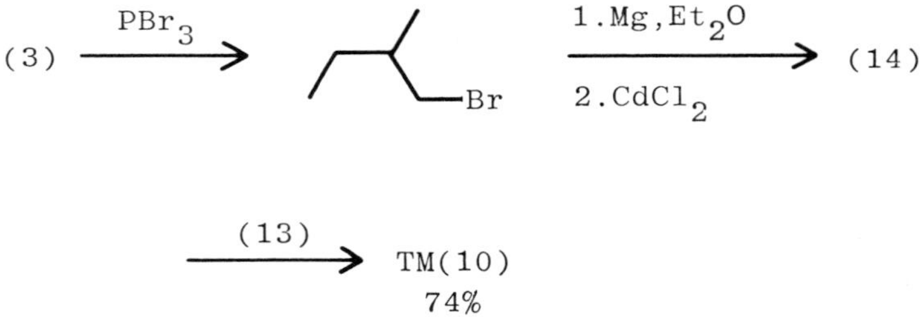

2. Faced with the problem of making simple vinyl ketones (15) and (16) as intermediates in a longer synthesis, chemists[197] found they needed both strategies outlined here:

Analysis

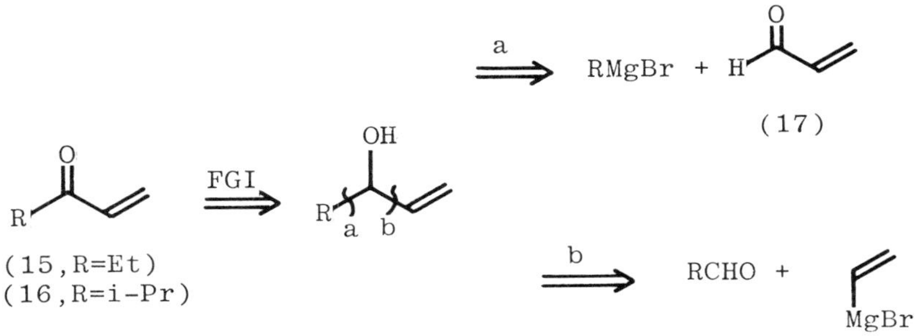

[Scheme showing disconnections of (15, R=Et) and (16, R=i-Pr) via FGI to allylic alcohol, then disconnection a → RMgBr + acrolein (17), and disconnection b → RCHO + CH₂=CH-MgBr]

Ethyl Grignard added chemoselectively to acrolein (17) in good yield, but iso-propyl Grignard failed to give the reaction. The alternative strategy (6) was successful in this case.

Synthesis[197]

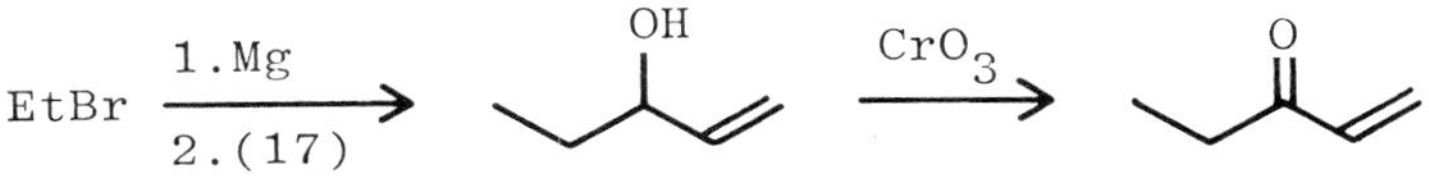

EtBr $\xrightarrow{\text{1. Mg} \atop \text{2. (17)}}$ [allylic alcohol] $\xrightarrow{\text{CrO}_3}$ TM(15)

CH₂=CH-Br $\xrightarrow{\text{1. Mg} \atop \text{2. i-PrCHO}}$ [allylic alcohol] $\xrightarrow{\text{CrO}_3}$ TM(16)

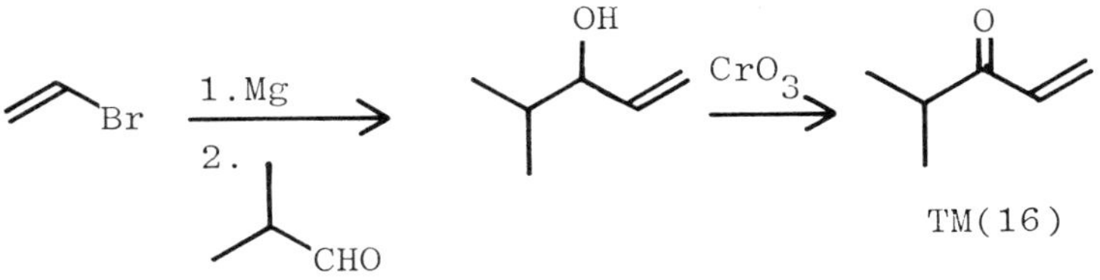

More Advanced Examples :

1. Frontalin (18), the pheromone of the western pine beetle, is an acetal (atom ● has two single bonds to oxygen). Disconnection reveals diol ketone (19).

Analysis 1

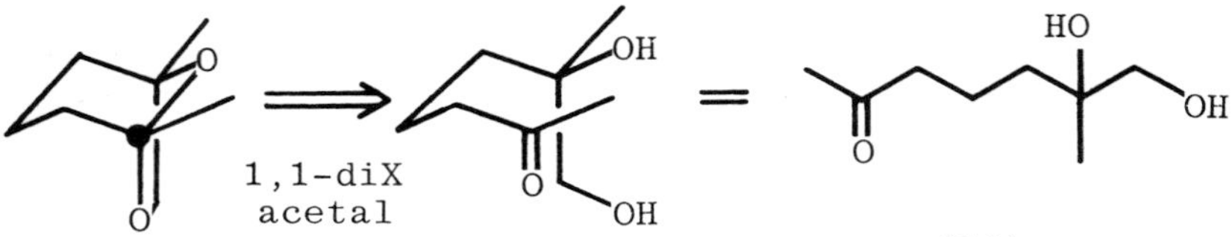

[Structure (18): bicyclic acetal with O, O] ⟹ [1,1-diX acetal structure with OH, OH] = [open-chain structure (19)]

(18)   1,1-diX acetal   (19)

The diol part of (19) could come from hydroxylation of the double bond in (20). This looks like an alkylation product of acetone, but the required halide (21) is not easy to make. An alternative is to make (20) from Grignard attack on nitrile (22).

Analysis 2

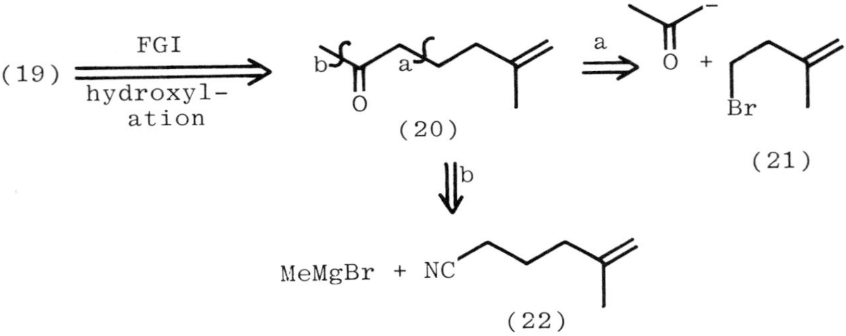

(19) $\xrightarrow[\text{hydroxylation}]{\text{FGI}}$ (20) $\overset{a}{\Longrightarrow}$ [acetone enolate] + [allyl bromide (21)]

↓b

MeMgBr + NC~~~(22)

Cyanide displacement on (23) gives (22) and we can make (24) easily by alkylation with symmetrical allylic halide (25).

Analysis 3

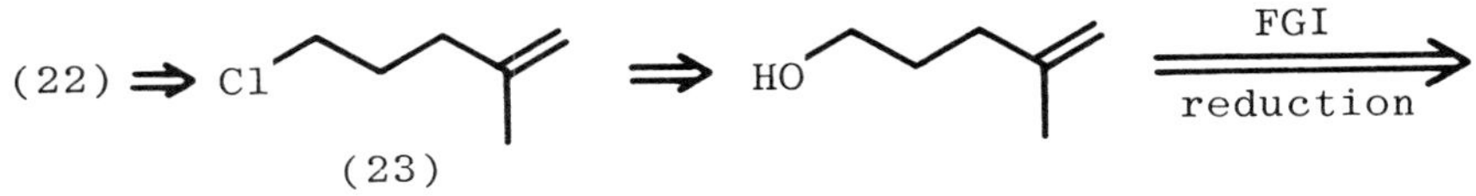

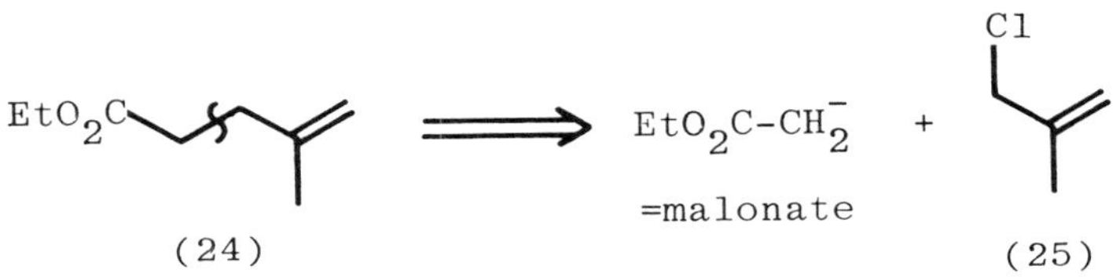

(22) ⇒ Cl~~~~~Y ⇒ HO~~~~~Y ═FGI/reduction═⇒

(23)

EtO$_2$C~~~~~Y ⇒ EtO$_2$C-CH$_2^-$ + Cl~~Y

(24)   =malonate   (25)

This is a long synthesis for such a simple compound but every step is easy to carry out. Hydroxylation was replaced by epoxidation in the published synthesis.

Synthesis[198]

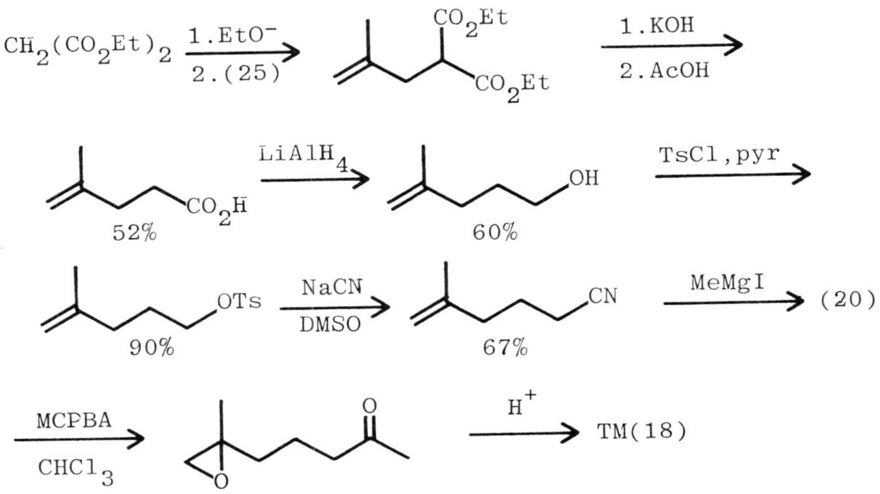

CH$_2$(CO$_2$Et)$_2$ —1.EtO⁻ / 2.(25)→ [CH$_2$=C(CH$_3$)CH$_2$CH(CO$_2$Et)$_2$] —1.KOH / 2.AcOH→

[CH$_2$=C(CH$_3$)CH$_2$CH$_2$CO$_2$H] 52% —LiAlH$_4$→ [CH$_2$=C(CH$_3$)CH$_2$CH$_2$CH$_2$OH] 60% —TsCl, pyr→

[CH$_2$=C(CH$_3$)CH$_2$CH$_2$CH$_2$OTs] 90% —NaCN / DMSO→ [CH$_2$=C(CH$_3$)CH$_2$CH$_2$CH$_2$CN] 67% —MeMgI→ (20)

—MCPBA / CHCl$_3$→ [epoxide-ketone] —H$^+$→ TM(18)

2. In the search for analogues of tetracyclines, compounds with sulphur in the second ring were investigated and (26) was needed for this work.[199]

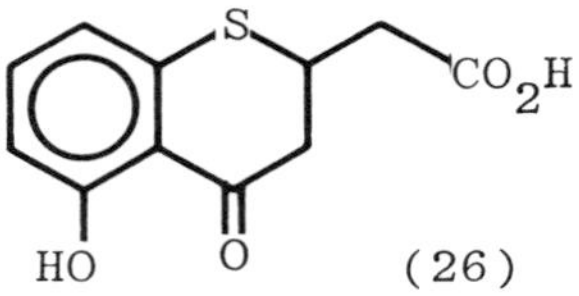

(26)

The starting material was thiol (27) whose synthesis appears on page 25. Michael addition to available (28) and hydrolysis gives diacid (29) which was cyclised in acid (HF) to (30). Removal of the chlorine atom and the methyl group gave (26).

Synthesis[199]

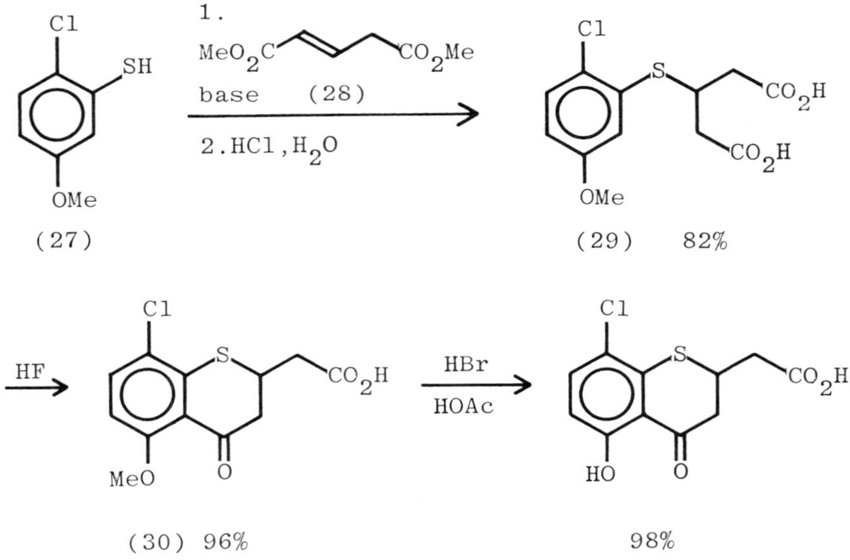

(27) → (29) 82%

(30) 96% → 98%

$$\xrightarrow[\text{H}_2\text{O},\text{MeOH}]{\text{H}_2,\text{Pd},\text{C}} \text{TM}(26) \quad 75\%$$

The final sulphur-containing antibiotic proved to be a highly active compound, 'superior in its anti-bacterial spectrum to all known tetracyclines'.

More Advanced Problems :

1. In the synthesis of (26) above, why does the cyclisation of (29) to (30) give such a high yield? What is the purpose of the chlorine atom in (27) since it is not present in (26)?

Answer : (a) Only one site is available for cyclisation and the two carboxylic acid side chains are identical. The only possible cyclisation product is (30). Reaction at OH is blocked by the MeO group.

(b) The chlorine atom blocks the alternative *para* cyclisation which is favoured when (31) is treated under the same conditions.[199]

2. δ-Halo ketone (32) was required for a study of the photochemistry of this class of compounds. Suggest a synthesis offering full stereo and regio-control.

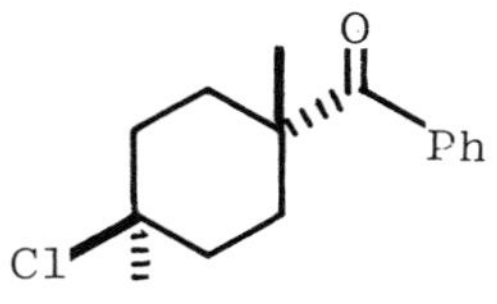

(32)

Answer : The chlorine atom most obviously comes from an OH group but a better choice is an alkene since the starting material (33) is then a Diels-Alder adduct.
Analysis

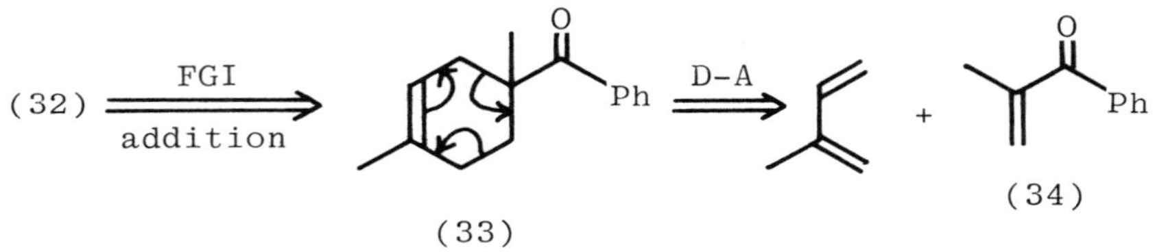

(32) $\xrightarrow[\text{addition}]{\text{FGI}}$ (33) $\xrightarrow{\text{D-A}}$ + (34)

The Diels-Alder orientation is correct ('para') and the stereochemistry of (33) → (32) should be right since the best conformation of the carbonium ion intermediate (35), with the large PhCO group equatorial, allows much better approach for Cl⁻ from the correct side.

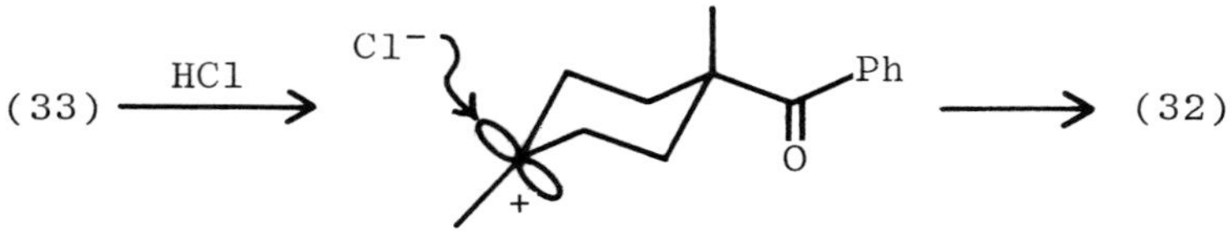

(33) $\xrightarrow{\text{HCl}}$ (35) ⟶ (32)

In practice (24) is not available and is replaced by (36) which is. The free acid (37) adds PhLi to give (33) (p T 107).

Synthesis[200]

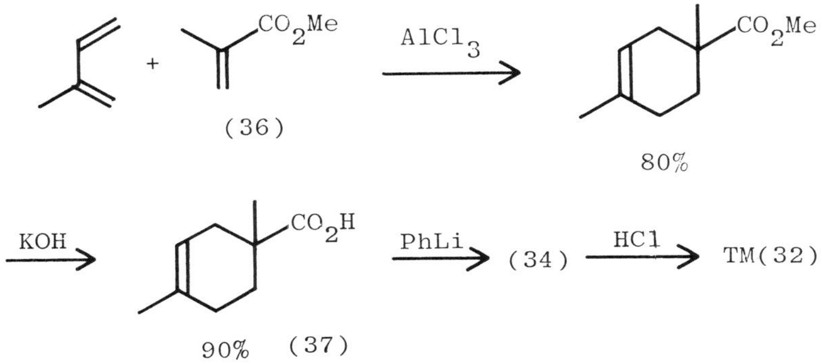

isoprene + CH$_2$=C(Me)CO$_2$Me (36) $\xrightarrow{\text{AlCl}_3}$ 1,4-dimethylcyclohex-3-ene-1-carboxylate (80%)

$\xrightarrow{\text{KOH}}$ (37) 90% $\xrightarrow{\text{PhLi}}$ (34) $\xrightarrow{\text{HCl}}$ TM(32)

CHAPTER 19

Two-Group Disconnections II: 1,3-Difunctionalised Compounds and α, β-Unsaturated Carbonyl Compounds

Do not despise these first three problems! Skill at their disconnections is vital.

Simple Problems : Make disconnections on these TMs and write the starting materials.

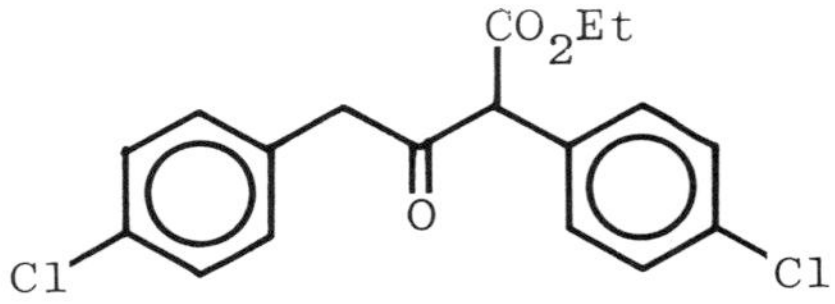

(1)

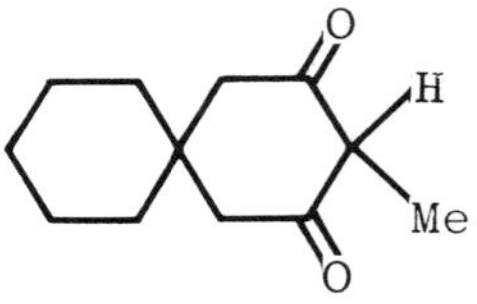

(2)

Answer : Two disconnections are possible for (1) giving either symmetrical ketone (3) or two molecules of the same ester (4).

Analysis

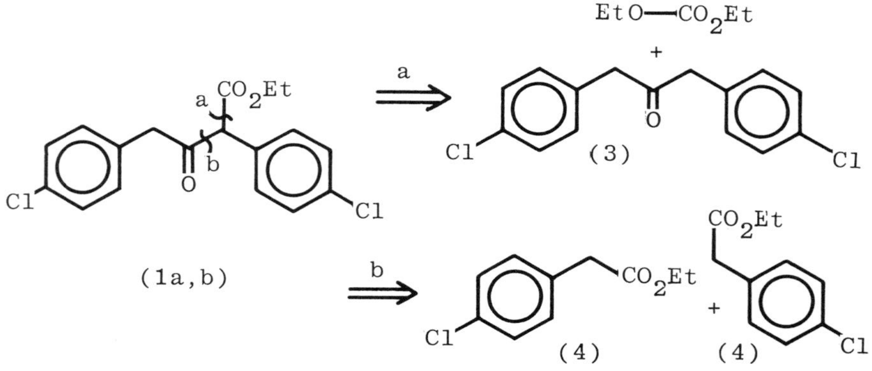

(1a,b)

(3)

(4) (4)

200

Condensation of (4) with i-PrMgBr as base gives (1) in 93% yield.[201]

Diketone (2) is symmetrical so only one disconnection (2a) is possible.

Analysis

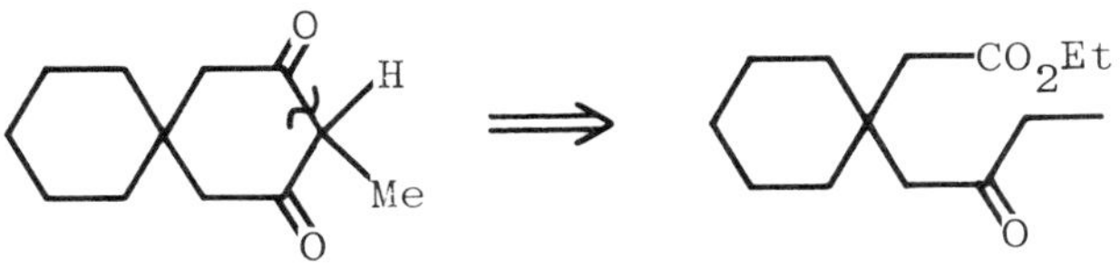

(2a) ⟹ (5)

Treatment[202] of (5) with NaOEt gives a quantitative yield of (2) : note that (2) can be formed as a stable enolate anion (p T 145) as there remains a proton between the two carbonyl groups.

2. Write disconnections and starting materials for these β-hydroxy carbonyl compounds.

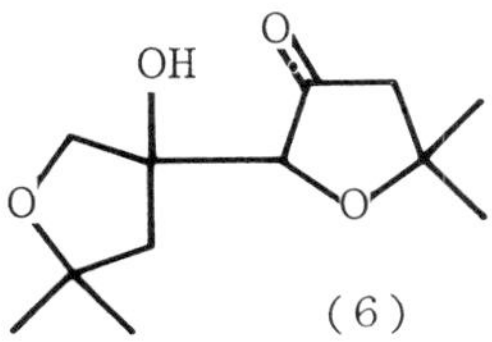

(6) 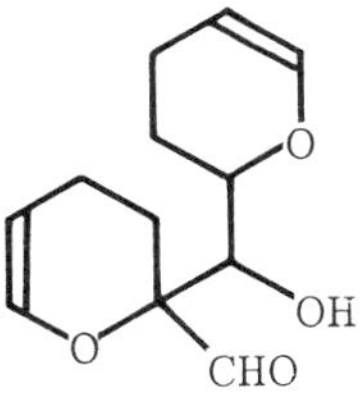   (7)

Answer : We must disconnect the bond next to the hydroxyl group on the carbonyl side. This gives two molecules of (8) for TM(6).

Analysis

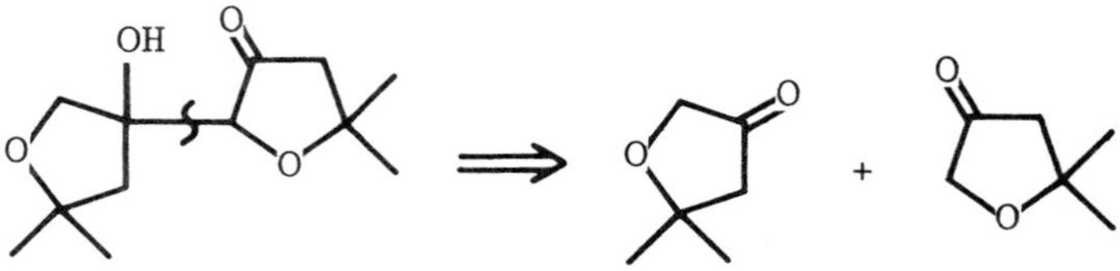

(6a) ⇒ (8) + (8)

Treatment of (8) with potassium in ether gives (6) in 50% yield.²⁰³

The same disconnection on (7) gives two molecules of aldehyde (9). Condensation of (9) with sodium hydroxide gives²⁰⁴ a good yield (76%) of (7).

Analysis

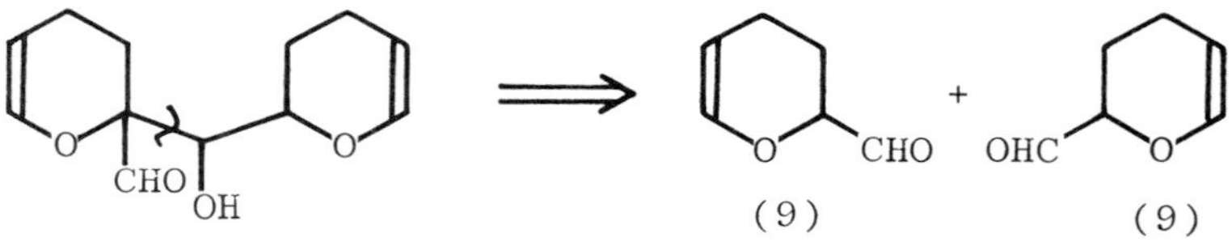

(7a) ⇒ (9) + (9)

3. The most important of these three types of target molecule is the α,β-unsaturated carbonyl compound. Disconnect TMs (10) and (11) and provide starting materials.

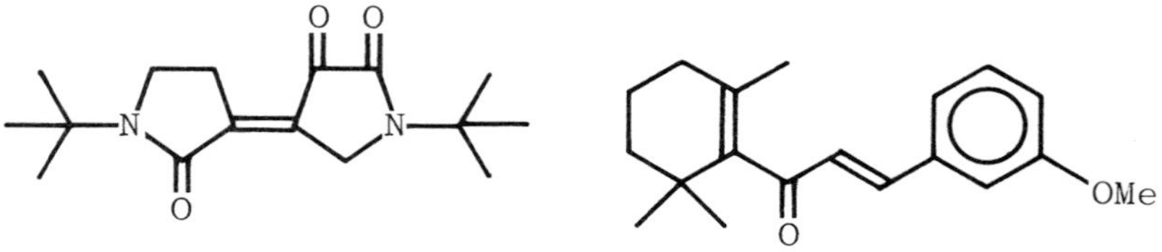

(10)   (11)

Answer : Two alternative disconnections of the double bond in (10) are available since there is a carbonyl group at each end of it.

Analysis

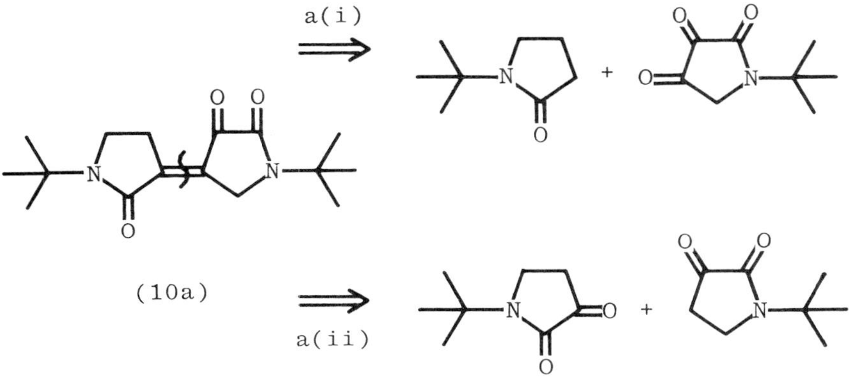

(10a) ⇒ a(i) gives pyrrolidinone + dioxo-pyrrolidinone

(10a) ⇒ a(ii) gives (12) + (12)

We prefer the second as it gives two molecules of the same keto-amide. Condensation here is very easy: standing (12) in the refrigerator with pyridine as catalyst[205] gives 96% of (10).

TM (11) again offers two alternative disconnections as it has two double bonds, each α,β to the carbonyl group.

Analysis

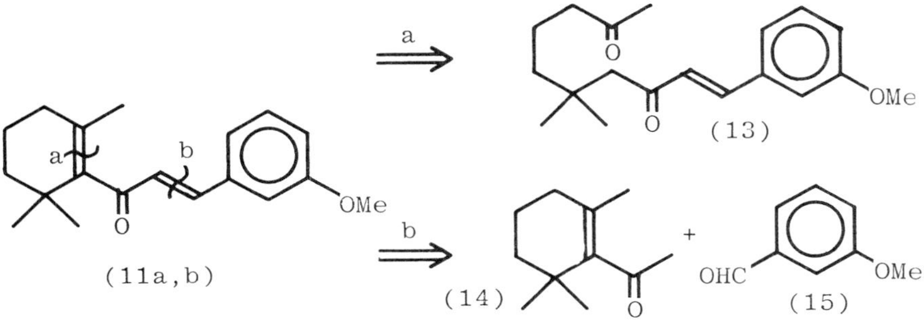

(11a,b) ⇒ a gives (13)

(11a,b) ⇒ b gives (14) + OHC-C₆H₄-OMe (15)

No doubt (13) would cyclise to (11), but disconnection (b) offers greater simplification. A mixture of (14) and (15) heated under reflux in ethanol with KOH gave[206] 60% of (11) which was used both to synthesise the natural product torreyol and to confirm its structure.

Example : The starting material (12) for one of these problems can itself be synthesised by carbonyl condensations. It contains ketone and amide groups and we can disconnect the ketone by the process on page T 147. The CO_2Et activating group can be added on one side only to give (16) which is a 1,3-dicarbonyl compound. Disconnection in the ring gives (17) which has two straightforward C-N disconnections to very simple starting materials.

Analysis

$$(12) \xrightarrow{\text{add } CO_2Et} (16) \xrightarrow{\text{1,3-diCO}}$$

$$(17) \xrightarrow[\text{1,3-diX}]{\text{C-N amide}} \quad \text{+ } (18)$$

The published synthesis does the Michael addition first to give (19) which is not isolated but combined immediately with symmetrical (18). Amide formation to give (17) and condensation to give (16) occur under the same conditions and decarboxylation is carried out in the usual way.

Synthesis[205]

$$\text{t-Bu-NH}_2 \xrightarrow{\diagup\!\!\diagdown\text{CO}_2\text{Et}} \text{t-Bu-NH-CH}_2\text{CH}_2\text{CO}_2\text{Et}$$

(19)

$$\xrightarrow[\text{EtOH}]{(18)\ \text{EtO}^-} (16) \xrightarrow[\text{2. HCl, heat}]{\text{1. NaOH, H}_2\text{O}} \text{TM}(12) \quad 57\% \text{ overall}$$

Example : Diol (20) was used on page 45 to make an acetal. It is a 1,3-diol so could be derived from β-hydroxy aldehyde (21) and a simple disconnection follows.

Analysis

(20) ⟹ (FGI reduction) ⟹ (21) ⟹ i-PrCHO + CH$_2$O

The synthesis is very easy : condensation of i-PrCHO with CH$_2$O in base gives a 96% yield of (20), not (21). The reason for this is that CH$_2$O and base is a reducing system by a Cannizzaro reaction (22).

Synthesis[207]

$$\text{i-PrCHO} \xrightarrow[\text{NaOH}]{\text{CH}_2\text{O}} \text{TM}(20)$$

$$\text{CH}_2\text{O} \xrightarrow{\text{HO}^-} \text{(22)}$$

Problem : Compounds of type (23) show promise as perfumes. Suggest a general synthesis for this structure.

(23)

Answer : Disconnecting the acetal reveals a 1,3-dihydroxy compound (24) which can be made by the reaction we have just met from CH_2O and aldehyde (25), an obvious Diels-Alder product.

Analysis

(23a) $\xrightarrow{\text{1,1-diO acetal}}$ (24) $\xrightarrow{\text{FGI reduction}}$

$\xrightarrow{\text{1,3-diO}}$ (25) + CH_2O $\xrightarrow{\text{D-A}}$ +

Again, the Cannizzaro reduction occurs under the condensation conditions. The ortho ester $HC(OEt)_3$ was used as a dehydrating agent in the final acetal formation.

Synthesis[208]

+ $\xrightarrow[\text{autoclave}]{100°C}$ (25) $\xrightarrow[\substack{\text{NaOH} \\ H_2O}]{CH_2O}$ (24) $\xrightarrow[\substack{\text{TsOH} \\ HC(OEt)_3}]{RCHO}$ TM(23)
90%   92%   80-90%

Problem : A recent synthesis[209] of the enzyme inhibitor elasnin (28) used (26) and (27) as starting materials. The structure of elasnin should make it clear which atoms were supplied by each starting material. Suggest syntheses for (26) and (27).

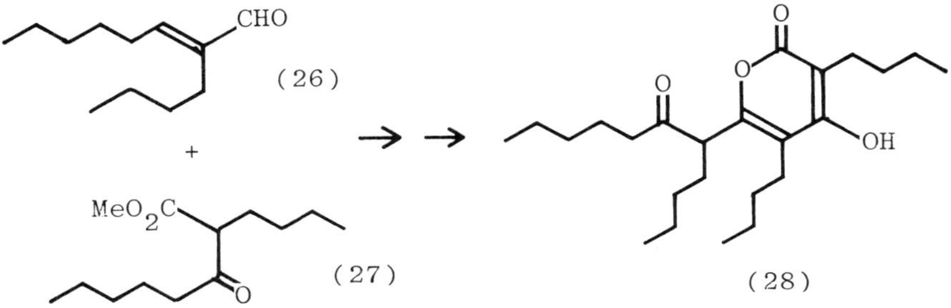

Answer : Enal (26) must be disconnected at the α,β double bond which reveals that it is formed from two molecules of hexanal (29).

Analysis

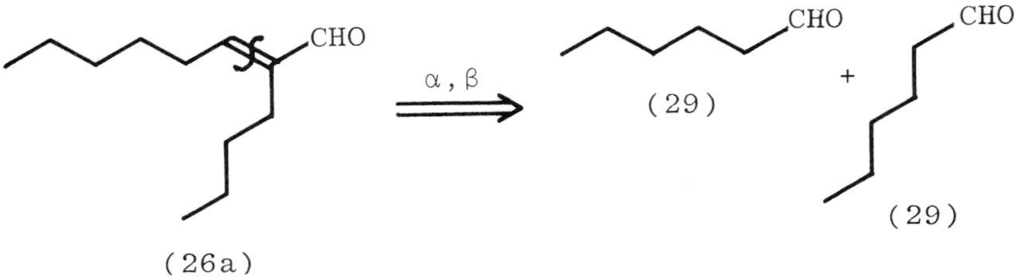

Acid-catalysed condensation with removal of water by xylene distillation gives a good yield.

Synthesis[209]

(29) $\xrightarrow[\text{xylene \quad distil}]{\text{boric acid}}$ (26) 77%

Keto-ester (27) is a 1,3-dicarbonyl compound: simplification and symmetry suggest disconnection (27a).
Analysis

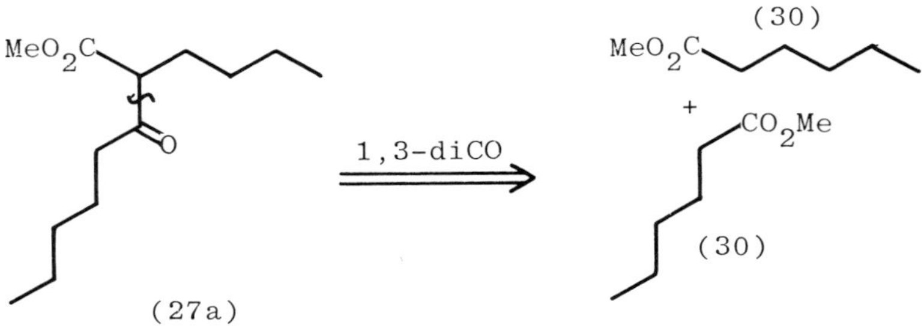

[Structure (27a): MeO$_2$C-CH(butyl)-C(=O)-pentyl] ⟹ (1,3-diCO) [MeO$_2$C-pentyl (30) + pentyl-CO$_2$Me (30)]

Condensation of simple ester (30) with NaH as base gives an excellent yield of (27).
Synthesis[209]

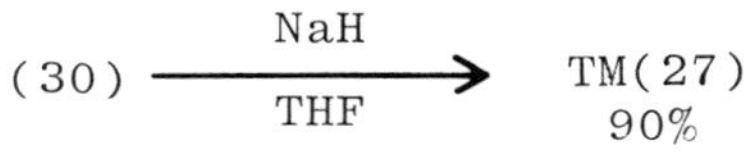

$$(30) \xrightarrow[\text{THF}]{\text{NaH}} \text{TM}(27) \quad 90\%$$

CHAPTER 20

Strategy IX: Control in Carbonyl Condensations

This Chapter also starts with some deceptively simple problems. These are worth doing if only to confirm your understanding of the basic reasons which underlie selectivity in carbonyl condensations.

Problems :

1.

PhCHO + MeCO.CO$_2$H (1) $\xrightarrow{\text{KOH}}$ Ph-CH=CH-CO$_2$H (2)   80% ref. 210

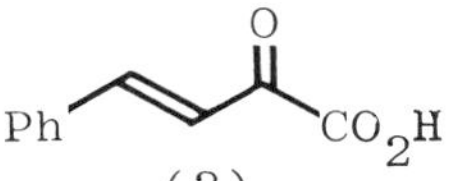

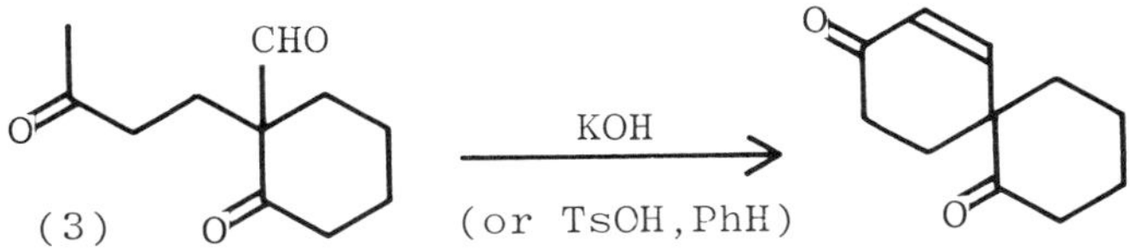

(3) $\xrightarrow{\text{KOH (or TsOH, PhH)}}$ (4)   ref. 211

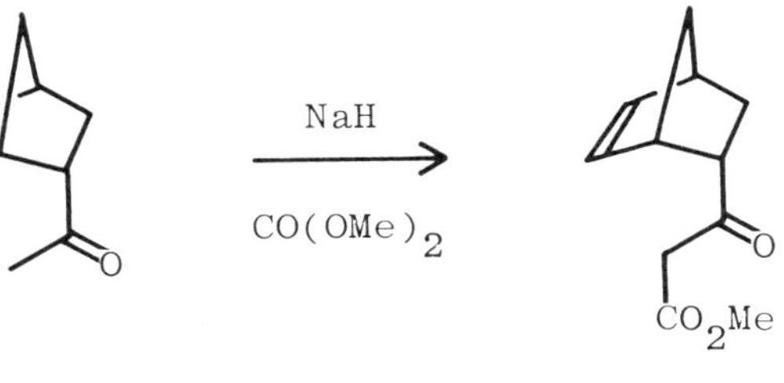

(5) $\xrightarrow[\text{CO(OMe)}_2]{\text{NaH}}$ (6) 80-85%   ref. 212

Answer : (a) PhCHO has no enolisable protons, but has the most electrophilic carbonyl group. The keto-acid (1) forms enolate anion (7) and attacks PhCHO.

$$(1) \xrightarrow{KOH} \underset{(7)}{\overset{O^-}{\underset{CO_2^-}{\diagup\!\!\!\diagdown}}} \xrightarrow{PhCHO} (2)$$

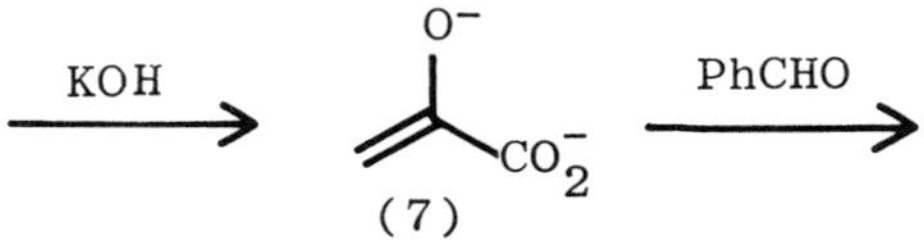

(b) Control is mainly achieved because this is an intramolecular reaction forming a stable six-membered ring. Though enolisation at three sites is possible (3a), all reactions give unstable products (cf page T 155) except attack by enolate (8) on either of the other carbonyl groups. The aldehyde is more reactive than the ketone, so (4) is formed.

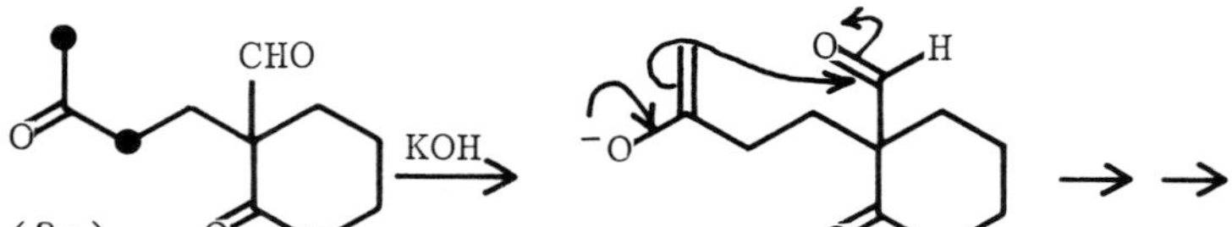

(3a) $\xrightarrow{KOH}$ (8) $\rightarrow \rightarrow$ (4)

● = site for enolisation

(c) Only (5) can enolise, and must react as it does to give a product (6) which can form a stable enolate anion (9). The alternative product (10) has no remaining protons at atom (●).

(6) 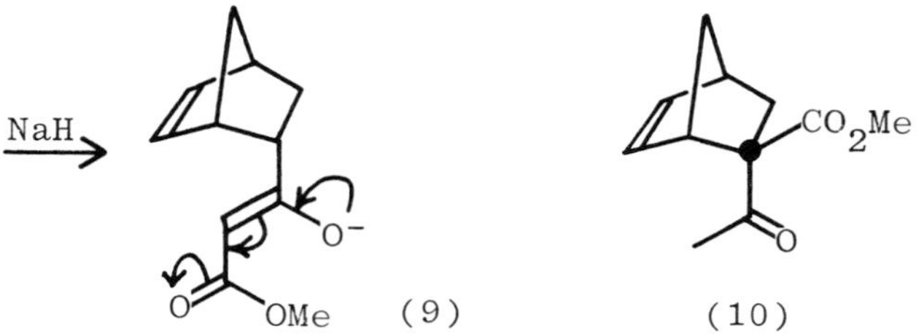 $\xrightarrow{NaH}$ (9)   (10)

Problem 2 : Again, only one product is formed in high yield from each of these condensations. Predict what it is, giving reasons.

(11) 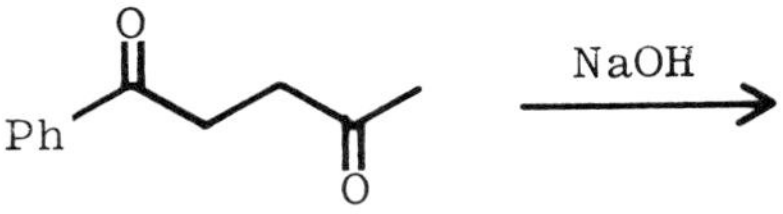 Ph-CO-CH₂-CH₂-CO-CH₃ —NaOH→ (12)   ref. 213

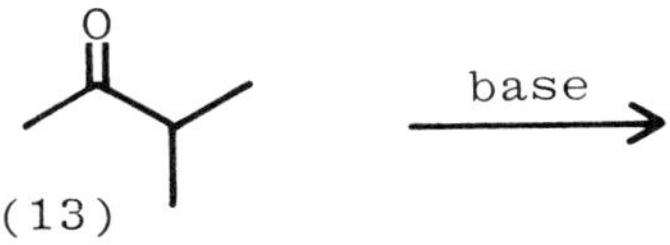

(13) (CH₃)₂CH-CO-CH₃ —base→ (14)   ref. 214

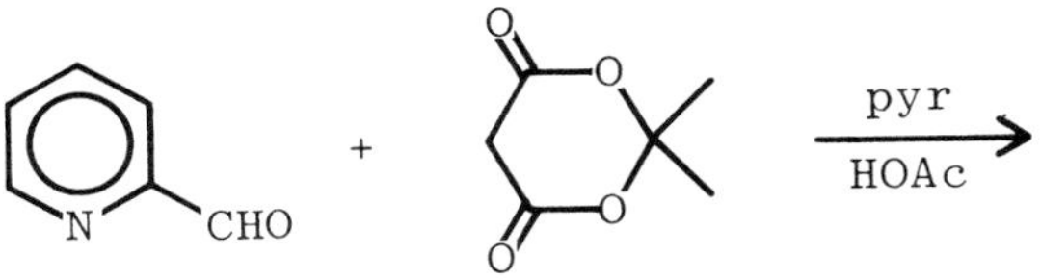

(15) 2-pyridinecarboxaldehyde + (16) Meldrum's acid (2,2-dimethyl-1,3-dioxane-4,6-dione) —pyr/HOAc→ (17) 76%   ref. 215

Answer : (a) Diketone (11) can form a stable ring if enolisation occurs on the methyl group. Other possible enolates can give only three-membered rings (p T 155).

(11) 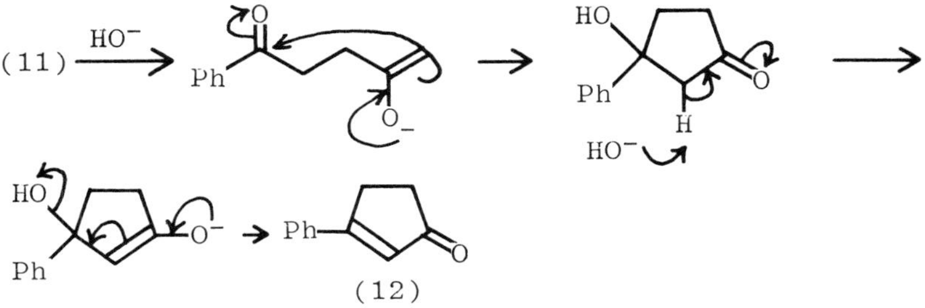 —HO⁻→ [enolate mechanism] → [cyclised intermediate] →

→ Ph—[cyclopentenone] (12)

(b) In this self-condensation of an unsymmetrical ketone, the only question is which side will enolise. Base catalysis favours enolisation at the methyl group to give (18). With some bases, (19) is the final product, with others dehydration to (14) occurs.

(13) 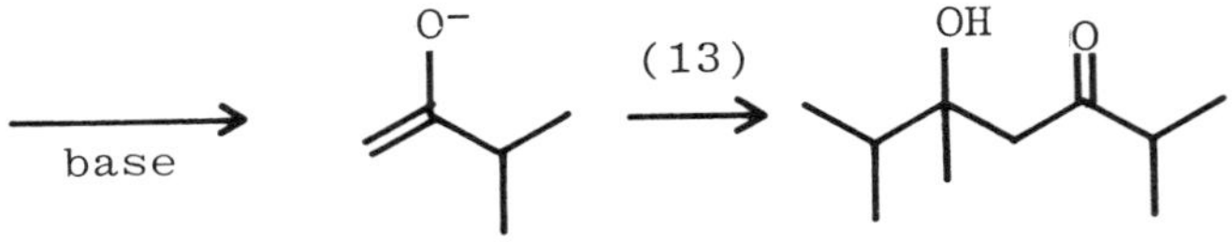 —base→ (18) —(13)→ (19)

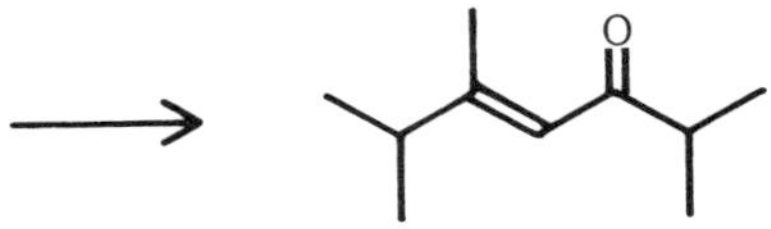

→ (14)

(c) Compound (16) is a cyclic malonate ester, able to form a specific enolate (20), while aldehyde (15) cannot enolise. This is a typical Knoevenagel reaction (p T 161).

(16) 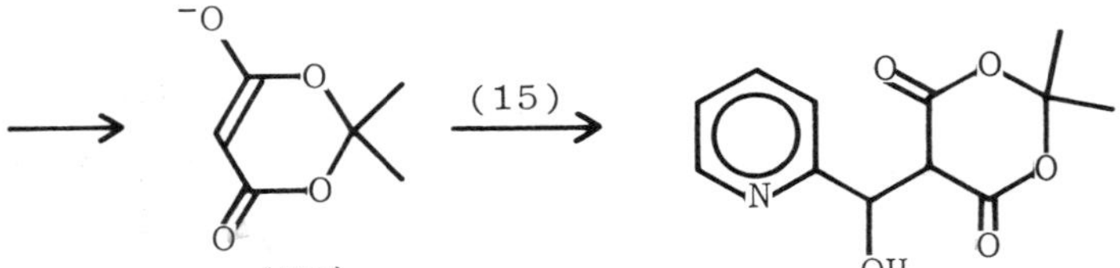 → (20) —(15)→

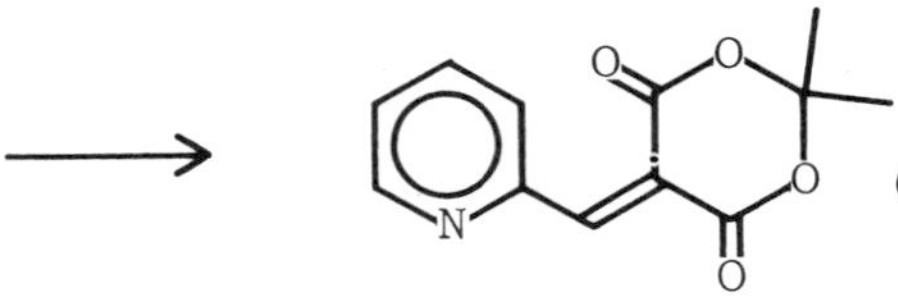

→ (17) 76%

Examples and Problems from Each Section of the Chapter
Self-Condensation : There have been many examples of self-condensations in Chapter 19 so two more will be enough to remind you about regioselectivity of enolisation (p T 154).

Example : Compound (21) is a β-hydroxy ketone. Disconnection of the usual bond reveals two molecules of enone (22), disconnected in turn to two molecules of acetone.

Analysis :

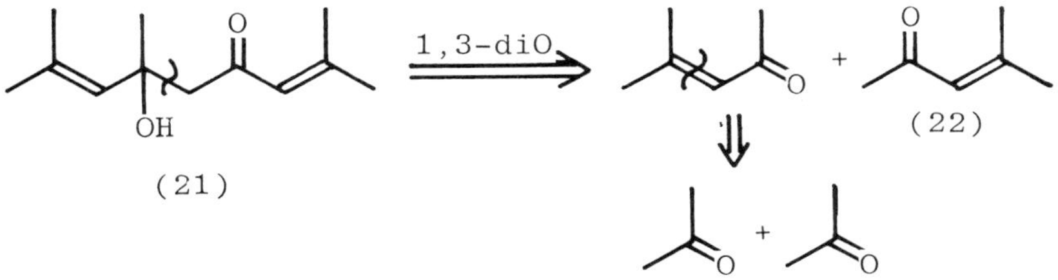

Acetone is symmetrical so its condensation to give (22) is straightforward.[216] Enolisation of (22) to (23) may look unambiguous, but the alternative removal of the γ-proton to give (24) is thermodynamically more favourable. A strong base (PhMeN⁻) secures kinetic control and a good yield of (21).

Synthesis[216,217]

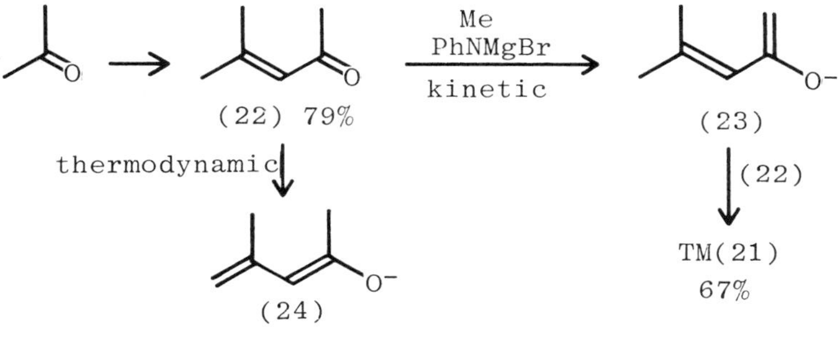

Problem : What product will be formed from (25) under base catalysis?

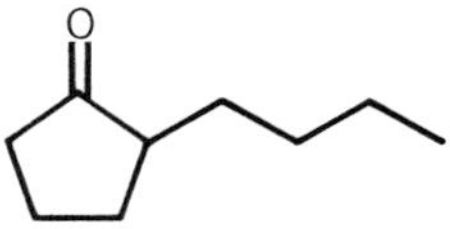

(25)

Answer : Kinetic enolisation occurs on the less substituted side to give (26) and hence (27) which dehydrates under the reaction conditions to (28). The alternative product (29) cannot dehydrate.
Synthesis[218]

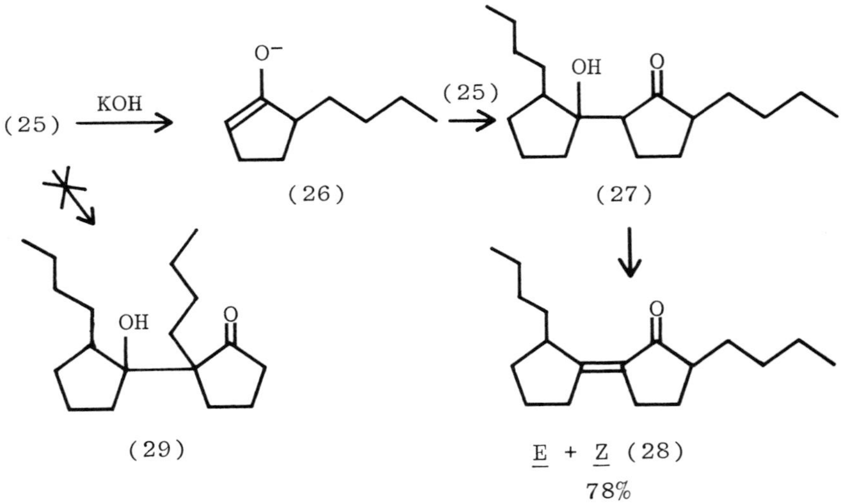

(25) $\xrightarrow{\text{KOH}}$ (26) $\xrightarrow{(25)}$ (27)

(29)

E + Z (28)
78%

(b) *Intramolecular Reactions*

Problems :

(a) Predict the cyclisation product from (30) in base.

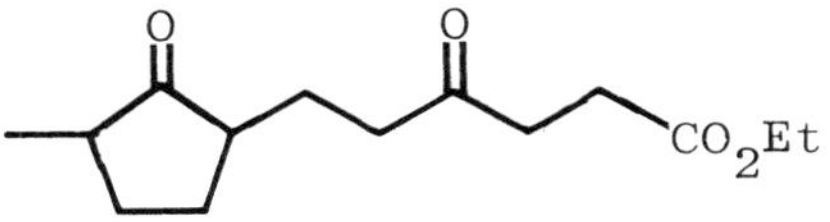

(30)

Answer : This diketo-ester has five sites for enolisation (● in 30a). Attack of these enols on the other carbonyl groups could produce ring sizes from three to nine but only one stable non-bridged cyclic

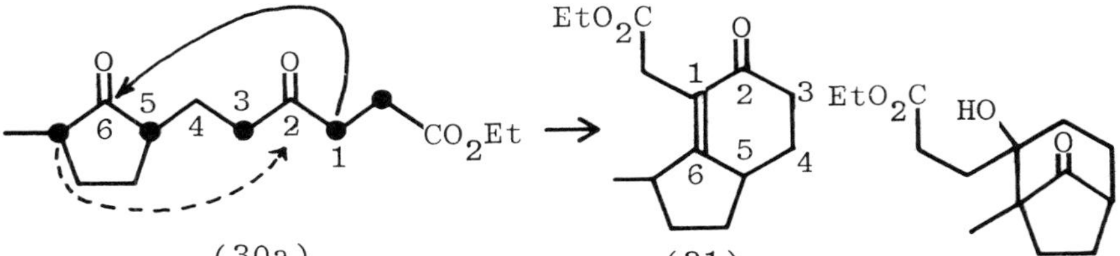

(30a) (31) (32)

compound : a six-membered enone (31). This is formed in vigorous basic (MeO^-) or acid (BF_3 or TsOH) conditions. Under mild acidic conditions (HCl, room temperature) the bridged compound (32) is formed : this was used in a synthesis of guaiol.[219]

2. Suggest a synthesis for TM(33).

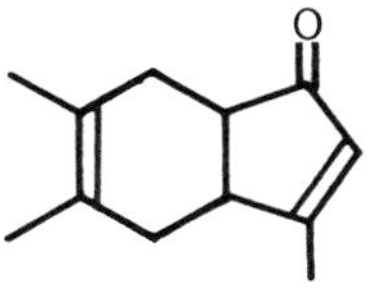

(33)

Answer : Enone disconnection (33a) reveals symmetrical diketone (34), an obvious Diels-Alder adduct.

Analysis

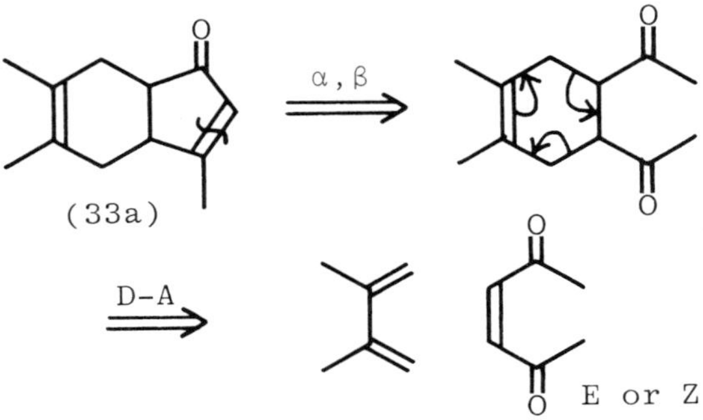

[Structures: (33a) bicyclic enone; ⇒ α,β ⇒ diketone (34); D-A ⇒ diene + E or Z enedione]

The compound was made by this route in model experiments for a steroid synthesis.

Synthesis[220]

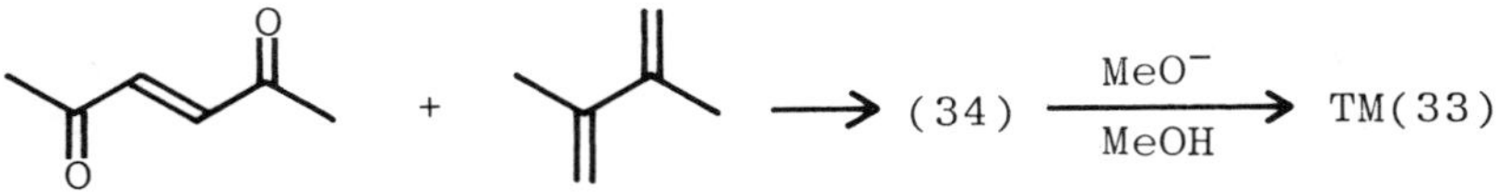

[Scheme: hexenedione + methylenebutenone → (34) → MeO⁻/MeOH → TM(33)]

Cross Condensations I
Use of Compounds which cannot enolise
Examples : Compound (35) is a double diene, capable of Diels-Alder reactions on the simple diene and on the furan ring and it was required to try out a route to polycyclic compounds using both these reactions. Wittig disconnection direct to available aldehyde (36) and easily made (37) is possible, but the alternative Wittig disconnection to (38) takes advantage of the known simple and high yielding condensation of acetone with (36).

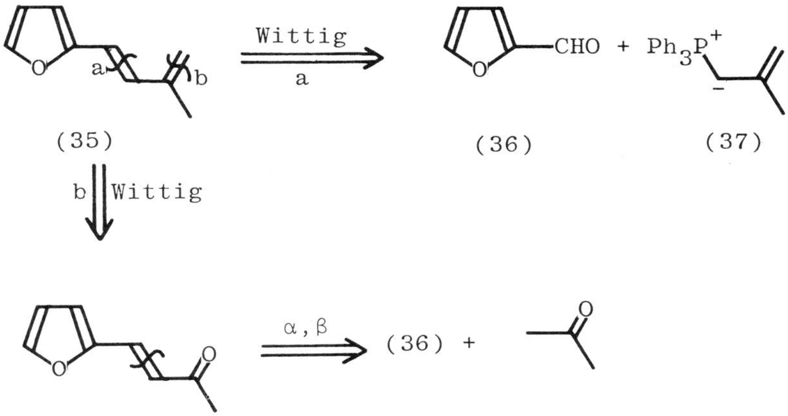

(35) → Wittig a → (36) furan-CHO + Ph₃P⁺-CH₂-C(Me)=CH₂ (37)

b ↓ Wittig

(38) furan-CH=CH-CO-Me ⇒ α,β (36) + acetone

The second route gave 84% of TM(35). There is no problem with the cross condensation as only acetone can enolise and (36) has the more reactive carbonyl group. *Synthesis*[221]

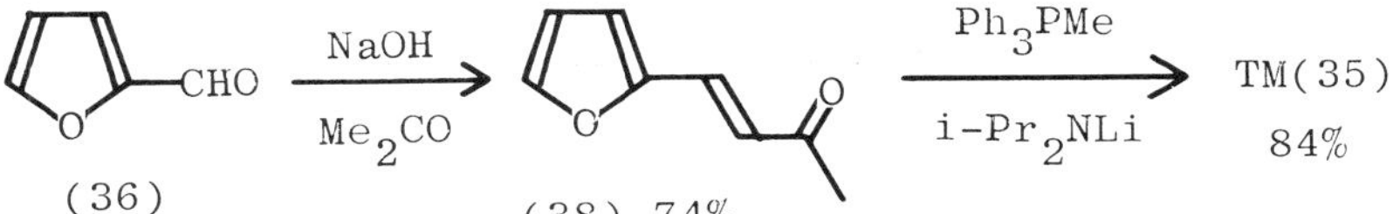

(36) →[NaOH, Me₂CO]→ (38) 74% →[Ph₃PMe, i-Pr₂NLi]→ TM(35) 84%

Problem : A search for anti-tubercular drugs suggested that condensation of aldehydes such as (39) with the successful anti-tubercular agent isoniazide (40) might yield some useful drugs. Suggest how (39) might be synthesised.

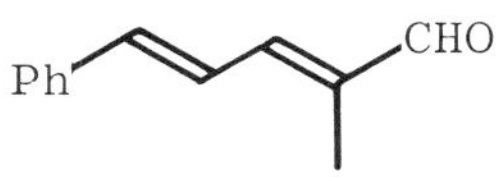

(39) Ph-CH=CH-CH=C(Me)-CHO

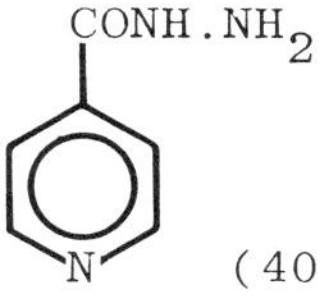

(40) 4-pyridyl-CONH·NH₂

Answer : Normal α,β disconnection reveals another enal and hence another α,β disconnection.

Analysis

$$Ph\diagdown\diagup_a\diagdown\diagup^{CHO} \xRightarrow{\alpha,\beta} Ph\diagdown\diagup\diagdown_{CHO} + \diagup\diagdown_{CHO}$$

(39a)   (40)

$$\Downarrow \alpha,\beta$$

PhCHO + MeCHO

Each step in this synthesis requires the condensation of an aldehyde (MeCHO or EtCHO) which can enolise with one which cannot (PhCHO or 40). These condensations give reasonable yields if an excess of the enolisable aldehyde is used to compensate for self-condensation.

Synthesis[222]

$$PhCHO + MeCHO \xrightarrow{base} (40) \xrightarrow[NaOH]{EtCHO} TM(39)$$

The Mannich Reaction

Examples : Over-reaction and reduction are the problems with formaldehyde condensations in base (page T 158). In acid solution over-reaction and acetal formation, e.g. to give (41), occur.[223] The Mannich Reaction avoids all these problems.

$$\text{MeCOEt} \xrightarrow[H^+]{CH_2O} (41)$$

As a model for his synthesis of vitamin D, Lythgoe[224] made triene (42). Disconnection (Wittig) of the central double bond is likely to give the greatest simplification and an α-methylene ketone (43) is one of the starting materials.

Analysis

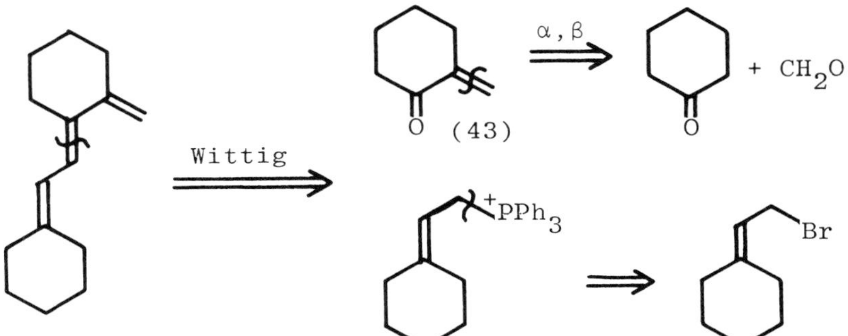

The Mannich reaction must be used to make (43). Allylic bromide (45) can be made from alcohol (46) by the strategy shown.

Analysis 2

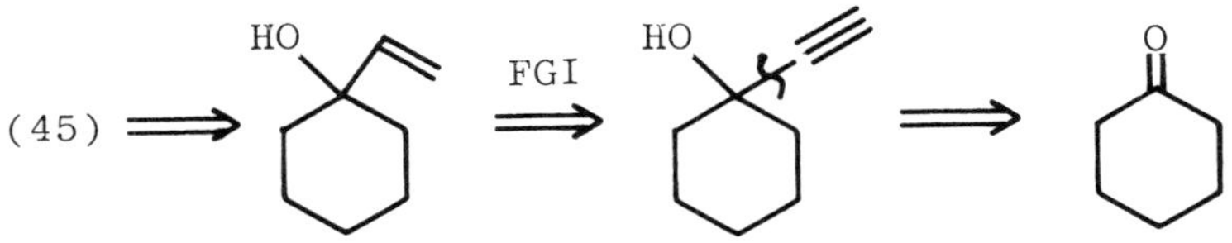

The starting material for both halves of the molecule is cyclohexanone. The Mannich reaction presented no problems and the NR_2 group was retained until the end of the synthesis to avoid releasing the sensitive exomethylene ketone (43).

Synthesis[224]

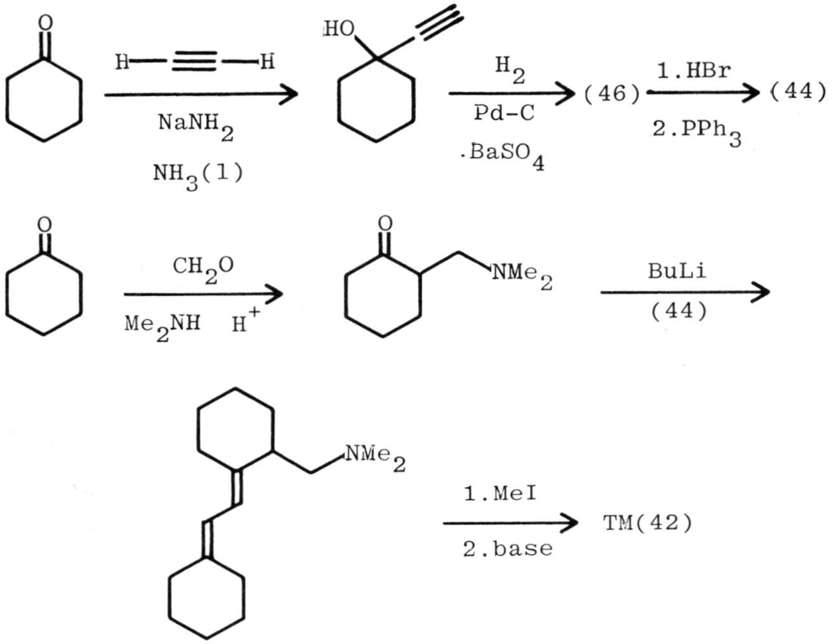

Problem : Suggest a synthesis of Darvon (47), used as an analgesic for mild pain.

(47)

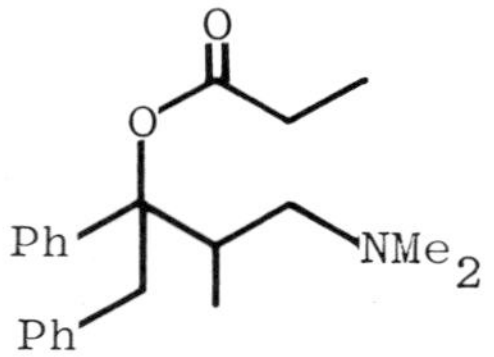

Answer : Ester disconnection gives a tertiary alcohol (48). Of the three possible Grignard disconnections, (a) is most helpful as it requires the Mannich product (49) of an aryl ketone (50) available by the Friedel-Crafts reaction.

Analysis

(47) $\xrightleftharpoons[\text{ester}]{\text{C-O}}$ Ph–C(OH)(Ph)–CH(a)(Me)–CH$_2$–NMe$_2$ (bonds b, c marked) $\xrightarrow{a}$ PhMgBr

+

PhH + EtCOCl $\xleftarrow{\text{F-C}}$ Ph–CO–CH$_2$–CH$_3$ (50) $\xleftarrow{\text{Mannich}}$ Ph–CO–CH(Me)–CH$_2$–NMe$_2$ (49)

Synthesis[225]

PhH $\xrightarrow[\text{AlCl}_3]{\text{EtCOCl}}$ (50) $\xrightarrow[\substack{\text{HCl} \\ \text{EtOH}}]{\substack{\text{CH}_2\text{O} \\ \text{Me}_2\text{NH}}}$ (49) $\xrightarrow{\text{PhCH}_2\text{MgBr}}$

(48) $\xrightarrow[\text{pyr}]{(\text{EtCO})_2\text{O}}$ TM(47)

There are two chiral centres in (47) so a mixture of diastereoisomers is produced in 75 and 15% yields. Fortunately the major isomer is the analgesic. In fact only one enantiomer of this diastereo isomer is analgesic and so (48) is resolved with camphor sulphonic acid before esterification. The other enantiomer is a useful cough suppressant.

Problem : Chemists studying cyclisation reactions wanted a series of nitrogen-containing tertiary alkyl chlorides of general structure (51). How might they be made?

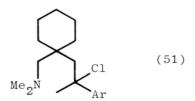

(51)

Hint : The chloride could be made from an alcohol, but this isn't very helpful. How else might the chloride atom be added?

Answer : If the chloride is made by HCl addition to (52) then a Wittig disconnection leads back to a Mannich product (53).

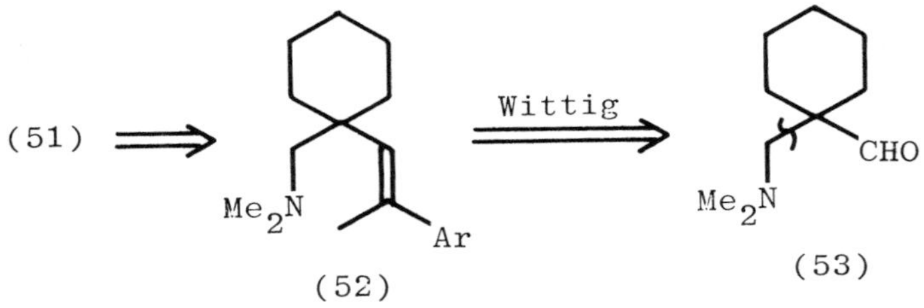

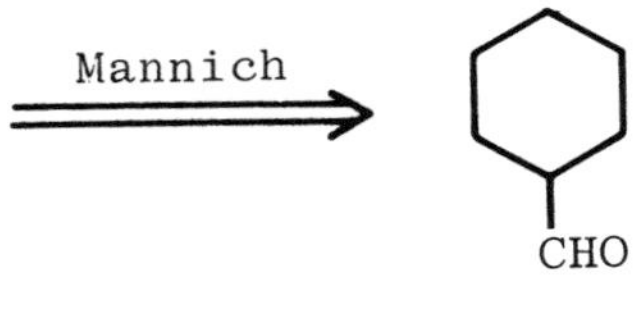

(51) ⟹ (52) ⟹ Wittig ⟹ (53)

⟹ Mannich ⟹ cyclohexane-CHO

Synthesis[226]

cyclohexane-CHO $\xrightarrow{\underset{H^+}{Me_2NH,\ CH_2O}}$ (53) $\xrightarrow[base]{Ph_3P^+-CH(Ar)}$ (52) $\xrightarrow{HCl}$ TM(51)

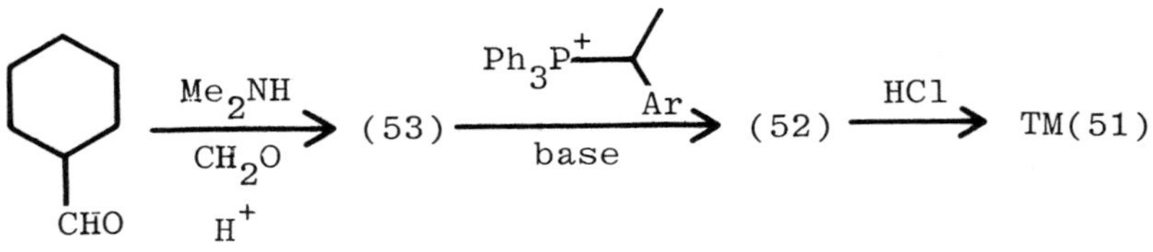

Specific Enol Equivalents

Example : The correct structure for bullatenone (cf p T 163) is (54). Disconnection of the ether bond gives the enol of α-haloketone (55).

Analysis 1

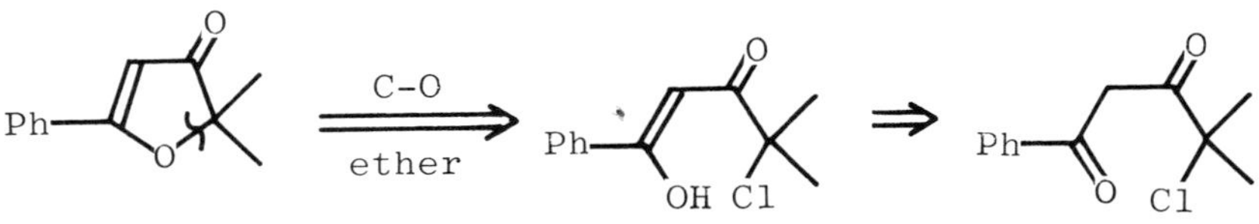

[Scheme: (54) ⟹ (C-O ether) Ph-C(OH)=C-CO-C(CH$_3$)$_2$Cl ⟹ Ph-CO-CH$_2$-CO-C(CH$_3$)$_2$Cl (55)]

Disconnection of the 1,3-diketone obviously comes next but it will be better to add an activating group to control the reaction. One possibility is to acylate specific enol equivalent (56) with acid chloride (57).

Analysis 2

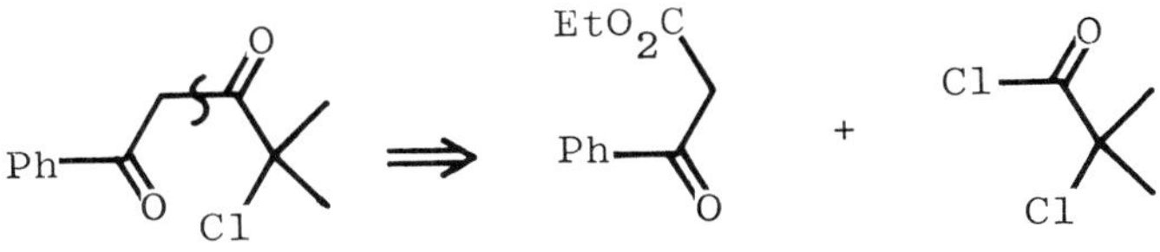

[Scheme: (55a) ⟹ (56) EtO$_2$C-CH$_2$-CO-Ph + (57) Cl-CO-C(CH$_3$)$_2$Cl]

Ketoester (56) is also a 1,3-dicarbonyl compound and can be made by two routes.

Analysis 3

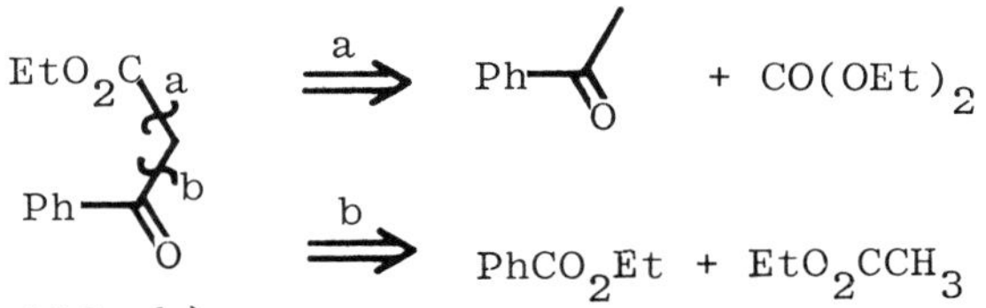

[Scheme: (56a,b) EtO$_2$C-CH$_2$-CO-Ph
 a ⟹ Ph-CO-CH$_3$ + CO(OEt)$_2$
 b ⟹ PhCO$_2$Et + EtO$_2$CCH$_3$]

Route (b) gives reasonable yields of (56). It is easier to use the bromo-acid bromide (58) available by phosphorus-catalysed bromination of the acid (page T 53). The acylation of (56) with (58) requires the magnesium salt as described on page T 163.

Synthesis[227]

$$\text{>-CO}_2\text{H} \xrightarrow{\text{red P, Br}_2} \text{>-COBr} \quad (58)$$
(with Br on the central carbon)

$$\text{PhCO}_2\text{Et} \xrightarrow[\text{EtO}^-]{\text{MeCO}_2\text{Et}} (56) \; 70\% \xrightarrow[\text{2.(58)}]{\text{1. Mg, EtOH}}$$

$$\text{EtO}_2\text{C-}\underset{\text{Ph}}{\overset{\text{O}}{\diagdown}}\text{-Br} \xrightarrow{\text{Et}_3\text{N}} \text{Ph-}\underset{\text{O}}{\overset{\text{EtO}_2\text{C}}{\diagdown}} \xrightarrow[\text{2. H}^+,\text{heat}]{\text{1. HO}^-,\text{H}_2\text{O}} \text{TM(54)}$$

Example : Ester (59) was needed for a photochemical synthesis of chrysanthemate ester (60), a component of the pyrethrin insecticides.[228] The α,β disconnection (59a) gives synthon (61) and aldehyde (62). This β,γ-unsaturated compound could be made by dehydration of (63) as the double bond can appear in only the required position. On page T 149 we discussed the synthesis of (62) by the aldol dimerisation of (64). An alternative strategy is to work at the ester oxidation level (65) which means synthon (66) is needed to combine with (64).

$$(59) \quad \text{structure-CO}_2\text{Me} \xrightarrow{h\nu} \text{structure-CO}_2\text{Me} \quad (60)$$

Analysis

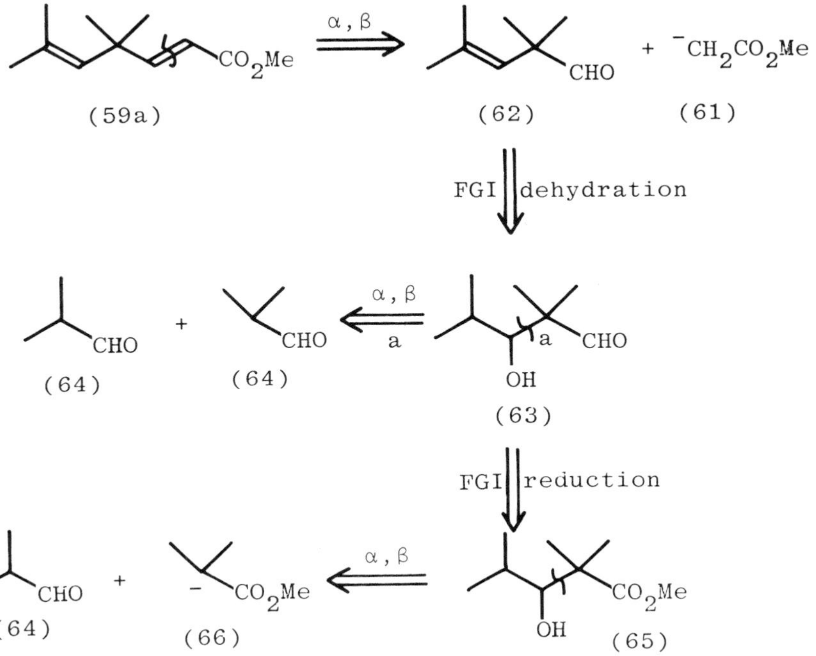

(59a) ⟹ (α,β) (62) + ⁻CH₂CO₂Me (61)

FGI ⇓ dehydration

(64) CHO + (64) CHO ⇐(α,β, a)= (63) [with OH, CHO]

FGI ⇓ reduction

(64) CHO + (66) ⁻CO₂Me ⇐(α,β)= (65) [with OH, CO₂Me]

Specific enol equivalents will be needed for both synthons (61) and (66). Since (61) is to give a double bond but (66) is to give an alcohol, the logical choices are a Wittig reagent - actually (67) - for (61) and a Reformatsky reagent for (66). The ester to aldehyde conversion (65 → 63) is easiest by over-reduction and re-oxidation after dehydration.

Synthesis[228]

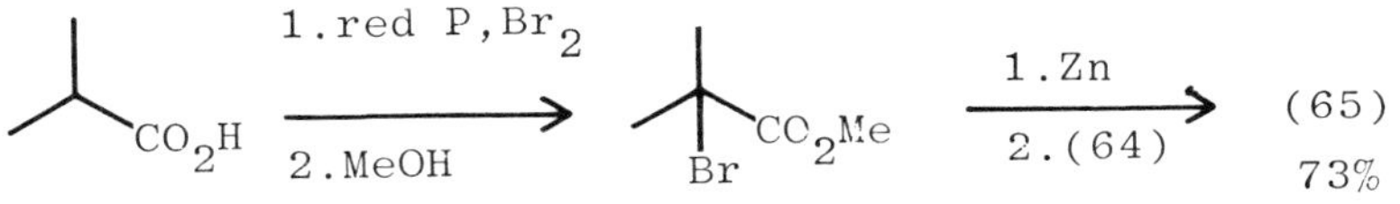

CO₂H →(1. red P, Br₂; 2. MeOH)→ (Br, CO₂Me) →(1. Zn; 2.(64))→ (65) 73%

$$\xrightarrow[\text{pyr}]{\text{POCl}_3} \quad \text{[alkene-CO}_2\text{Et]} \quad \xrightarrow{\text{LiAlH}_4} \quad \text{[alkene-OH]}$$

60%   90%

$$\xrightarrow[\text{TFA}]{\substack{\text{DCC}\\ \text{DMSO}\\ \text{pyr}}} \quad (62) \quad 62\% \quad \xrightarrow{\text{NaH}} \quad \text{TM}(59) \quad 50\%$$

$(\text{EtO})_2\text{P}(=\text{O})\text{CH}_2\text{CO}_2\text{Me}$

(67)

Problem :

Devise a synthesis for TM(68).

[structure of (68): cyclohexene with CH=C(CN)CO₂Et side chain]   (68)

Answer : Of the two double bonds, only one is α,β to a carbonyl group so we must disconnect that. Aldehyde (70) is an obvious Diels-Alder adduct.

Analysis

[structures: (68a) → (69) CH₂(CN)CO₂Et + cyclohexene-CHO (70) ⟹ acrolein + butadiene]

No control is needed for the condensation as there are already two activating groups on (69).

Synthesis[229]

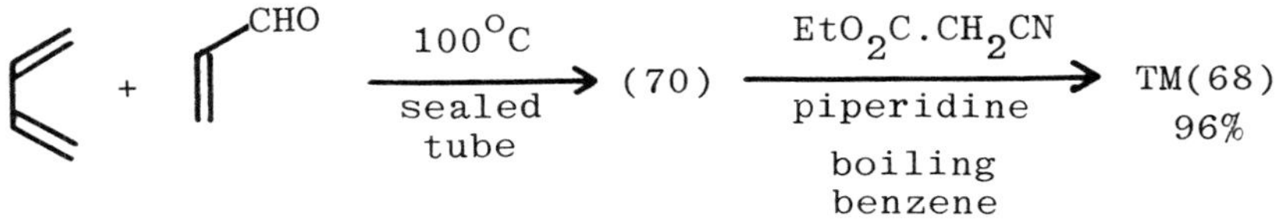

The product (68) is a useful herbicide and parasiticide.

Example : The synthesis of aryl cyclohexanones (71) illustrates some of the problems in this area. Of the two possible 1,3-diCO disconnections (71a and b) we may dismiss (a) as it leads to no simplification of the problem. The ring-chain disconnection (b) suggests the acylation of a symmetrical cyclohexanone and so is very promising.

Analysis 1

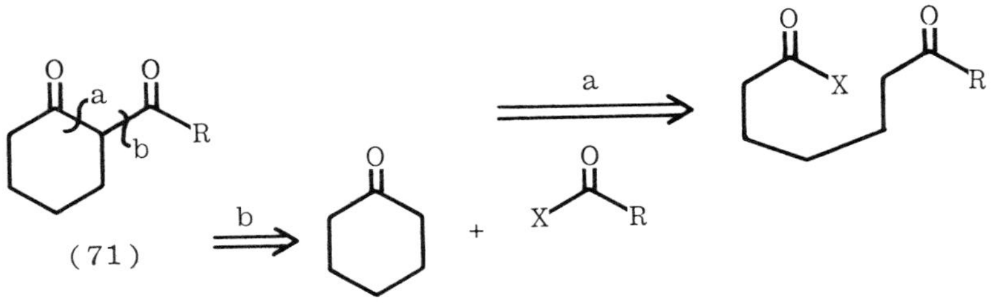

The problem arises in controlling the reaction. Cyclohexanone enolises easily but attacks itself easily too. Direct acylation with RCO_2Et and EtO^- catalyst gives poor yields.[230] Clearly a specific enol is needed. An activating group could easily be added, as in (72), but acylation of this, to replace the last remaining proton, is not a good idea (cf p 211).

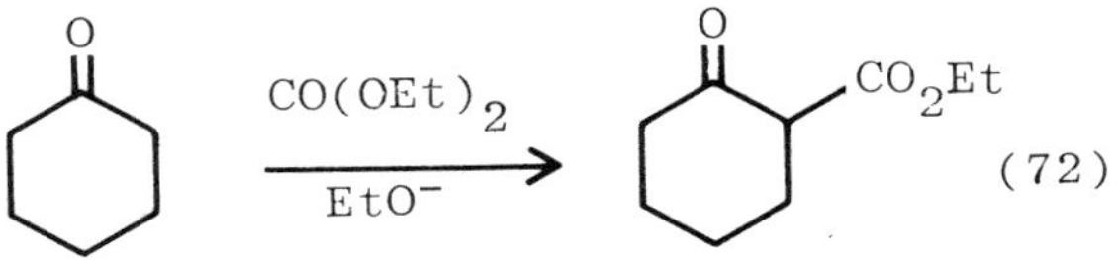

The answer is to use an enamine. Work by Hünig[231] has established that the best combination is the morpholine enamine (73) as the nucleophile and an acid chloride as the electrophile. The yields are generally excellent.

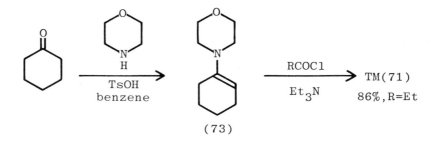

CHAPTER 21

Two-Group Disconnections III: 1,5-Difunctionalised Compounds: Michael Addition and Robinson Annelation

Problem : Carry out both the reverse Michael disconnections on TM(1), continue the analysis to find available starting materials, and choose which route you prefer.

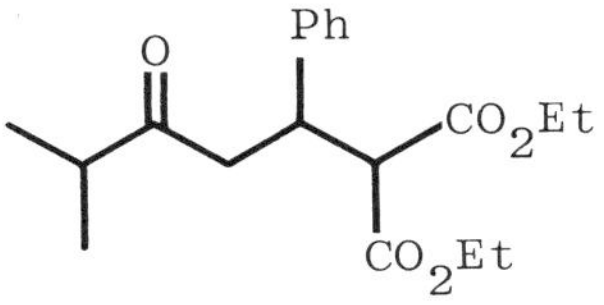

O
Ph
CO_2Et
CO_2Et

(1)

Answer : Good syntheses can be developed from both disconnections. Make sure you have appreciated the need for control in the reactions used.

Analysis

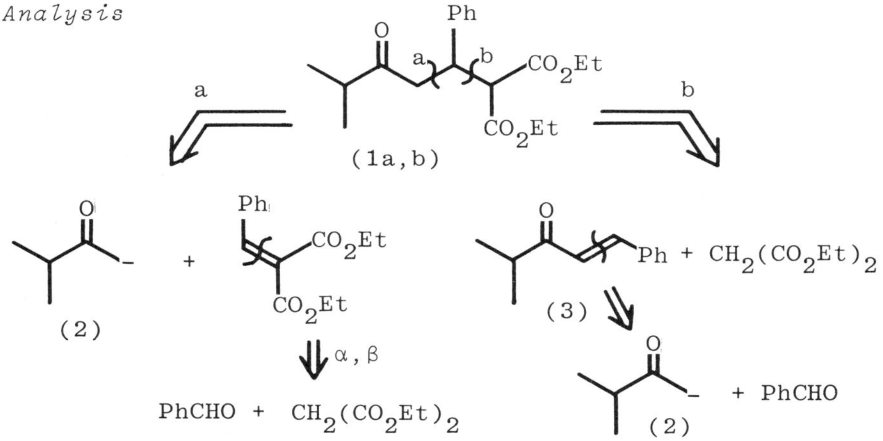

O
Ph
a
b
CO_2Et
CO_2Et
a
b
(1a,b)
O
−
(2)
+
Ph
CO_2Et
CO_2Et
α, β
PhCHO + $CH_2(CO_2Et)_2$
O
Ph + $CH_2(CO_2Et)_2$
(3)
O
−
(2)
+ PhCHO

The analysis uses the same starting materials whichever bond you disconnect first : malonate ester, PhCHO, and the specific enolate (2). This last could be an enamine, or an activated version (4) (cf p T 160 ff).

$$\text{(4)} \quad \underset{\text{(4)}}{\text{Me}_2\text{CH-CO-CH}_2\text{-CO}_2\text{Et}} \Longrightarrow \text{Me}_2\text{CH-CO-CH}_3 + \text{CO(OEt)}_2$$

Route (b) offers a short cut since the reaction between (5) and PhCHO under dehydrating conditions needs no control as (3) is the only possible enone from a ketone enolate attacking the more reactive aldehyde (p T 167). The Michael reaction is also better by this route as explained on p T 171. This is the published synthesis.

Synthesis[232]

$$(5) \xrightarrow[\text{H}_2\text{O}]{\text{PhCHO, NaOH}} (3) \; 57\% \xrightarrow[\text{EtOH}]{\text{CH}_2(\text{CO}_2\text{Et})_2, \text{NaOEt}} \text{TM}(1) \; 79\%$$

Example : The lactone (8), needed for a natural product synthesis, might be made from (6) via epoxide (7) and so a synthesis for (6) was required. Wittig disconnection reveals a 1,5-dicarbonyl compound (9), best made by Michael addition of a substituted malonate (11) to enone (10). The enone was made by the simple but reliable Grignard route rather than risking a Mannich reaction of unknown regioselectivity.

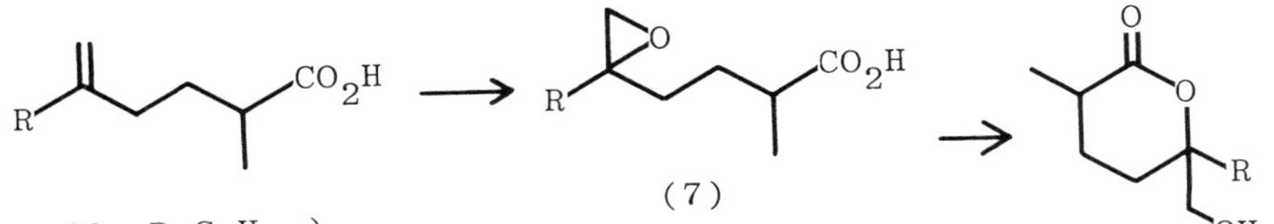

Analysis

(6) $\xrightarrow{\text{Wittig}}$ (9) $\Rightarrow$ (10) + (11)

(10) $\underset{\text{FGI}}{\Downarrow}$

RCHO + BrMg–CH=CH$_2$ $\rightleftharpoons$ allyl alcohol

(11) CH(Me)(CO$_2$Et)$_2$

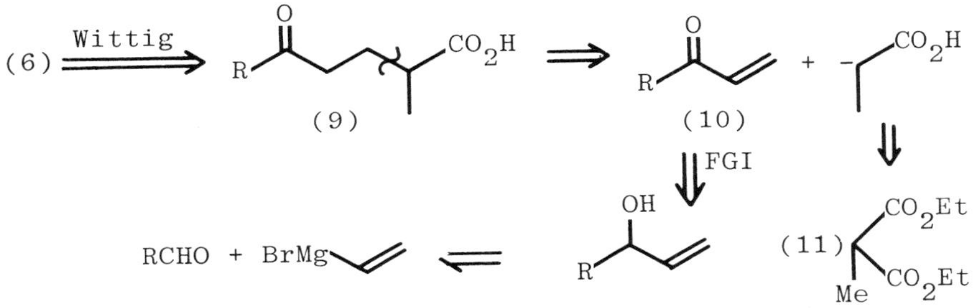

Synthesis[2,3,3]

$$CH_2(CO_2Et)_2 \xrightarrow[\text{2. MeI}]{\text{1. EtO}^-} (11)$$

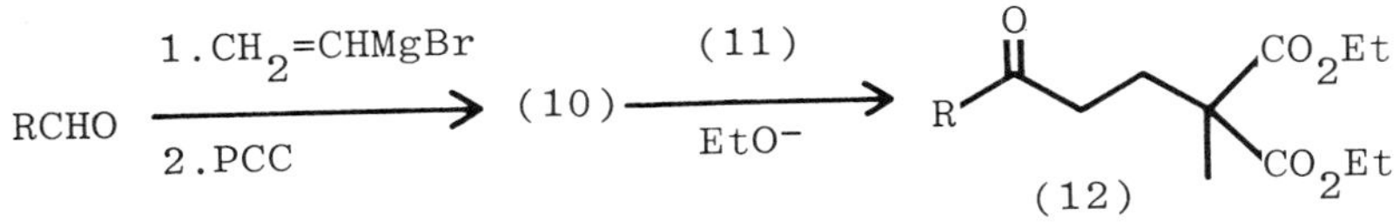

$$RCHO \xrightarrow[\text{2. PCC}]{\text{1. CH}_2=CHMgBr} (10) \xrightarrow[\text{EtO}^-]{(11)} (12)$$

$$\xrightarrow{\text{AcOH}} (9) \xrightarrow{2Ph_3P=CH_2} TM(6)$$

Hydrolysis and decarboxylation of (12) occur in hot acetic acid. The CO$_2$H group in (9) is not protected: the first mole of Ph$_3$P=CH$_2$ removes the acid proton and the second does the Wittig reaction.

Problem : Suggest a synthesis for (13).

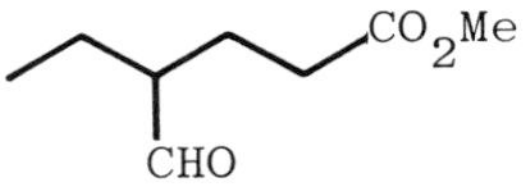

(13) — structure: CH₃CH₂-CH(CHO)-CH₂-CH₂-CO₂Me

Answer : Disconnection at the branchpoint is best giving acrylate (15) and specific enol (14).

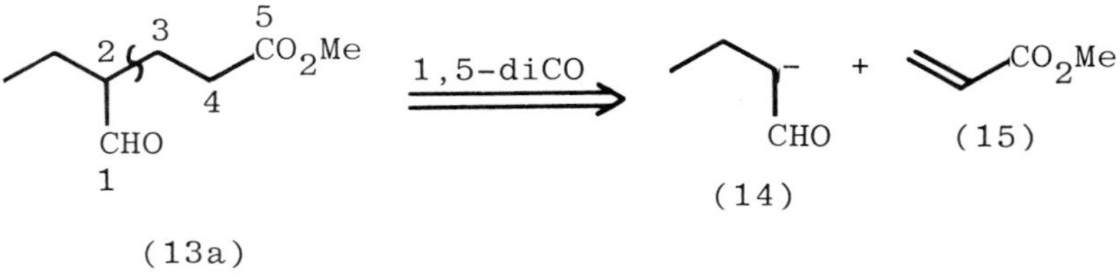

(13a) numbered 1(CHO)–2–3–4–5(CO₂Me) ⇒ 1,5-diCO ⇒ enolate at C2 (14, CH₃CH₂CH⁻CHO) + acrylate (15, CH₂=CH-CO₂Me)

The enamine (16) is the best reagent for synthon (14) : this was one of Stork's original enamine syntheses.

Synthesis[2,3,4]

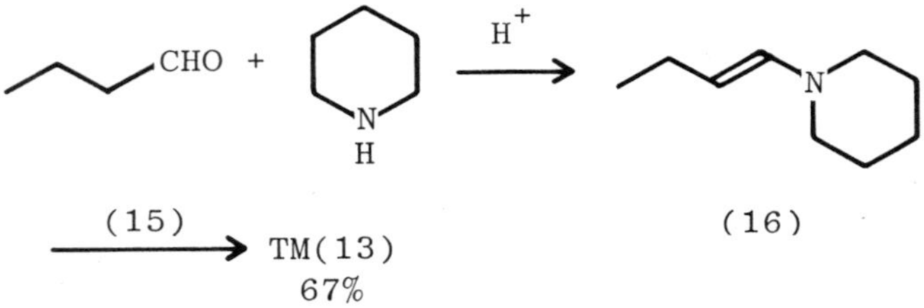

CH₃CH₂CH₂CHO + piperidine →(H⁺) enamine (16)

(16) + (15) ⟶ TM(13), 67%

This product has been used to make (17) and (18), both intermediates in indole alkaloid chemistry. How might (13) be converted into (17) and (18)?

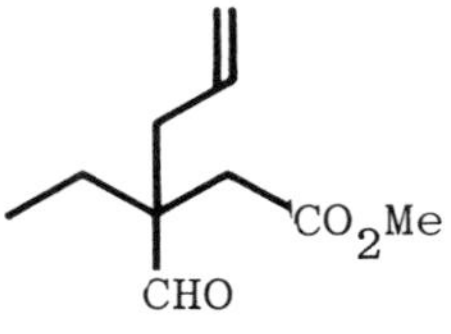

(17): quaternary carbon bearing ethyl, CHO, CH₂-CH=CH₂ (allyl), and CH₂CO₂Me

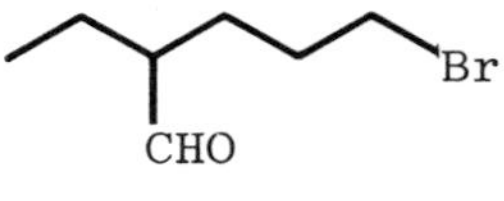

(18): CH₃CH₂CH(CHO)CH₂CH₂CH₂Br

Answer : Alkylation of the specific enol of the aldehyde in (13) will give (17). The enamine is again the best way to do this.

Synthesis[235]

(13) $\xrightarrow[\text{H}^+]{\text{pyrrolidine NH}}$ [enamine with CO$_2$Me chain] $\xrightarrow[\text{2. H}^+, \text{H}_2\text{O}]{\text{1. allyl Br}}$ TM(17)

To make (18), we must reduce the ester in the presence of the aldehyde : protection is necessary and an acetal the obvious way.

Synthesis[236]

(13) $\xrightarrow[\text{H}^+]{\text{MeOH}}$ [dimethyl acetal with CO$_2$Me] $\xrightarrow{\text{LiAlH}_4}$ [dimethyl acetal with OH]

$\xrightarrow[\text{2. LiBr, DMF}]{\text{1. MsCl, Et}_3\text{N}}$ [dimethyl acetal with Br] 84% $\xrightarrow[\text{H}_2\text{O}]{\text{HCl}}$ TM(18)

This particular OH → Br conversion avoids hydrolysing the acetal with HBr or other acidic reagent.

Problem : How would you make TM(19) whose cyclisation was discussed on page 211?

(19) [bicyclic lactone with CHO substituent]

Answer : The better Michael disconnection is (19a) (branchpoint and more stable anion) and the intermediate (20) can be made by direct formylation of cyclohexanone.

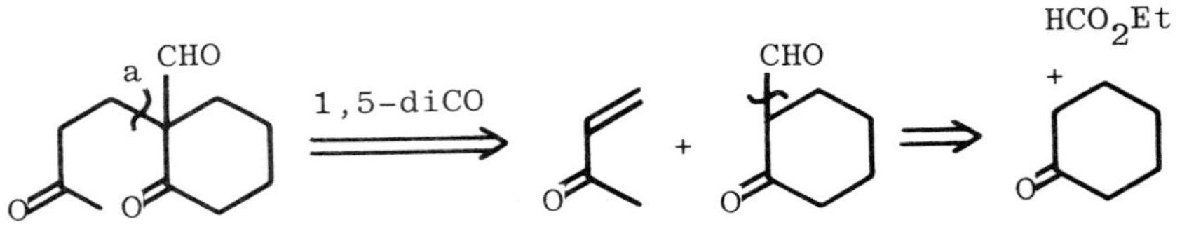

(19a) ⟹ [1,5-diCO] + (20) ⟹ HCO₂Et + cyclohexanone

Synthesis[211]

cyclohexanone →[HCO₂Et / EtO⁻] (20) →[KOH, methyl vinyl ketone] TM(19)

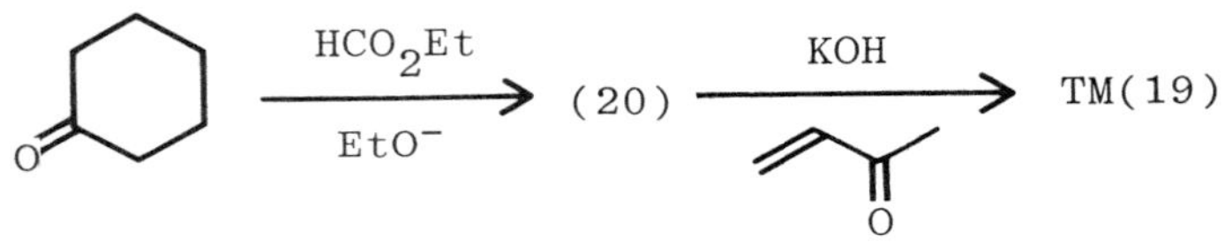

Robinson Annelation

Example : Compound (21) may not look like a Robinson annelation product, but it is certainly an enone so α,β disconnection gives a 1,5-di-carbonyl compound A reverse Michael reaction disconnecting the ring from the chain gives enone (22).

Analysis

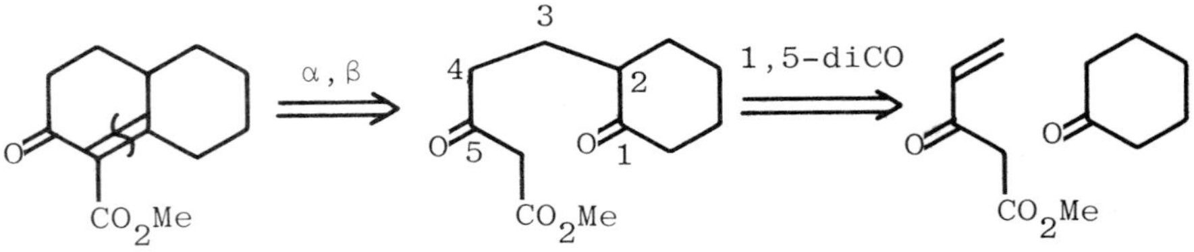

(21) ⟹[α,β] (intermediate) ⟹[1,5-diCO] (22)

Compound (23), which we met on page 210, was made so that it could be decomposed to give (22) on heating, by a reverse Diels-Alder reaction. An enamine again controls the Michael addition, and cyclisation is spontaneous.

Synthesis[212]

[Structure (23): norbornene with -C(O)-CH$_2$-CO$_2$Me substituent] $\xrightarrow{\text{heat}}$ (22) 80%

cyclohexanone + pyrrolidine $\xrightarrow{H^+}$ [enamine] $\xrightarrow{(22)}$ TM(21)

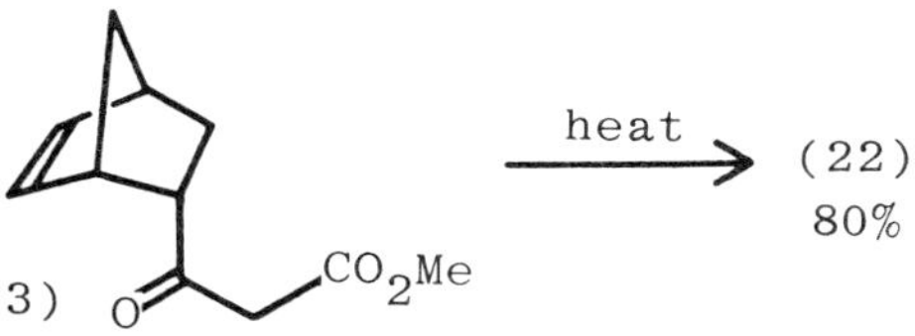

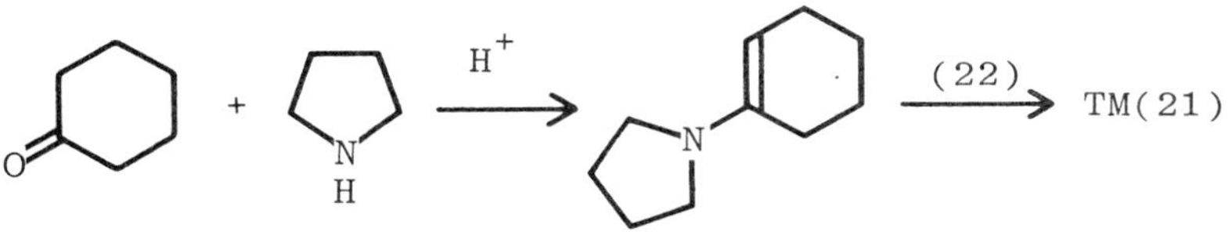

Problem : Suggest a synthesis for TM(24).

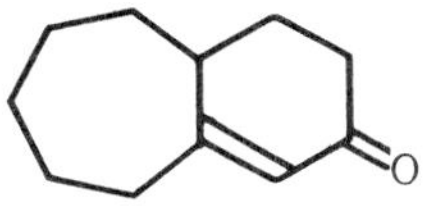

[Structure (24): bicyclic enone with fused 7- and 6-membered rings]

Answer : Following the Robinson annelation disconnection is simple here.

Analysis

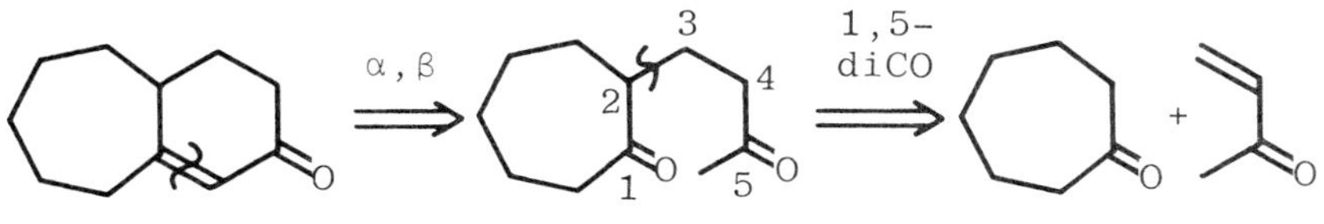

(24a) $\xRightarrow{\alpha,\beta}$ [1,5-diketone intermediate with positions 1-5 labelled] $\xRightarrow{\text{1,5-diCO}}$ cycloheptanone + methyl vinyl ketone

An enamine can again be used to control the reaction.

Synthesis[237]

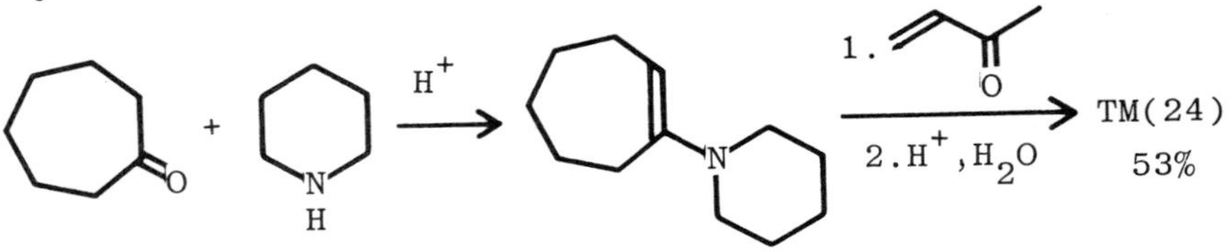

[Scheme: cycloheptanone + piperidine, H⁺ → enamine; 1. methyl vinyl ketone (CH₂=CH-COMe); 2. H⁺, H₂O → TM(24), 53%]

Problem : Compounds like (25) have been used in alkaloid syntheses. How might they be made?

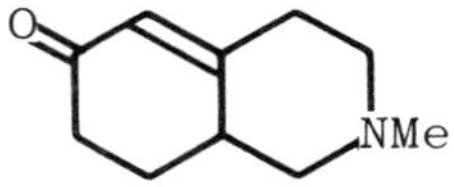

(25)

Answer : The cyclohexenone is a clue to a Robinson annelation : disconnection reveals symmetrical amino ketone (26) as starting material (see page T 147 for its synthesis). An enamine is again the best control.

Analysis

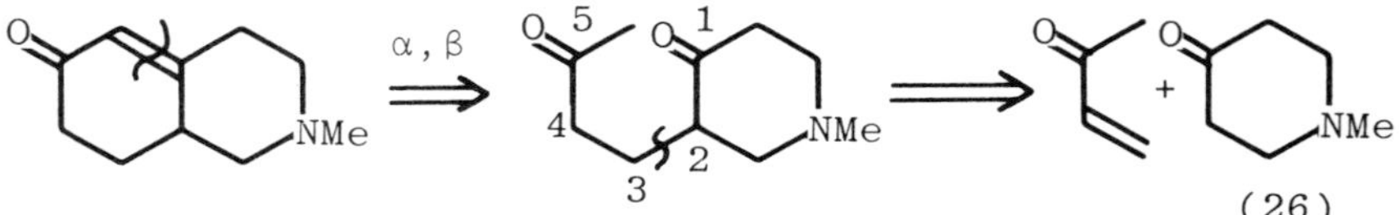

(25a) ⇒ α,β ⇒ [diketoamine numbered 1-5] ⇒ methyl vinyl ketone + (26)

Synthesis[238]

(26) + piperidine → enamine; 1. CH₂=CH-COMe; 2. H⁺, H₂O → TM(25), ~80%

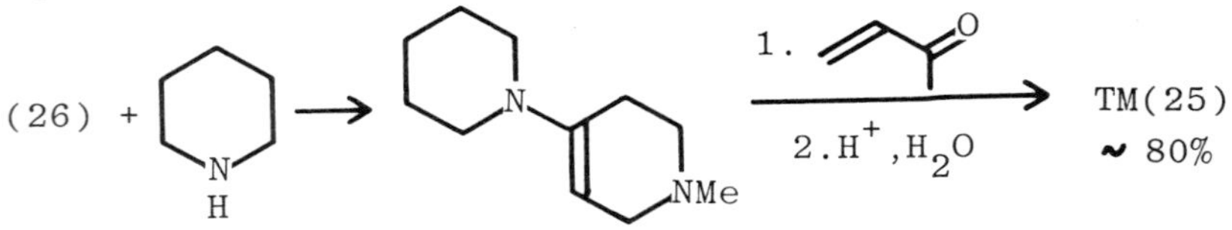

Example : Compound (27) was needed for synthesis of analogues of vernolepin, an anti-tumour compound.[239] Robinson disconnection suggests unsymmetrical ketone (28)

as starting material but it would be very difficult to control the regioselectivity of this reaction. A change in oxidation level to (29) makes control a simple matter.

Analysis

[Structures: (27) decalone with OH and ketone, α,β ⇒ diketone with OH ⇒ methyl vinyl ketone + cyclohexanone with OH (28); FGI ⇓ β-ketoester (29) ⇐ cyclohexanone + CO(OEt)₂]

The Mannich method was used to release the vinyl ketone into the reaction mixture from (30).

Synthesis[240]

[Scheme: acetone + CH₂O / Et₂NH / H⁺ → Mannich base with NEt₂ → MeI → quaternary ammonium salt (30)]

[Scheme: cyclohexanone → (EtO)₂CO / EtO⁻ → (29) → (30) / EtO⁻ → (31) bicyclic enone with CO₂Et, 70%]

Reduction of the ester group in (31) requires protection of the ketone : the double bond wanders during this step but returns to conjugation in (27).

Synthesis[241]

[Reaction scheme showing compound (31) (a diketone) + HOCH₂CH₂OH, TsOH, benzene → bicyclic mono-ketal with CO₂Et group (95%); then LiAlH₄, Et₂O → bicyclic mono-ketal with CH₂OH group (90%); then H⁺, H₂O → TM(27), a bicyclic enone with CH₂OH group (65%).]

Cyclic 1,3-diketones

Problem : Suggest a general approach to cyclohexadiones of type (32).

[Structure (32): cyclohexane-1,3-dione with R substituent at the 5-position.]

Answer : Comparison with dimedone (p T 177) suggests a similar analysis, starting with the 1,3-dicarbonyl disconnection to give keto-ester (33), and then disconnecting next to the branchpoint to give enone (34).

Analysis

[Scheme: (32a) ⇒ (1,3-diCO) ⇒ open-chain keto-ester (33) with CO₂Et at C1, R at C3 ⇒ (1,5-diCO) ⇒ CH₂(CO₂Et)₂ + enone (34) R–CH=CH–C(O)–CH₃, which comes from RCHO + acetone.]

The synthesis of enone (34) requires an aldol condensation between acetone and RCHO : this may not give a good yield as RCHO may prefer to condense with itself if it has enolisable protons. The alternative disconnection (33b) avoids this problem as we can use acetoacetate for the synthon (34) and a specific enol equivalent for (35).

Analysis

[Scheme: 1,5-diCO disconnection of R-CH(CO₂Et)-CH₂-C(=O)-CH₃ giving acetone synthon (34) ≡ ethyl acetoacetate; plus RCHO + ⁻CH₂CO₂Et ⇌(α,β) R-CH=CH-CO₂Et (35)]

Both methods have been used successfully. The first route gives (32, R=Pr-i) in good yield as enone (34, R=Pr-i) is available from condensation (though not in very good yield) as the aldol dimer of (36) cannot dehydrate (control by removal of one of the products).

Synthesis[242] of (32, R=Pr-i)

[Scheme: acetone + (CH₃)₂CHCHO (36) —base→ (CH₃)₂CH-CH=CH-C(=O)CH₃, 37%; then CH₂(CO₂Et)₂ / EtO⁻ → cyclohexanedione with CO₂Et and i-Pr substituents → 1. KOH, H₂O; 2. H⁺, heat → 5-isopropylcyclohexane-1,3-dione, 70%]

The product was used in the synthesis of terpenes. Compound (32, R=Me) has been made by the alternative strategy from readily available ester (37) *Synthesis*[243] of (32, R=Me)

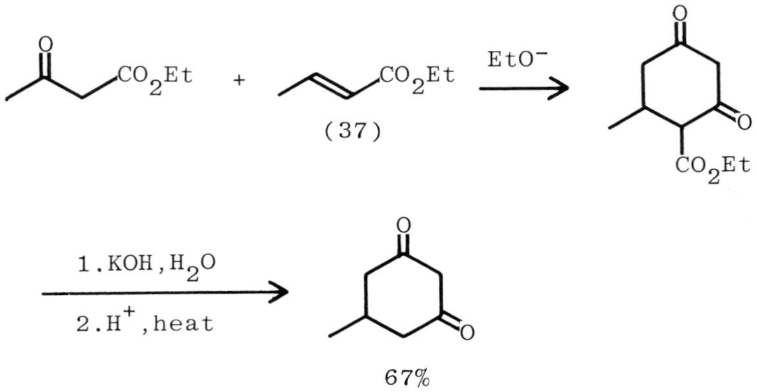

[reaction scheme: ethyl acetoacetate + ethyl crotonate (37), EtO⁻ → methyl-substituted cyclohexanedione with CO₂Et; then 1. KOH, H₂O; 2. H⁺, heat → 5-methylcyclohexane-1,3-dione, 67%]

CHAPTER 22

Strategy X: Use of Aliphatic Nitro Compounds in Synthesis

Example :

Amide (1) was needed for a synthesis of an isoquinoline. Disconnecting the amide reveals the amine (2) which could be made by reduction of an unsaturated nitro compound. The α,β-disconnection is simple.

Analysis

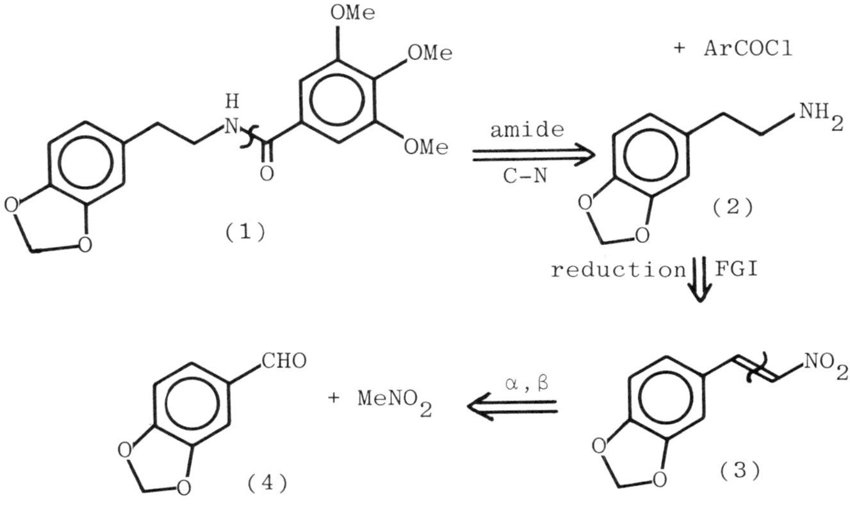

Aldehyde (4) can be made by chloromethylation (p T 9), the condensation with nitromethane with mild base gives an excellent yield of crystalline (3) and $LiAlH_4$ can be used for the reduction.

Synthesis[244]

(4) $\xrightarrow[\text{EtNH}_2]{\text{MeNO}_2}$ (2) $\xrightarrow{\text{LiAlH}_4}$ 97%

(2) $\xrightarrow{\text{ArCOCl}}$ TM(1)
60%   80%

Problem :

How might an aliphatic nitro compound be used in the synthesis of amphetamine analogues such as (5)?

(5) [structure: 3,4-methylenedioxy-substituted benzene with CH₂CH(NH₂)CH₃ side chain]

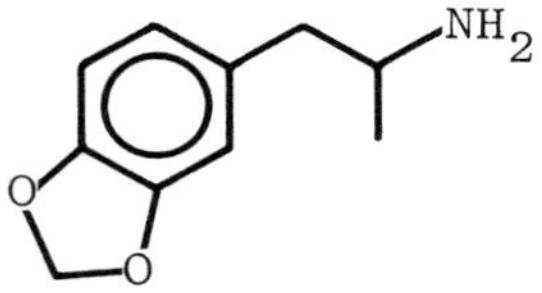

Answer :

Reversing the reduction (FGI) restores both the nitro group and the double bond (6). Disconnection reveals aldehyde (4) and nitroethane.

Analysis

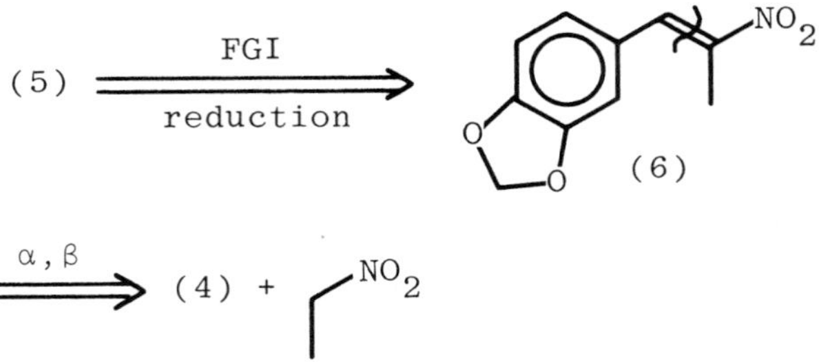

(5) $\xrightarrow[\text{reduction}]{\text{FGI}}$ (6) [methylenedioxyphenyl-CH=C(CH₃)NO₂]

$\xRightarrow{\alpha,\beta}$ (4) + CH₃CH₂NO₂

This has proved an excellent route to amphetamine analogues. The condensation gives good yields with a variety of weak bases,[245] and the reduction goes well with LiAlH$_4$ or one of the modern metal hydrides, e.g. Redal (7).

Synthesis[246]

$$(4) \xrightarrow[\text{NH}_4\text{OAc}]{\text{EtNO}_2} (6) \quad 80\%$$

$$\xrightarrow[\substack{\text{NaAlH}_2(\text{OCH}_2\text{CH}_2\text{OMe})_2 \\ (7)}]{} \text{TM}(5) \quad 85\%$$

Example :

A famous example of the use of nitro compounds in synthesis was the original synthesis[247] of the antibiotic chloramphenicol (8), which is still used to treat tropical diseases. This synthesis also confirmed the structure of chloramphenicol and established that the (-)-threo compound was the biologically active stereoisomer.

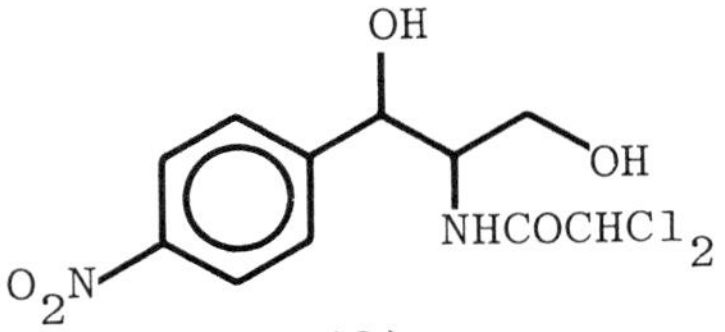

(8)

Disconnecting the amide reveals an amine (9) with two hydroxyl groups both positioned so that the corresponding nitro compound (10) could be made from nitromethane and two aldehydes.

Analysis

(8) $\xrightarrow{\text{C-N amide}}$ [4-O$_2$N-C$_6$H$_4$-CH(OH)-CH(NH$_2$)-CH$_2$OH] (9) $\xrightarrow{\text{FGI reduction}}$

[4-O$_2$N-C$_6$H$_4$-CH(OH)-CH(NO$_2$)-CH$_2$OH] (10) $\Longrightarrow$ 4-O$_2$N-C$_6$H$_4$-CHO + CH$_2$O + CH$_3$NO$_2$

There is no problem of chemoselectivity here : it is not possible to reduce the aliphatic NO$_2$ group in (10) without reducing the aromatic NO$_2$ group too. This is easily solved by introducing the aromatic nitro group *after* the reduction.

Condensations with formaldehyde pose a problem and it turns out that the best way to proceed is to condense nitromethane with formaldehyde and use the unpurified product to react with benzaldehyde.

Synthesis[247]

CH$_3$NO$_2$ $\xrightarrow[\text{KOH}]{\text{CH}_2\text{O}}$ O$_2$N-CH$_2$-CH$_2$-OH $\xrightarrow[\substack{\text{MeOH}\\\text{NaOMe}}]{\text{PhCHO}}$

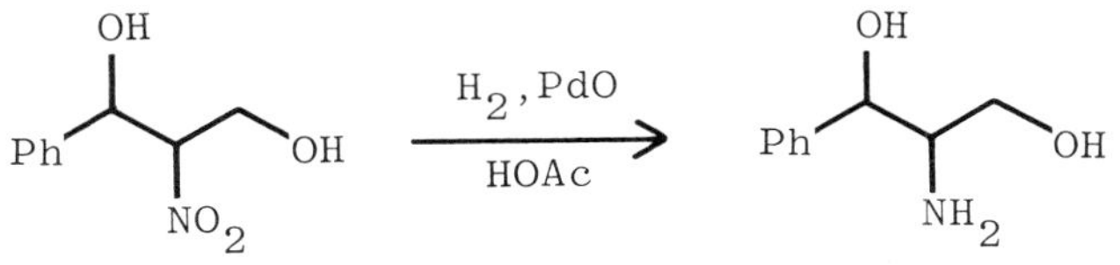

[Scheme: PhCH(OH)CH(NO$_2$)CH$_2$OH → (H$_2$, PdO, HOAc) → PhCH(OH)CH(NH$_2$)CH$_2$OH (11)]

The unwanted erythro isomer of (11) crystallised out at this point, so the synthesis continues with the mother liquors. We must now protect the NH$_2$ and both OH groups during the nitration - this also allows purification of crystalline threo (12).

Synthesis 2

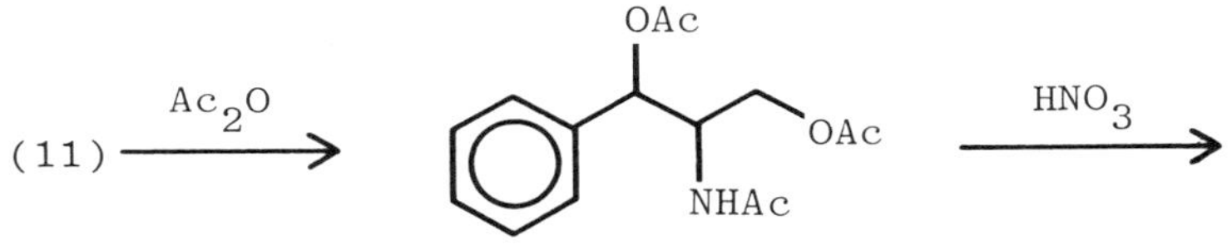

[Scheme: (11) → (Ac$_2$O) → PhCH(OAc)CH(NHAc)CH$_2$OAc (12) → (HNO$_3$) →]

[Scheme: O$_2$N-C$_6$H$_4$-CH(OAc)CH(NHAc)CH$_2$OAc → (1. H$^+$, H$_2$O; 2. Cl$_2$CH.COCl) → TM(8)]

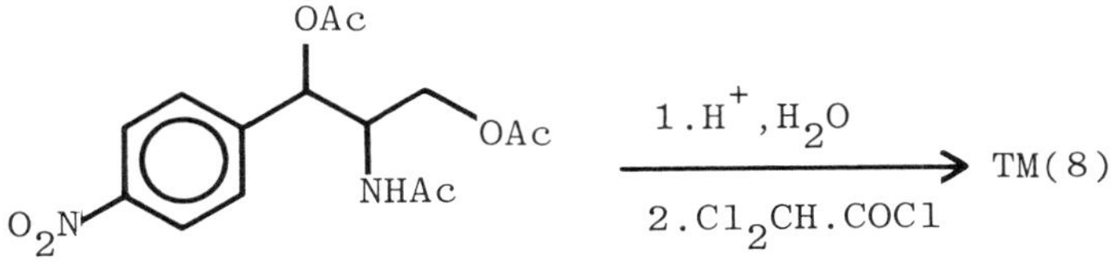

Example :
A more reliable way to control formaldehyde condensations is to use the Mannich reaction:

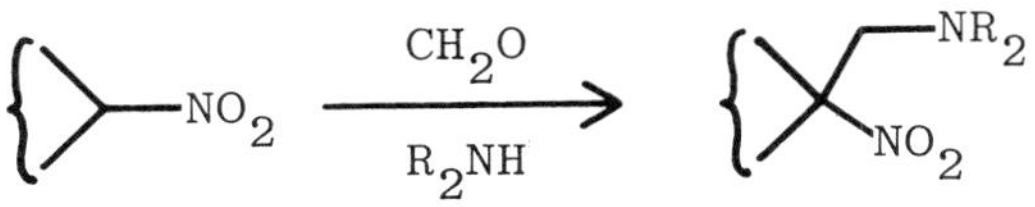

[Scheme: R-CH$_2$-NO$_2$ → (CH$_2$O, R$_2$NH) → R-C(NR$_2$)(NO$_2$)-]

Problem :
Suggest a synthesis for diamine (13).

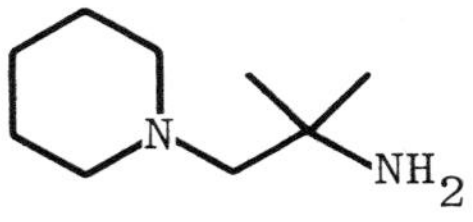

(13)

Answer :
FGI from NH_2 to NO_2 reveals a Mannich disconnection.

Analysis

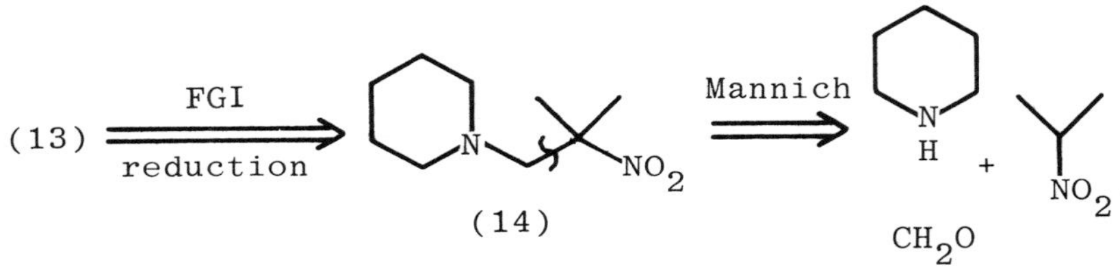

(13) $\xrightarrow[\text{reduction}]{\text{FGI}}$ (14) $\xRightarrow{\text{Mannich}}$ piperidine-NH + CH_2O + $Me_2C(NO_2)$

Catalytic reduction over Raney nickel gives high yields of diamines by this route.

Synthesis[248]

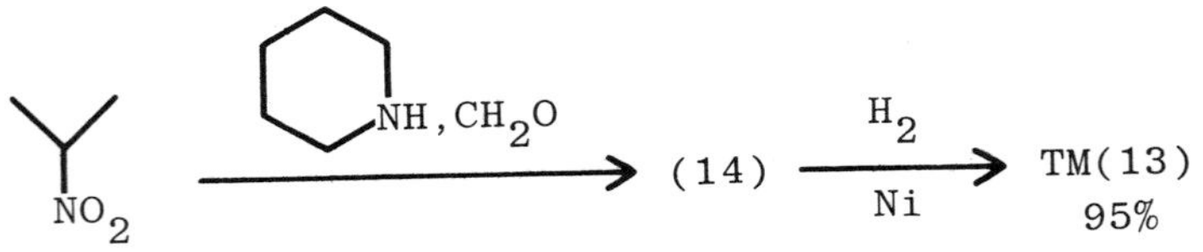

Me_2CHNO_2 $\xrightarrow{\text{piperidine-NH, } CH_2O}$ (14) $\xrightarrow[\text{Ni}]{H_2}$ TM(13) 95%

Example :
The rather restricted range of available aliphatic nitro compounds has been extended by the discovery[249] that oximes of ketones, e.g. (15) can be oxidised to secondary nitro compounds (16) with the per-acid CF_3CO_3H.

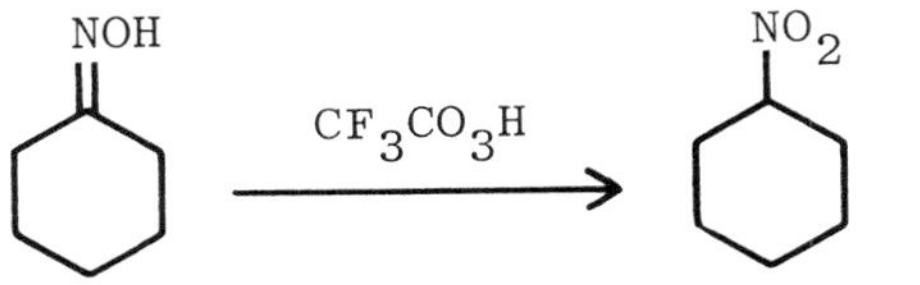

(15) →[CF$_3$CO$_3$H] (16) 62%

Problem :

Suggest a synthesis of spiro-lactam (17), needed to provide an authentic sample for comparison with a rearrangement product.

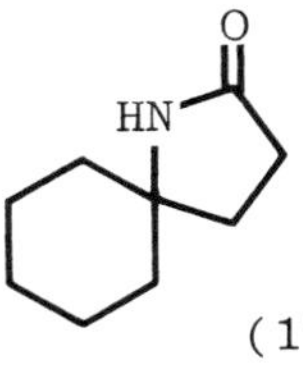

(17)

Answer :

Amide disconnection reveals (18) and FGI (amino to nitro) gives (19) which could be made by Michael addition of nitro compound (16) to an acrylate ester.

Analysis

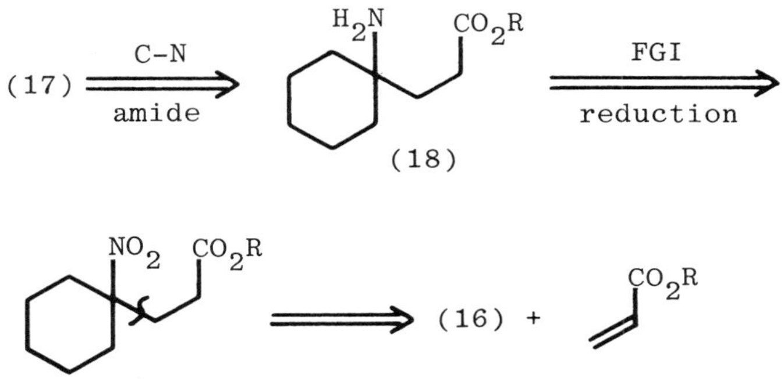

(17) ⟹[C-N, amide] (18) ⟹[FGI, reduction]

(19) ⟹ (16) + CH$_2$=CHCO$_2$R

This sequence gives high yields when hydrogenation over Raney nickel is used. A quaternary ammonium hydroxide was used to catalyse the Michael addition. Cyclisation is spontaneous.

Synthesis[250]

(16) $\xrightarrow[R_4N^+OH^-]{\diagdown CO_2Et}$ [1-nitro-1-(2-ethoxycarbonylethyl)cyclohexane] 92% $\xrightarrow[Ni]{H_2}$ TM(17) 96%

Conversion of Nitro Compounds to Ketones

Example :
Sodium borohydride in methanol selectively reduces the double bond of nitrocompounds (20) while leaving the nitro group intact.[251] This provides another method of synthesising nitro alkanes.

Ar—CH=CH—NO$_2$ $\xrightarrow[MeOH]{NaBH_4}$ Ar—CH$_2$—CH$_2$—NO$_2$

(20)

With this reaction available, a simple synthesis of unsymmetrical dibenzyl ketones (21) can be planned. The carbonyl group can be derived from a nitro group (22) by TiCl$_3$-catalysed hydrolysis (p T 183). Reversing the selective reduction gives (23) and α,β-disconnection separates this into aldehyde (24) and nitro compound (25), available by reduction of (20).

Analysis

$$Ar^1\text{-CO-CH}_2\text{-}Ar^2 \quad (21) \xRightarrow[\text{hydrolysis}]{\text{FGI}} Ar^1\text{-CH}_2\text{-CH(NO}_2\text{)-CH}_2\text{-}Ar^2 \quad (22)$$

$$\xRightarrow[\text{reduction}]{\text{FGI}} Ar^1\text{-CH}_2\text{-C(NO}_2\text{)=CH-}Ar^2 \quad (23) \xRightarrow{\alpha,\beta}$$

$$Ar^2\text{CHO} \quad (24) \quad + \quad Ar^1\text{-CH}_2\text{-CH}_2\text{-NO}_2 \quad (25)$$

One of the examples used was compound (3), made by an improved technique in higher yield, so we can illustrate the method with it.

Synthesis[251]

Piperonal (3,4-methylenedioxybenzaldehyde) $\xrightarrow[\substack{\text{MeNH}_3\text{Cl} \\ \text{KOAc} \\ \text{HC(OMe)}_3 \\ \text{MeOH}}]{\text{MeNO}_2}$ Ar^1-CH=CH-NO$_2$ (3) 94%

$\xrightarrow[\text{MeOH}]{\text{NaBH}_4}$ (25) 82% $\xrightarrow{Ar^2\text{CHO}}$ (23) 75%

Ar^1 = 3,4-methylenedioxyphenyl

Ar^2 = 3-methoxy-4-benzyloxyphenyl (MeO, PhCH$_2$O substituents)

$$\xrightarrow[\text{MeOH}]{\text{NaBH}_4} (22) \xrightarrow[\substack{\text{NaOH} \\ \text{MeOH}}]{\text{TiCl}_3} \text{TM}(21) \quad 67\% \text{ from } (23)$$

Problem :

Suggest a synthesis for cyclic ketone (26) needed as a synthetic intermediate.

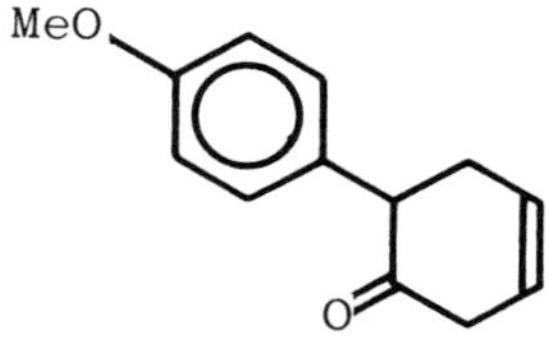

(26) — 2-(4-methoxyphenyl)cyclohex-3-en-1-one

Answer :

If the ketone group comes from a nitro group, a Diels-Alder disconnection is possible to give a simple condensation product (27) from nitromethane and aldehyde (28).

Analysis

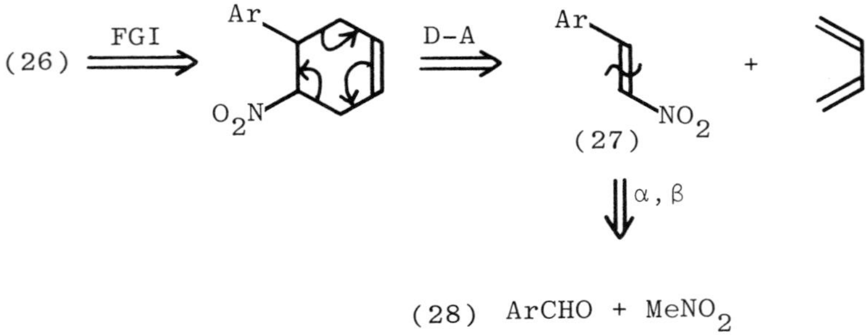

$(26) \xRightarrow{\text{FGI}}$ Ar, O_2N-cyclohexene $\xRightarrow{\text{D-A}}$ Ar-CH=C(NO$_2$)- (27) + diene

$\Downarrow \alpha,\beta$

(28) ArCHO + MeNO$_2$

Various catalysts, mostly amines, have been used[252] to give high yields of (27). In the published synthesis, the Nef reaction was used to transform the nitro group into the ketone (26). No doubt $TiCl_3$ would be at least as good.

Synthesis[252,253]

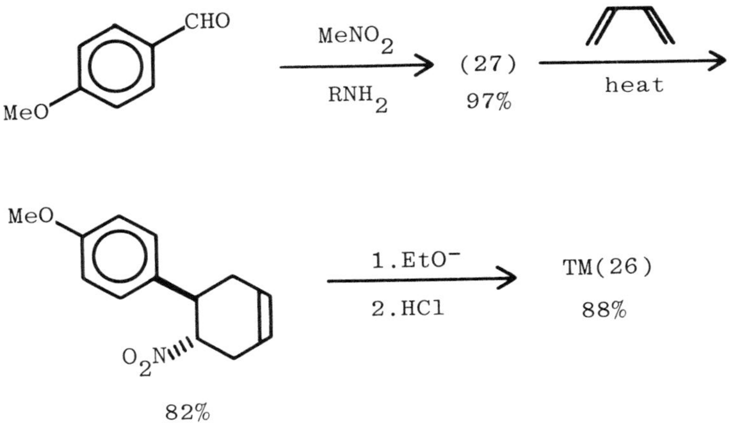

MeO-C₆H₄-CHO $\xrightarrow[RNH_2]{MeNO_2}$ (27) 97% $\xrightarrow{\text{heat}}$

MeO-C₆H₄-(cyclohexene with O₂N) 82% $\xrightarrow[2.HCl]{1.EtO^-}$ TM(26) 88%

CHAPTER 23

Two-Group Disconnections IV: 1,2-Difunctionalised Compounds

(a) *Methods using acyl anion equivalents*
Examples :
In an early attempt[254] to synthesise camphor, diol (1) was an important intermediate. We shall want to disconnect the ring from the chain, so a preliminary disconnection of a methyl group gives an α-hydroxy ketone (2) which can be made from ketone (3) and an acyl anion equivalent.
Analysis

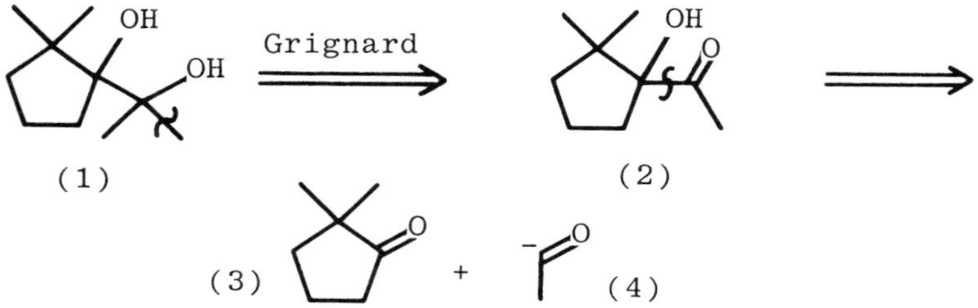

Acetylene anion serves for synthon (4), the hydration with mercury ion catalysis being unambiguous.
Synthesis[254]

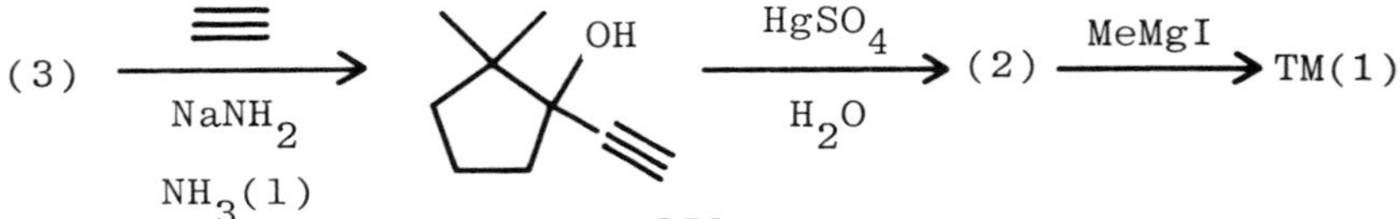

Example 2 :
In a study[255] of stereochemistry, the half ether (5) of an unsymmetrical diol (6) was required. There is little prospect of making (5) from (6) as chemoselectivity presents a formidable problem. Grignard disconnection from the tertiary alcohol would leave α-methoxyl ketones (7) or (8).

Analysis 1

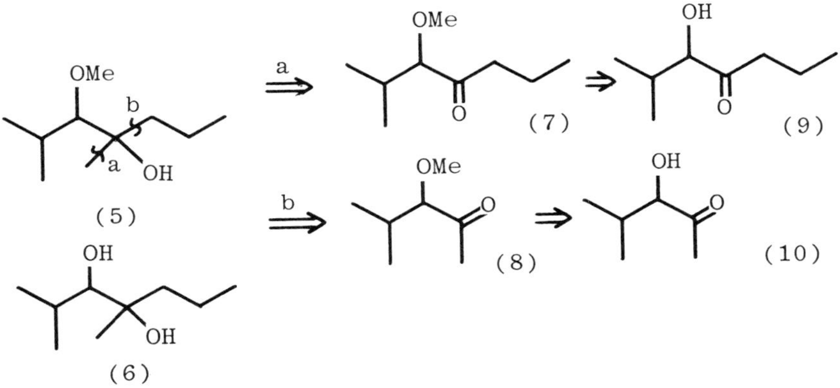

Disconnection of the ether is now a good idea as it leaves an α-hydroxy ketone to be made by acyl anion equivalent addition. We prefer (10) as the acetylene anion can then serve as synthon (4).

Analysis 2

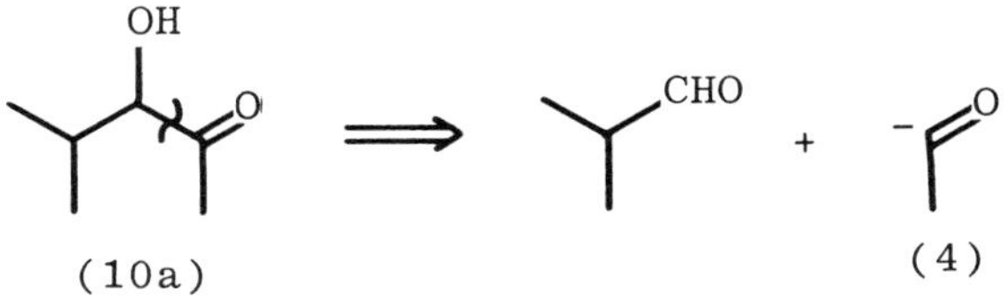

The synthesis goes according to plan with the methylation best completed before hydration of the acetylene.

Synthesis[255]

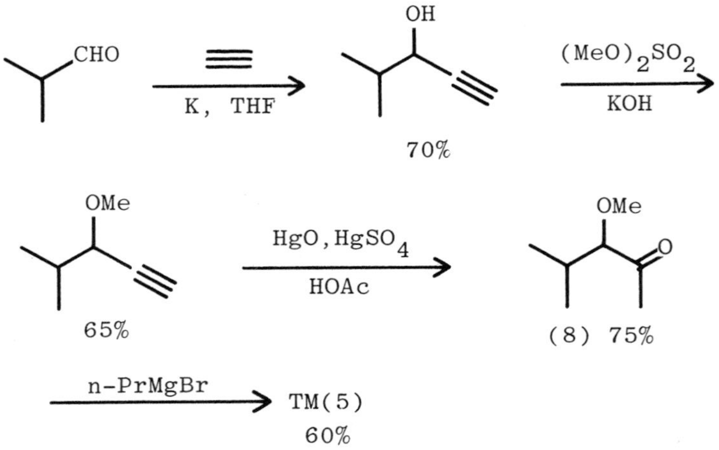

[Scheme: iPr-CHO + HC≡CH / K, THF → iPr-CH(OH)-C≡CH (70%) → (MeO)₂SO₂ / KOH → iPr-CH(OMe)-C≡CH (65%) → HgO, HgSO₄ / HOAc → iPr-CH(OMe)-C(=O)-CH₃ (8) 75% → n-PrMgBr → TM(5) 60%]

Problem :

Suggest a synthesis of compounds of the general class (11).

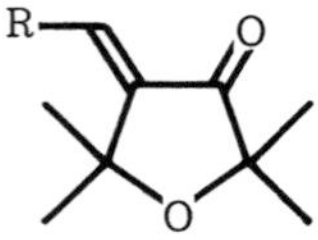

[Structure (11): tetrahydrofuran-3-one with 2,2-dimethyl, 5,5-dimethyl, and 4-(=CHR) exocyclic alkene]

Answer :

Compound (11) contains enone and ether functional groups but only disconnection of the enone will lead to simplification.

Analysis 1

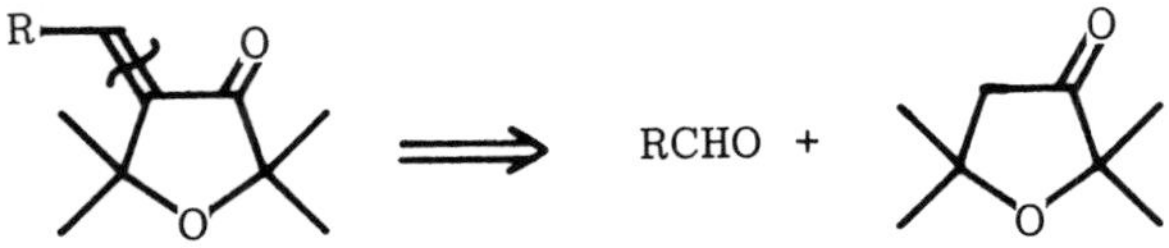

[Retrosynthesis: (11a) ⟹ RCHO + (12)]

(11a)   (12)

Now the ether may be disconnected, diol (13) being the obvious starting material. This contains an α-hydroxy ketone so we might consider disconnecting an acyl anion equivalent from a ketone. If we use an acetylene, the starting material (14) is symmetrical so hydration presents no problems.

Analysis 2

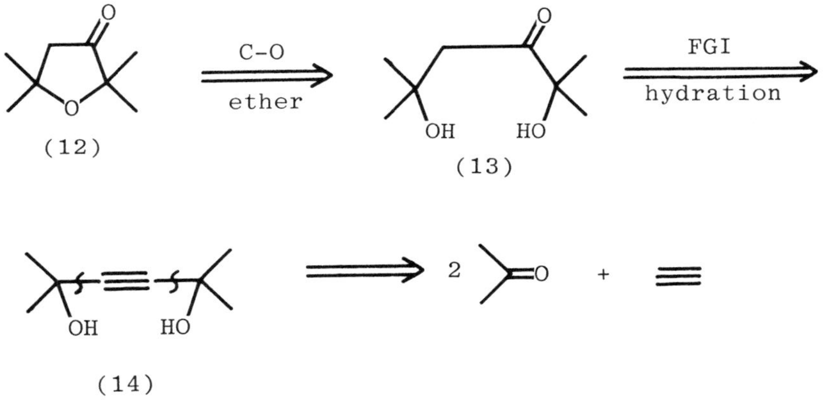

In adding two molecules of acetone to acetylenes, there are no problems of chemoselectivity as the di-anion (15) of the monoadduct reacts preferentially on carbon. Diol (13) cannot be isolated as it cyclises under the hydration conditions.

Synthesis[256]

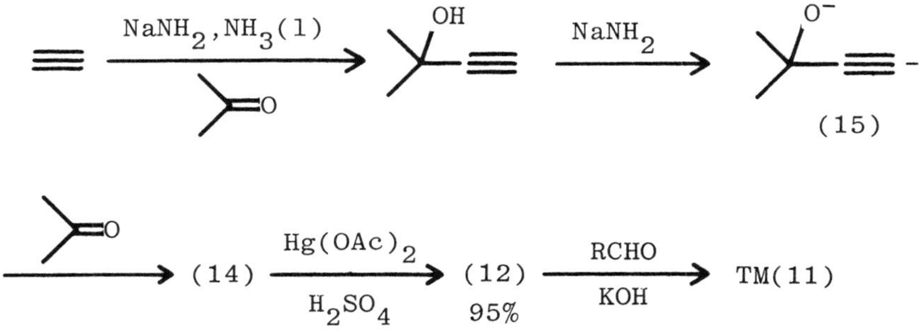

More Advanced Examples :

Reagents are available nowadays for acyl anions other than (4). Thus when Heathcock made the ketone (16), which he used in stereoselective aldol reactions,[257] he needed α-hydroxy ketone (17). This required synthon (18) for which an acetylene is not a good choice as there are as yet no means of controlling the regio-selectivity of hydration of (19).

Analysis

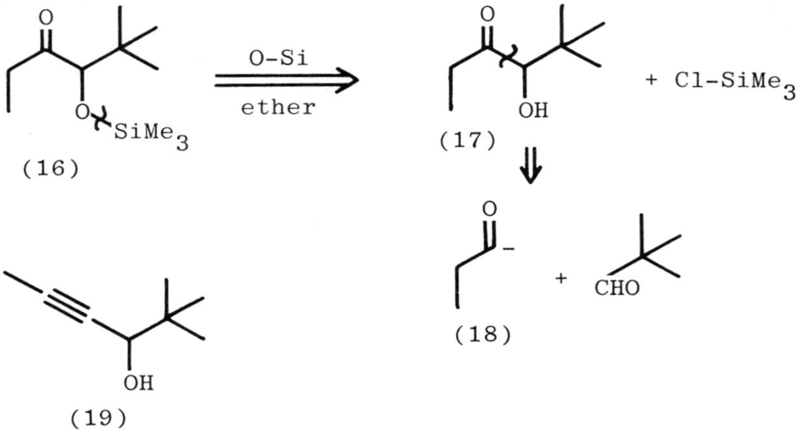

(16) ⇒ (17) + Cl-SiMe₃   (O-Si ether)

⇓

(18) + CHO-tBu

(19)

The anion from vinyl ether (20) (this vinyl anion is stabilised by the inductive effect of the oxygen atom) added to the aldehyde to give (21) which hydrolysed to (17). The anion is a reagent for synthon (18).

Synthesis[257]

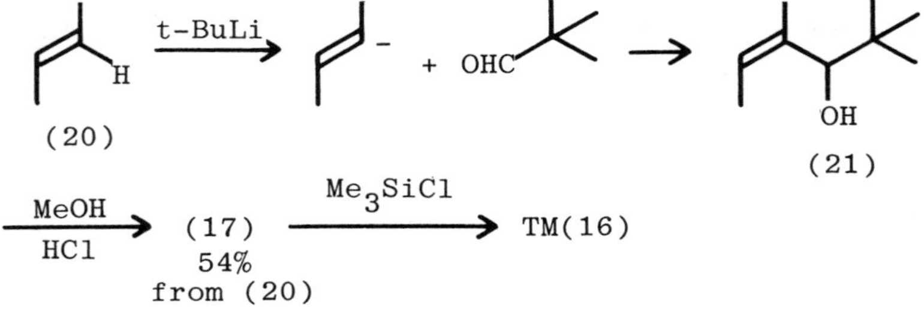

(20) →(t-BuLi)→ [anion] + OHC-tBu → (21)

→(MeOH, HCl)→ (17) 54% from (20) →(Me₃SiCl)→ TM(16)

(b) *Methods from Alkenes*

Examples :

Some insect pheromones are internal ketals. We have already mentioned multistriatin (pp T 2 and 99) and frontalin (p 193). Brevicomin (22) is another example. Disconnection of the ketal gives (23) containing a 1,2-diol. Among other syntheses, hydroxylation of protected enone (24) by epoxidation and acid catalysed rearrangement gives brevicomin stereospecifically.

Analysis

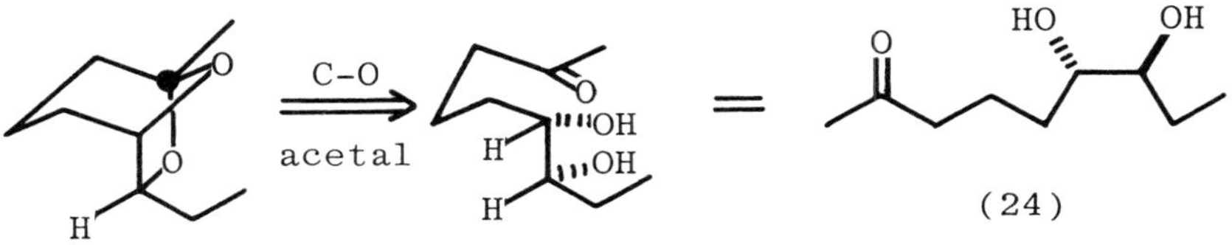

Synthesis[258]

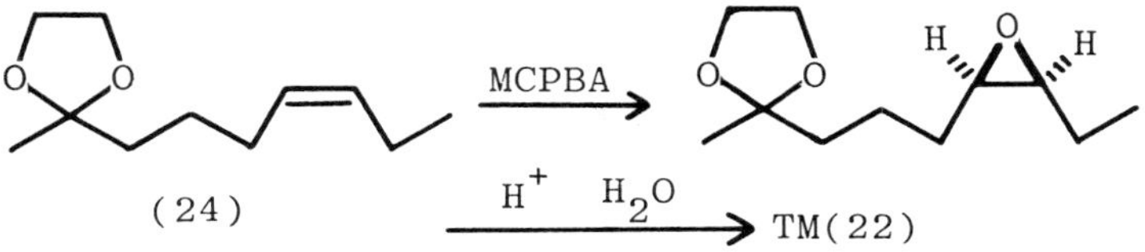

Problem :

Suggest a synthesis of unsymmetrical diol (25) needed to study rearrangement reactions.

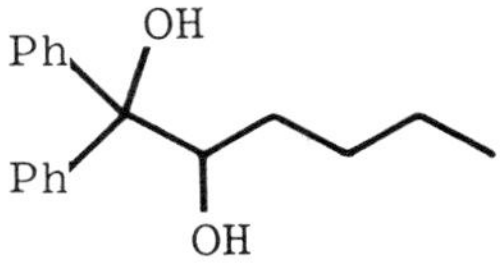

Answer :

Problems of acyl anion equivalents met above in the synthesis of similar TMs disappear if (25) is made from the alkene (26). A Wittig is the obvious method to make (26) and reaction between (27) and Ph_2CO will probably give (26). An alternative is the dehydration of (28), made by Grignard addition to ester (29). Osmium tetroxide was used for the hydroxylation.

Analysis

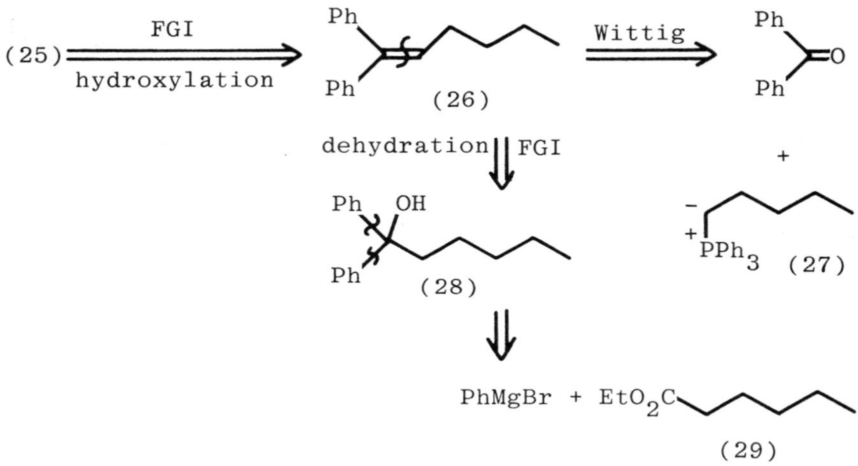

$$(25) \xrightarrow[\text{hydroxylation}]{\text{FGI}} \underset{(26)}{Ph_2C=CH\text{-}C_4H_9} \xrightarrow{\text{Wittig}} Ph_2C=O$$

$$\text{dehydration} \Updownarrow \text{FGI} \qquad + \qquad \overset{-}{\underset{PPh_3}{C_5H_{11}^+}} \text{ (27)}$$

$$\underset{(28)}{Ph_2C(OH)\text{-}C_5H_{11}}$$

$$\Downarrow$$

$$PhMgBr + EtO_2C\text{-}C_5H_{11} \text{ (29)}$$

Synthesis[259]

$$\underset{(29)}{n\text{-}C_5H_{11}CO_2Et} \xrightarrow{PhMgBr} (28) \xrightarrow[Et_2O]{HCl}$$

$$(26) \xrightarrow{OsO_4} TM(25)$$

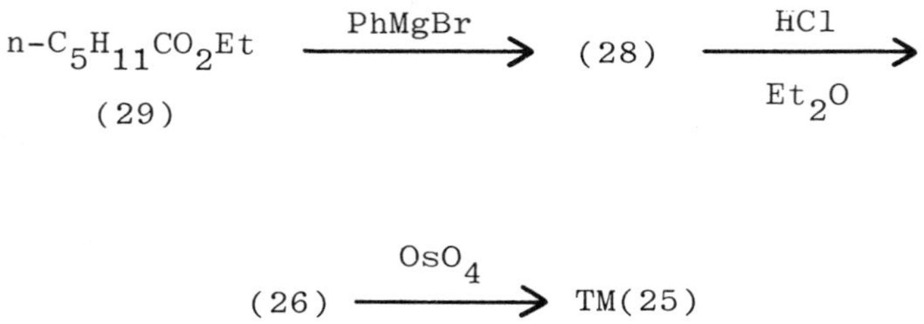

(c) α-*Functionalisation of Carbonyl Compounds*

Example :

Workers studying the cyclisation of acetylenic alcohols[260] decided to make (30). Disconnection of the acetylene leaves α-hydroxy ketone (31). Since this ketone is blocked on one side by the aromatic ring it is reasonable to make (31) from (33) via bromoketone (32).

Analysis

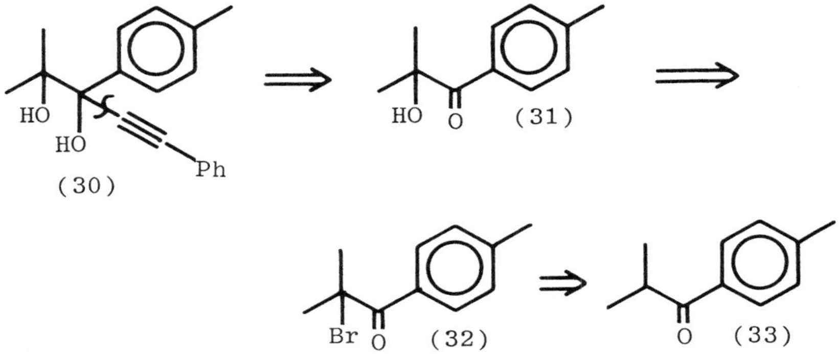

They avoided the obvious Friedel-Crafts synthesis of (33), presumably to avoid the separation of *o* and *p* isomers, and used instead either Grignard route to (34).

Synthesis[260]

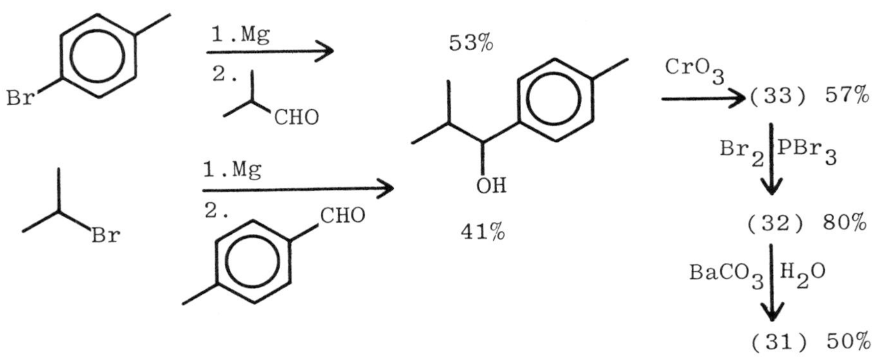

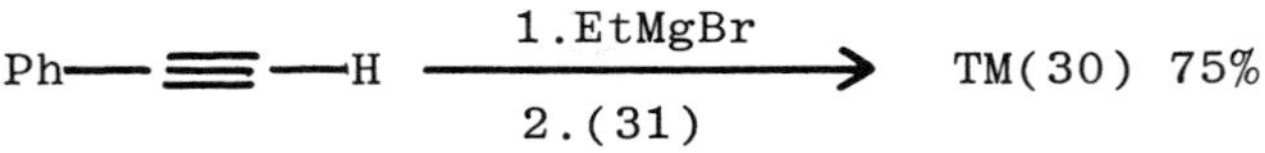

Ph—≡—H $\xrightarrow{\text{1.EtMgBr} \atop \text{2.(31)}}$ TM(30) 75%

Problem :

Devise a synthesis for the beta blocker butoxamine (34).

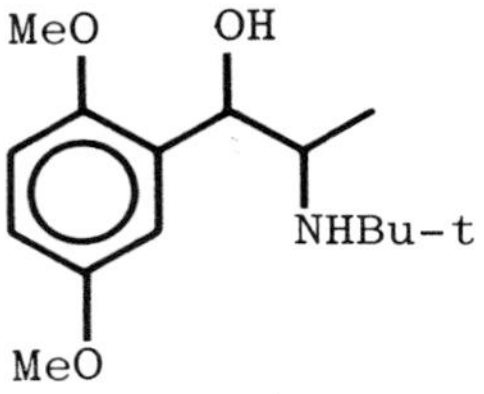

MeO, OH, NHBu-t, MeO (34)

Answer :

By similar argument, bromoketone (36) is a good intermediate. There is no problem of *o,p*-isomers here, so the Friedel-Crafts synthesis is preferred.

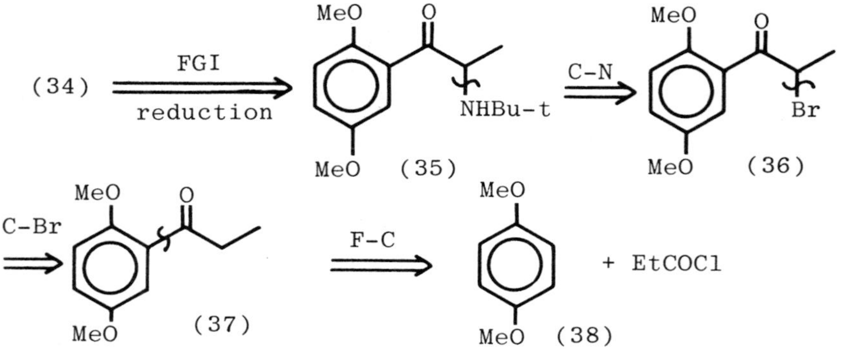

(34) $\xRightarrow{\text{FGI} \atop \text{reduction}}$ (35) [MeO, O, NHBu-t, MeO] $\xRightarrow{\text{C-N}}$ (36) [MeO, O, Br, MeO]

$\xRightarrow{\text{C-Br}}$ (37) [MeO, O, MeO] $\xRightarrow{\text{F-C}}$ (38) [MeO, MeO] + EtCOCl

In practice, the anhydride gave a better Friedel-Crafts reaction. Note that the hindered amine will cause no problems of over-reaction. Only the erythro isomer of (34) is biologically active so this must be separated from the threo which can be recycled by oxidation to (35).

Synthesis[261]

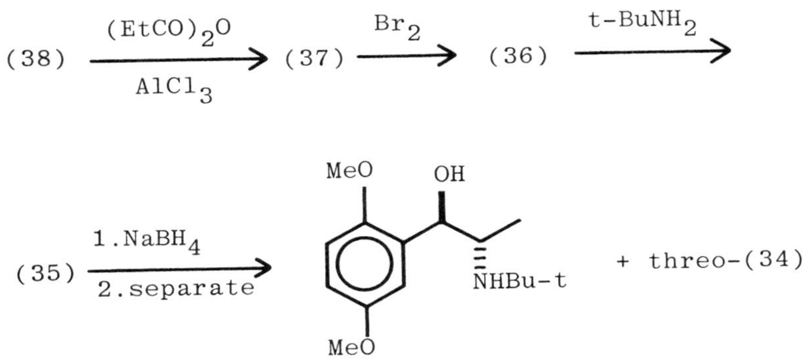

$$(38) \xrightarrow[\text{AlCl}_3]{(\text{EtCO})_2\text{O}} (37) \xrightarrow{\text{Br}_2} (36) \xrightarrow{t\text{-BuNH}_2}$$

$$(35) \xrightarrow[\text{2.separate}]{1.\text{NaBH}_4} \text{erythro-}(34) + \text{threo-}(34)$$

(where erythro-(34) is the 2,5-dimethoxyphenyl compound with CH(OH)–CH(NHBu-t)–CH₃ side chain)

(d) *Strategy of Available Starting Materials*

Example :

The naturally-occurring amino acids are good sources of 1,2-difunctionalised compounds so that tertiary alcohol (40) can easily be made[252] from proline ester (39).

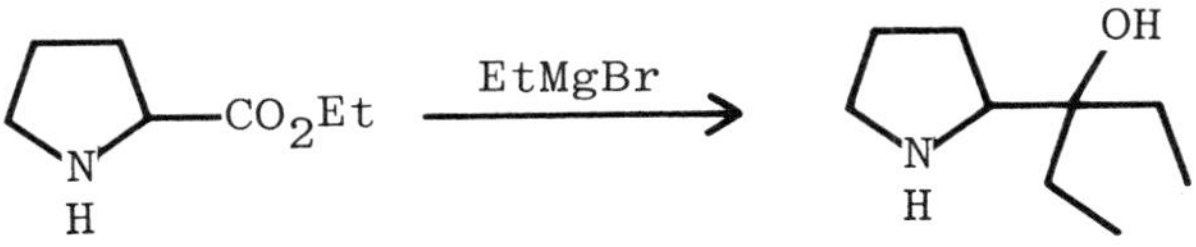

(39) proline-CO₂Et $\xrightarrow{\text{EtMgBr}}$ (40) proline-C(OH)(Et)₂

Problem :

Suggest a synthesis of (41).

$$\underset{(41)}{\text{Ph}\diagup\!\!\diagdown\!\!\diagup\!\!\overset{\text{O}}{\underset{\text{OH}}{\diagdown\!\!\diagup}}}$$

Answer:

Enone disconnection gives hydroxyketone (42) used as a starting material on p T 195.

Analysis

$$\underset{(41a)}{\text{Ph}\diagup\!\!\diagdown\!\!\diagup\!\!\overset{\text{O}}{\underset{\text{OH}}{\diagdown\!\!\diagup}}} \xRightarrow{\alpha,\beta} \text{PhCHO} + \underset{(42)}{\overset{\text{O}}{\underset{\text{OH}}{\diagdown\!\!\diagup}}}$$

Synthesis[263]

$$(42) \xrightarrow[\text{base}]{\text{PhCHO}} \text{TM}(41) \quad 89\%$$

Problem:

Suggest a synthesis of tosylate (43), needed to study nitrogen participation in substitution reactions.

$$\underset{\text{Me}}{\overset{\bigcirc\!\!\!\!\text{N}}{\diagdown\!\!\diagup}}\!\!\diagdown\!\text{OTs} \quad (43)$$

Answer :

The tosylate must come from alcohol (44). Disconnection to an epoxide (45) is no good as the amine will attack the wrong atom. Change of oxidation level to (46) is more hopeful as the α-halo acid (47) is easily made. Another possibility is to use naturally occurring alanine (48).

Analysis

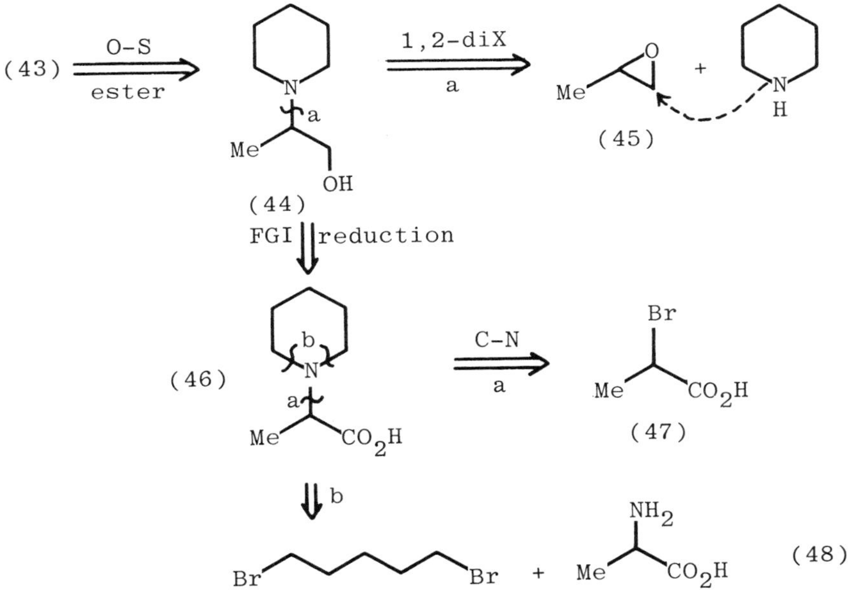

The compound has been made by the last route from alanine.

Synthesis[264]

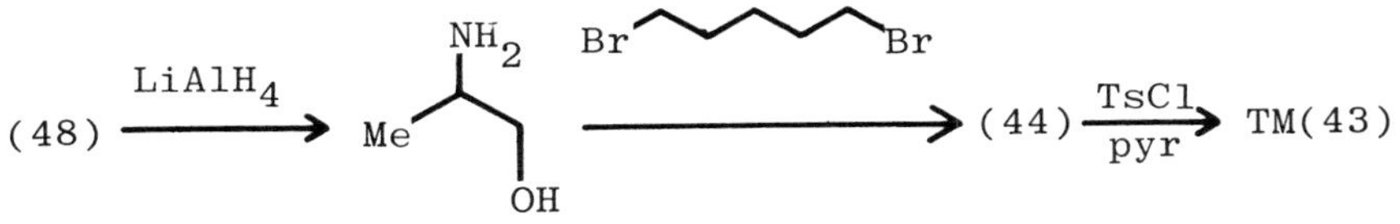

General Examples :

The 1,2-difunctionalised compounds are difficult to classify and in studying a TM it is not always obvious which approach is best. We shall end with two unclassified examples to emphasise that each TM must be taken on its own merits and an individual synthesis designed.

When Pattenden wished to make the penicillium metabolite multicholanic acid (50) he decided anhydride (49) was a key intermediate. The 1,2-difunctionalisation is marked ●.

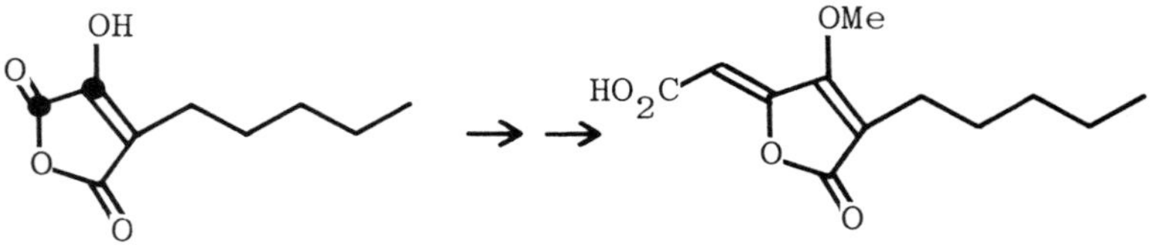

(49)   (50)

Opening the anhydride and writing the enol as a ketone reveals that we are trying to make an α-ketoacid (51). The best disconnection is (51a) which uses the 1,3-dicarbonyl relationship, is next to a branchpoint, and produces a symmetrical starting material (52).

Analysis

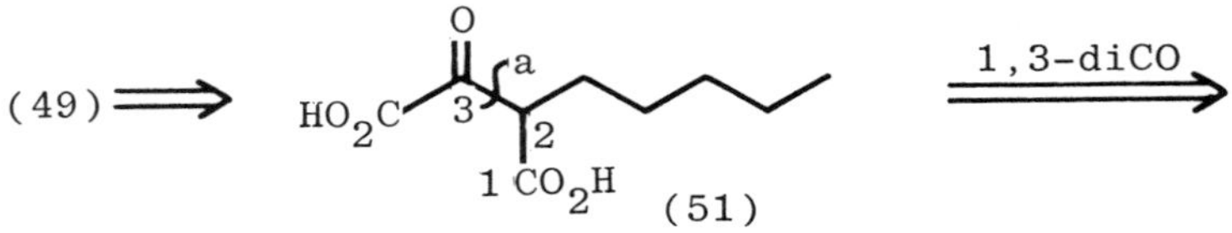

(49) ⟹ [structure (51) with labels 1 CO₂H, 2, 3, a] $\xrightarrow{1,3-diCO}$

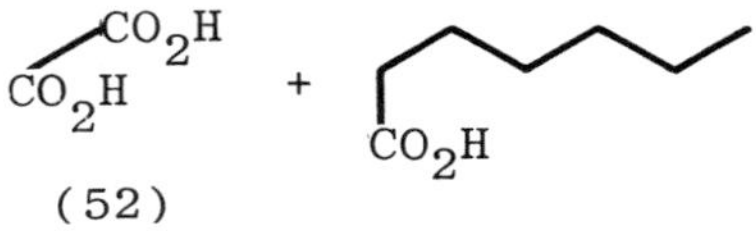

(52)

Naturally we shall need to esterify all the carboxylic acid groupings and we then have an unambiguous condensation between enolisable ester (53) and unenolisable but more electrophilic (α-diCO) diethyl oxalate (54). Hydrolysis of the esters in (55) and cyclisation occur under the same conditions.

Synthesis[265]

[Reaction scheme: hexyl-CO$_2$Et (53) + CH$_2$(CO$_2$Et)$_2$ (54) →[EtO$^-$] EtO$_2$C-CO-CH(CO$_2$Et)-pentyl (55) →[H$_2$SO$_4$] TM(49)]

The synthesis of a local anaesthetic required amino lactone (56) as intermediate. The amino groups could be made from a ketone, but if we use a nitro group instead (57) a reverse Michael disconnection gives a simple condensation product (58).

Analysis

[Scheme: isobenzofuranone with butyl-CH(NH$_2$)- side chain (56) ⇒ (FGI reduction) nitro analogue (57) ⇒ (1,3-diX C-O) o-(CO$_2$H)-C$_6$H$_4$-CH=C(NO$_2$)-butyl (58) ⇒ o-(CO$_2$H)-C$_6$H$_4$-CHO (59) + CH$_3$(CH$_2$)$_3$CH$_2$NO$_2$ (60)]

In practice, the condensation of nitroalkane (60) and aldehyde (59) gives nitro lactone (57) directly. *Synthesis*[266]

$$(59) \xrightarrow[\text{base}]{(60)} (57) \xrightarrow[\text{cat}]{H_2} TM(56)$$

CHAPTER 24

Strategy XI: Radical Reactions in Synthesis: FGA and its Reverse

The mechanism of allylic bromination by N-bromosuccinimide (NBS, 1) (p T 198) is so often given wrongly in textbooks that the correct route is worth repeating here.[267] NBS is usually used in refluxing carbon tetrachloride and the initation step is either thermolysis

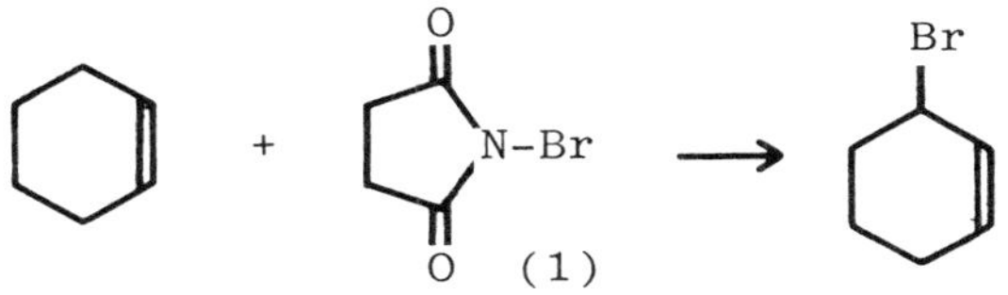

of the N-Br bond or of an added initiator such as dibenzoyl peroxide $(PhCO_2)_2$.

Initiation

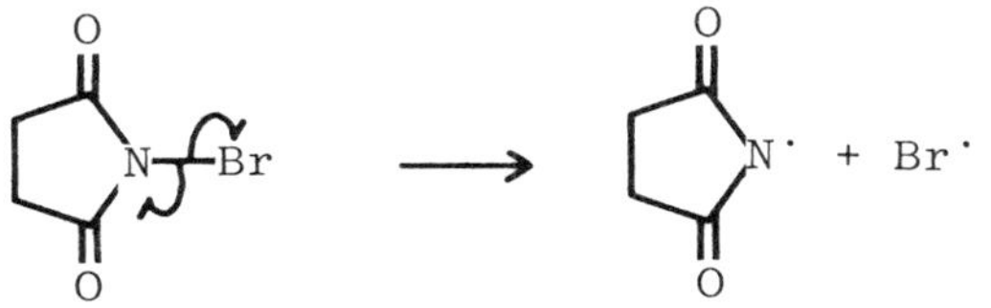

Chain propogation is by hydrogen abstraction by bromine radicals. The weakest C-H bond is broken, that is the allylic bond, to give the most stable radical (2).

Propagation 1

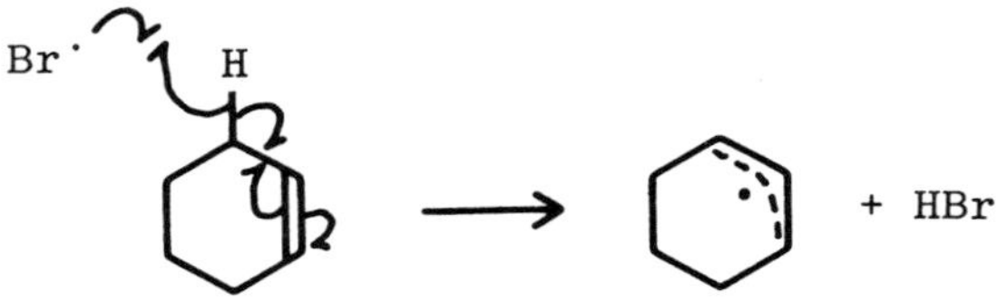

[Br· abstracts H from cyclohexene to give radical (2) + HBr]

(2)

This radical (2) captures molecular bromine to give the product and a new bromine radical so that the cycle continues.

Propagation 2

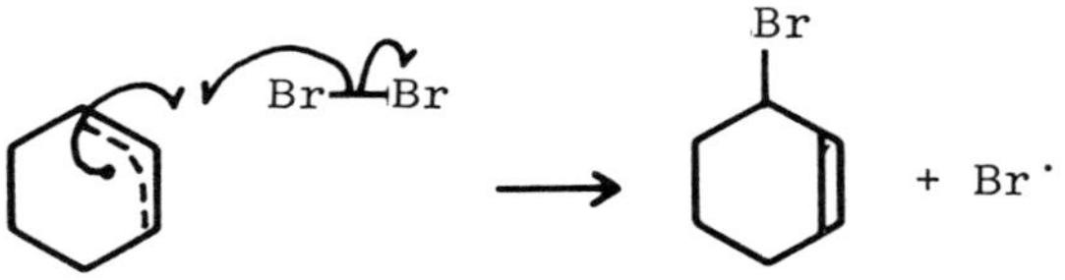

[Radical (2) + Br—Br → 3-bromocyclohexene + Br·]

(2)

The origin of the molecular bromine is at first puzzling. It is formed from NBS by an ionic reaction with the HBr released in propagation 1.

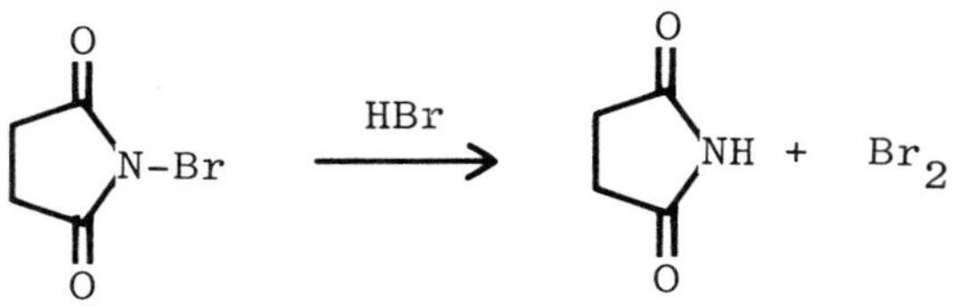

[succinimide N-Br + HBr → succinimide N-H + Br_2]

The amount of bromine produced must therefore be exactly enough to combine with radical (2). Any more and ionic bromination of the double bond would occur.

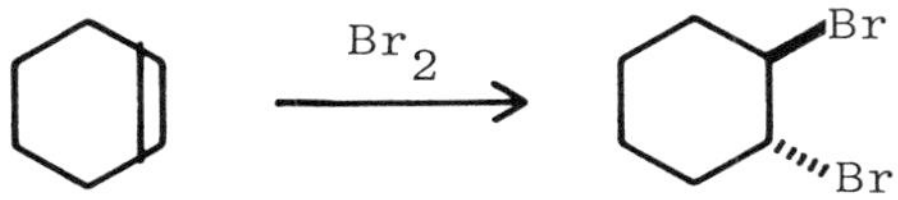

Example :
Allylic bromination of unsymmetrical alkenes may give many products. Occasionally one product is formed in reasonable yield, e.g. (3), but this is a matter for trial and error.
Synthesis[268]

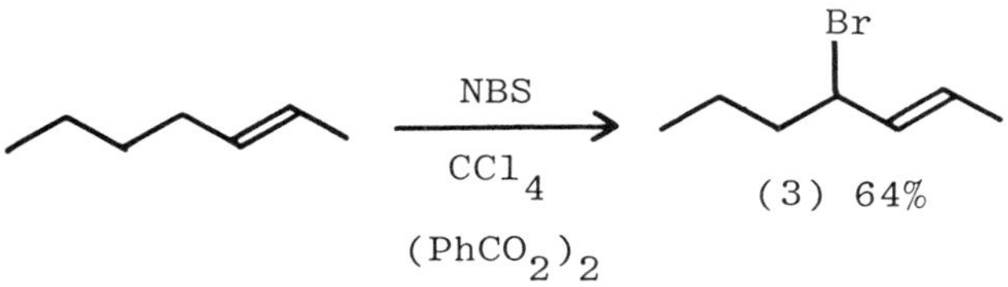

Problem :
For syntheses of carotenoids (4), where R is a large group, disconnection (a) achieves the most simplification. How would you make the phosphonium salt (5) needed[269] for this strategy?

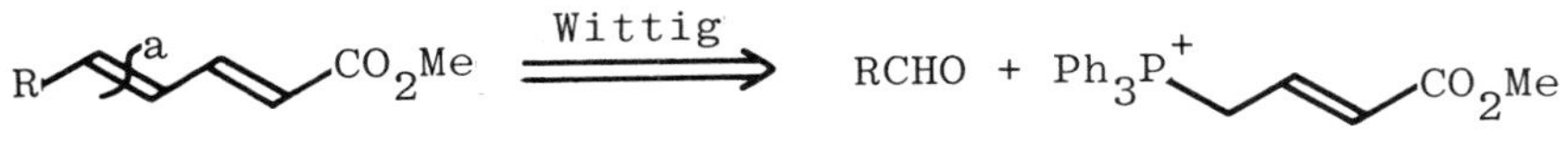

Answer :
Bromide (6) is clearly needed and can be made[270] by allylic bromination of ester (7). Crotonic acid (8) is

available or can be made by a malonic acid condensation.[271]

Analysis

$Ph_3\overset{+}{P}$∨∨CO_2Me ⟹ Br∨∨CO_2Me ⟹
(5a)   (6)

∨∨CO_2Me ⟹ ∨∨CO_2H ⟹ MeCHO + $^-CH_2CO_2H$
(7)   (8)   =malonate

The Knoevenagel method (p T 161) gives high yields of crotonic acid. The allylic bromination uses NBS in CCl_4 and the ylid (9) is stable enough to be isolated.

Synthesis[269-271]

MeCHO $\xrightarrow[\text{pyr,pip}]{CH_2(CO_2H)_2}$ (8) $\xrightarrow[H^+]{MeOH}$ (7) $\xrightarrow[CCl_4]{NBS}$
            75%

(6) $\xrightarrow{Ph_3P}$ TM(5) $\xrightarrow{NaOH}$ $Ph_3\overset{+}{P}$∨∨CO_2Me
              83%   (9) 71%

Benzylic Halogenation. Example:
The starting materials (10) and (11) for the Palanil synthesis described on page T 123 are both made by benzylic halogenation.

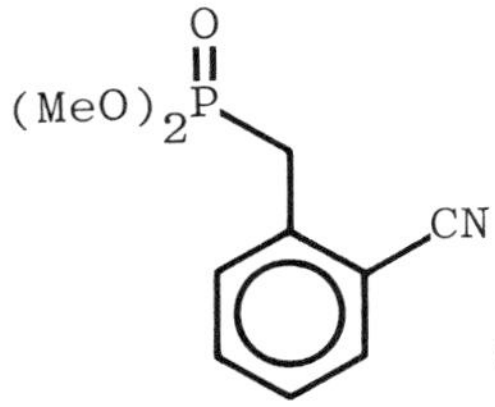

(MeO)₂P(=O)-CH₂-C₆H₄(o-CN) (10)

OHC-C₆H₄-CHO (para) (11)

Phosphonate ester (10) is made from chloride (12) available by direct chlorination with photochemical generation of radicals.

Synthesis[272]

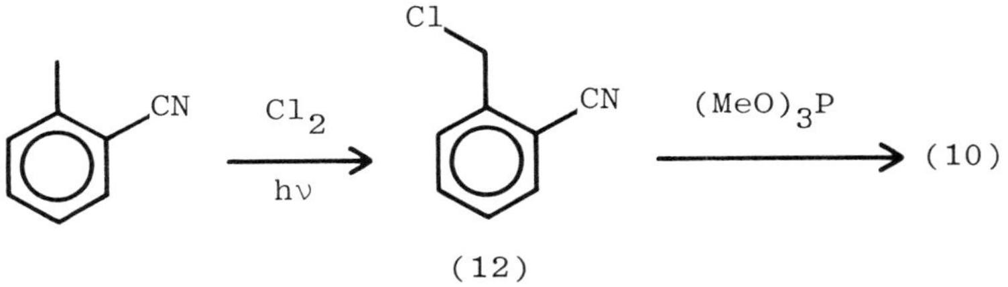

o-tolunitrile →[Cl₂, hν] (12) →[(MeO)₃P] (10)

Dialdehyde (11) is not such an obvious halogenation product, but bromination of p-xylene can be controlled to give a reasonable yield of tetrabromide (13) which is hydrolysed to (11) in acid.

Synthesis[273]

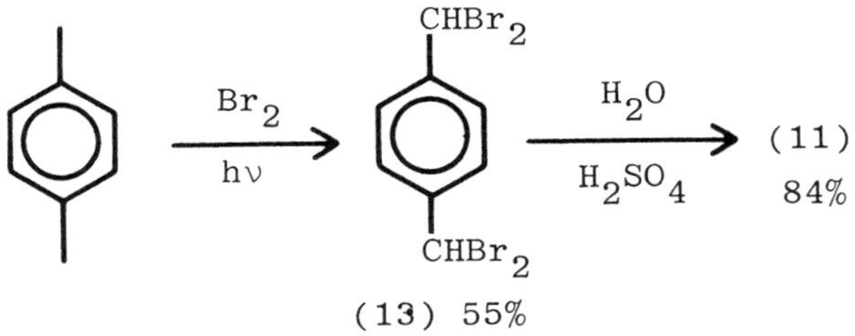

p-xylene →[Br₂, hν] (13) 55% [p-C₆H₄(CHBr₂)₂] →[H₂O, H₂SO₄] (11) 84%

Problem:
Substituted styrenes, such as acid (14) are polymerised with styrene itself to make resins with functional groups

attached.[274] Suggest a synthesis for (14).

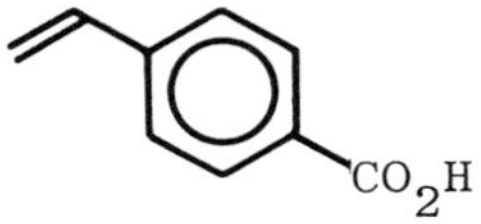

$$\text{(14)} \quad \text{CH}_2=\text{CH-C}_6\text{H}_4\text{-CO}_2\text{H}$$

Answer :

The Wittig reaction can be used for the double bond and with benzylic bromination in mind we prefer phosphonium salt (15), bromide (16), and hence available acid (17) as starting materials.

Analysis

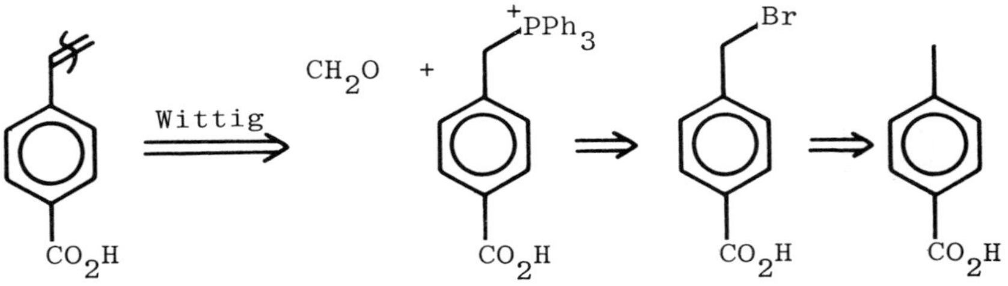

(14a) $\xRightarrow{\text{Wittig}}$ CH$_2$O + (15) [4-(Ph$_3$P$^+$CH$_2$)-C$_6$H$_4$-CO$_2$H] $\Rightarrow$ (16) [4-(BrCH$_2$)-C$_6$H$_4$-CO$_2$H] $\Rightarrow$ (17) [4-CH$_3$-C$_6$H$_4$-CO$_2$H]

Bromination of (17) with NBS and peroxide gives (16). The Wittig reaction can be carried out with the CO$_2$H group unprotected using enough base to make the anion (18).

Synthesis[275]

$$(17) \xrightarrow[\text{CCl}_4]{\text{NBS},\ (\text{PhCO}_2)_2} (16) \xrightarrow[\text{Me}_2\text{CO}]{\text{Ph}_3\text{P}} (15) \xrightarrow[\text{H}_2\text{O}]{\text{NaOH}}$$

66%

(18) [structure: benzene ring with CH₂⁺PPh₃ and CO₂⁻ substituents] $\xrightarrow{CH_2O}$ TM(14)

Oxidation of Allylic Positions
Example :
Introduction of a carbonyl group next to a double bond seems to be affected by steric factors since the cycloheptene (19) gives only ketone (20) on allylic oxidation with SeO_2.
Synthesis[276]

(19) $\xrightarrow[EtOH]{SeO_2}$ (20)

C-C Bond Formation
Radical Dimerisation
Example :
We make 1,6-dicarboxylic acids and their derivatives in the laboratory by cleaving cyclohexenes (Chapter 27) and this was the industrial method too. The need for vast quantities of amine (21) for nylon manufacture suggested dinitrile (22) as a starting material and, as this is symmetrical, coupling of radicals (23) must be a possible synthesis. Acrylonitrile (24) is a cheap starting material and might give (23) under the right conditions.

Analysis

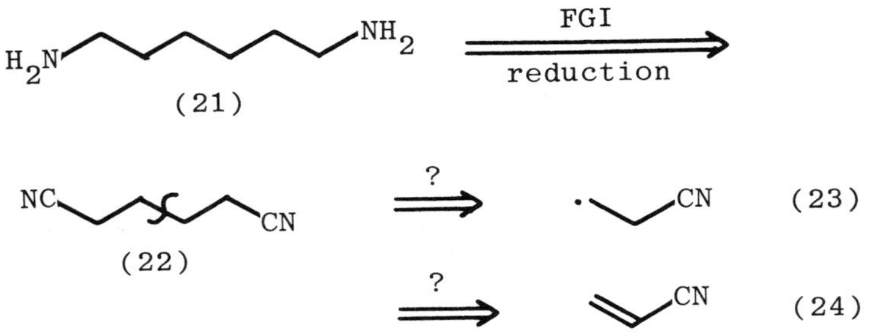

$H_2N\frown\frown\frown NH_2$ (21) $\xrightarrow[\text{reduction}]{\text{FGI}}$

$NC\frown\frown CN$ (22) $\Longrightarrow$ $\cdot\frown CN$ (23)

$\Longrightarrow$ $\diagdown CN$ (24)

These ideas were developed into a working industrial process at Monsanto.[277]

The final method was an electrochemical reductive dimerisation working extremely efficiently and capable of dimerising many electron-deficient olefins.

Synthesis[277]

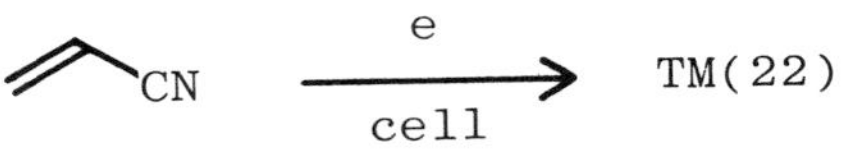

$\diagdown CN \xrightarrow[\text{cell}]{e} TM(22)$

Problem :
Suggest some ideas for a radical synthesis of diacid (25).

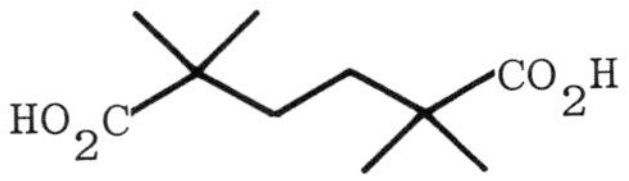

$HO_2C\diagup\diagdown\diagup CO_2H$

(25)

Answer :
The molecule is symmetrical so we clearly need a supply of radical (26). This cannot be made from an olefin, but

acid (27) looks a good starting material as all its nine CH bonds are identical so some radical abstraction reaction might be found to remove H· and leave (26).
Analysis

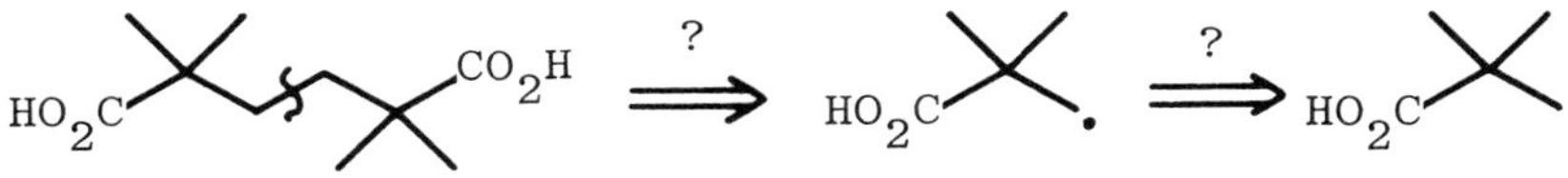

$$(25a) \implies (26) \xrightarrow{?} (27)$$

Experiments showed that Fe(II) catalysed oxidation of (27) with hydrogen peroxide gave a modest yield of (25), presumably *via* (26).
Synthesis[278]

$$(27) \xrightarrow[\substack{H_2SO_4 \\ H_2O}]{\substack{Fe(II) \\ H_2O_2}} TM(26) \quad 37\%$$

Example :
Pinacols, e.g. (28), are much easier to make as the two hydroxyl groups come from reductive dimerisation of carbonyl groups (p T 201). A modern method uses V(II) as the reducing agent.[279]

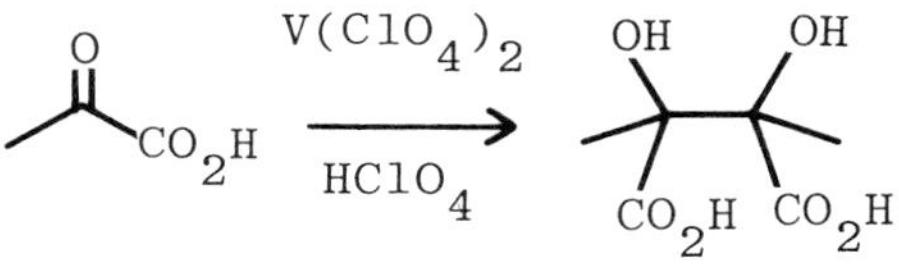

$$\text{MeCOCO}_2\text{H} \xrightarrow[HClO_4]{V(ClO_4)_2} (28)$$

(28) 80%

Problem :

Suggest a synthesis for (29), needed to explore geometrical constraints on Diels-Alder reactions.

(29)

Answer :

The compound is an acetal and disconnection of this reveals pinacol (30), made from available acrolein.

Analysis

$$(29) \underset{\text{acetal}}{\overset{1,1\text{-diO}}{\Longrightarrow}} \text{(30)} \overset{\text{pinacol}}{\Longrightarrow} \text{CHO} + \text{CHO}$$

(30)

Zinc was used as the metal for the dimerisation. Diol (30) is formed as a mixture of diastereoisomers, but this is unimportant as (29) was isomerised in strong base to (31). This compound refused to undergo any Diels-Alder reactions!

Synthesis[280]

CHO $\xrightarrow[\text{HOAc}]{\text{Zn, EtOH}}$ (30) $\xrightarrow[\text{CH}_2\text{Cl}_2]{\text{TsOH}}$ TM(29)
                         78%

$\xrightarrow{\text{KOBu-t}}$ (31)

Example:
The acyloin reaction is so useful for making medium and large rings that cyclic ketones are often made this way. Reduction of acyloin (32) with zinc in acid solution or HI removes the unwanted hydroxyl group to give cyclodecanone (33) in good yield.
Synthesis[281]

[cyclodecane-1,2-dicarboxylic acid diethyl ester (with CO₂Et, CO₂Et groups)] → [2-hydroxycyclodecanone (32)] $\xrightarrow{HI}$ [cyclodecanone (33) 78%]

Hydrocarbon Synthesis

The structure of alkanes can be difficult to establish as their NMR spectra are difficult to interpret. The stereochemistry as well as the structure of (34) was in doubt, but synthesis by three different routes established both.

[2,6-dimethylheptane (34)]

Natural products citronellal (35) and linalool (36) have the same skeleton as (34). Hydrogenation of linalool gave alcohol (37). Dehydration and hydrogenation would be the obvious way to make (34) from (37) but an alternative was used here. The halide (38) was reduced with sodium in liquid ammonia.

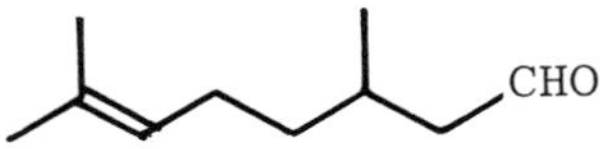
CHO

(35)

Synthesis 1[282]

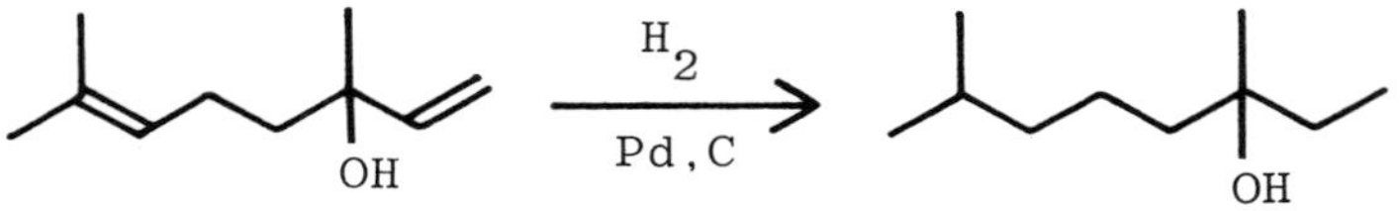
OH
H2
Pd,C
OH

(36) (37)

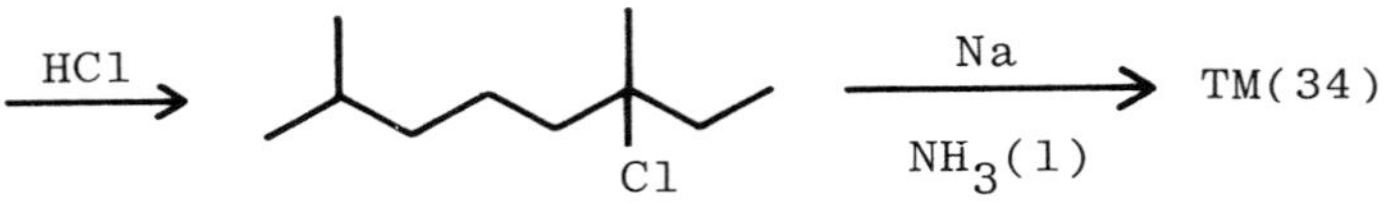
HCl
Cl
Na
NH3(l)
TM(34)

(38)

The product did indeed have the same structure as (34) but the stereochemistry was still unknown. If a carbonyl group is added (FGA) to give ketone (39), disconnection via standard Grignard routes to available optically active (40) is possible.

Analysis

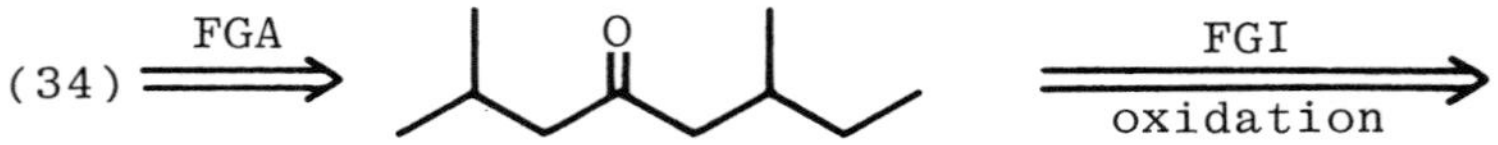
(34)
FGA
O
FGI
oxidation

(39)

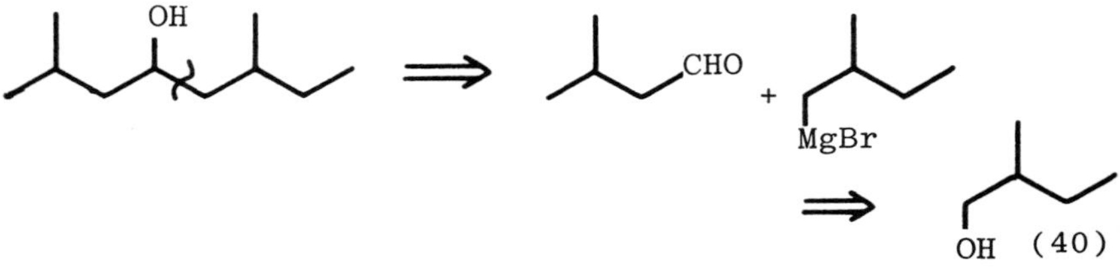
OH
CHO
+
MgBr
OH
(40)

Synthesis by this route from (-)-(S)-(40) gave (34) with the same rotation as the original sample. The carbonyl group in (39) was removed by the Wolf-Kishner method (table T 24.1).

Synthesis[282]

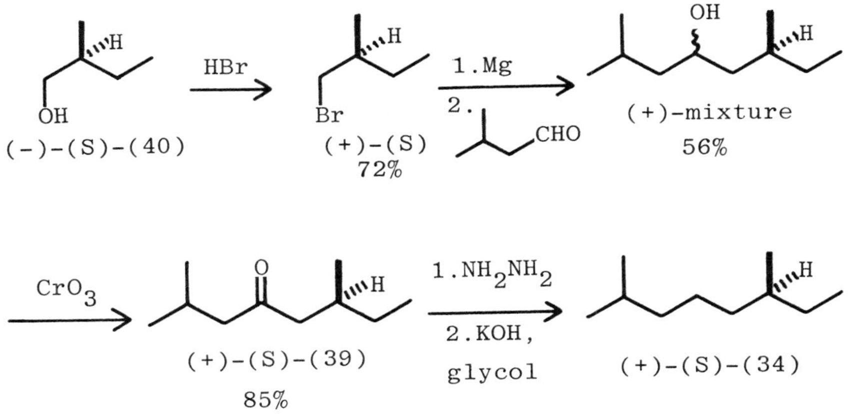

(-)-(S)-(40) →[HBr] (+)-(S)-Br 72% →[1. Mg; 2. CHO-CH(CH₃)₂] (+)-mixture 56% (OH compound)

→[CrO₃] (+)-(S)-(39) 85% →[1. NH₂NH₂; 2. KOH, glycol] (+)-(S)-(34)

Problems :
How would you convert citronellal (35) into (34)?

Answer :
Hydrogenation and Wolf-Kishner reduction in either order, e.g.:

Synthesis[282]

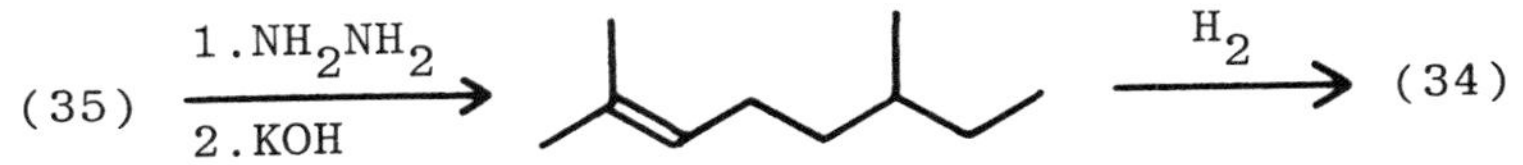

(35) →[1. NH₂NH₂; 2. KOH] (alkene intermediate) →[H₂] (34)

Problem :
Suggest a synthesis of (41) to provide an authentic specimen for g.l.c. comparison with oil fractions.

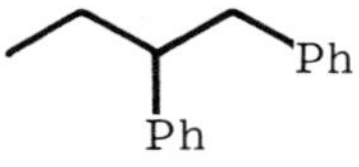

(41) [structure]

Answer :

Hydrogenation of an alkene such as (42) or (43) is a good method, and these could both be derived from branchpoint alcohol (44).

Analysis

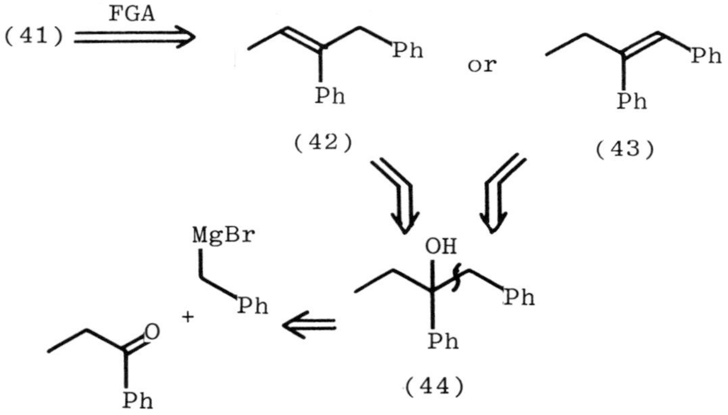

(41) ⟹ᶠᴳᴬ (42) or (43) ⟹ (44) ⟹ PhCOEt (ketone) + PhCH₂MgBr

In the dehydration of (44) in acid, a mixture of alkenes was formed which was hydrogenated directly to (41).

Synthesis[108]

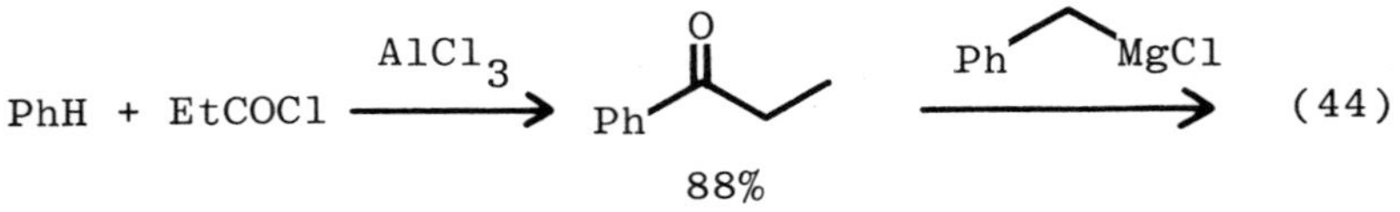

PhH + EtCOCl $\xrightarrow{AlCl_3}$ PhCOEt (88%) $\xrightarrow{PhCH_2MgCl}$ (44)

(44) $\xrightarrow{H_2SO_4}$ alkenes $\xrightarrow[\text{Raney Ni}]{H_2}$ TM(41)

If you preferred to add a carbonyl group, the most obvious place is next to the benzene ring (45).

Analysis

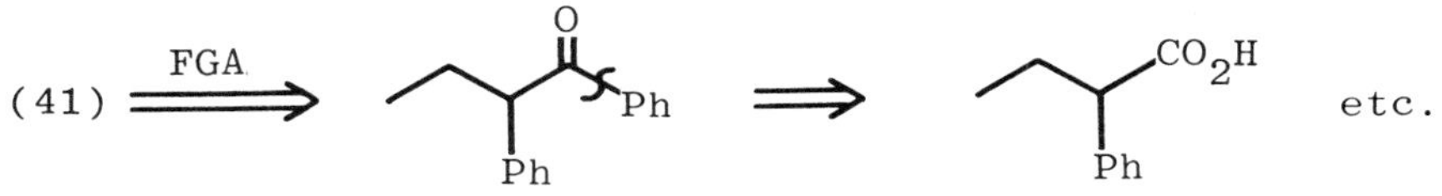

(41) $\xrightarrow{FGA}$ [structure: CH$_3$CH$_2$CH(Ph)C(O)Ph] $\Rightarrow$ [structure: CH$_3$CH$_2$CH(Ph)CO$_2$H] etc.

There are many other approaches based on FGA.

More Advanced Problem :

Suggest a synthesis of the oestrogen benzoestrol (45).

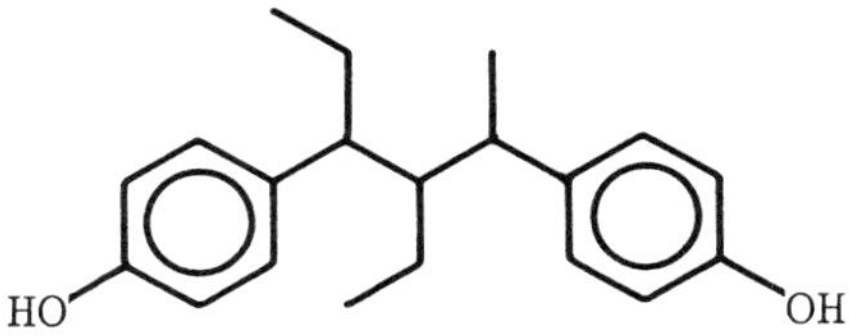

(45)

Answer :

Alkenes (46) or (47) could be hydrogenated to (45) and these could be made by Wittig reactions or from the three branchpoint alcohols (48 - 50).

Analysis 1

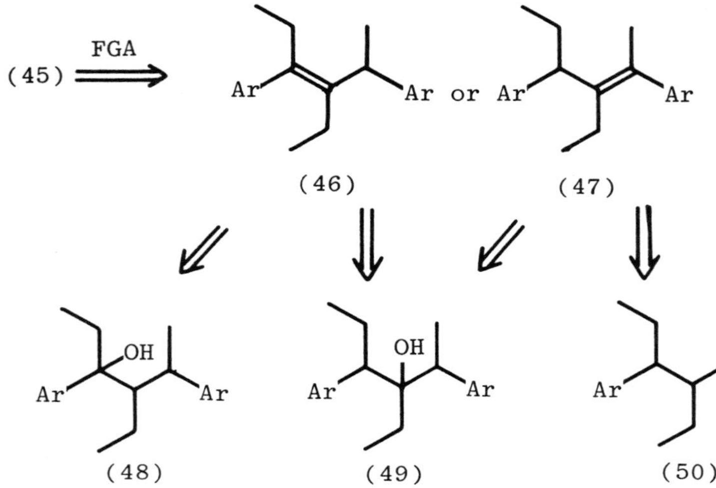

(45) $\xrightarrow{FGA}$ (46) or (47)

⇓

(48)   (49)   (50)

There are many possibilities and I shall simply analyse the published synthesis. Removal of the methyl group from (50) gives ketone (51) which can be made from enone (52) by Michael addition of an ethyl group. Unambiguous cross-condensation between enolisable (54) and reactive (53) gives (52).

Analysis 2

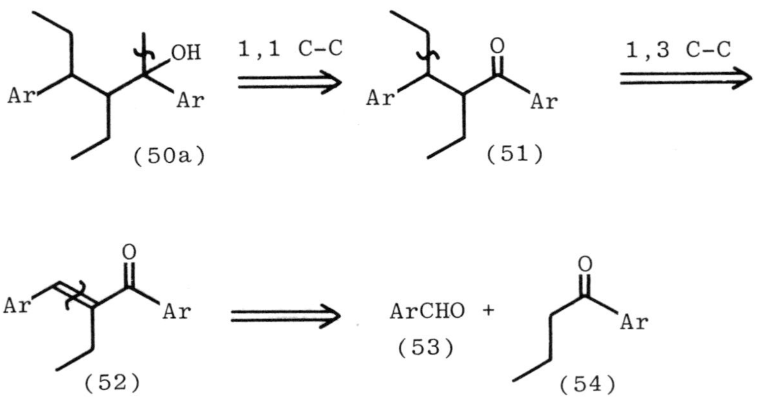

(50a) $\xrightarrow{1,1 \text{ C-C}}$ (51) $\xrightarrow{1,3 \text{ C-C}}$

(52) ⇒ ArCHO + (54)
      (53)

No copper is necessary for the Michael addition. Alcohol (50) is dehydrated to a mixture of alkenes, all giving (45) on hydrogenation. The two phenol groups are protected throughout as their methyl ethers. Synthesis[283]

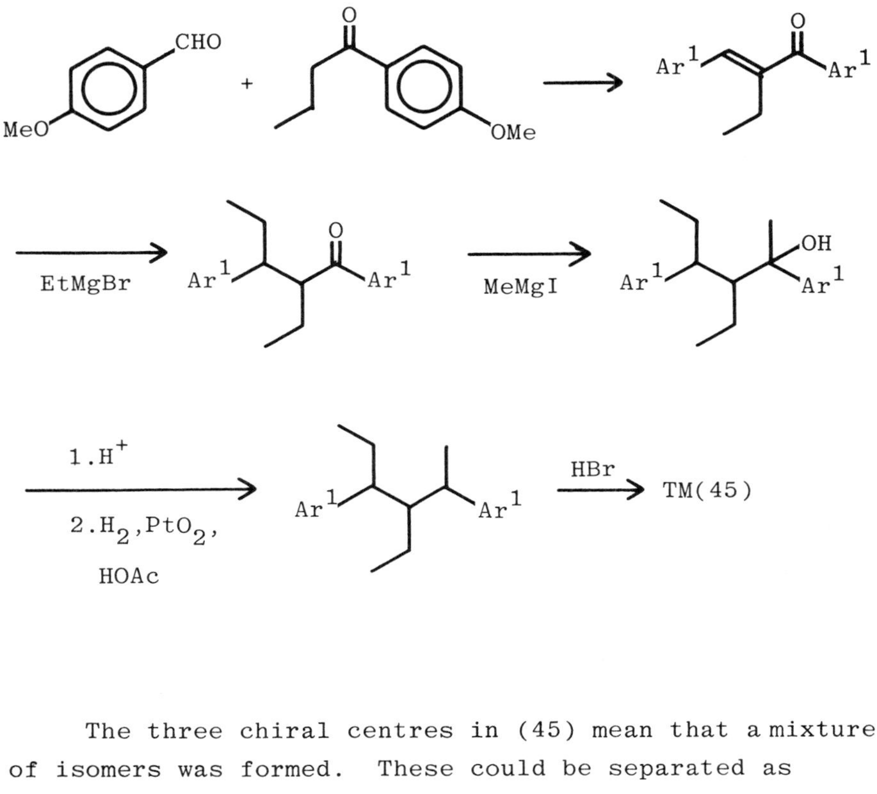

The three chiral centres in (45) mean that a mixture of isomers was formed. These could be separated as their crystalline benzoates.

CHAPTER 25

Two-Group Disconnections V: 1,4-Difunctionalised Compounds

(a) *Methods using unnatural electrophilic synthons*
Example :
Chemists wishing to study the thermal rearrangement of (1) - it gives (2) by a [3,3] sigmatropic shift (p T 288) - decided it could be made by a double Wittig reaction on (3).
Analysis

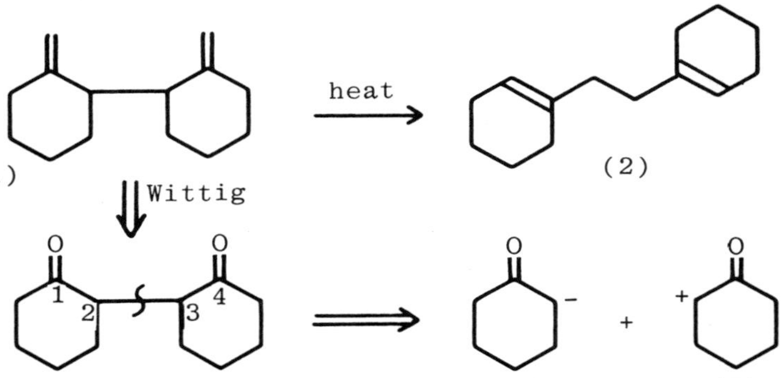

This is a 1,4-diketone and disconnection of the central bond separates the two rings. We require a specific enol equivalent for (4) - they used activated ketone (6) - and a reagent for unnatural synthon (5) - they used α-chloroketone (7).

Synthesis[284]

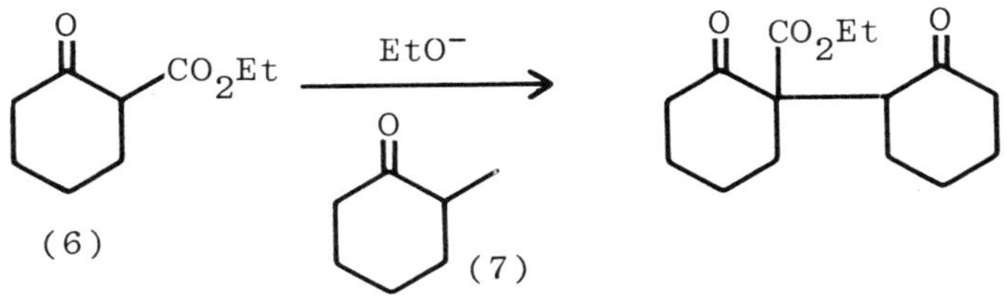

[Scheme: compound (6) 2-carbethoxycyclohexanone + EtO⁻, with 2-methylcyclohexanone (7) → coupled product; then H⁺, H₂O → (3); then Ph₃P=CH₂ → TM(1), 54%]

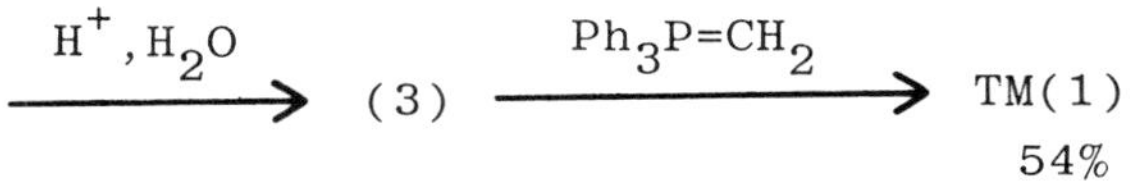

Problem :
Suggest a synthesis for spiroketone (8), an intermediate in a synthesis of acorone.

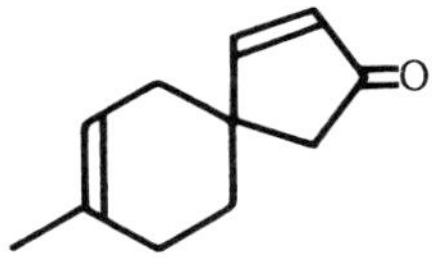

[Structure (8)]

Answer :
Enone disconnection (8a) reveals 1,4-dicarbonyl compound (9) best disconnected at the central bond (a) to sever the longer chain from the ring. Aldehyde (11) is a Diels-Alder product.

Analysis

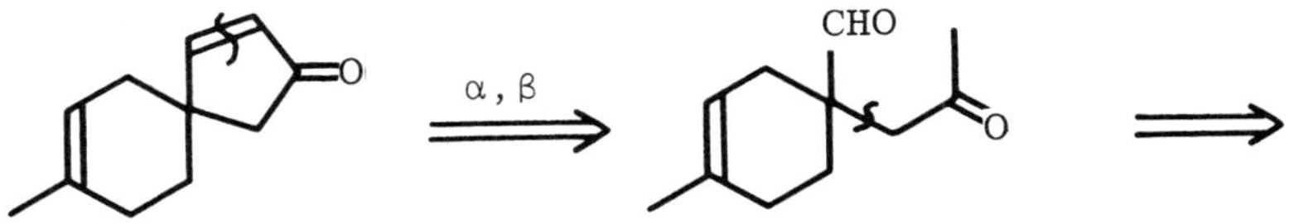

[(8a) ⟹ α,β (9) ⟹]

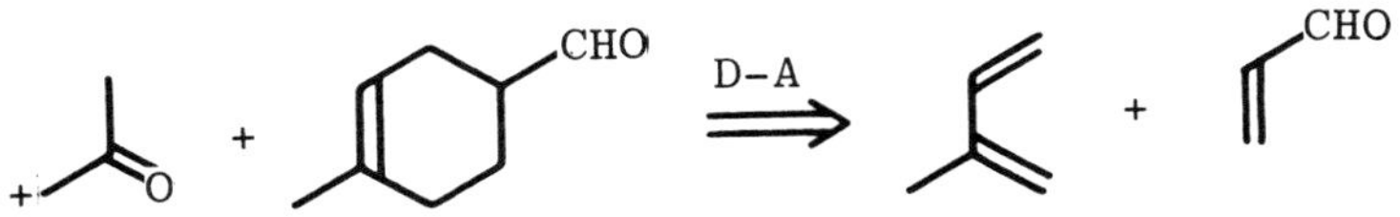

[(10) + (11) ⟹ D-A diene + acrolein CHO]

Reagent (12) was used for synthon (10), though no doubt bromoacetone would also add to the enamine of (11). Mercury catalysed hydrolysis of vinyl chloride released (9) which duly cyclised to (8) in base.

Synthesis[285]

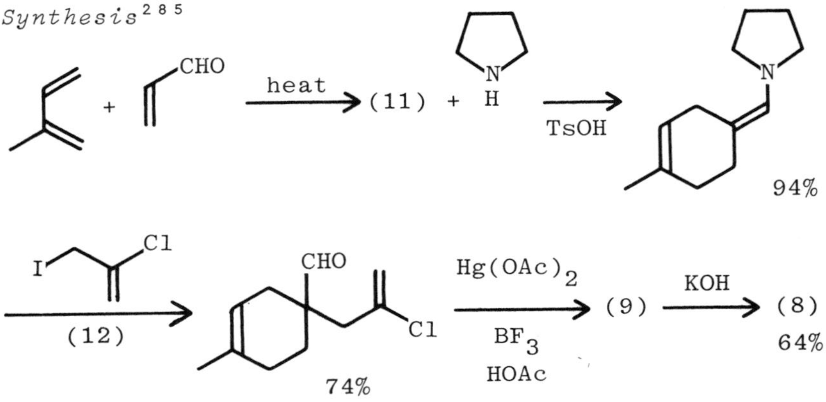

Use of Epoxides : Examples :
Lactone (13) is clearly derived from (14) : branchpoint disconnection gives acid (15) and synthon (16). Epoxide (17) has the right regioselectivity for this synthon.

Analysis

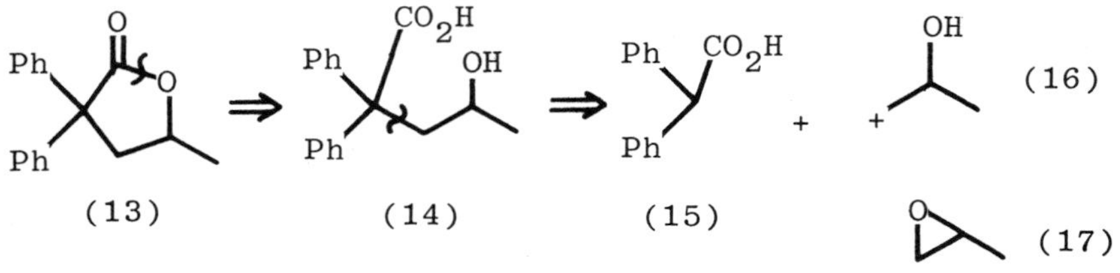

There is no room for an activating group on (15) but nitrile (18) gives a suitable anion with NH_2^- as base. No intermediates can be isolated, the lactone being formed in one step.

Synthesis[286]

$$\underset{(18)}{\underset{Ph}{\overset{Ph}{>}}\!\!-CN} \quad \xrightarrow[2.\,(17)]{1.\,NaNH_2} \quad \underset{69\%}{TM(13)}$$

This method* has been made more general by use of modern reagents: low temperatures and the strong hindered base i-Pr$_2$NLi allow the deprotonation of many nitriles and their capture by a variety of epoxides. Acid hydrolysis gives lactones.[287]

$$\underset{R^2}{\overset{R^1}{>}}\!\!-CN \quad \xrightarrow[R^3\!\!-\!\!\triangle\!\!-\!\!O]{i-Pr_2NLi} \quad \underset{R^2}{\overset{R^1}{>}}\!\!\underset{R^3}{\overset{CN}{\underset{OH}{\bigvee}}} \quad \xrightarrow{HCl} \quad \underset{R^2}{\overset{R^1}{\underset{R^3}{\text{lactone}}}}$$

General Problem :
Lactones (19) and (20) both have peach odours and are used in perfumes. Suggest syntheses.

(19)   (20)

Answer :
Hydroxy acids (21) and (22) can be disconnected in many ways but they are particularly suitable for epoxide routes.
* A third route to (13) appears on p T 222.

Analysis

(19) ⇒ [structure (21): HO-CH₂CH₂-CH(CO₂H)-C₇H₁₅] ⇒ [epoxide] + [⁻CH(CO₂H)-C₇H₁₅] (23)

(20) ⇒ [structure (22): C₇H₁₅-CH(OH)-CH₂-CH₂-CO₂H] ⇒ [epoxide (24)] + ⁻CH₂CO₂H (25)

Malonates are good reagents for synthons (23) and (25). Lactones (26) and (27) are formed on addition of the epoxides and give the perfumery lactones on hydrolysis and decarboxylation.

Synthesis[288]

$$CH_2(CO_2Et)_2 \xrightarrow[\text{2. n-}C_7H_{15}Br]{\text{1. EtO}^-} (EtO_2C)_2CH \cdot C_7H_{15}\text{-n} \xrightarrow[\text{2. }\triangle O]{\text{1. EtO}^-}$$

[structure (26): tetrahydrofuranone with EtO₂C and C₇H₁₅ substituents]

$$\xrightarrow[\text{2. H}^+,\text{heat}]{\text{1. HO}^-,\text{H}_2\text{O}} \text{TM(19)}$$

$$\text{n-}C_7H_{15}CH=CH_2 \xrightarrow{\text{PhCO}_3\text{H}} (24)\ 100\%$$

$$CH_2(CO_2Et)_2 \xrightarrow[\text{2. (24)}]{\text{1. EtO}^-}$$ [structure (27): lactone with CO₂Et and C₇H₁₅ substituents]

$$\xrightarrow[\text{2. H}^+,\text{heat}]{\text{1. HO}^-,\text{H}_2\text{O}} \text{TM(20)}\ 70\%$$

Since keto acids (30) are ideal intermediates, another route for (20) is to use available 1,4-starting material succinic anhydride (28), convert it into (29) (page T 39) and add an alkyl cadmium reagent. Reduction of keto-ester (30) gives (20).

Synthesis[289]

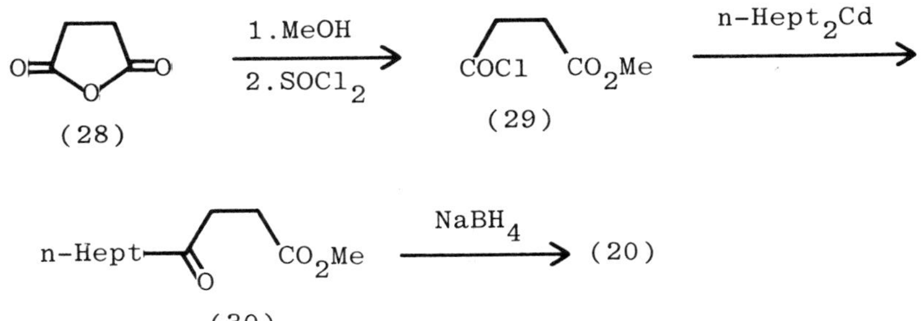

(28) →[1. MeOH / 2. SOCl$_2$]→ COCl–CH$_2$CH$_2$–CO$_2$Me (29) →[n-Hept$_2$Cd]→

n-Hept–CO–CH$_2$CH$_2$–CO$_2$Me (30) →[NaBH$_4$]→ (20)

Other routes you might have suggested include cyanide addition to enones, e.g. (31), or nitro alkane addition to unsaturated acids (32).

Analysis

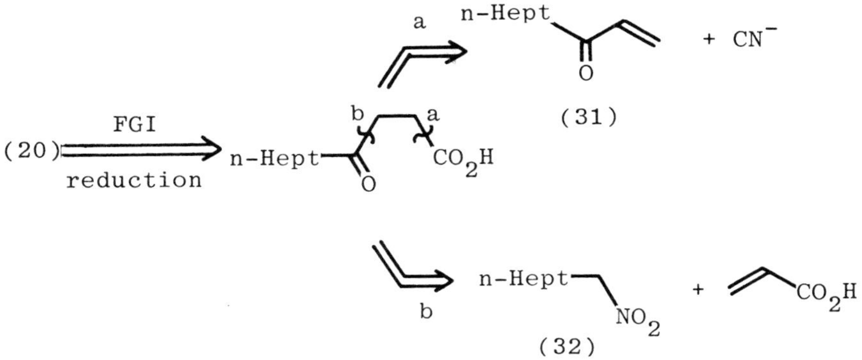

(20) ⇒[FGI reduction] n-Hept–CO–CH$_2$CH$_2$–CO$_2$H

a ⇒ n-Hept–CO–CH=CH$_2$ + CN⁻ (31)

b ⇒ n-Hept–CH$_2$–NO$_2$ + CH$_2$=CH–CO$_2$H (32)

Unnatural Nucleophilic Synthons

Problem :

Suggest a synthesis of 1,4-diketone (33).

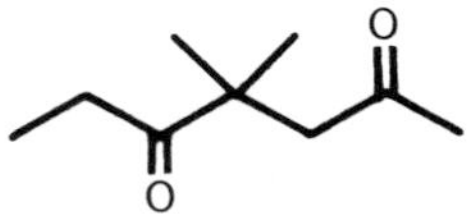

(33)

Answer :

Disconnection of the central bond is unhelpful : neither specific enolate (34) nor α-haloketone (35) will be easy to make.

Analysis 1

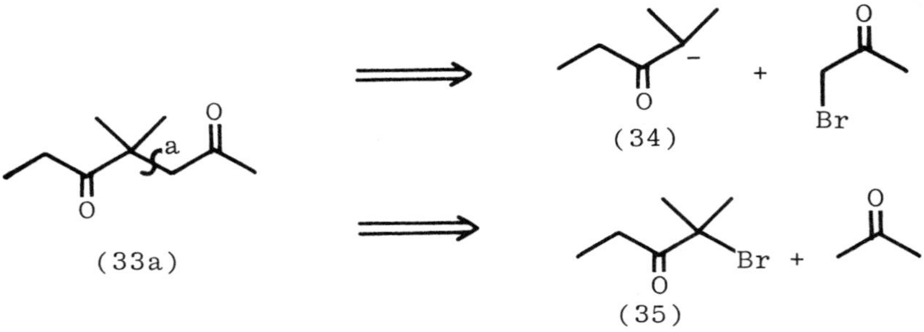

(33a) ⟹ (34) + Br-CH₂-CO-CH₃

⟹ (35) Br + acetone

The only possible acyl anion disconnection is (33b) which requires (36). A nitroalkane is ideal for this.

Analysis 2

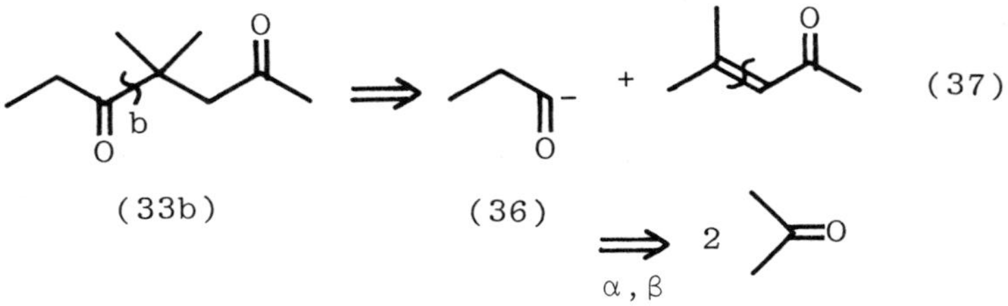

(33b) ⟹ (36) + (37)

⟹ 2 acetone
α,β

Enone (37) is mesityl oxide, the aldol dimer of acetone (p T 150). In this synthesis, the Nef hydrolysis proved satisfactory: no doubt $TiCl_3$ would have been at least as good.

Synthesis[290]

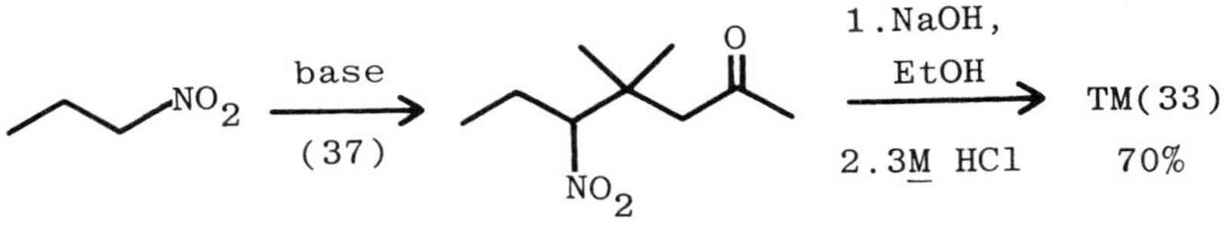

$CH_3CH_2CH_2NO_2$ —base→ (37) —1. NaOH, EtOH; 2. 3M HCl→ TM(33) 70%

Example :

In attempts to prevent a sheep disease, Australian workers wanted (38), a carbocyclic analogue of the natural product equol.[291]

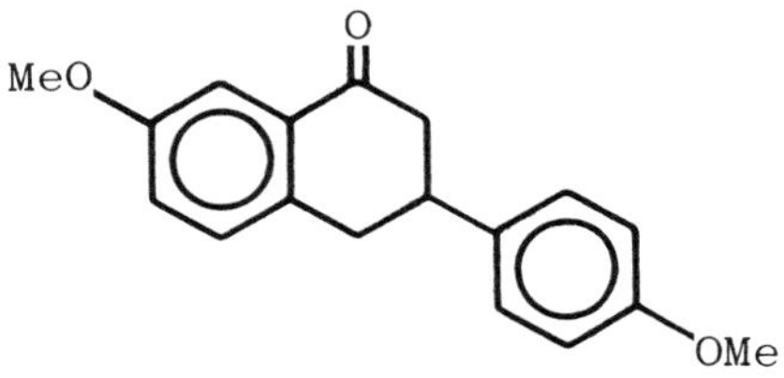

(38) [a 6-methoxy-tetralone with a 4-methoxyphenyl substituent]

Friedel-Crafts disconnection (38a) is unambiguous because of the symmetry of (39). Further disconnection requires FGA. A carbonyl group next to the aromatic ring gives a 1,4-dicarbonyl compound (40) and allows disconnection of an acyl anion equivalent to give an enone (41). This can be made by Mannich reaction from (42).

Analysis

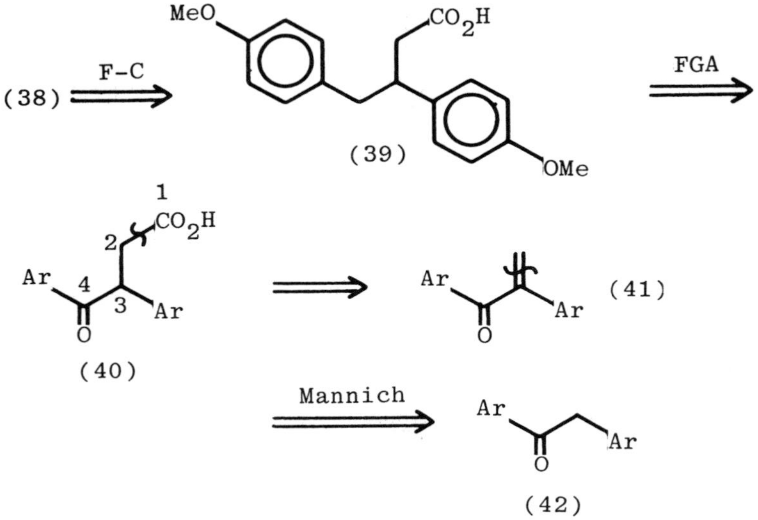

(38) $\xrightarrow{\text{F-C}}$ [structure (39): MeO-C6H4-CH2-CH(CO2H)-C6H4-OMe] $\xrightarrow{\text{FGA}}$

[structure (40): Ar-C(=O)-CH(Ar)-CH2-CO2H with positions 1,2,3,4 labeled] $\Longrightarrow$ [structure (41): Ar-C(=O)-C=C-Ar with double bond crossed]

$\xrightarrow{\text{Mannich}}$ [structure (42): Ar-C(=O)-CH2-Ar]

Ketone (42) could be made by a Friedel-Crafts reaction, but because the two aromatic rings are the same, another FGI provides a short cut. Hydroxy ketone (43) is the product of a benzoin dimerisation (p 188) of (44).

Analysis 2

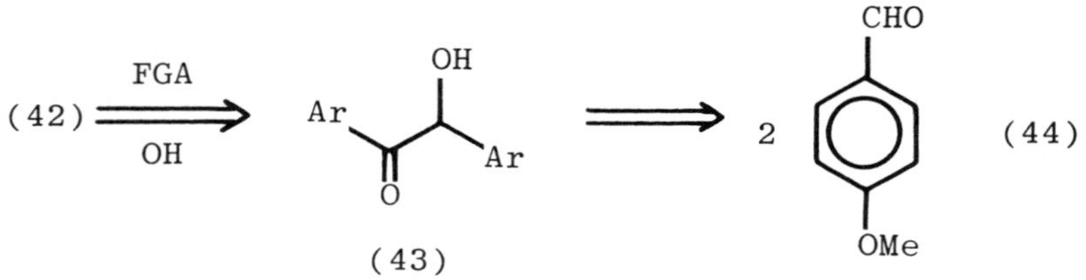

(42) $\xrightarrow[\text{OH}]{\text{FGA}}$ [structure (43): Ar-C(=O)-CH(OH)-Ar] $\Longrightarrow$ 2 [structure (44): 4-MeO-C6H4-CHO]

Tin and HCl reduce out the benzylic OH from (43) in high yield.[292] The Mannich base (45) decomposes to (41) simply on heating. Cyanide addition gives (46) which can be hydrolysed to (40), but a short cut is to hydrolyse to amide (47) and reduce out the carbonyl group by the Clemmensen method (Table T 24.1). Under these conditions the amide is hydrolysed to the acid. Cyclisation to (38) occurs with strong acid, acid anhydrides, or by $AlCl_3$-catalysed reaction of the acid chloride.

Synthesis[291-2]

(44) $\xrightarrow{NaCN}$ (43) $\xrightarrow[HCl]{Sn}$ (42) 92% $\xrightarrow[\text{H-N(piperidine)}]{CH_2O}$

(45) Ar-CO-CH(Ar)-CH$_2$-N(piperidine) $\xrightarrow{\text{heat}}$ (41) 100% $\xrightarrow[\text{MeOH HOAc}]{NaCN}$ (46) Ar-CO-CH(Ar)-CH$_2$-CN

$\xrightarrow[\text{HOAc}]{H_2SO_4}$ Ar-CO-CH(Ar)-CH$_2$-CONH$_2$ $\xrightarrow[\text{HCl}]{Zn-Hg}$ (39) 71% $\xrightarrow[\text{PPA}]{\text{e.g.}}$ TM(38) ~70%

Strategy of Available Starting Materials
(Consult Table T 25.1)

Problem :

Lactone (47) was used in a synthesis of β-vetivone.[293] How might it be made?

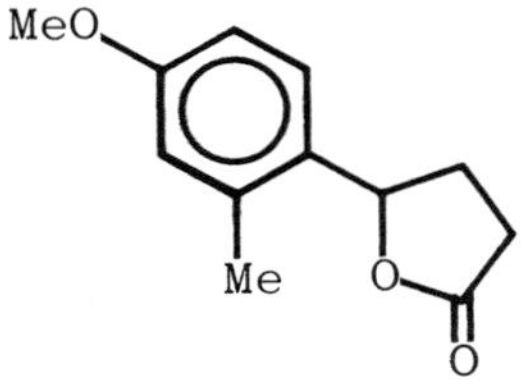

(47)

Answer :

Disconnecting the lactone gives hydroxyacid (48). With succinic anhydride in mind as an available 1,4-dicarbonyl starting material, we can write keto acid (49) as an intermediate and the orientation is then correct for a Friedel-Crafts disconnection as both MeO and Me activate this position.

Analysis

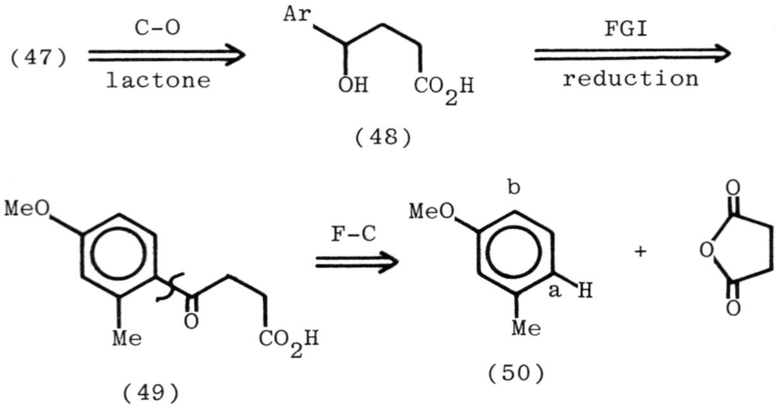

(47) $\underset{\text{lactone}}{\overset{\text{C-O}}{\Longrightarrow}}$ Ar–CH(OH)–CH$_2$–CH$_2$–CO$_2$H (48) $\underset{\text{reduction}}{\overset{\text{FGI}}{\Longrightarrow}}$

(49) $\overset{\text{F-C}}{\Longrightarrow}$ (50) + succinic anhydride

Two positions in (50) (a and b) are equally activated electronically. Experiments showed that (49) was in fact the major product from the Friedel-Crafts reaction.[294] The starting material is made from available *m*-cresol (51).

Synthesis[293-4]

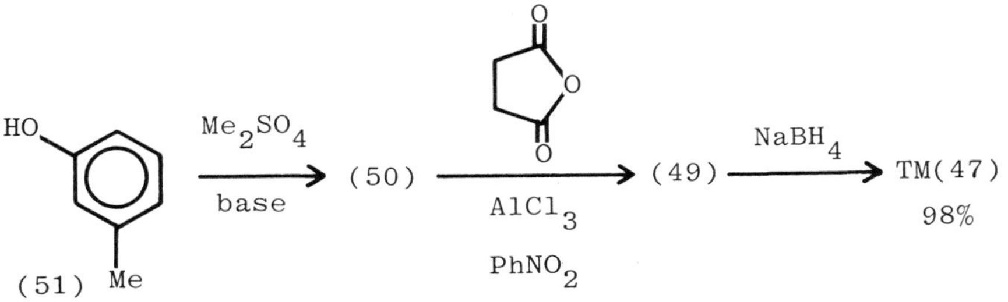

HO-[m-cresol with Me] (51) → Me$_2$SO$_4$/base → (50) → [succinic anhydride]/AlCl$_3$/PhNO$_2$ → (49) → NaBH$_4$ → TM(47) 98%

Example :

Laevulinic acid (52) appears in the list of readily available 1,4-difunctionalised compounds. It is in fact one of the cheapest as, like furfuraldehyde, it comes from the acid-catalysed degradation of sugars. Laevulinic acid can be isolated in good yield from the degradation of even low-grade cellulose, like sawdust, and so it is a good starting material for organic syntheses.[295]

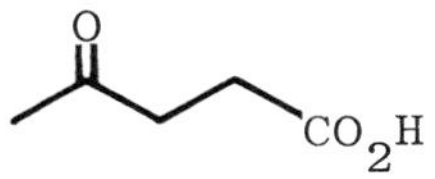

Problem :

Ester (53) was used in a synthesis of trisporic acid[296] (cf p T 350). How might it be made?

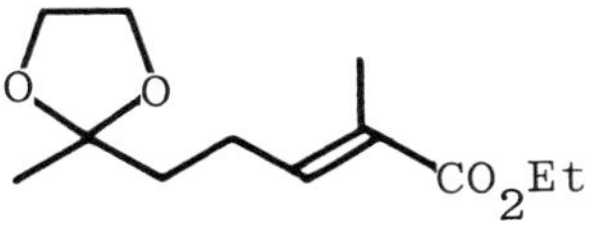

(53)

Answer :

The disconnection (53a) needed to find the skeleton of laevulinic acid corresponds to the Wittig reaction and the rest is straightforward.

Analysis

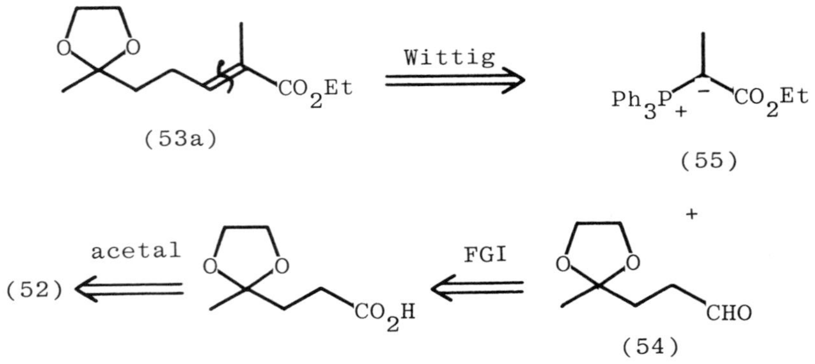

[Scheme: (53a) ⟹ (Wittig) Ph₃P⁺–C(CH₃)(CO₂Et)⁻ (55)]

(52) ⟸ (acetal) [acetal-CH₂CH₂-CO₂H] ⟸ (FGI) [acetal-CH₂CH₂-CHO] (54)

The adjustment of the oxidation level is most easily achieved by reducing the protected ester (56) to the alcohol and re-oxidising. The Wittig reaction with a stabilised ylid (55) gives mostly *E*-(53).

Synthesis[296]

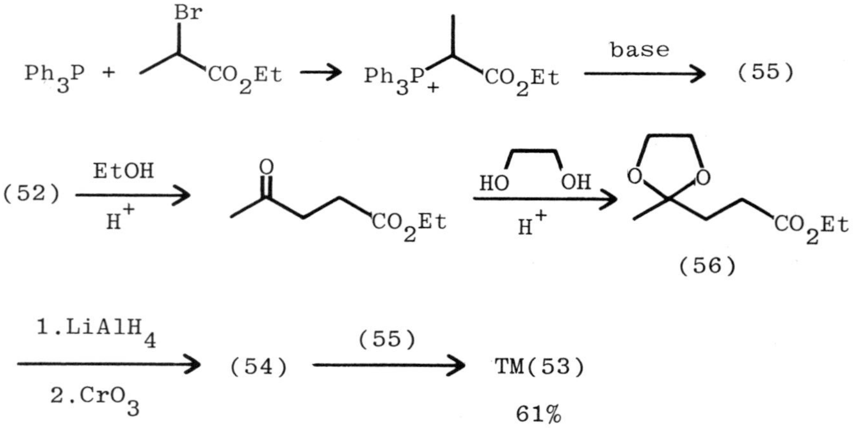

Ph₃P + BrCH(CH₃)CO₂Et → Ph₃P⁺CH(CH₃)CO₂Et $\xrightarrow{\text{base}}$ (55)

(52) $\xrightarrow[\text{H}^+]{\text{EtOH}}$ CH₃COCH₂CH₂CO₂Et $\xrightarrow[\text{H}^+]{\text{HOCH}_2\text{CH}_2\text{OH}}$ (56)

$\xrightarrow[\text{2.CrO}_3]{\text{1.LiAlH}_4}$ (54) $\xrightarrow{(55)}$ TM(53) 61%

FGA Strategy
Example :
Chalcogran (57), the aggregation pheromone of a pest on

Norwegian spruce,[297] is an acetal. Disconnection reveals two 1,4-relationships (58), either of which could be connected by an acetylene.

Analysis 1

(57) $\xrightarrow{\text{acetal}}$ (58)

We prefer to disconnect the right hand half of the molecule in this way because optically active lactones (59) can be made from glutamic acid (60), another available 1,4-difunctionalised compound (p T 96).

Analysis 2

(58) $\xrightarrow{\text{FGA}}$ $\Rightarrow$

$\Rightarrow$ (59) $\rightarrow$ (60)

The lactone (59) is made by the method described on page T 96. Addition of the protected acetylene (61) gives the hemiacetal (62) which can be hydrogenated directly to (63). Cyclisation occurs as soon as the protecting group is removed.

Synthesis[298]

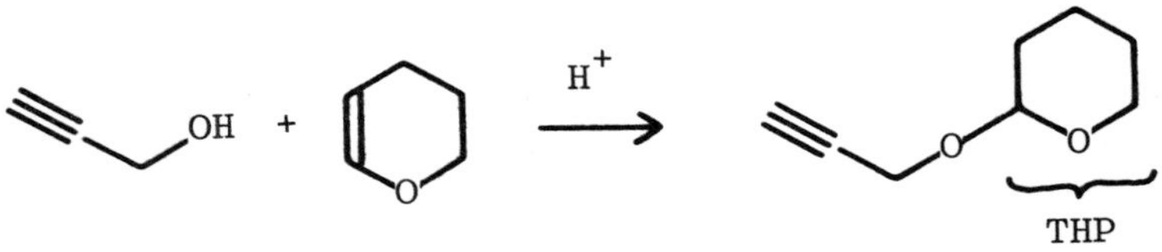

HC≡C-CH₂-OH + (dihydropyran) →[H⁺] HC≡C-CH₂-O-THP

(61)

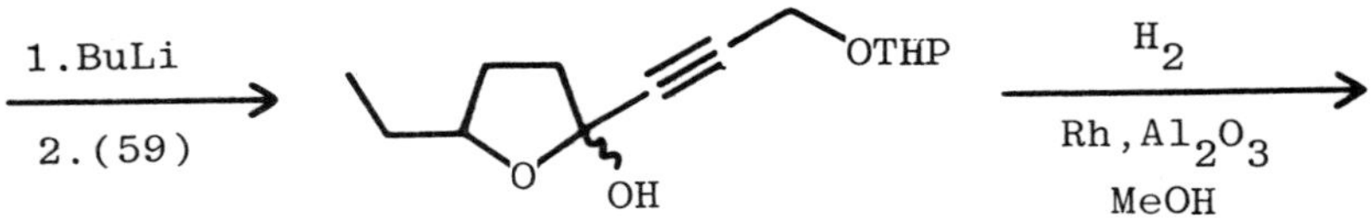

→[1. BuLi][2. (59)] (62) →[H₂][Rh, Al₂O₃][MeOH]

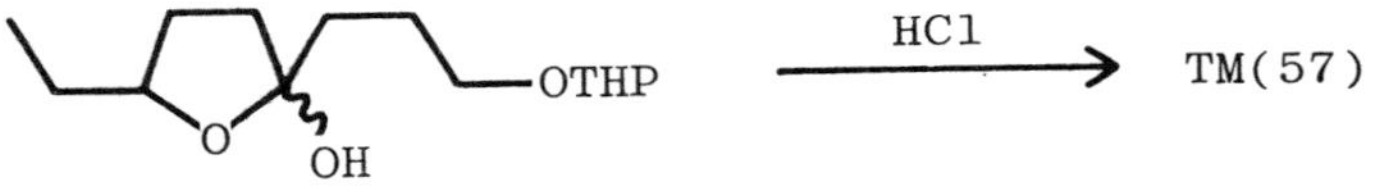

(63) →[HCl] TM(57)

CHAPTER 26

Strategy XII: Reconnections: Synthesis of 1,2- and 1,4-Difunctionalised Compounds by C=C Cleavage

Problem :
Suggest a synthesis of (1).

MeCONH–CH(Me)–CH2–CH2–CHO (1)

Answer :
This is a 1,4-difunctionalised compound containing amide and aldehyde groups. Disconnection of the amide gives (2) which would instantly cyclise if we made it. The amino group logically comes from a ketone suggesting 1,4-dicarbonyl compound (3) as starting material. The aldehyde in (2) and (3) would have to be protected against attack from nitrogen so one strategy is to have a double bond present in (2) and (3) which can be cleaved to give (1). We should *reconnect* (3) to give (4) as the true starting material.

Analysis

(1) $\xRightarrow{\text{C-N amide}}$ H2N–CH(Me)–CH2–CH2–CHO (2) $\xRightarrow{\text{FGI reduction}}$

HON=C(Me)–CH2–CH2–CHO $\xRightarrow{\text{C=N oxime}}$

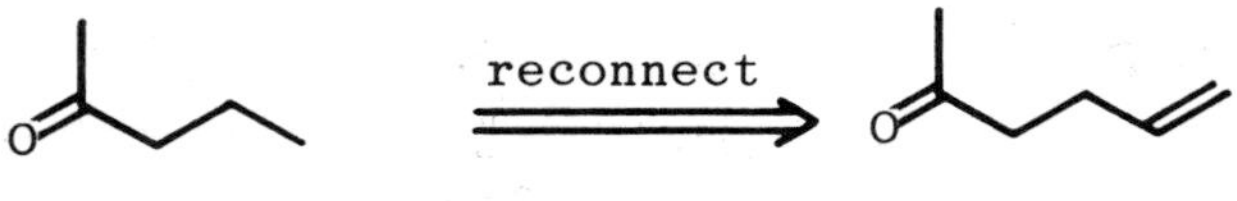

[Structures: (3) butan-2-one-like ketone; (4) with allyl group]

Since ketone (5) is available (T1) and the extra methyl groups will be removed in the ozonolysis, this was the starting material for the published synthesis. Reductive work-up (zinc was used) for the ozonolysis is necessary to preserve the aldehyde.

Synthesis[299]

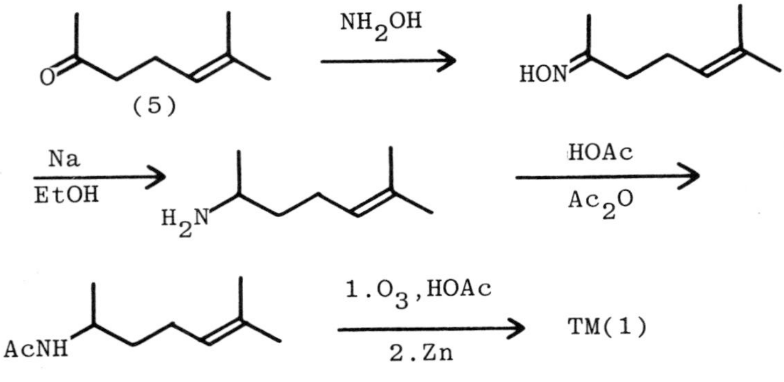

[Scheme: (5) 6-methylhept-5-en-2-one → NH$_2$OH → oxime (HON=) → Na/EtOH → amine (H$_2$N-) → HOAc/Ac$_2$O → acetamide (AcNH-) → 1. O$_3$, HOAc 2. Zn → TM(1)]

Example :
A synthesis of aspartic acid is based on this strategy. Disconnection (a) is attractive since acylamino malonate (7) is a reagent for synthon (6). Synthon (8) can be represented by allyl bromide.

Analysis

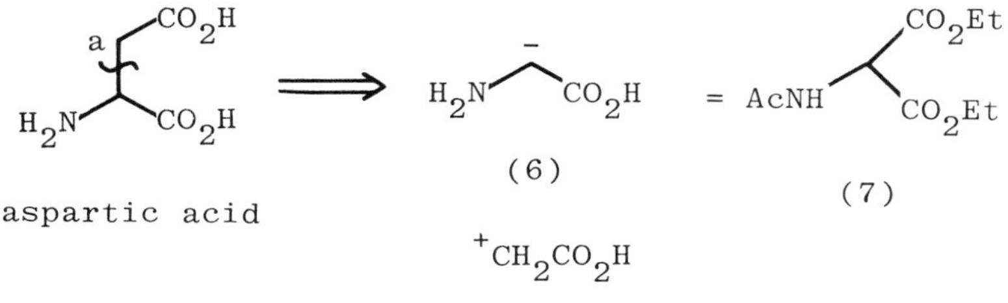

[Scheme: aspartic acid (H₂N-CH(CO₂H)-CH₂-CO₂H, bond "a") ⟹ H₂N-CH⁻-CO₂H (6) + ⁺CH₂CO₂H (8); (6) = AcNH-CH(CO₂Et)₂ (7)]

In the published synthesis the ozonolysis is performed on the protected product (9) and aldehyde (10) isolated before oxidation, hydrolysis and decarboxylation give aspartic acid.

Synthesis[300]

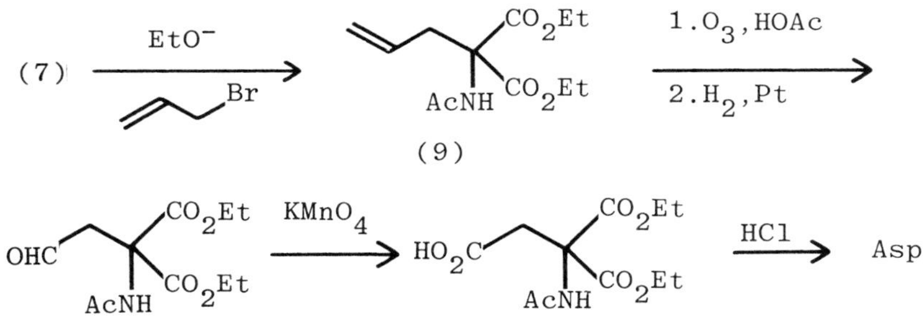

(7) $\xrightarrow[\text{Br}\diagdown\diagup]{\text{EtO}^-}$ CH₂=CH-CH₂-C(AcNH)(CO₂Et)₂ (9) $\xrightarrow[\text{2. H}_2,\text{Pt}]{\text{1. O}_3,\text{HOAc}}$

OHC-CH₂-C(AcNH)(CO₂Et)₂ (10) $\xrightarrow{\text{KMnO}_4}$ HO₂C-CH₂-C(AcNH)(CO₂Et)₂ $\xrightarrow{\text{HCl}}$ Asp

Reagent (7) is made by the α-functionalisation strategy (p T 191). The amino group comes from oxime (12) which can be made by direct nitrosation of malonate.

Analysis

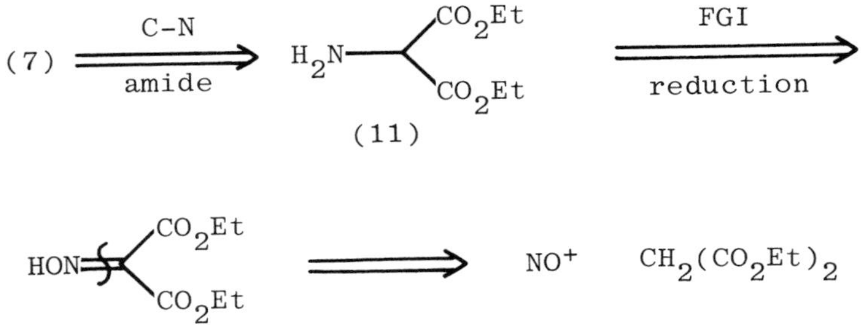

(7) $\underset{\text{amide}}{\overset{\text{C-N}}{\Longrightarrow}}$ $H_2N-\overset{CO_2Et}{\underset{CO_2Et}{\diagup\!\!\!\diagdown}}$ $\overset{\text{FGI}}{\underset{\text{reduction}}{\Longrightarrow}}$

(11)

$HON=\overset{CO_2Et}{\underset{CO_2Et}{\diagup\!\!\!\diagdown}}$ $\Longrightarrow$ NO^+ $CH_2(CO_2Et)_2$

(12)

The reduction is carried out in acetic anhydride so that amine (11) is acetylated as it is formed. This prevents the free amino group attacking the esters.

Synthesis[301]

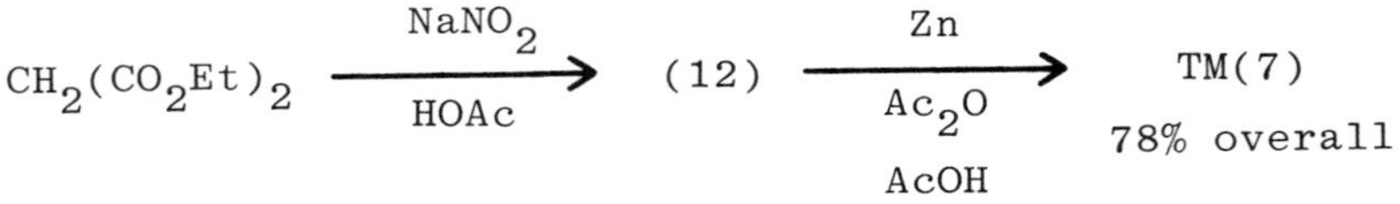

$CH_2(CO_2Et)_2 \xrightarrow[\text{HOAc}]{\text{NaNO}_2} (12) \xrightarrow[\substack{Ac_2O \\ AcOH}]{Zn} TM(7)$
78% overall

Example :

Ozonolysis of condensation products can provide a route to 1,2-difunctionalised compounds. Camphor (13) condenses with many carbonyl compounds (HCO_2Et is used here) at its only free α-CH_2 group to give, e.g. (14). Ozonlysis releases the α-diketone (15).

Synthesis[302]

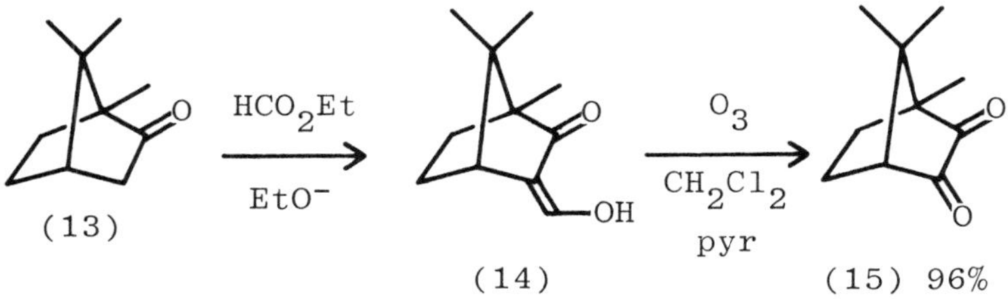

(13) →(HCO₂Et, EtO⁻)→ (14) →(O₃, CH₂Cl₂, pyr)→ (15) 96%

Problem :
Suggest a synthesis of keto-ester (16), by a reconnection strategy.

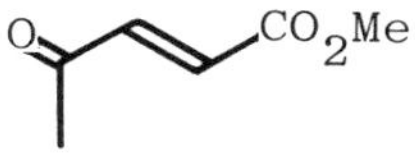

(16) — structure: CH₃-CO-CH=CH-CH₂-CO₂Me (as drawn)

Answer :
The keto group must come from double bond cleavage so we may add anything we please as R in (17). The obvious choice is hydrogen as (17) can then be made by a Wittig reaction on available aldehyde (18).

Analysis

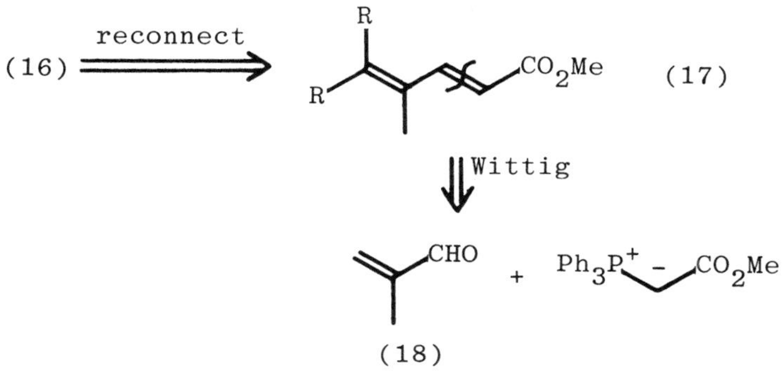

(16) ⇒ reconnect ⇒ R₂C=C(Me)-CH=CH-CO₂Me (17)

⇓ Wittig

CH₂=C(Me)-CHO (18) + Ph₃P⁺-CH⁻-CO₂Me

The chemoselectivity of the ozonolysis is all right because ozone attacks the most electron-rich double bond, that is the one furthest from the carbonyl group in (17, R=H). Reductive work-up is again needed after the ozonolysis.

Synthesis[303]

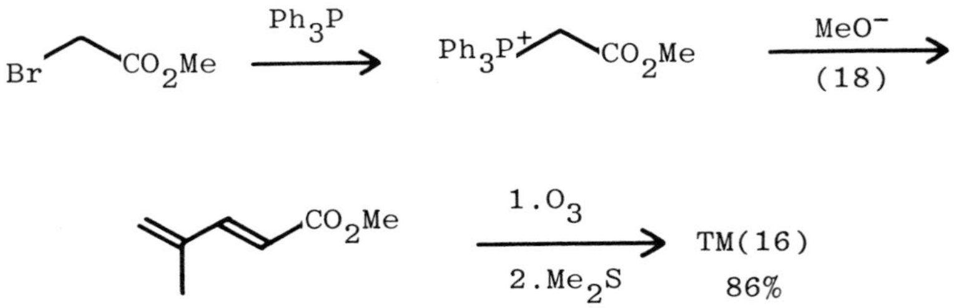

$$Br\diagup\diagdown CO_2Me \xrightarrow{Ph_3P} Ph_3\overset{+}{P}\diagup\diagdown CO_2Me \xrightarrow{MeO^-} (18)$$

$$\diagup\diagdown\diagup CO_2Me \xrightarrow[2.Me_2S]{1.O_3} TM(16) \quad 86\%$$

1,4-Functionalisation without Reconnection
Problem :
Suggest a synthesis for bicyclic ketone (19).

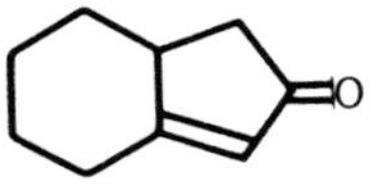

(19)

Answer :
Enone disconnection reveals 1,4-diketone (20) and ring-chain disconnection requires synthon (21). Allyl or propargyl halides (p T 221) could fill this role.

Analysis

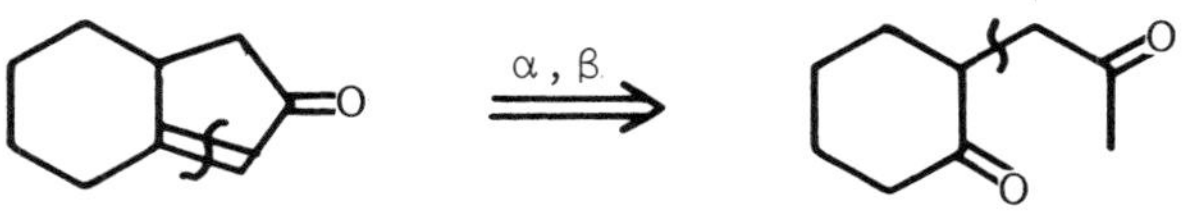

(19a) ⟹ α,β (20)

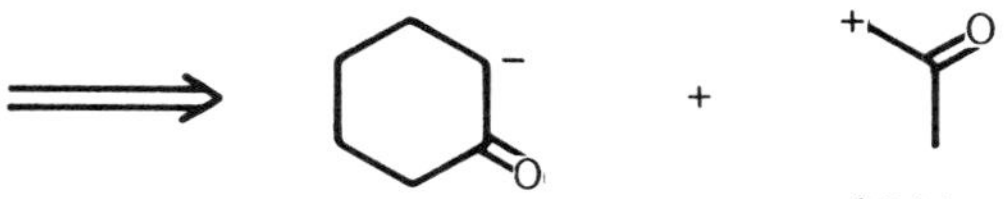

⟹ [cyclohexanone enolate] + (21)

One published synthesis uses a propargyl bromide for (21) and an activating group to provide a specific enol equivalent (22).

Synthesis[304]

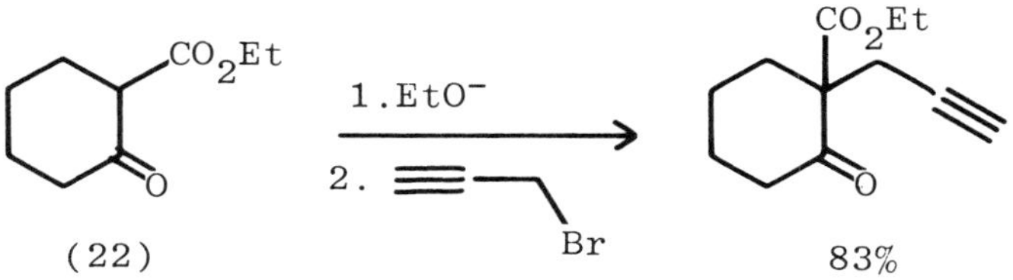

(22) →(1. EtO⁻; 2. ≡—Br) 83%

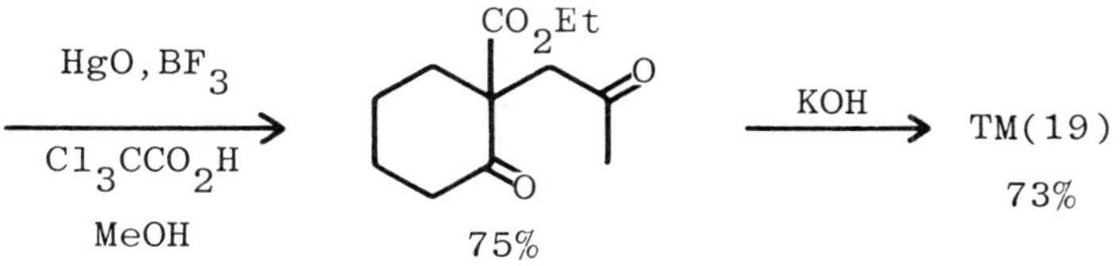

→(HgO, BF$_3$; Cl$_3$CCO$_2$H; MeOH) 75% →(KOH) TM(19) 73%

Hydrolysis, decarboxylation, and cyclisation all occur under the same conditions.

Problem:

Suggest a synthesis for the analgesic methadone (23).

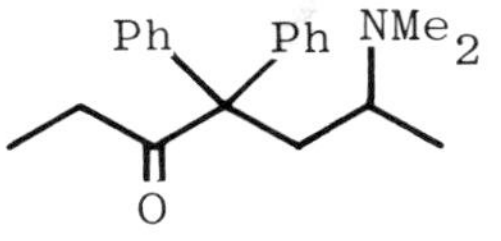

(23)

Answer :

This 1,4-difunctionalised compound cannot easily be made from diketone (24) because of the chemoselectivity problem. The α-halo ketone (25) is a better starting material and this could be made by HBr addition to (26).

Analysis 1

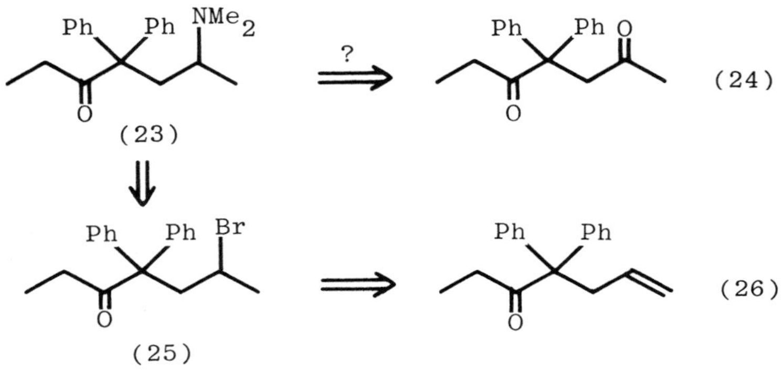

(23) (24) (25) (26)

Addition of an allyl halide regiospecifically to (27) would give (26), but a more reliable route is to allylate nitrile (28), as we did on page T 222 .

Analysis 2

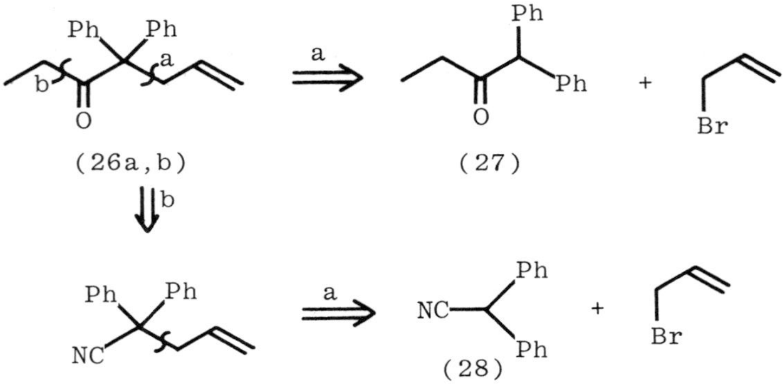

(26a,b) (27) (28)

One published route uses nitrile (28) as starting material, assembles the amino group in (30), and introduces the ethyl group at the end. The synthesis of (29) is on p T 222 .

Synthesis[305]

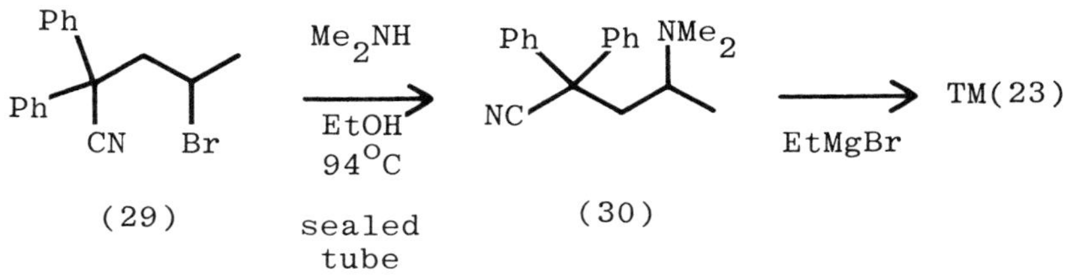
Ph
Ph
CN
Br
(29)
Me_2NH
EtOH
94°C
sealed tube
Ph
Ph
NMe_2
NC
(30)
EtMgBr
TM(23)

More Advanced Examples :

It might seem that allylic functionalisation can be used only on terminal alkenes such as (26) or trisubstituted alkenes, such as (31)[306] when the orientation of addition is unambiguous.

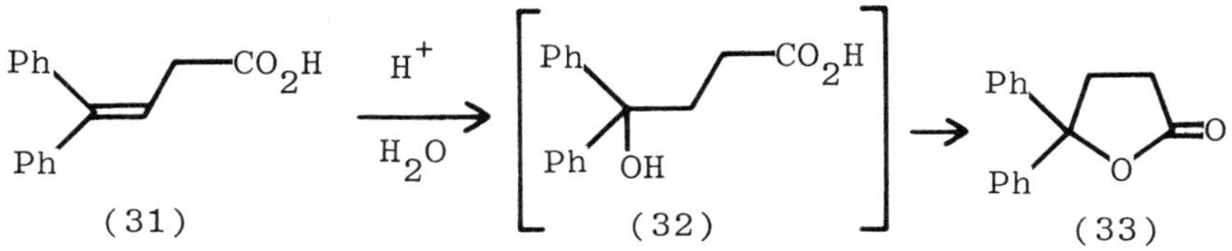
Ph
Ph
CO_2H
(31)
H^+
H_2O
Ph
Ph
OH
CO_2H
(32)
Ph
Ph
O
O
(33)

However, (32) is probably not an intermediate in the formation of lactone (33) as the carbonium ion (34) is probably captured intramolecularly by CO_2H rather than by water so that (33) is formed directly from (34).

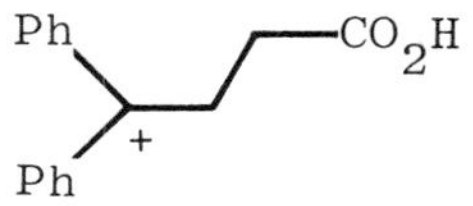
Ph
Ph
+
CO_2H

(34)

Since five-membered rings are formed faster than four or six-membered rings, an otherwise electronically symmetrical double bond (35) or (36) will be captured intramolecularly to give a γ-lactone (37) in acid.

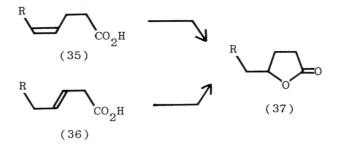

Disconnection of (37) can therefore be made to a double bond in either position (35) or (36) whichever is more convenient. Lactone (39) could come from either (40) or (41). We prefer (41) because (40) is an impossibly strained compound and because (41) can easily be made by a Diels-Alder reaction.

Analysis

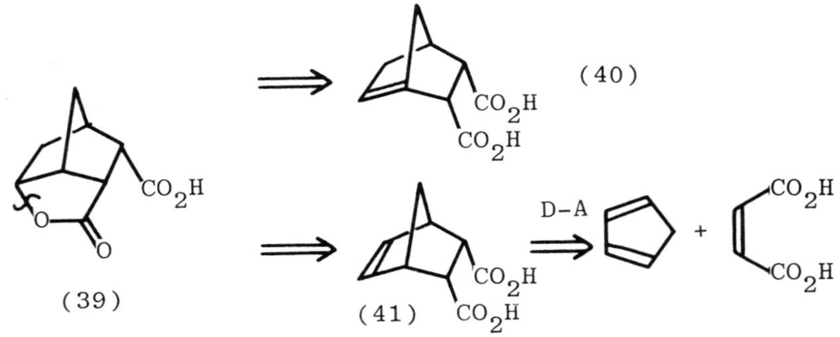

It is easier to use maleic anhydride for the Diels-Alder reaction and hydrolyse to (41) before cyclising to (39) in strong acid.

Synthesis[307]

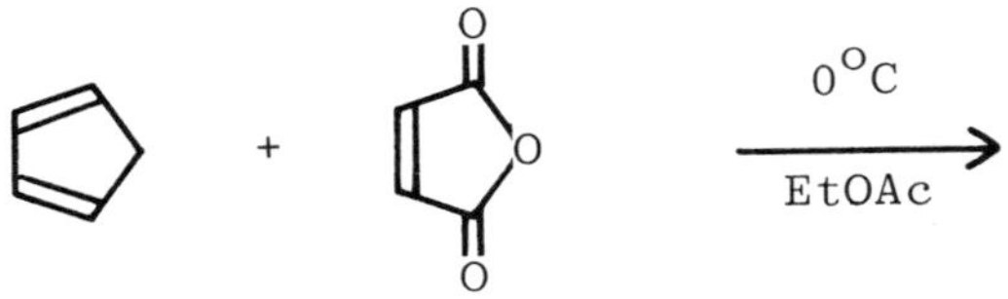

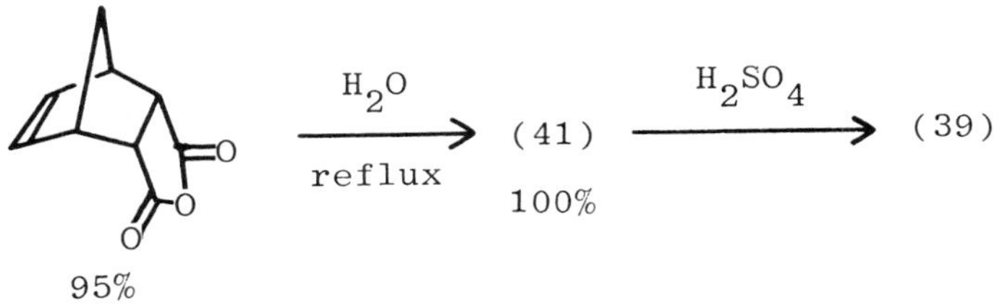

Problem :
Suggest a synthesis for the bicyclic lactone (42).

(42)

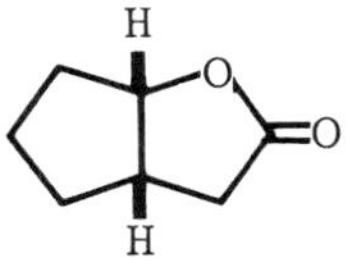

Answer :
Either double bond position (43) or (44) is satisfactory but only (44) leads to an easy disconnection.
Analysis

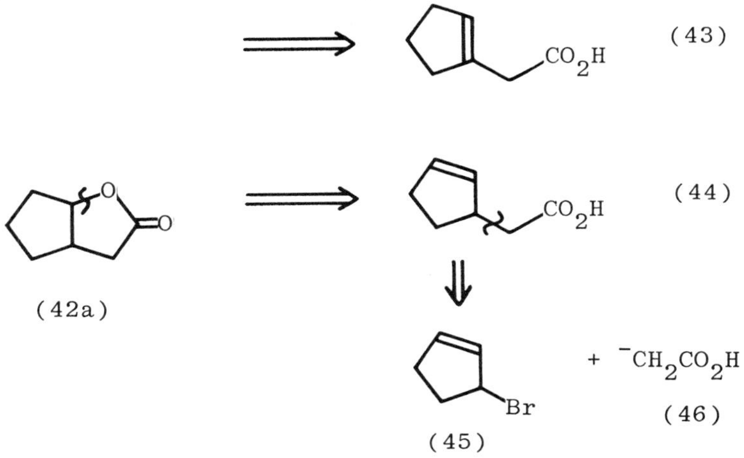

(43)

(44)

(42a)

(45) Br   + $^-CH_2CO_2H$
            (46)

We can use malonate for synthon (46) and prepare allylic halide (45) from available cyclopentadiene (47).

Synthesis[308]

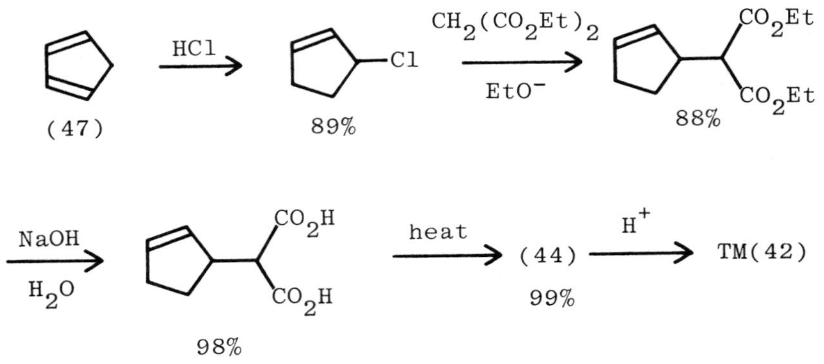

(47) →HCl→ [cyclopentenyl-Cl] 89% →CH$_2$(CO$_2$Et)$_2$ / EtO$^-$→ [cyclopentenyl-CH(CO$_2$Et)$_2$] 88%

→NaOH/H$_2$O→ [cyclopentenyl-CH(CO$_2$H)$_2$] 98% →heat→ (44) 99% →H$^+$→ TM(42)

Example:
The two Diels-Alder adducts (48) and (49) (regioselectivity is poor here in the absence of Lewis acid catalysis) give lactones (50) and (51).[309]

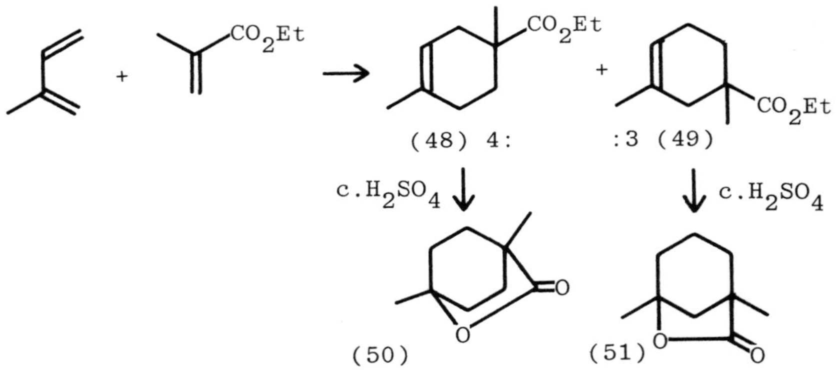

(48) 4: :3 (49)

↓ c.H$_2$SO$_4$ ↓ c.H$_2$SO$_4$

(50)   (51)

Cyclisation has occurred to the tertiary carbonium ions as kinetic preference for a five-membered ring cannot overcome thermodynamic preference for a six-membered ring in this cyclisation of trisubstituted olefin.

CHAPTER 27

Two-Group Disconnections VI: 1,6-Difunctionalised Compounds

Problem:

Suggest a synthesis for TM (1).

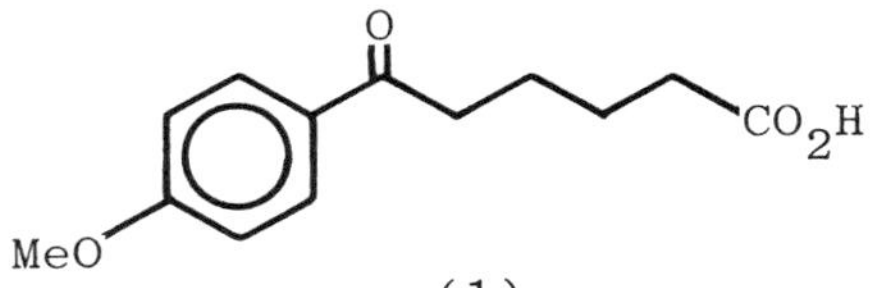

Answer:

First thoughts might be of a Friedel-Crafts disconnection using anhydride (2) which could be made from cyclohexene. An alternative is to reconnect (1) immediately to (3) and make this by standard Grignard methods. Both routes use the same disconnections.

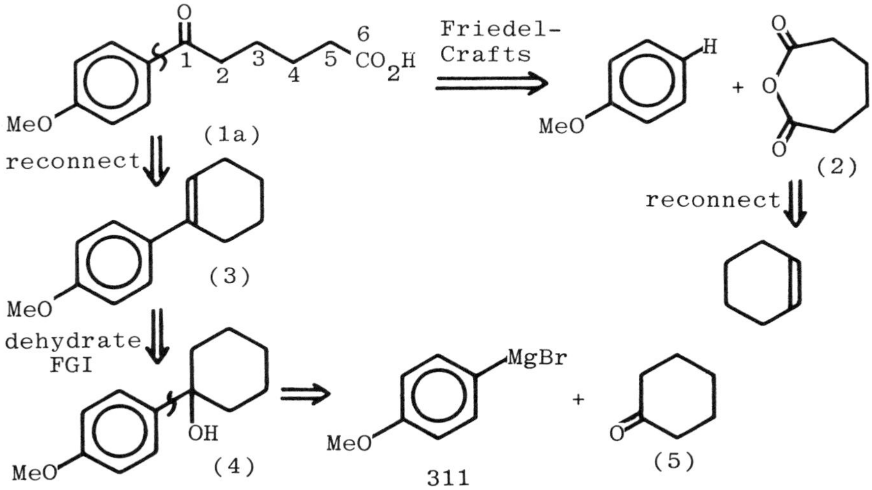

The compound has been made by the second route using CrO_3 oxidation rather than the more usual ozonolysis. No doubt the alternatives would work too.

Synthesis[310]

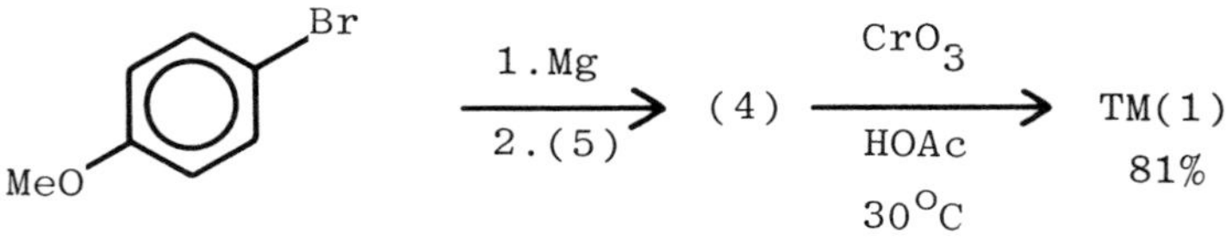

MeO—C6H4—Br $\xrightarrow{\text{1. Mg}}_{\text{2. (5)}}$ (4) $\xrightarrow[\text{30°C}]{\text{CrO}_3 \text{ HOAc}}$ TM(1) 81%

Problem 2:

Non-conjugated diene (6) was needed in a search for oils with certain viscosity properties. Suggest a synthesis.

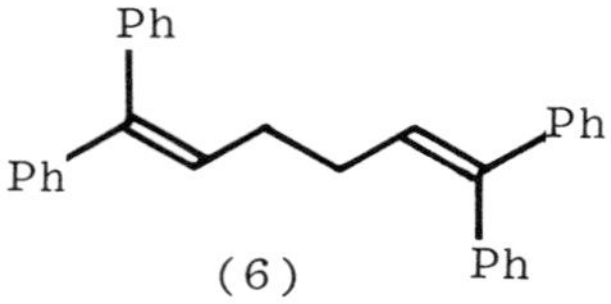

Ph₂C=CH-CH₂-CH₂-CH=CPh₂ (6)

Answer:

Diol (7) can dehydrate only to (6): it is a 1,6-difunctionalised compound which can be made from an adipic diester (8), and hence from cyclohexene. Dehydration occurs readily in acid solution to give a high yield of TM (6).

Analysis

(6) 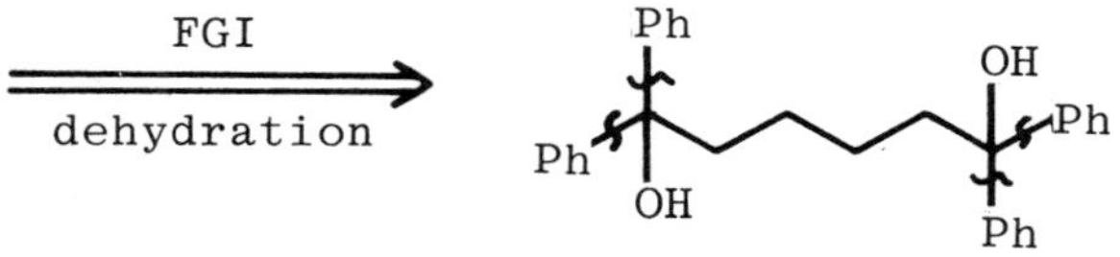 $\xrightarrow[\text{dehydration}]{\text{FGI}}$ Ph₂C(OH)-CH₂-CH₂-CH₂-CH₂-C(OH)Ph₂ (7)

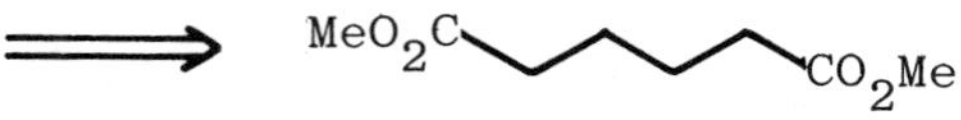

$\Longrightarrow$ MeO₂C-(CH₂)₄-CO₂Me (8)

Synthesis[311]

PhBr $\xrightarrow[\text{2.(8)}]{\text{1.Mg}}$ (7) $\xrightarrow{\text{H}^+}$ TM(6) 90%

Example :

In one synthesis[312] of the queen substance of honey bees (9) (cf p   ) the double bond is derived from an acetylene so that disconnections (a) and (b) are possible.

Analysis 1

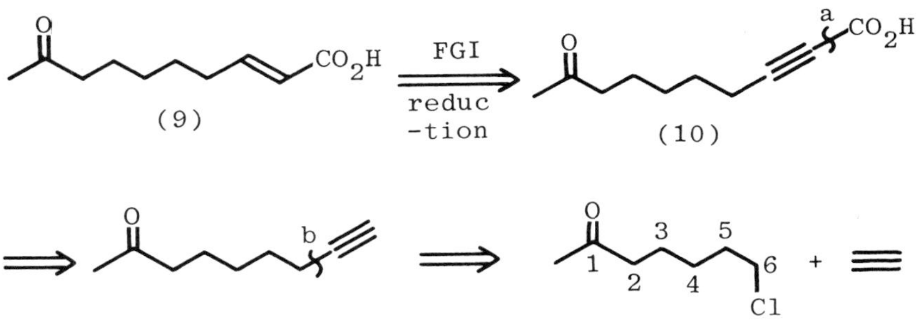

(9) $\xRightarrow[\text{reduction}]{\text{FGI}}$ (10)

$\Rightarrow$ (with disconnection b) $\Rightarrow$ (11) + ≡

Chloroketone (11) is a 1,6-difunctionalised compound and could be made from keto acid (12) which reconnects to (13).

Analysis 2

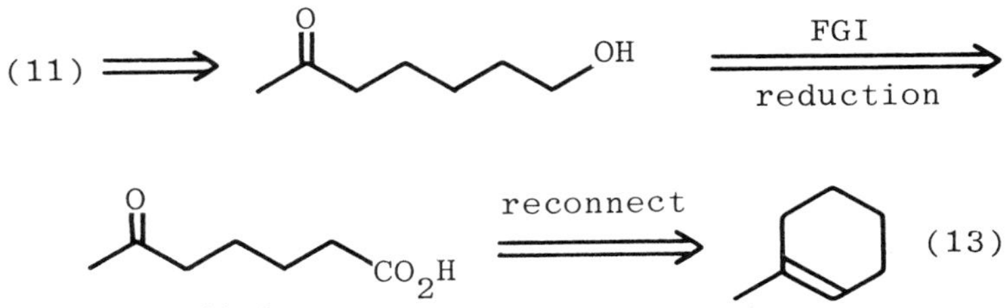

(11) $\Rightarrow$ hydroxy ketone $\xRightarrow[\text{reduction}]{\text{FGI}}$

(12) $\xRightarrow{\text{reconnect}}$ (13)

The discovery that photolysis of hypochlorite (14) gave chloroketone (11) directly provided a short cut in this synthesis. The ketone in (11) will need to be protected during reaction with acetylene.

Synthesis[312]

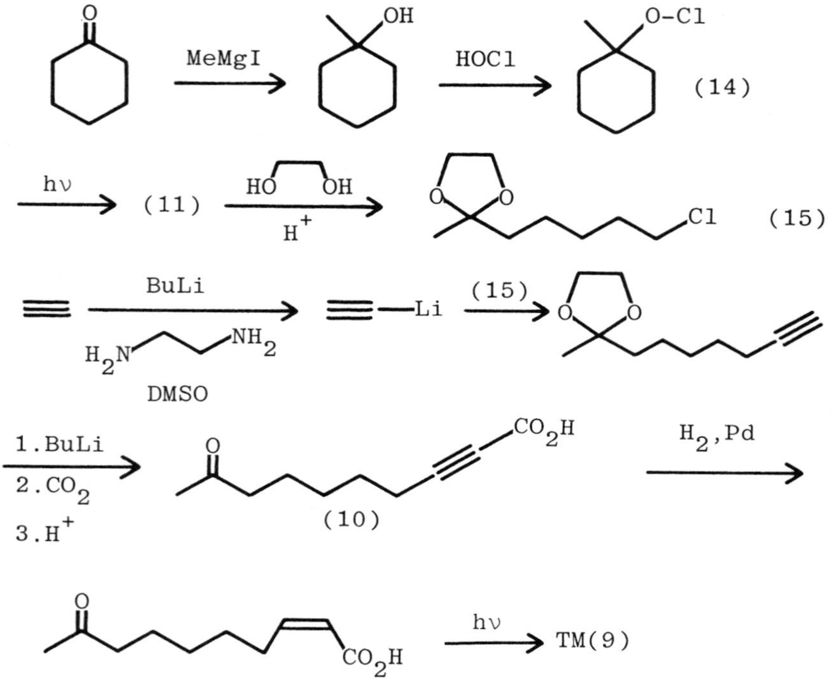

[reaction scheme: cyclohexanone →(MeMgI) 1-methylcyclohexanol →(HOCl) hypochlorite (14) →(hν) (11) →(HOCH₂CH₂OH, H⁺) ketal chloride (15); acetylene →(BuLi, H₂N−CH₂CH₂−NH₂, DMSO) ≡−Li →(15) ketal alkyne; →(1. BuLi, 2. CO₂, 3. H⁺) (10) →(H₂, Pd) cis enone acid →(hν) TM(9)]

Rather than reduce (10) with sodium in liquid ammonia (p T 127), these workers preferred to make the *cis* compound and isomerise it to the more stable *trans* (9) by light.

Example:

Optically active diacid (16) was needed as a synthetic intermediate. Reconnection gives (17) with no obvious disconnections. The FGA strategy provides a solution as a carbonyl group next to the ring (18) allows a Diels-Alder disconnection.

Analysis

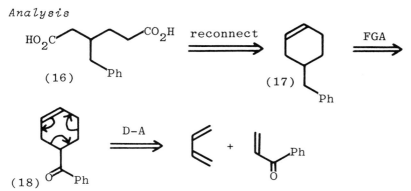

However, none of these compounds can be easily resolved except TM (16) itself. We prefer to resolve as early as possible, (page T 94), so it is better to carry out the Diels-Alder reaction with acrylic acid and resolve acid (19) before adding the phenyl group by a Grignard reaction. The benzylic alcohol group in (20) can be taken out by metal-ammonia reduction.

Synthesis[313]

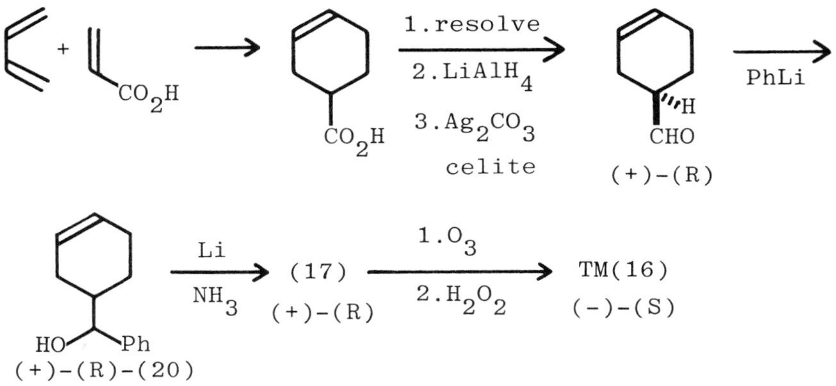

Problem :

Suggest a synthesis for (21), needed for prostaglandin synthesis.

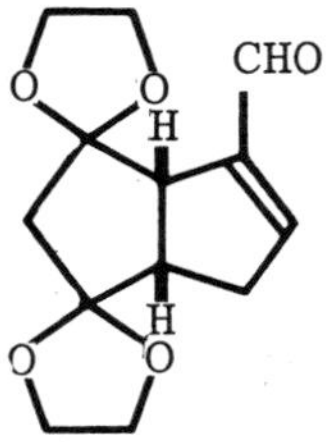

(21)

Answer :

Disconnection of the acetals leads nowhere but α,β disconnection (21a) gives symmetrical (22). This is a 1,6-dicarbonyl compound and can be reconnected to (23). Removal of the acetals now reveals a Diels-Alder adduct (24).

Analysis

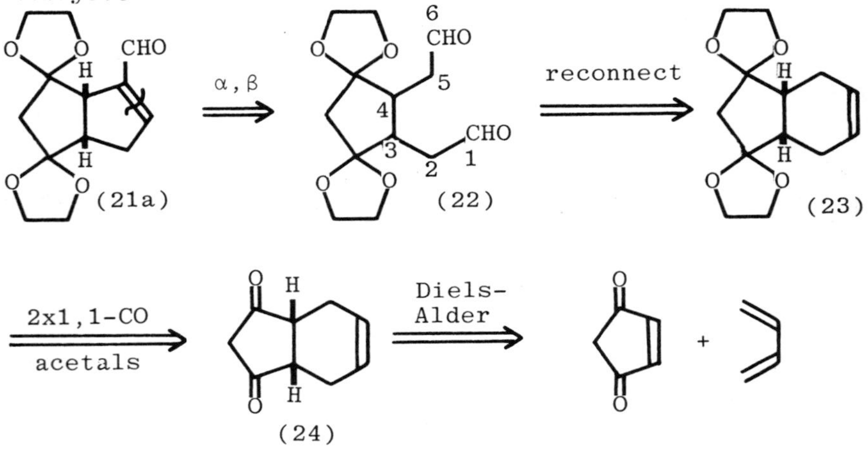

Synthesis[3,14]

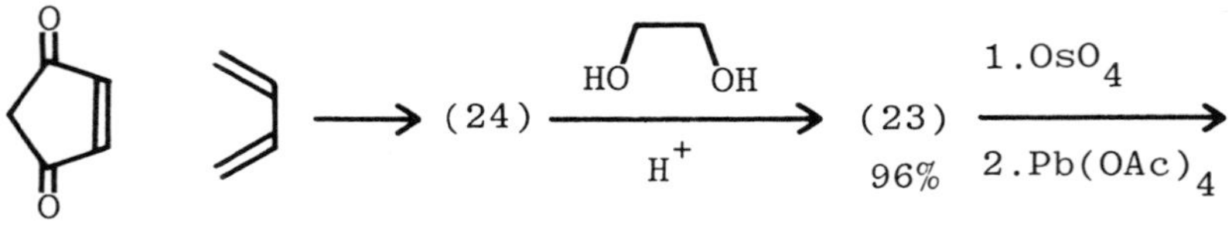

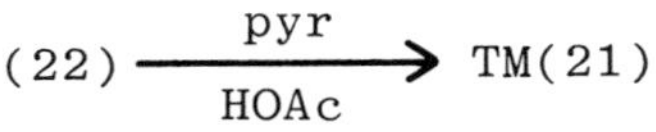

b) *Baeyer-Villiger Cleavage*

Example :

The cyclic amino acid (25) could be made from α,ε-dibromo acid (26) and this can be made from (27) by α-bromination and replacement of hydroxyl by Br. This 1,6-difunctionalised compound (27) is simply the ring opened form of lactone (28), the Baeyer-Villiger (p T 226) cleavage product from cyclohexanone.

Analysis

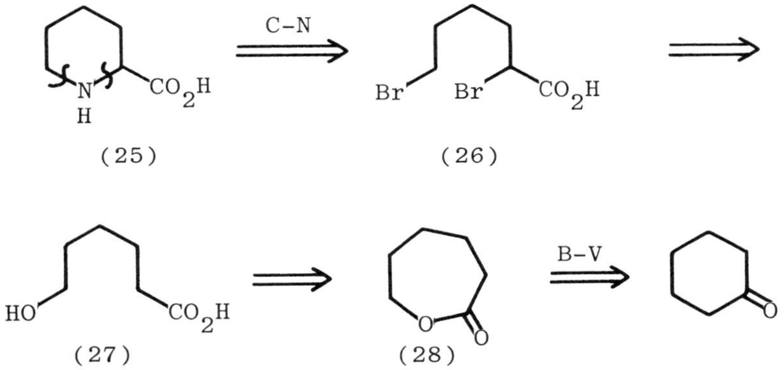

Direct bromination of lactone (28) gives methyl ester (29), which can be hydrolysed to (26). These intermediates have also been used in a synthesis of lysine.

Synthesis[315]

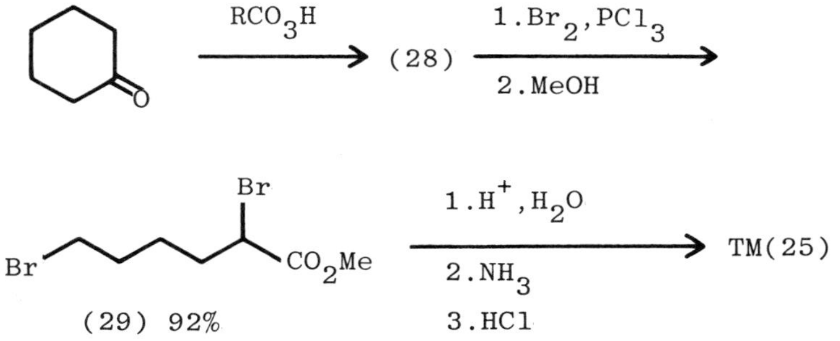

Regio and Stereoselectivity :
The per acid first adds to the ketone to give adduct (30) which rearranges *via* a transition state (31) which is electron-deficient around the former carbonyl group. Consequently, the group which can best supply electrons to combat the deficiency migrates best. It does so with retention as it is a one step reaction in which the chiral centre (● in 30) never becomes detached.

$$R\overset{O}{\underset{}{\text{-C-}}}R \xrightarrow{R^1CO_3H} \text{(30)} \longrightarrow \text{(31)} \longrightarrow R\text{-O-C(O)-}R^1$$

Problem :
Given these facts, what would be the structure of the hydroxy-acid formed by hydrolysis of the Baeyer-Villiger product from bicyclic ketone (32)?

(32) $\xrightarrow{\text{MCPBA}}$ lactone (33) $\xrightarrow[H_2O]{HO^-}$ hydroxy-acid (34)

Answer :
The more substituted group, that is the bridgehead, migrates with retention of configuration. Bicyclic

lactone (33) and the final product therefore both have *cis* arrangements of the substituents around the five-membered ring.

Synthesis[316]

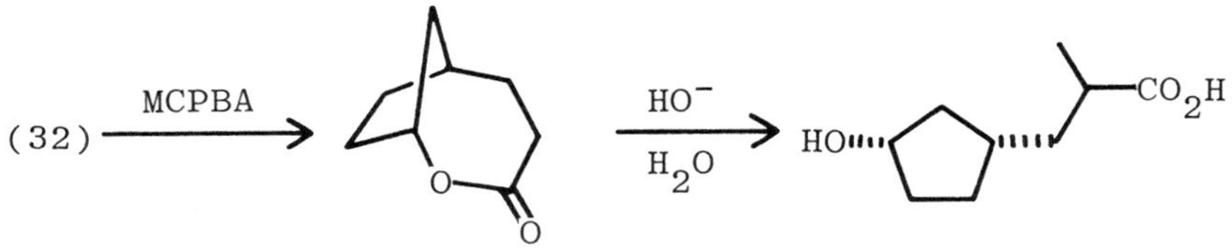

(32) —MCPBA→ (33) —HO⁻/H₂O→ (34)

Other Approaches to 1,6-Difunctionalised Compounds
Example:
The reconnection strategy for (35) via olefin (36) was ignored in the book (p T 228) and is not very helpful because dehydration of the obvious intermediate (37) gives a mixture of olefins.

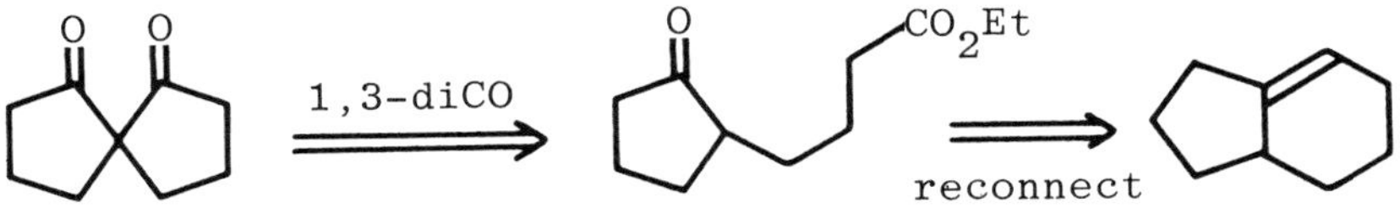

⇒ 1,3-diCO ⇒ (35) ⇒ reconnect ⇒ (36)

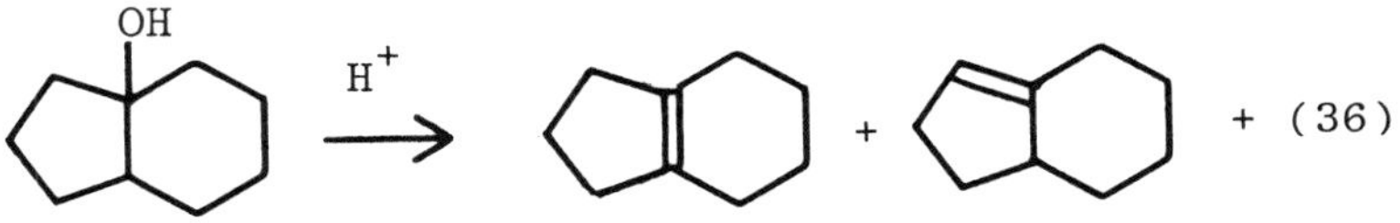

(37) —H⁺→ + + (36)

Problem:
Consider the superficially similar diketone (38). What is wrong with the reconnection strategy here?

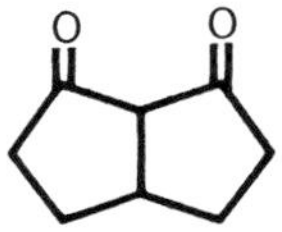

(38)

Answer :

1,3-Dicarbonyl disconnection of symmetrical (38) again reveals a suitable 1,6-dicarbonyl starting material (39) but reconnection gives an impossibly strained alkene (40).

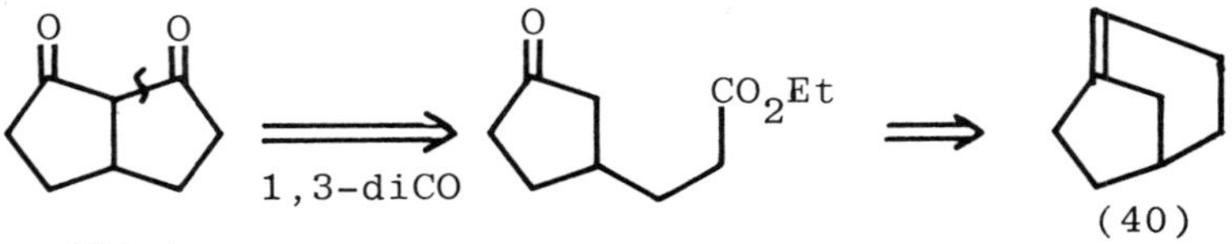

(38a) $\Longrightarrow$ (39) $\Longrightarrow$ (40) impossible
  1,3-diCO

Problem :

Given the impossibility of reconnection, what next best disconnection for (39) can you find? What is the difficulty with it?

Answer :

Disconnecting the ring from the chain gives Michael acceptors (41) or (44) but it also gives difficult synthons (42) or (43).

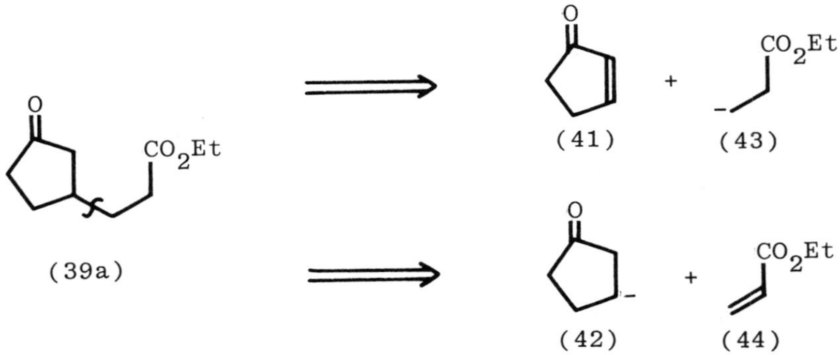

(39a) $\Longrightarrow$ (41) + (43)

$\Longrightarrow$ (42) + (44)

Problem :

Suggest a general approach to a reagent for synthons (42) or (43).

Answer :

Two approaches look reasonable. One is to prepare a protected Grignard reagent such as (45) or (46). It is necessary to change the oxidation level of the ester in (43) as esters cannot be protected as acetals.

(45) 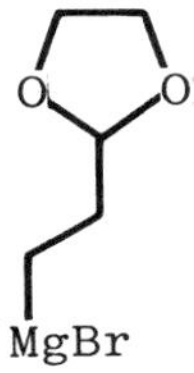 (46)

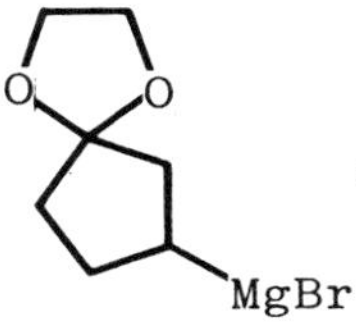

An alternative is to carry out a normal Michael addition and extend the chain by cyanide displacement, though this does not strictly produce a reagent for either (42) or (43).

Synthesis[317]

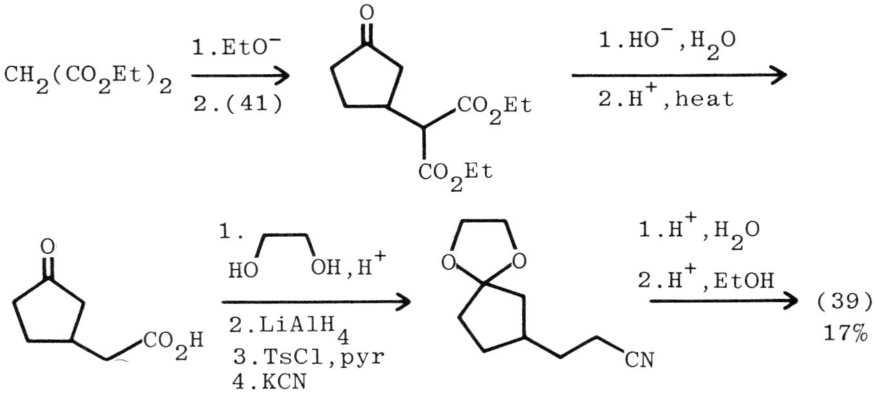

$$CH_2(CO_2Et)_2 \xrightarrow[2.(41)]{1.EtO^-} \text{[cyclopentanone-CH(CO_2Et)_2]} \xrightarrow[2.H^+,heat]{1.HO^-,H_2O}$$

$$\text{[cyclopentanone-CH_2CO_2H]} \xrightarrow[\substack{2.LiAlH_4 \\ 3.TsCl,pyr \\ 4.KCN}]{1.HO\overset{}{\frown}OH,H^+} \text{[acetal-CH_2CH_2CN]} \xrightarrow[2.H^+,EtOH]{1.H^+,H_2O} (39) \quad 17\%$$

Another approach is to use the Baeyer-Villiger reaction we have just carried out to make (34), esterify, and oxidise to the ketone.

Synthesis³¹⁶

(34) 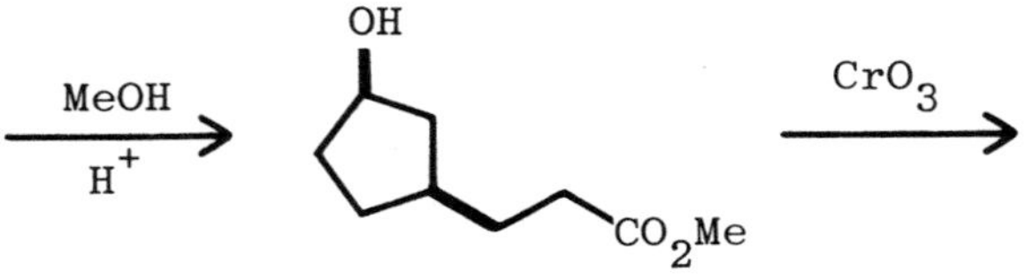 →[MeOH, H⁺] [cyclopentanol with CH₂CH₂CO₂Me substituent] →[CrO₃] (39) 40%

The highest yield comes from a modification of (43) in which the oxidation level is changed to that of an alcohol and the copper reagent from protected (47) added to (41).

Synthesis³¹⁶

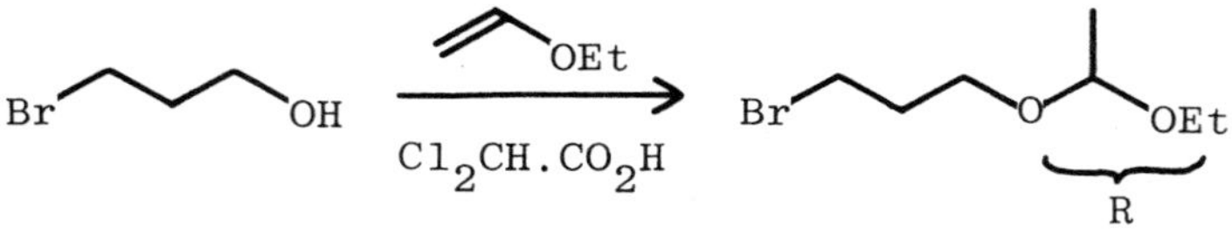

Br~~~OH →[CH₂=CHOEt, Cl₂CH.CO₂H] Br~~~O-CH(CH₃)-OEt (47) 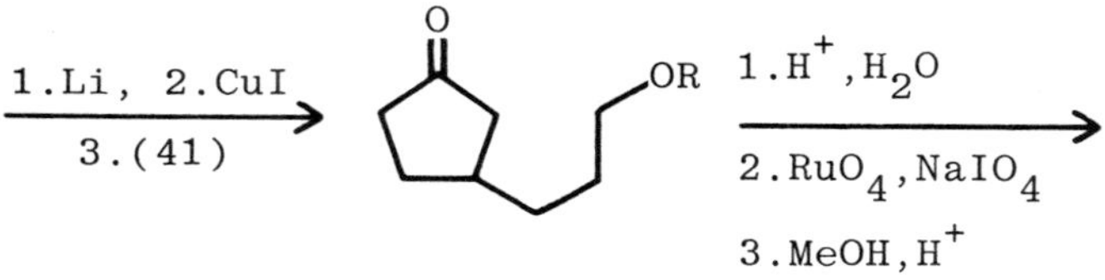
                                              ⎵⎵⎵⎵⎵⎵
                                                 R

→[1. Li, 2. CuI, 3. (41)] [3-substituted cyclopentanone with CH₂CH₂CH₂OR] →[1. H⁺, H₂O; 2. RuO₄, NaIO₄; 3. MeOH, H⁺] (39) 50%

However (39) is made, it cyclises to (38) in excellent yield. and this diketone was used in the synthesis of an 'architectural' compound, peristylane.

Synthesis³¹⁶

(39) 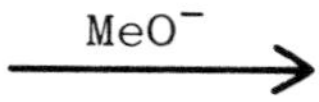 →[MeO⁻] TM(38) 85-95%

CHAPTER 28

General Strategy B: Strategy of Carbonyl Disconnections

Problems in this area tend to be quite difficult so you will find more of a mixture of example and problem than usual.

Problem :

A triketone can be made in moderate yield by the base-catalysed cyclisation of (1). Count up the relationships in (1) and suggest possible outline approaches, on the lines of the discussion on page T

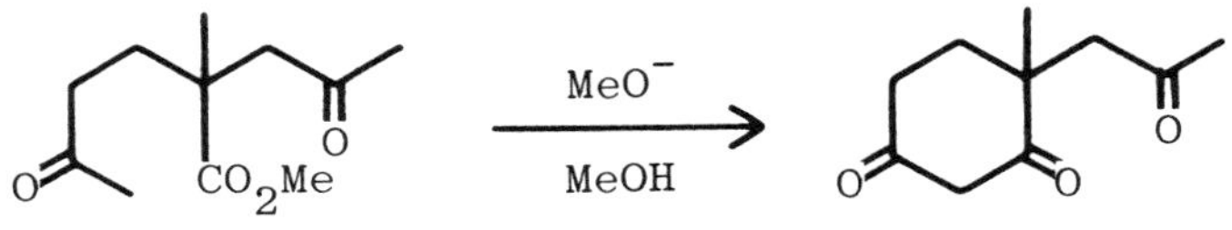

Answer :

The diketo-ester (1) has 1,4 1,5 and 1,6-dicarbonyl relationships. Both the 1,4 and 1,5 relationships can be disconnected at the branchpoint in the middle of the molecule, to give sensible intermediates such as (3) and (4), but difficult synthons (2) and (5). The 1,6 reconnection requires symmetrical Diels-Alder adduct (6).

Analysis

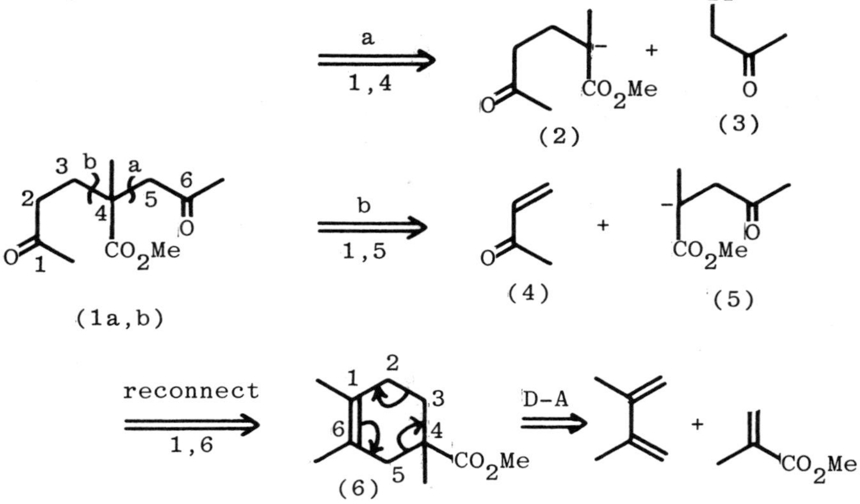

The synthesis was carried out by the Diels-Alder addition and ozonolysis with reductive work-up. No doubt other syntheses could have been based on the 1,4 and 1,5 strategies.

Synthesis[318]

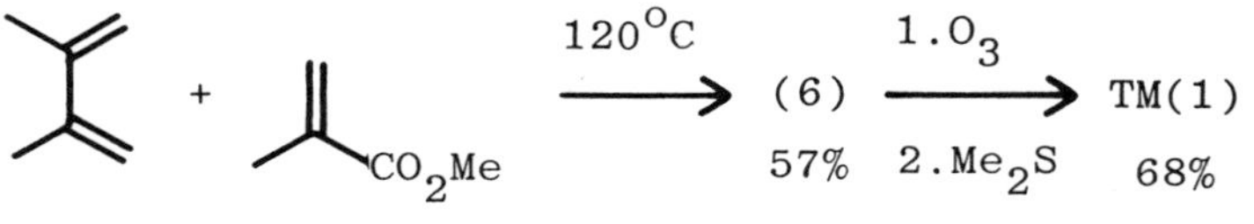

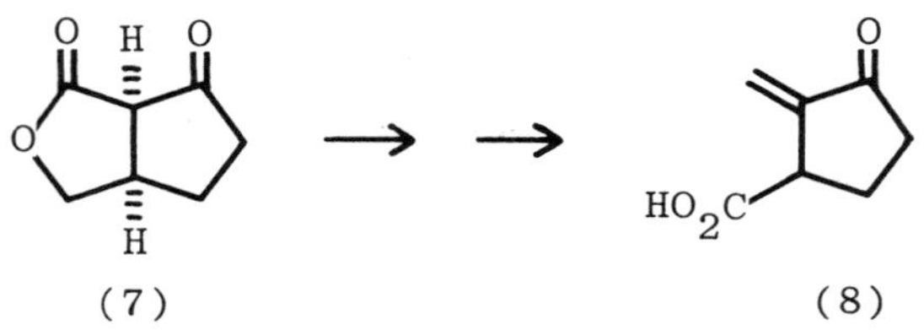

Example :

The antibiotic sarkomycin (8) has been synthesised *via* intermediate (7). There are three sensible initial disconnections: the lactone (C-O) and the two 1,3-dicarbonyls, followed by the C-O disconnection.

Analysis 1

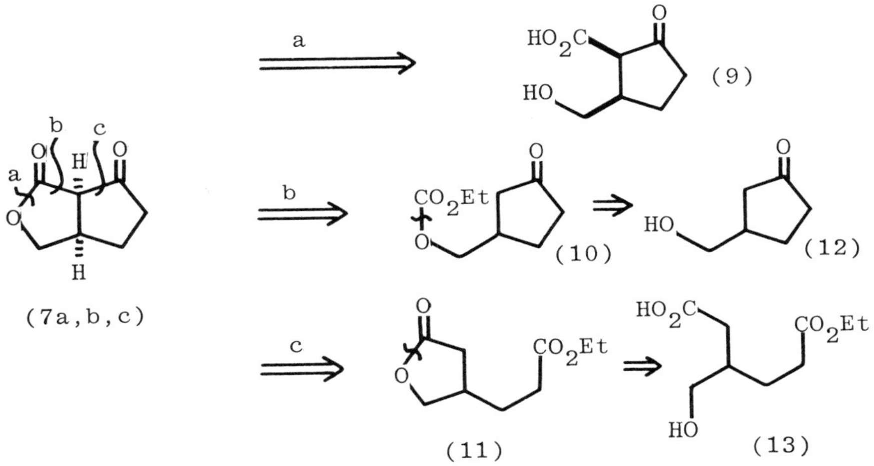

The *cis* stereochemistry of (7) is irrelevant to (10) or (11), as cyclisation can give only *cis* fusion of two five-membered rings, but it might be a problem for (9). In any case, no disconnection, other than removing single carbon atoms, can easily be made on (9). Much the same is true of (12), but reconnection of the 1,6-dicarbonyl in (13) provides a good starting material (14).

Analysis 2

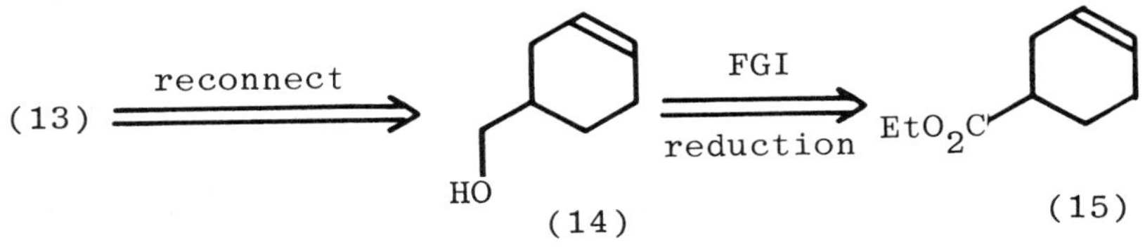

Synthesis[319]

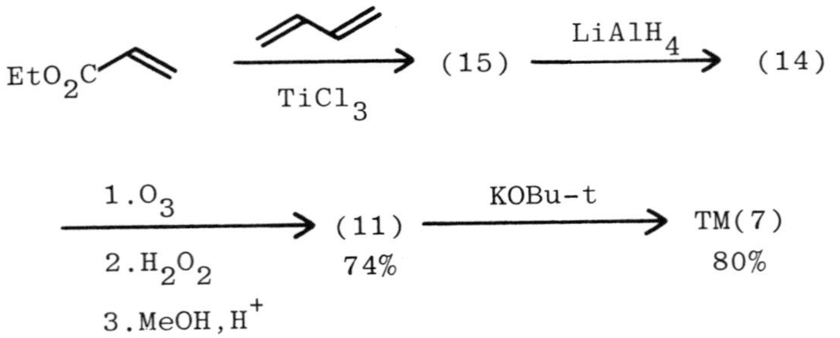

EtO_2C—CH=CH₂ →[TiCl₃, CH₂=CH-CH=CH₂] (15) →[LiAlH₄] (14)

→[1.O_3 / 2.H_2O_2 / 3.MeOH,H^+] (11) 74% →[KOBu-t] TM(7) 80%

Cyclisation to (11) is spontaneous during the work-up from the ozonolysis.

Example :

FGA and carbonyl strategy : the synthesis of long chain fatty acids such as (16) can be achieved in a few steps by suitable FGA. Adding a carbonyl group in a 1,6 relationship allows reconnection to (17) and hence the disconnection of the six carbon atoms in the ring.

Analysis

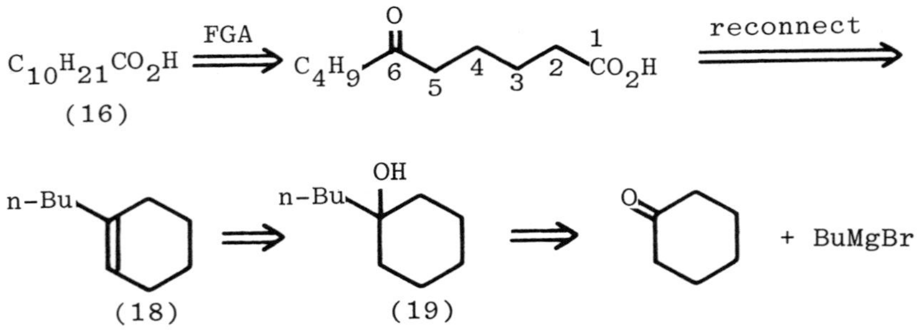

$C_{10}H_{21}CO_2H$ ⇒[FGA] C_4H_9-CO-CH₂-CH₂-CH₂-CH₂-CO_2H (positions 6,5,4,3,2,1) ⇒[reconnect]

(16)

n-Bu-cyclohexene (18) ⇒ n-Bu,OH-cyclohexane (19) ⇒ cyclohexanone + BuMgBr

Synthesis[320]

cyclohexanone →[BuMgBr] (19) →[H^+] (18) 40%

→[1.O_3 / 2.H_2O_2] (17) 63% →[1.NH_2NH_2 / 2.KOH] TM(16)

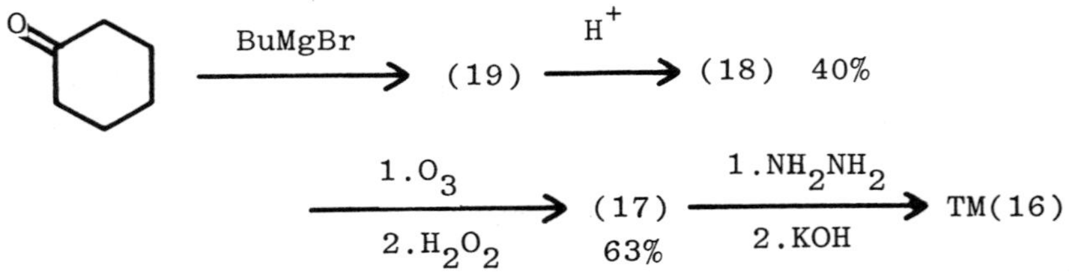

An interesting alternative strategy is to reconnect (17) to the 1,3-dicarbonyl compound (20) which can be made by one 1,3-dicarbonyl condensation and cleaved to (17) by the reverse of another.

Analysis 2

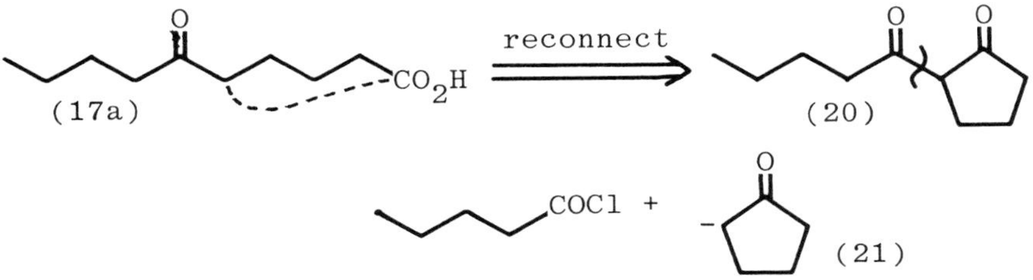

[Structure of (17a): linear chain with ketone and CO₂H, with dashed bond indicating disconnection] → reconnect → [Structure of (20): chain with ketone attached to cyclopentanone]

[Structure: BuCH₂COCl + cyclopentanone enolate (21)]

An enamine provides synthon (21) and the cleavage of (20) to (17) occurs in aqueous base. Attack on the slightly strained five-ring ketone is faster than on the exocyclic ketone.

Synthesis[320]

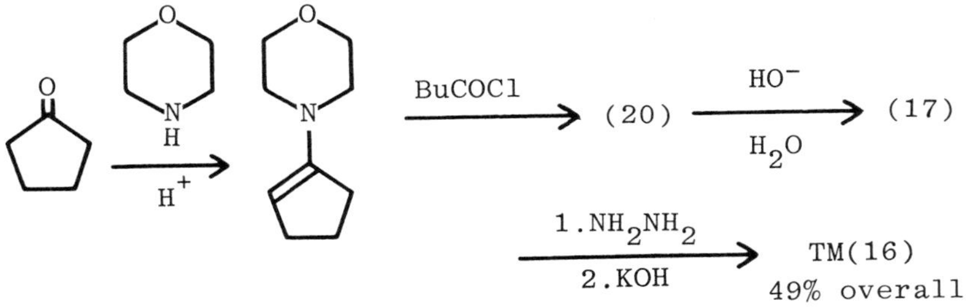

[Cyclopentanone + morpholine/H⁺ → morpholine enamine of cyclopentanone → BuCOCl → (20) → HO⁻/H₂O → (17)]

[(20) → 1. NH₂NH₂, 2. KOH → TM(16), 49% overall]

The acid chloride of (16) can be used to acylate another enamine and these processes repeated, adding five or six carbon atoms each time to the previously assembled chain.

Problem:

Consider general approaches to the synthesis of (22), the sex pheromone of the olive fly, a serious pest of olive trees.

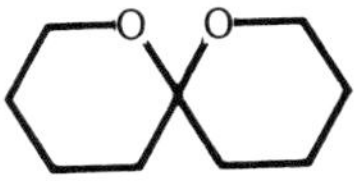

[Structure of (22): spiroketal, 1,7-dioxaspiro[5.5]undecane]

Answer :

The first disconnection must be of the acetal to reveal two identical 1,5-relationships (23). Disconnection of these would require a change of oxidation level to, say (24), and the addition of activating groups (25).

Analysis 1

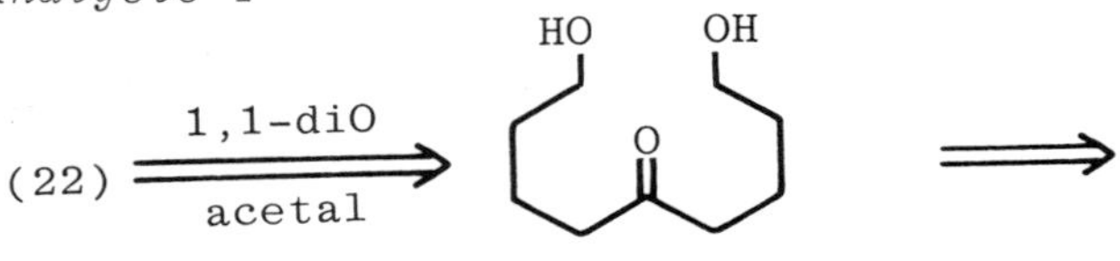

(22) $\xrightleftharpoons[\text{acetal}]{\text{1,1-diO}}$ [HO–(CH$_2$)$_3$–CO–(CH$_2$)$_3$–OH] (23) $\Longrightarrow$

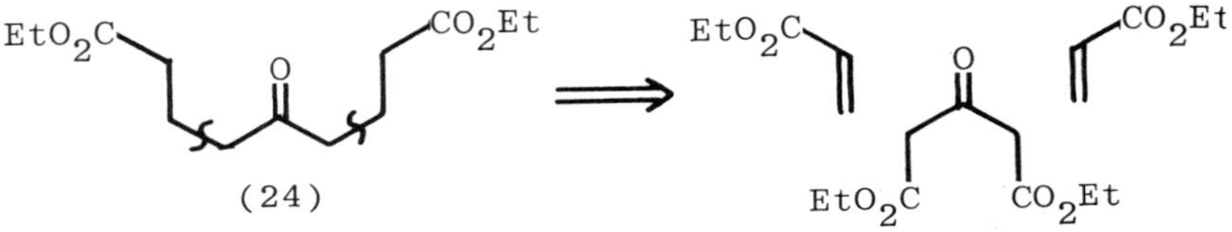

[EtO$_2$C–(CH$_2$)$_3$–CO–(CH$_2$)$_3$–CO$_2$Et] (24) $\Longrightarrow$ [EtO$_2$C–CH(CO$_2$Et)–CH$_2$–CO–CH$_2$–CH(CO$_2$Et)–CO$_2$Et] (25)

This synthesis has not been attempted, as far as I know. The compound has been made by the random method, oxidation of diol (26) giving 3 per cent of the product.

Synthesis[3][2][1]

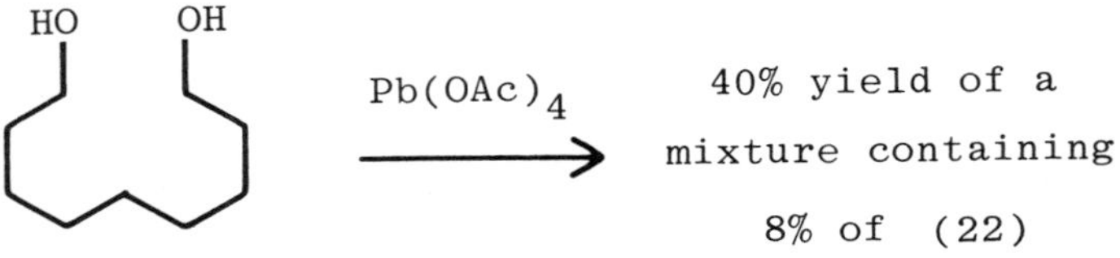

[cyclodecane-1,2-diol] $\xrightarrow{\text{Pb(OAc)}_4}$ 40% yield of a mixture containing 8% of (22)

A more logical approach was to adjust the oxidation level of (23) and to reconnect in the manner of (17a) to give symmetrical (26), easily made by a Michael addition.

Analysis 2

[Structure (24a): EtO₂C-(CH₂)₃-CO-(CH₂)₃-CO₂Et with reconnect shown]

$$\text{(24a)} \Longrightarrow$$

[Structure (26): 1,3-cyclohexanedione with CH₂CH₂CO₂Et substituent at C2] $\Longrightarrow$ [1,3-cyclohexanedione] + CH₂=CH-CO₂Et

(26)

This time the ketone cleavage was carried out in acid solution. Protection of (24) is necessary before reduction, and cyclisation to (22) is spontaneous on removal of the protecting group.

Synthesis[322]

[1,3-cyclohexanedione] $\xrightarrow[\text{base}]{\text{CH}_2=\text{CH-CO}_2\text{Et}}$ (26) $\xrightarrow{\text{conc. H}_2\text{SO}_4}$ EtO₂C~~~CO-~~~CO₂H 91%

$\xrightarrow[\text{H}^+]{\text{EtOH}}$ (24) 96% $\xrightarrow[\text{H}^+]{\text{HO-OH}}$ EtO₂C~~~[dioxolane]~~~CO₂Et 83%

$\xrightarrow{\text{LiAlH}_4}$ HO~~~[dioxolane]~~~OH 96% $\xrightarrow{\text{H}^+}$ TM(22) 87%

The best strategy is to leave (23) at the correct oxidation level, add one activating group and use the

1,3-dicarbonyl disconnection to reveal two molecules of (27).

Analysis 3

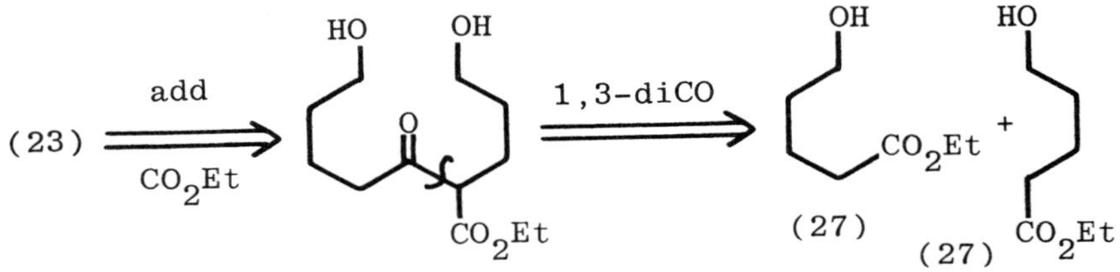

(23) $\xrightarrow[\text{CO}_2\text{Et}]{\text{add}}$ [diol keto diester] $\xrightarrow{\text{1,3-diCO}}$ (27) + (27)

In practice, lactone (28) is a better reagent for (27). Condensation gives an enol ether which, without isolation, is treated with acid to give (22) in high yield.

Synthesis[323]

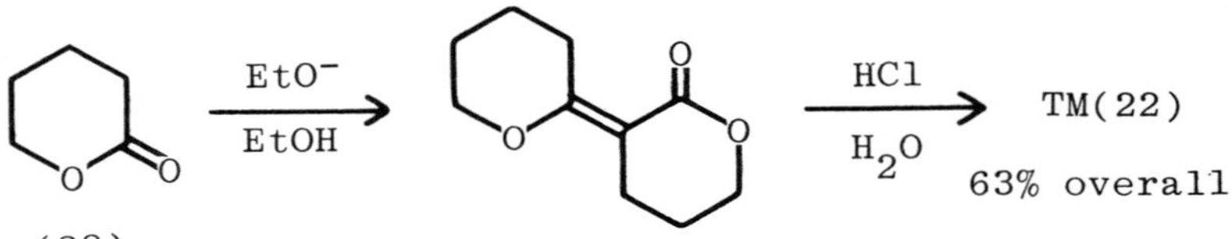

(28) $\xrightarrow[\text{EtOH}]{\text{EtO}^-}$ [enol ether] $\xrightarrow[\text{H}_2\text{O}]{\text{HCl}}$ TM(22)

63% overall

Furans occur widely in nature and many are important commercially. Thus alcohol (29) is used in various insecticides. The carbon atoms joined to the ring oxygen atom are at the carbonyl oxidation level so that (29) can be made by acid-catalysed cyclisation of (30).

Analysis

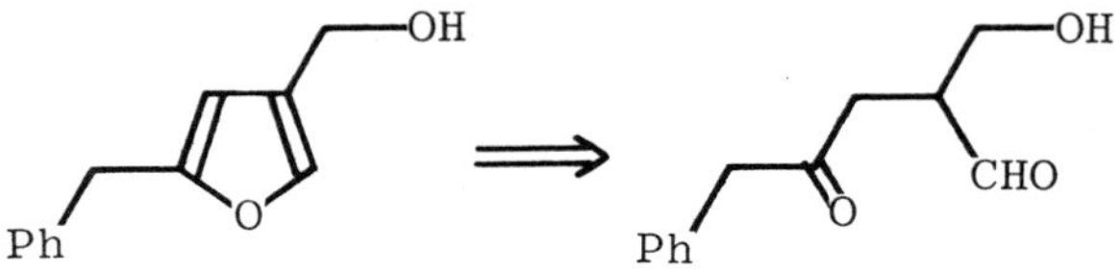

(29) $\Longrightarrow$ (30)

Problem :
Consider possible strategies for the synthesis of (30).
Answer :
Intermediate (30) contains one 1,3 and two 1,4-relationships. The 1,4 disconnections look unpromising : perhaps the best of them is (30a) but a synthesis for (31) and a reagent for (32) (an enamine?) might be hard to find.

Analysis 1

Ph—C(=O)—CH2—CH(CH2OH)—CHO (30a)

$\xrightarrow{1,4\text{-diCO}}$

Ph—C(=O)—CH(Cl)—CH2— (31) + ⁻CH(OH)—CHO (32)

The 1,3-disconnection can only be made at the dicarbonyl oxidation level which means going all the way to ester (33) to preserve the distinction between the two groups.

Analysis 2

(30) $\xrightarrow[\text{reduction}]{\text{FGI}}$ Ph—C(=O)—CH2—CH(CHO)—CH2—CO2Et (33)

$\Longrightarrow$ Ph—C(=O)—CH2—CH2—CH2—CO2Et (34) + HCO2Et

We must now disconnect the 1,4-relationship in (34) and there are many possibilities here. Disconnection of the central bond is possible, e.g. to give (31) and malonate, or cyanide could be added to (35). The best is probably to use an available 1,4-dicarbonyl compound such as (37).

Analysis 3

[Scheme showing retrosynthetic analysis:

(34a,b) ⟹ᵃ PhCH₂C(=O)CH=CH₂ (35) + CN⁻

(34a,b) ⟹ᵇ PhCH₂⁻ (36) + succinic anhydride (37)]

One published synthesis uses nitrile (38) for synthon (36) and the first step is essentially a 1,3-dicarbonyl synthesis. Removal of the cyanide by decarboxylation gives (34). Protection of the ketone is necessary for the condensation but not for the reduction if this is kept to the end after the furan is formed.

Synthesis[324]

[Synthesis scheme:

Ph-CH(CN)- (38) + EtO-CH(CO₂Et)CH₂CH₂CO₂Et → Ph-C(CN)(-)-C(=O)CH₂CH₂CO₂Et →(1. conc. HCl, HOAc, H₂O; 2. EtOH, H⁺) (34)

→(H⁺, HOCH₂CH₂OH) Ph-C(dioxolane)-CH₂CH₂CO₂Et →(NaH, HCO₂Et) Ph-C(dioxolane)-CH(CHO)CO₂Et

→(c. HCl) Ph-CH₂-furan-CO₂Et →(LiAlH₄) TM(29)]

Another interesting furan is ipomeamarone (39), a stress metabolite produced by sweet potatoes when they are attacked by fungi.

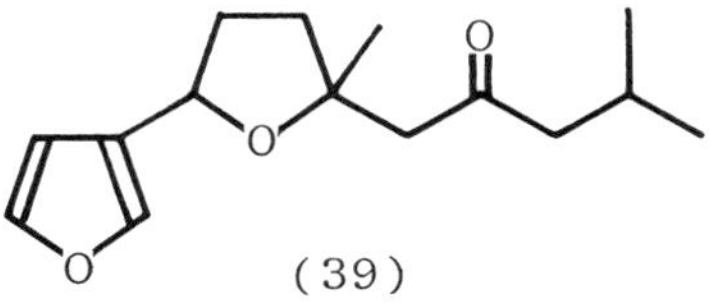

(39)

There is only one substituent on the furan so it will probably be best to disconnect within the side chain and come back to an intact furan as starting material. The side chain contains 1,3 1,4 and 1,6 relationships but the C-O disconnection (39a) is particularly helpful. Important as carbonyl chemistry is, it is also important to look at other possible disconnections.

Analysis 1

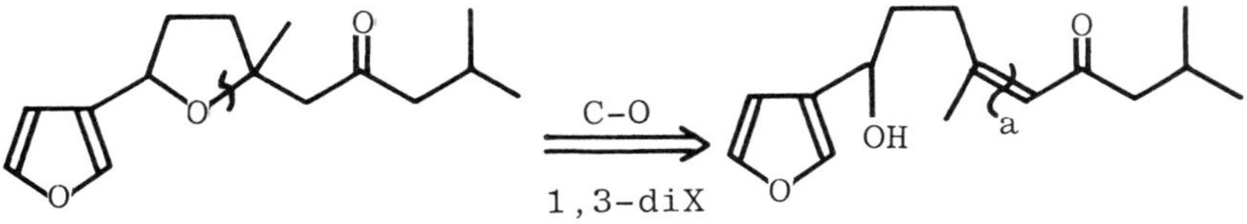

(39a) $\xrightarrow{\text{C-O}}$ (40)
1,3-diX

The α,β-disconnection (40a) now gives considerable simplification and leaves only the 1,4-relationship in (41) and the potentially difficult (regioselectivity) synthon (42).

Analysis 2

(40) 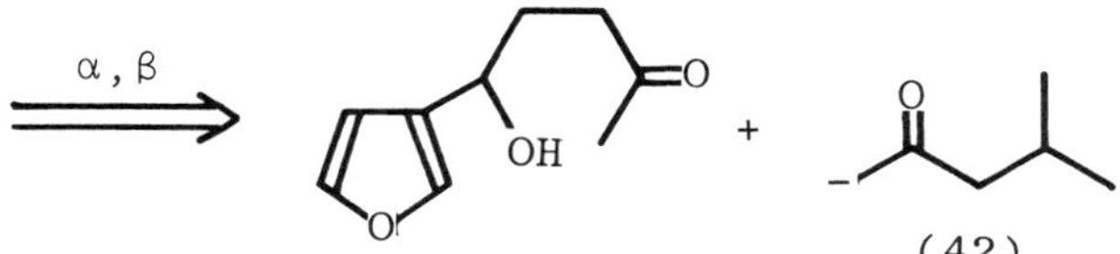 $\xrightarrow{\alpha,\beta}$ (41) + (42)

Disconnection of the central bond in (41) might require epoxide (43) or ketoester (45). Since furan acid (46) is one of the few relatively accessible 3-functionalised furans, the higher oxidation level is preferred.

Analysis 3

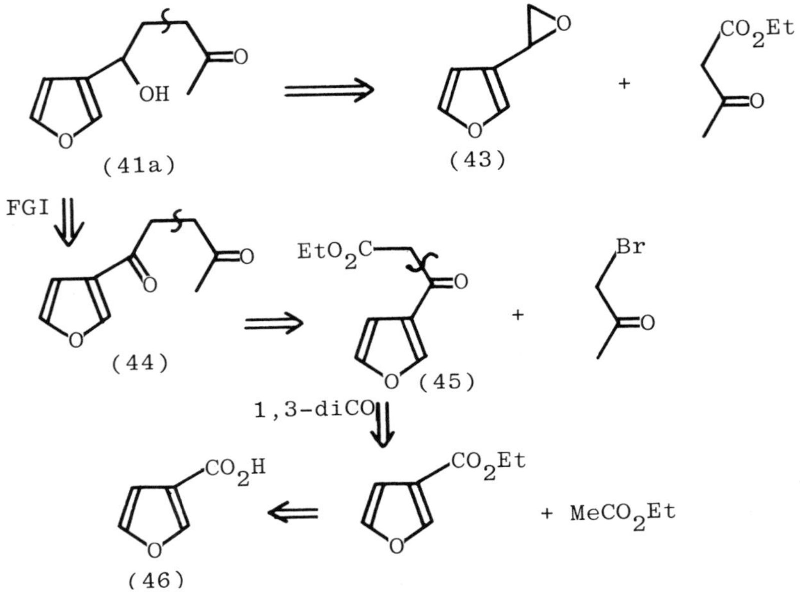

[Retrosynthetic scheme: (41a) ⇒ (43) + ethyl acetoacetate fragment; FGI down to (44); (44) ⇒ (45) + bromoacetone; 1,3-diCO down to (46) ⇐ ethyl furan-3-carboxylate + MeCO$_2$Et]

Ipomeamarone has been synthesised by a slight variant on this strategy in which the methyl group is added to lactone (47) and a Wittig reagent is used for synthon (42).[325]

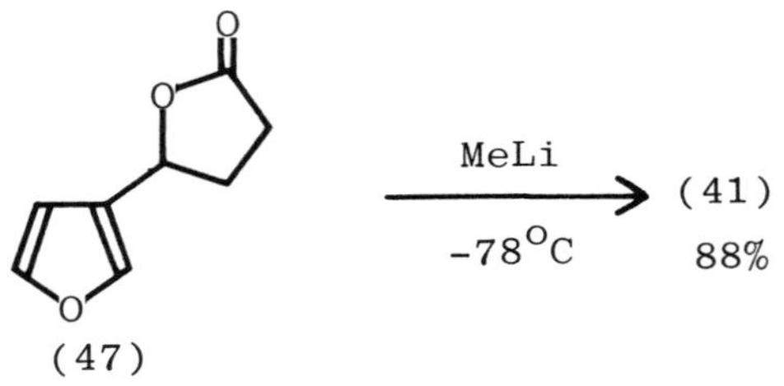

(47) $\xrightarrow[-78°C]{\text{MeLi}}$ (41)   88%

CHAPTER 29

Strategy XIII: Introduction to Ring Synthesis: Saturated Heterocycles

(a) *Cyclisations*

Example :
The kinetic advantages of five-membered rings over other sizes are well illustrated by the radical reaction leading to cyclic amine (3) used on p T 247 . Chlorination of secondary amine (1) gives N-chloro compound (2) which gives (3) on heating in acid solution.[326]

n-BuNHEt $\xrightarrow{Cl_2}$ n-BuNEt-Cl $\xrightarrow[H^+]{100°C}$ (pyrrolidine with N-Et)

(1) (2) (3) 70-80%

The secret of the cyclisation is that salt (4) decomposes to radical (5) (and Cl·) which abstracts a hydrogen atom from the fourth carbon atom along the chain (5, arrows) via transition state (6) to give radical (7). This recaptures Cl· (from Cl_2) to give (8) which cyclises to (3) very easily.

(2) $\xrightarrow{H^+}$ (4) $\rightarrow$ (5) $\rightarrow$

(4) (5)

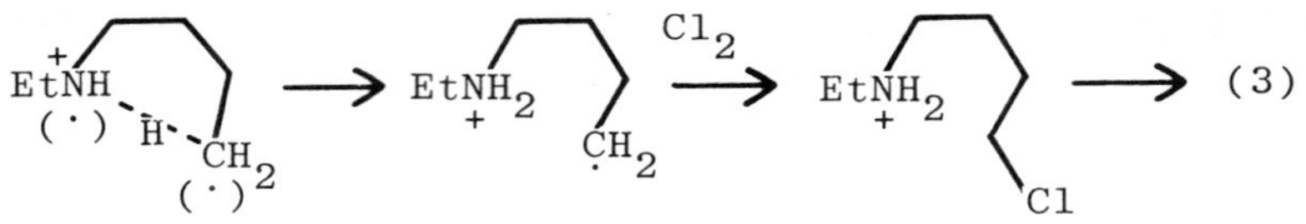

EtNH
(·)
H
CH2
(·)
EtNH2
CH2
Cl2
EtNH2
Cl
(3)

(6) (7) (8)

The site selectivity in this reaction is excellent - 70 - 80% yield of (3) - and it depends on the stability of cyclic transition state (6). This looks at first like a six-membered ring, but N--H--CH_2 is linear and shorter than two bonds so that it is somewhere between a five and a six-membered ring. Cyclisation to (3) is via a conventional five-membered cyclic transition state.

Problem :

Treatment[327] of (9) with strong base gives cyclic amine (10). Why is this particular product formed?

MeO
Cl
NHCH2Ph
PhLi
MeO
N
Ph

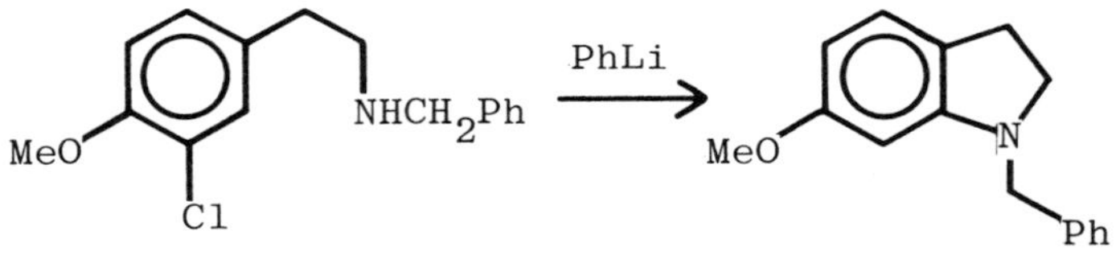

(9) (10)

Answer :

The clue is that displacement of chloride by nitrogen has occurred at the 'wrong' carbon atom. The strong base must therefore generate a benzyne (11) which cyclises easily to give a five-membered ring but which could cyclise to the other end of the triple bond only with great difficulty.

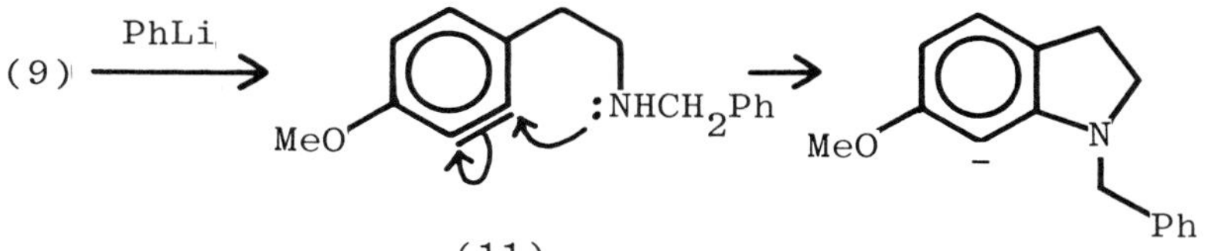

(9) —PhLi→ [structure (11) with MeO, :NHCH₂Ph] → [structure (10) with MeO, N-Ph] ⟶ (10)

Problem :

Cyclisation of (12) to a β-lactam (13) is an attractive reaction for those wanting to synthesise antibiotics. What other reaction(s) might compete with the cyclisation and why?

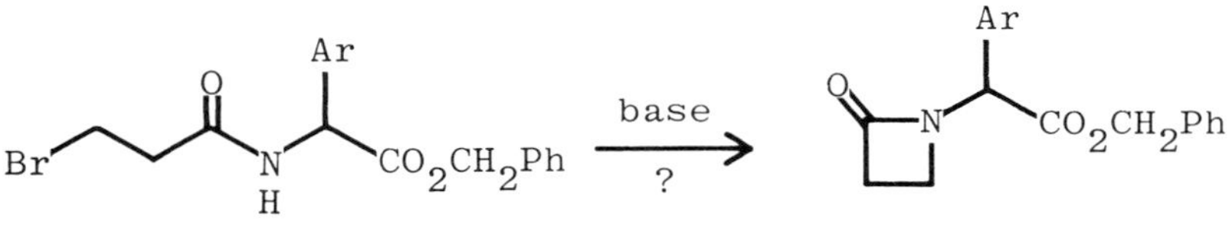

[structure (12): Br-CH₂CH₂-C(=O)-N(H)-CH(Ar)-CO₂CH₂Ph] —base ?→ [structure (13): β-lactam with Ar and CO₂CH₂Ph]

(12)   (13)

Answer :

The problem is that this is a cyclisation to give a four-membered ring with all the attendant kinetic and thermodynamic problems (p T 241). Cyclisation to give a three-membered ring (14) is not a serious problem as the reaction is reversible. Attack by the carbanion (15) to give stable (16) might be a problem but the carbanion is difficult to form. More serious problems are intermolecular reactions such as the nitrogen atom of one molecule displacing the bromide from another or elimination via carbanion (17).

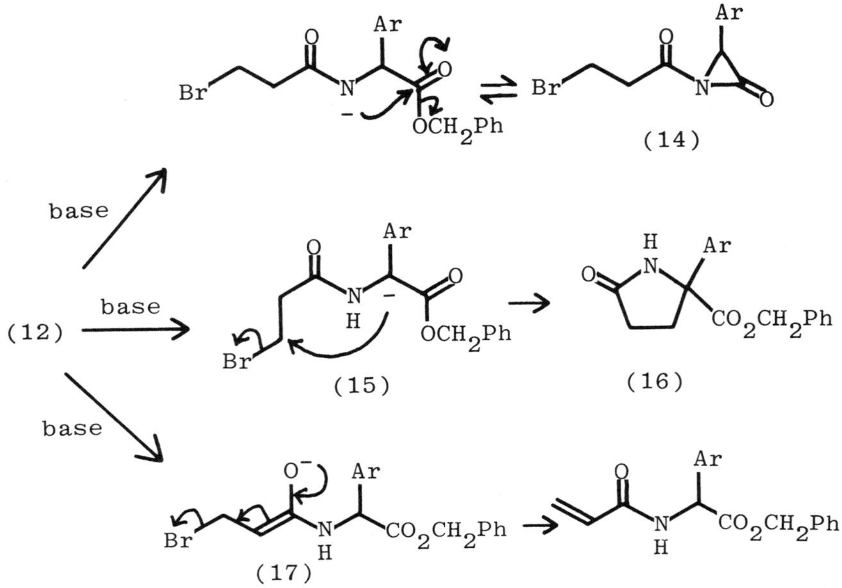

In the event, intramolecular reactions could be avoided by working at high dilution and conditions were found - adding (12) to NaH in a special solvent at controlled temperatures - under which elimination was minimised.[328] For example, (18) could be made in 80% yield.

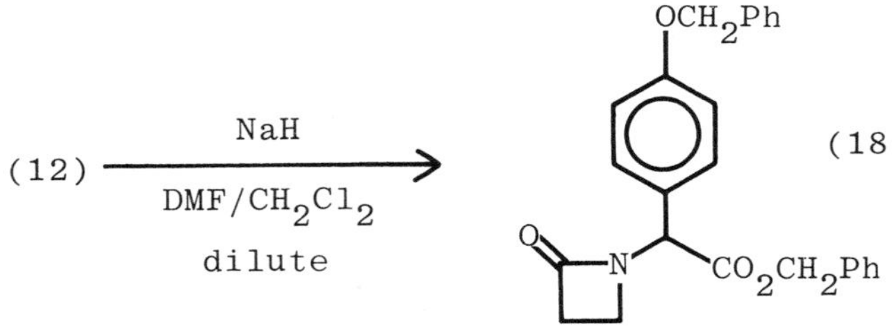

(b) *Saturated Heterocycles*

Epoxidation

Example :

A stereoselective synthesis[329] of the *cis*-isomer (23) (this isomer is used as a perfumery compound whereas the *trans* isomer has little smell, see p T 93) starts from β-pinene (19), a turpentine constituent.

Epoxidation occurs on the side away from the CMe$_2$ bridge to give (20). This fragments in water to give (21) which can be oxidised to (22) under mild conditions. Hydrogenation occurs preferentially on the opposite side to the CMe$_2$OH substituent to give (23).

Problem :
Epoxide (25) is needed as an intermediate for reductive alkylation. How would you make it from carvone (24)?

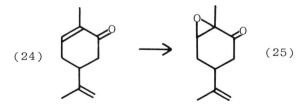

Answer :
Peracids would attack the other, more electron-rich, double bond. We need a nucleophilic epoxidising agent

which will attack Michael fashion (26) and the best is simply HOO⁻.

*Synthesis*³³⁰

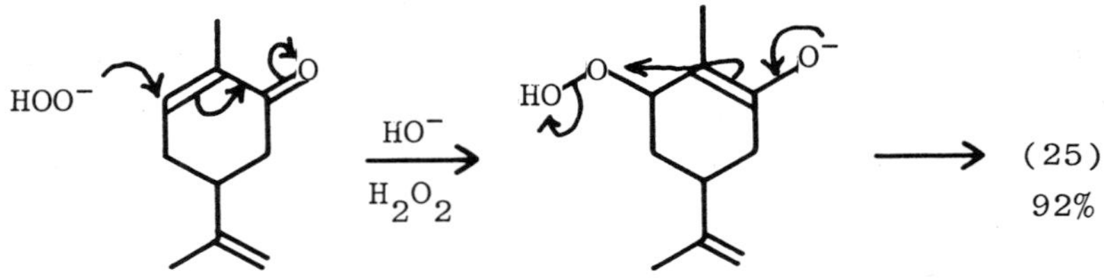

(25) 92%

(26)

Example :

The Darzens reaction,³³¹ which is described on p T 253, is a good example of cyclisation to give a three-membered ring. The enolate of an α-halo ester, e.g. (27), adds to a ketone to give (28) which cannot be isolated as it cyclises so readily.

*Synthesis*³³¹

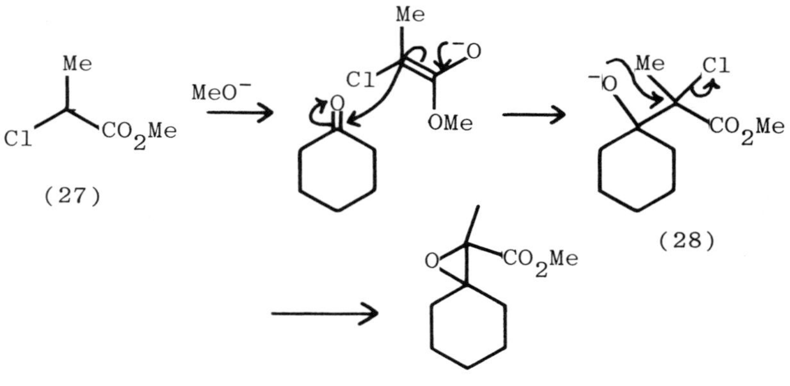

(27)

(28)

(ii) *Four-Membered Rings*

Example :

The attack of bromine or iodine on β,γ-unsaturated acids (29) usually gives five-membered lactones (30) even when the CO_2H has to attack the less substituted end of the double bond (cf p 308).

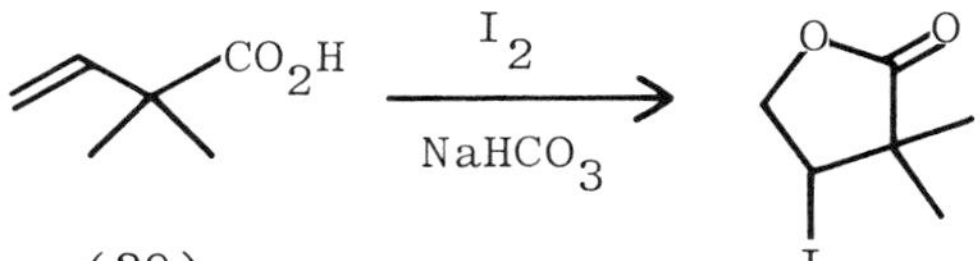

$$(29) \xrightarrow[NaHCO_3]{I_2} (30)$$

However, under carefully controlled conditions,[332] the product of attack at the more stable carbonium ion can be isolated even though this is an unstable four-membered lactone (31).

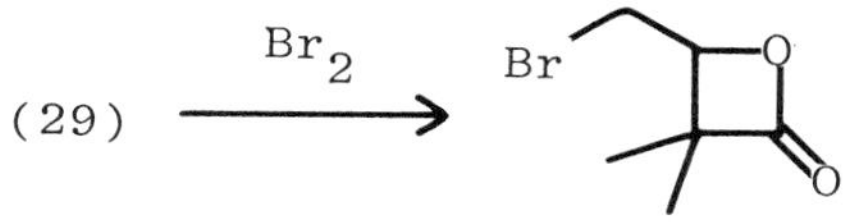

$$(29) \xrightarrow{Br_2} (31)\ 70\text{-}80\%$$

Problem :

Suggest how available oestrone ether (32) might be converted into potentially anti-oestrogenic (33).

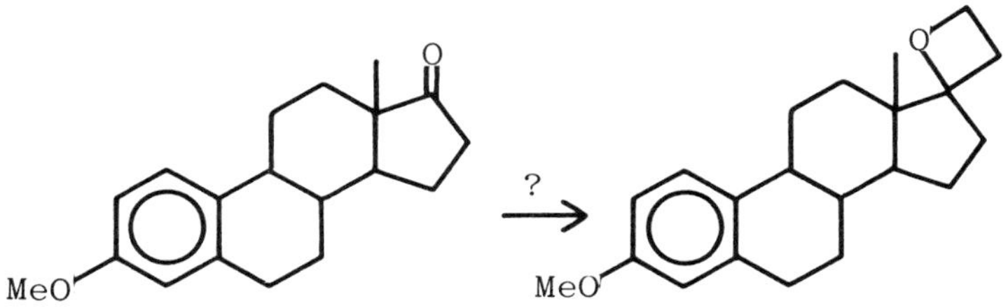

Answer :

Ether disconnection gives diol (34) which must be disconnected next to the oxygen to give (32) and synthon (35). A protected Grignard reagent such as $THPOCH_2CH_2MgBr$ might just be made to work, but changing the oxidation level to (36) is much easier.

Analysis

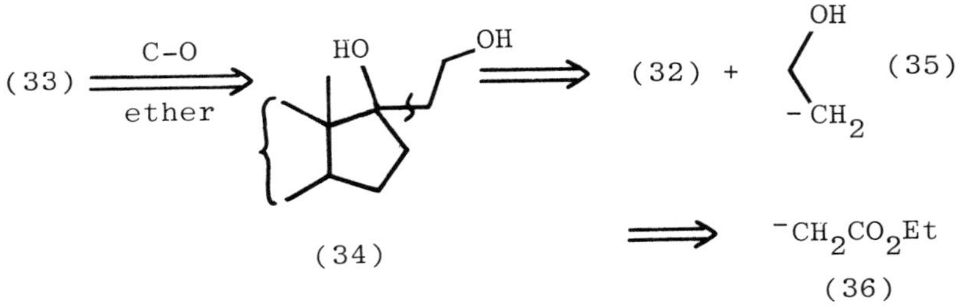

(33) $\xrightleftharpoons[\text{ether}]{\text{C-O}}$ (34) $\Longrightarrow$ (32) + (35)

(35) structure: $HOCH_2-CH_2^-$

$\Longrightarrow {}^-CH_2CO_2Et$ (36)

A Reformatsky reagent is ideal for (36) and the product can be reduced to (34). Treatment with acid will not be enough to make a four-membered ring, but mono-tosylation (cf p T 243) and treatment with strong base gives (33).

Synthesis[333]

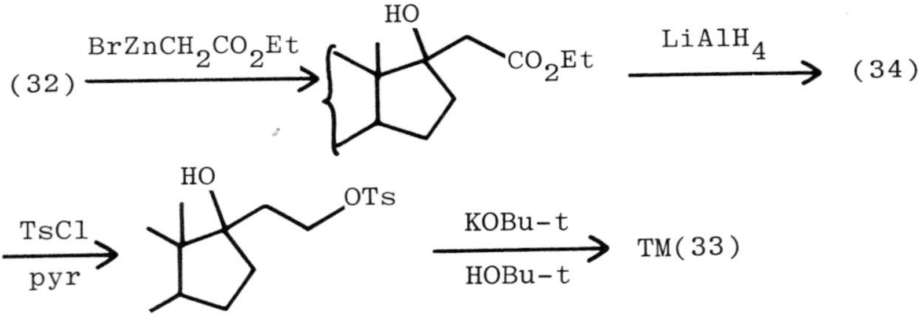

(32) $\xrightarrow{BrZnCH_2CO_2Et}$ [hydroxy-cyclopentane with CH$_2$CO$_2$Et] $\xrightarrow{LiAlH_4}$ (34)

$\xrightarrow[\text{pyr}]{TsCl}$ [hydroxy-cyclopentane with CH$_2$CH$_2$OTs] $\xrightarrow[\text{HOBu-t}]{KOBu-t}$ TM(33)

(iii) *Five-Membered Rings*

Problem :

Amino ketone (37) was needed for a study of the influence of nitrogen atoms on the Clemmensen reduction of ketones. How might it be synthesised?

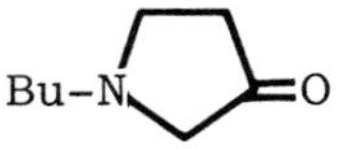

Bu-N[pyrrolidinone ring]=O

(37)

Answer :

C-N disconnection requires difficult (38) so it is better to look for C-C disconnections. The strategy of adding an activating group (39) to provide a 1,3-dicarbonyl disconnection is successful (cf p 204).

Analysis

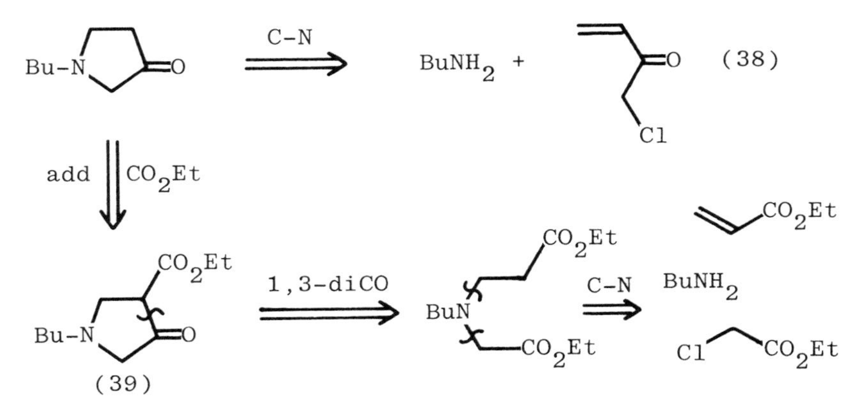

[Scheme: Bu-N pyrrolidinone (37) ⇒ (C-N) BuNH$_2$ + CH$_2$=C(–)–CH$_2$Cl (38); add CO$_2$Et ↓ gives (39) Bu-N pyrrolidinone with CO$_2$Et; (39) ⇒ (1,3-diCO) BuN(CH$_2$CH$_2$CO$_2$Et)(CH(CO$_2$Et)) ⇒ (C-N) BuNH$_2$ + CH$_2$=CHCO$_2$Et + ClCH$_2$CO$_2$Et]

Either electrophile could be added to BuNH$_2$ first; in the published synthesis the Michael addition was first (contrast p T 246). It is of no importance whether (39) or (41) is formed in the cyclisation : both decarboxylate to (37).

Synthesis[3,4]

[Scheme: CH$_2$=CHCO$_2$Et + BuNH$_2$ —EtOH→ BuNHCH$_2$CH$_2$CO$_2$Et (88%) —ClCH$_2$CO$_2$Et / K$_2$CO$_3$→ (40) —Na, xylene, 90%→ (39) + BuN-pyrrolidinone with CO$_2$Et (41) —1. HO$^-$, H$_2$O; 2. H$^+$, heat→ TM(37) (73% from (40))]

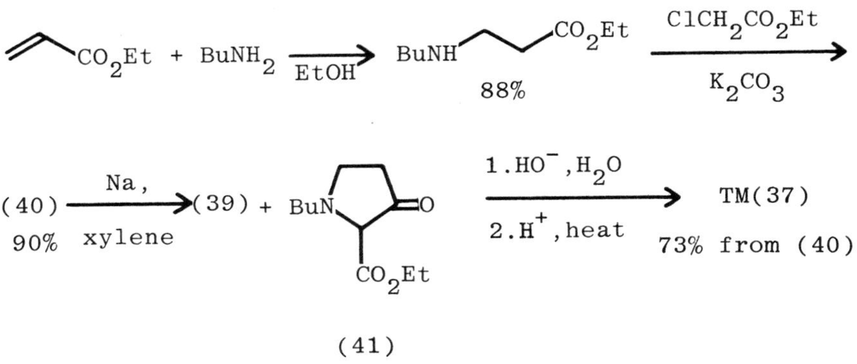

The analgesic and anti-inflammatory Indoprofene (42) can be made by several routes, two of which involve cyclisations. Treatment of phthalic anhydride with amino acid (43) gives imide (45) in one step - the cyclisation of (44) must be faster than the intermolecular reaction by which it is formed. Reduction of (45) gives Indoprofene as the reactive imide is easily reduced as far as the unreactive amide.

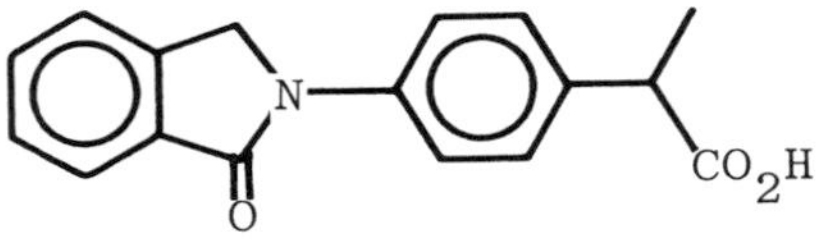

(42)

Synthesis[335]

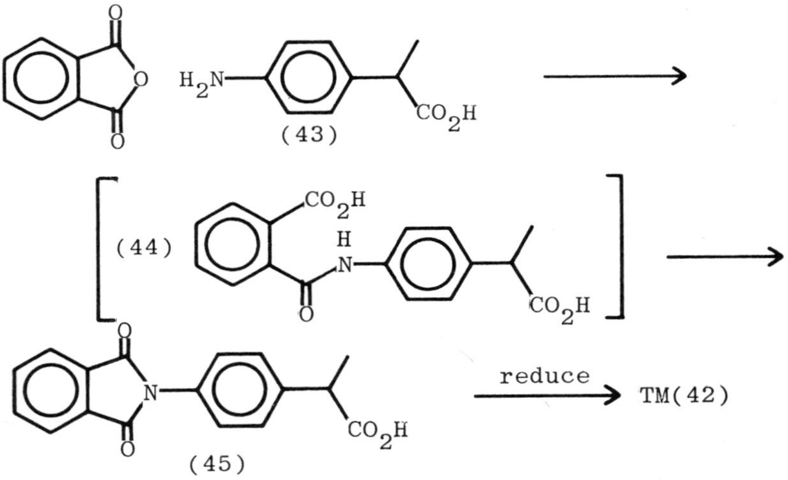

(45) reduce → TM(42)

Alternatively, secondary amine (46) can be treated with phosgene to give (47). This time a catalyst ($AlCl_3$) is necessary for cyclisation. Notice that cyclisation gives only (42) and not the alternative four-membered ring.

Synthesis[335]

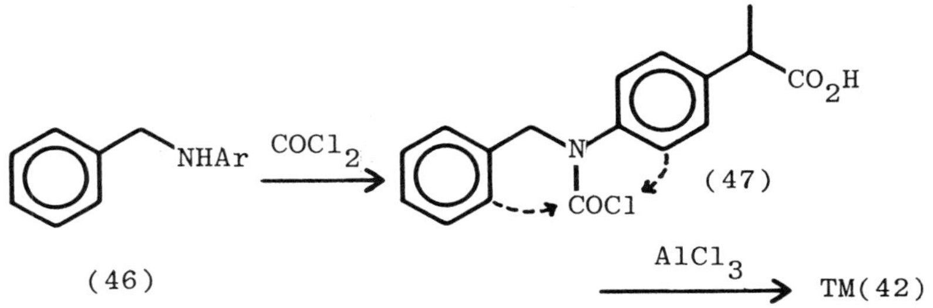

(46) PhCH$_2$NHAr $\xrightarrow{COCl_2}$ (47) $\xrightarrow{AlCl_3}$ TM(42)

Problem:
What general strategies can you suggest for the synthesis of symmetrical 2,3-di substituted tetrahydrofurans (48)?

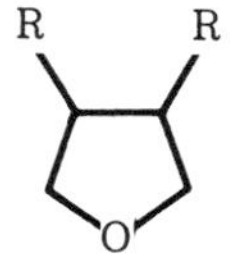

(48)

Answer:
The first disconnection reveals that this is a 1,4 problem (49). Changing the oxidation level to (50) allows disconnection of the central bond (a) or an acyl anion disconnection (b).

Analysis 1

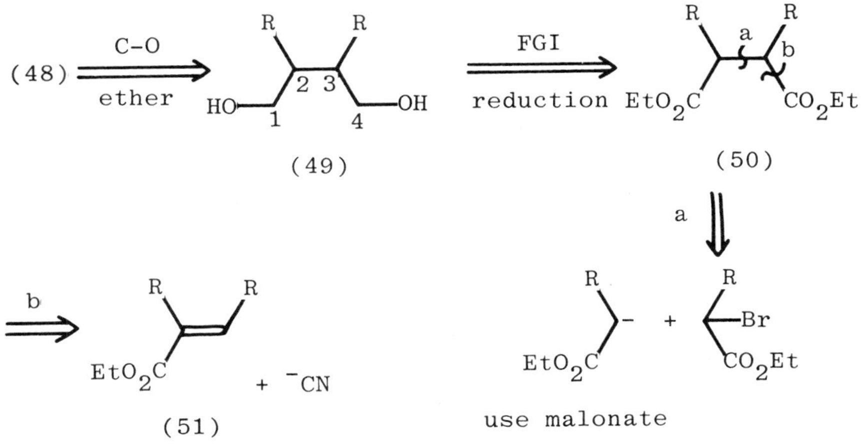

$(48) \underset{ether}{\overset{C-O}{\Longrightarrow}}$ HO—1—2(R)—3(R)—4—OH (49) $\underset{reduction}{\overset{FGI}{\Longrightarrow}}$ EtO$_2$C—(R)—a—(R)—b—CO$_2$Et (50)

a ⇓

EtO$_2$C—(R)—$^-$ + Br—(R)—CO$_2$Et

use malonate

b ⇒ EtO$_2$C—(R)=(R) + $^-$CN

(51)

Problem :
How might these specific examples be made?

(52)   [3,4-diphenyl tetrahydrofuran]   (53) [cis-fused bicyclic tetrahydrofuran]

Answer :
Either strategy will do for (52) which has been made by route (b) using cyanide (54) instead of an ester (51). Cyclisation of (49) merely needs acid, in contrast to the synthesis of (33).

Synthesis[336]

$$PhCHO + PhCH_2CN \xrightarrow{base} \underset{(54)\ 100\%}{\underset{NC}{Ph}\diagup\diagdown\underset{}{Ph}} \xrightarrow[H_2O]{KCN,\ EtOH} \underset{87\%}{\underset{NC}{Ph}\diagup\diagdown\underset{CN}{Ph}}$$

$$\xrightarrow[H_2SO_4]{H_2O} \underset{100\%}{\underset{HO_2C}{Ph}\diagup\diagdown\underset{CO_2H}{Ph}} \xrightarrow[Et_2O]{LiAlH_4} (49, R=Ph) \xrightarrow{H^+} \underset{97\%}{TM(52)}$$

Strategy (b) would also do for (53) but a much easier route is based on the Diels-Alder reaction.

Analysis 2

$$(53) \underset{FGA}{\Longrightarrow} [\text{cis-fused bicyclic anhydride (55)}] \underset{D-A}{\Longrightarrow} \text{butadiene} + \text{maleic anhydride}$$

Synthesis[336]

$$(55) \xrightarrow[2.LiAlH_4]{1.H_2, Pd} [\text{cis-1,2-bis(hydroxymethyl)cyclohexane}] \xrightarrow[DMSO]{heat} \underset{68\%}{TM(53)}$$

Two Heteroatoms

Problem :

Suggest a synthesis of antipyrine (56), one of the very earlier synthetic drugs, still used as a fever-reducing drug in some countries.

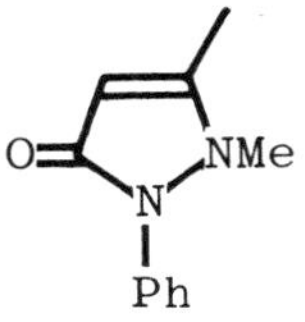

O=[ring]—NMe (56)
   |
   Ph

Answer :

Removing the substituted hydrazine (58) leaves acetoacetic ester (57). Phenylhydrazine is available, so it is easier to make (59) and methylate afterwards. This removes the chemoselectivity problem as the more nucleophilic NH_2 group attacks the more electrophilic ketone.

Analysis

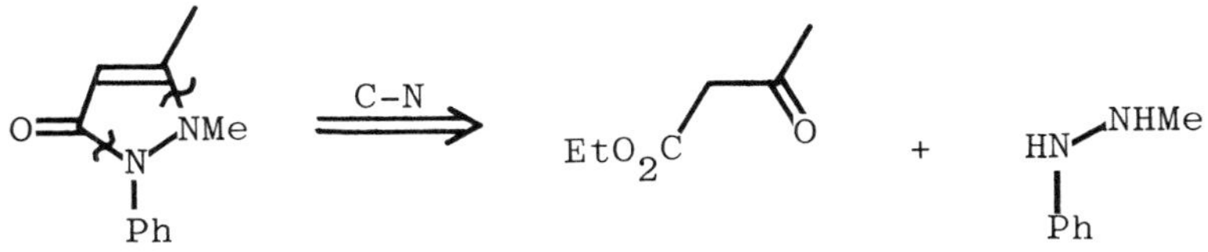

(56a) ⟹ C–N ⟹ EtO_2C-CH$_2$-CO-CH$_3$ (57) + HN(Ph)-NHMe (58)

Synthesis[337]

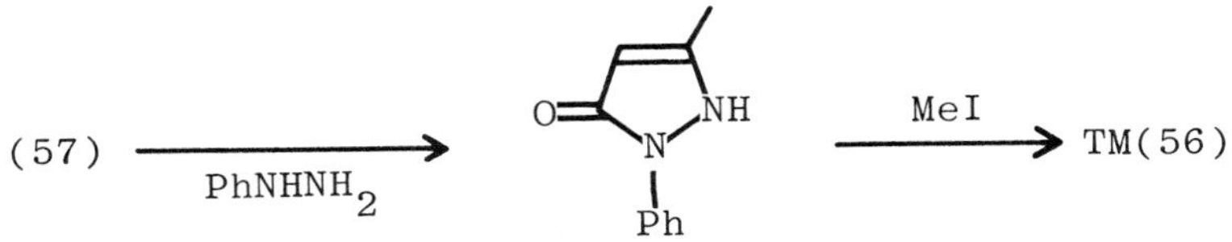

(57) —PhNHNH$_2$→ (59) —MeI→ TM(56)

where (59) is the pyrazolone with NH, N-Ph.

Example :

Diol (60) is used to make 'easy care' cotton fabrics as the two hydroxyl groups cross-link the cotton. It is obviously a formaldehyde adduct of (61) which can be disconnected to give two available compounds containing both heteroatoms.

Analysis

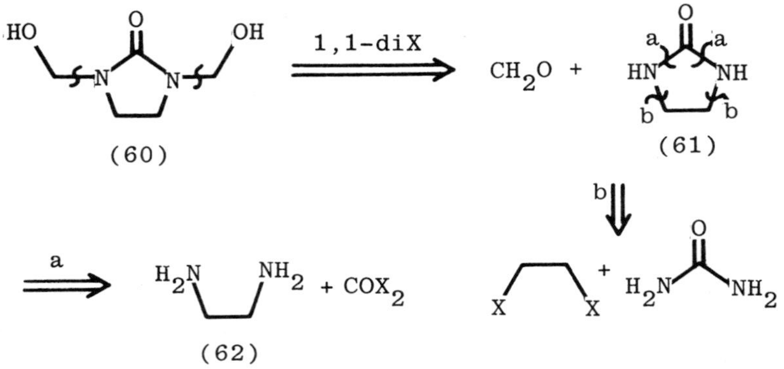

HO~N(C=O)N~OH (60) ⟹ CH_2O + HN(C=O)NH (61)
  (a,a bonds; b,b bonds marked)

(61) ⟹[b] $H_2N\frown NH_2$ (62) + COX_2

⟹[a] X~X + $H_2N-C(=O)-NH_2$

Both routes to (61) are successful,[338] though manufacture is by CO_2 addition to (2) under pressure, making (61) readily available.

Synthesis[340]

(61) $\xrightarrow[\text{mild alkali}]{CH_2O}$ TM(60)

1,3-Dipolar Cycloadditions

Problem :

Suggest how (63) might be made by a 1,3-dipolar cycloaddition (p T 247).

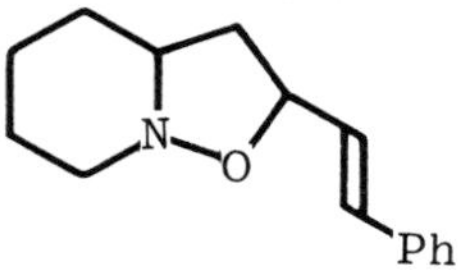

(63) — bicyclic isoxazolidine with vinyl-Ph substituent

Answer :

The disconnection must go back to nitrone (64) which gives us the other half of the molecule too.

Analysis

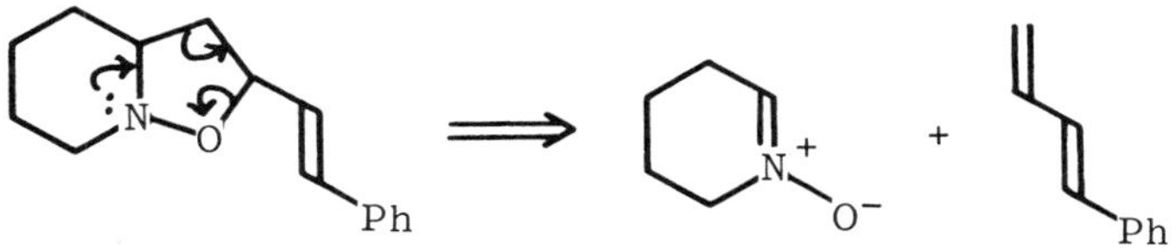

(63a) ⟹ (64) + (65)

The nitrone (64) does indeed add to the less hindered of the two double bonds in diene (65).

Synthesis[341]

(64) + (65) —heat→ TM(63)

(iv) *Six-Membered Rings*

Problem:
What first disconnection(s) are available on (66) using both amino and ketone groups?

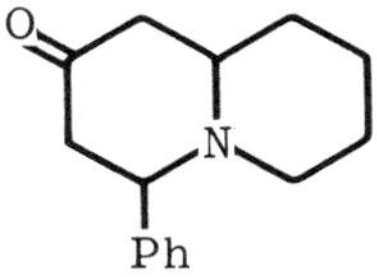

(66)

Answer:
There are two 1,3-diX disconnections (p T 48), that is reverse Michael reactions.

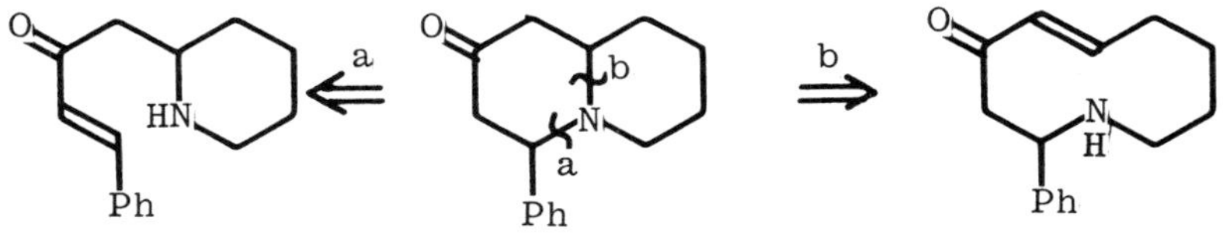

(67)   (66a,b)

Disconnection (a) produces greater simplification and (67) can be made from (63). How?

Answer:

We must cleave the N-O bond and then oxidise the allylic alcohol in (68). Cyclisation is spontaneous.

Synthesis[341]

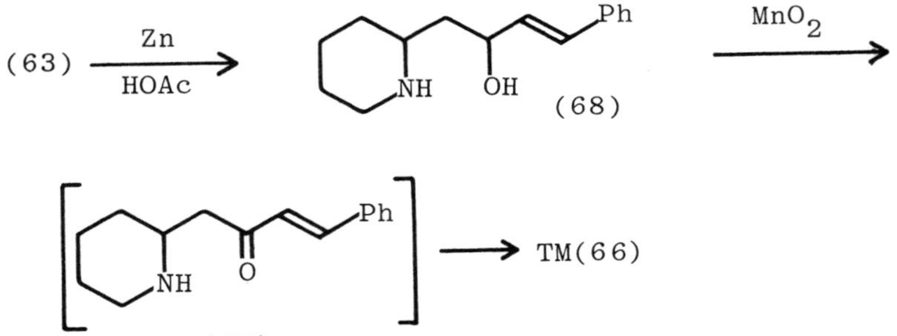

(63) $\xrightarrow[\text{HOAc}]{\text{Zn}}$ [piperidine-NH-CH$_2$-CH(OH)-CH=CH-Ph] (68) $\xrightarrow{\text{MnO}_2}$

[piperidine-NH-CH$_2$-CO-CH=CH-Ph] =(67) $\longrightarrow$ TM(66)

Example:

The potential analgesic (69) is clearly made from (70) and double Michael disconnection on the lines of p T 249 gives dienone (71). Double α,β-disconnection reveals the true starting materials.

Analysis

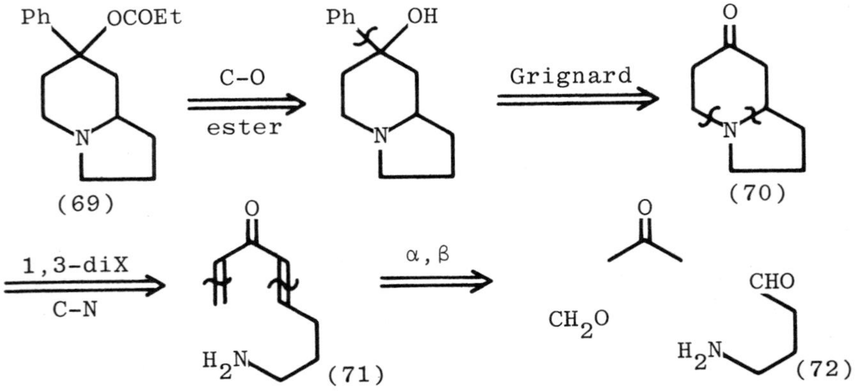

(69) $\xRightarrow[\text{ester}]{\text{C-O}}$ (Ph, OH indolizidine) $\xRightarrow{\text{Grignard}}$ (70)

$\xRightarrow[\text{C-N}]{\text{1,3-diX}}$ (71) $\xRightarrow{\alpha,\beta}$ CH$_2$O + acetone + H$_2$N-CH$_2$CH$_2$CH$_2$-CHO (72)

Putting this simple strategy into operation demands protection of the aldehyde group in (72) (otherwise it will cyclise) and activation of acetone as (74). The three-component cyclisation to form both rings of (75) works amazingly well.

Synthesis[342]

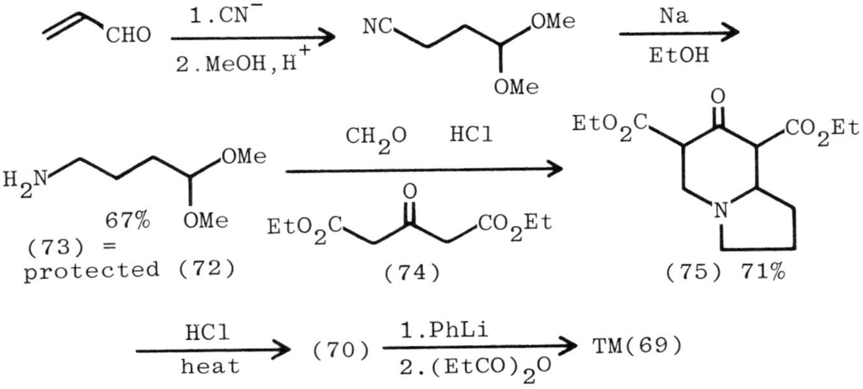

$CH_2=CH-CHO \xrightarrow[2.\ MeOH,\ H^+]{1.\ CN^-} NC-CH_2CH_2-CH(OMe)_2 \xrightarrow[EtOH]{Na}$

$H_2N-CH_2CH_2CH_2-CH(OMe)_2$
67%
(73) = protected (72)

$\xrightarrow{CH_2O \quad HCl}$ with $EtO_2C-CH_2-CO-CH_2-CO_2Et$ (74) → (75) 71%

(75): bicyclic indolizidine with EtO_2C and CO_2Et groups, ketone

$\xrightarrow[heat]{HCl}$ (70) $\xrightarrow[2.\ (EtCO)_2O]{1.\ PhLi}$ TM(69)

(v) *Seven-Membered Rings*

Problem:

Suggest a synthesis for (76) needed for the synthesis of enzyme inhibitors.

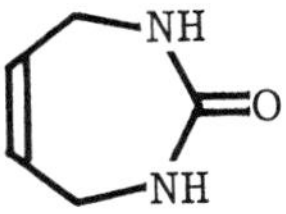

(76): seven-membered ring with -NH-C(=O)-NH- and a C=C double bond

Answer:

A molecule of urea is easily recognised and reaction with some derivative of available alcohol (77) (p T 128) looks reasonable.

Analysis

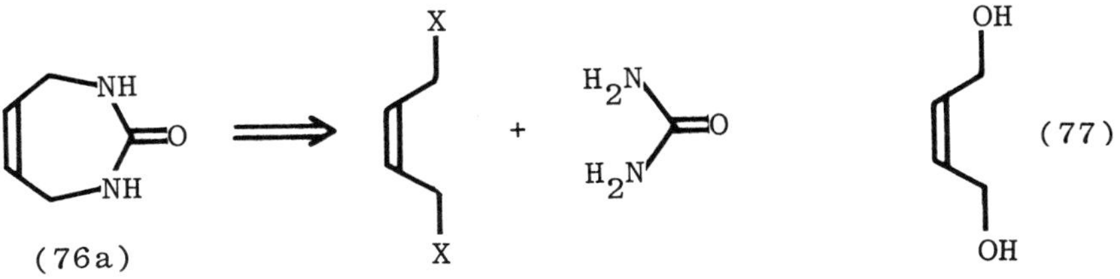

(76a) ⇒ X-CH_2-CH=CH-CH_2-X + H_2N-CO-NH_2 ; (77) = HOCH_2-CH=CH-CH_2OH

In practice, however, dichloride (78) was first converted into diamine (79) by the phthlimide route (p T 65) before cyclisation with COS. Perhaps urea is not reactive enough to combine with (78).
Synthesis[343]

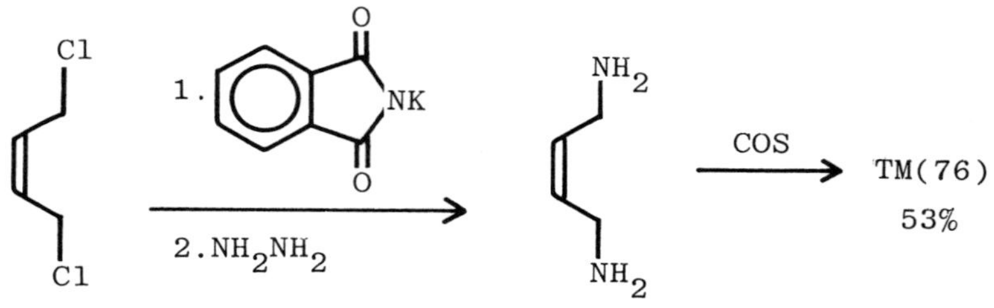

CHAPTER 30

Three-Membered Rings

(a) *Cyclisations*

Problem :
Suggest a synthesis of symmetrical ketone (1). The symmetry is a help, so disconnect by opening *both* three-membered rings at once!

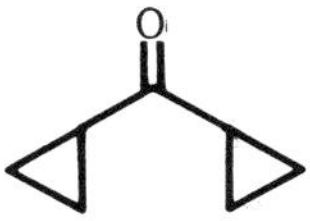

Answer :
Disconnection gives dihaloketone (2) made from (3). This compound is best approached by the strategy we have often used for symmetrical ketones; adding an activating group and using a 1,3-dicarbonyl disconnection. There is a very similar example on p 329.

Analysis

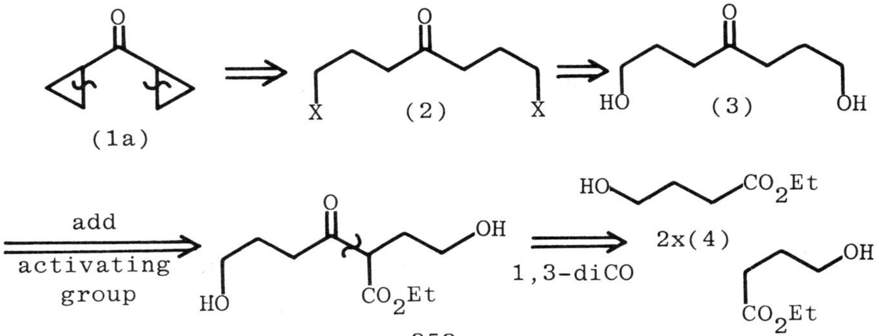

Available lactone (5) replaces (4) in the synthesis: there is no need to isolate any of the intermediates.

Synthesis[344]

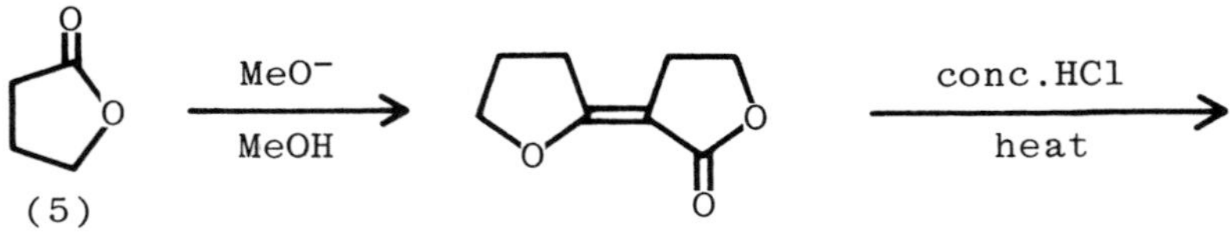

[Reaction scheme: (5) γ-butyrolactone → MeO⁻/MeOH → furan-lactone intermediate → conc. HCl, heat → (2, X=Cl) → NaOH → TM(1), 55% overall]

Example :

Simple cyclopropyl ketones such as (7) can be made from available acid (6) by standard methods.

Synthesis[345]

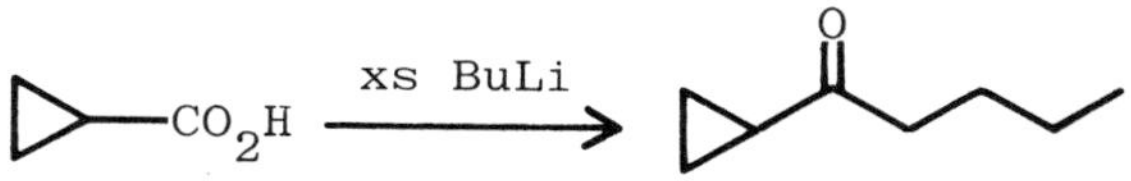

[Reaction scheme: (6) cyclopropyl-CO₂H → xs BuLi → (7) cyclopropyl ketone, 70%]

Acid (6) is available[346] from nitrile (8) which is made by cyclisation of (9). The statistical method of chemoselectivity is used to build (9) from cheap diol (10).

Analysis

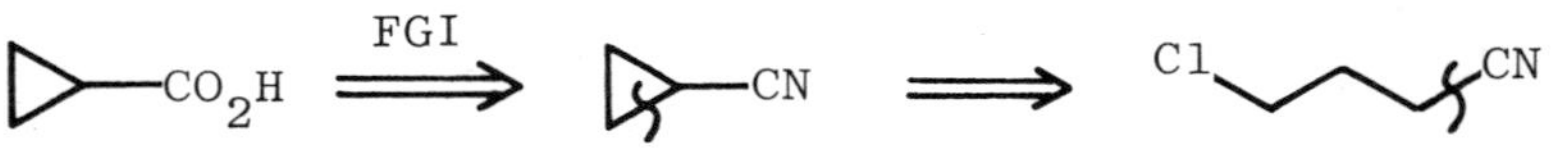

[Retrosynthetic scheme: (6) cyclopropyl-CO₂H ⇒ FGI ⇒ (8) cyclopropyl-CN ⇒ (9) Cl~~~CN ⇒ Cl~~~Br ⇒ Cl~~~OH ⇒ HO~~~OH (10)]

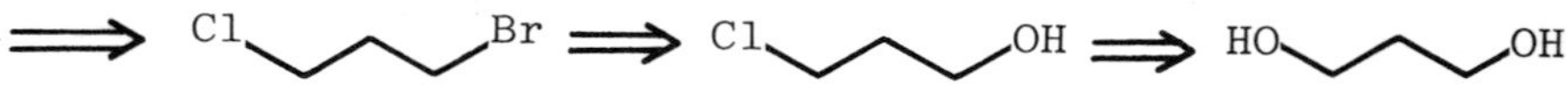

Synthesis[3,4,6]

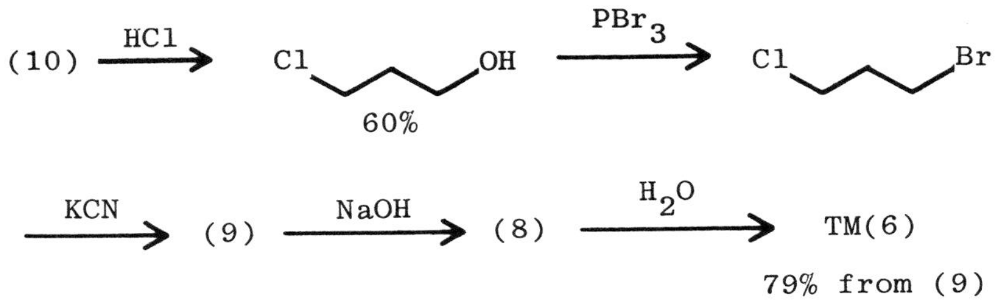

(10) →[HCl] Cl~~~OH (60%) →[PBr₃] Cl~~~Br

→[KCN] (9) →[NaOH] (8) →[H₂O] TM(6)
79% from (9)

Julia[3,4,5] synthesised ketone (11) by both cyclisation of (12) and addition of organometallic reagents to (6) or its acid chloride.

Analysis

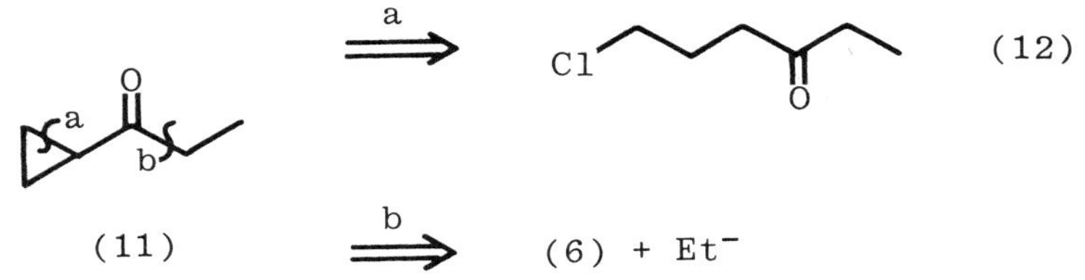

(11) ⇒[a] Cl~~~C(=O)~~ (12)

(11) ⇒[b] (6) + Et⁻

The cyclisation route gave the highest yield, but (12) is not available and had to be made by a low-yield radical coupling process.

Cyclisation Synthesis[3,4,5]

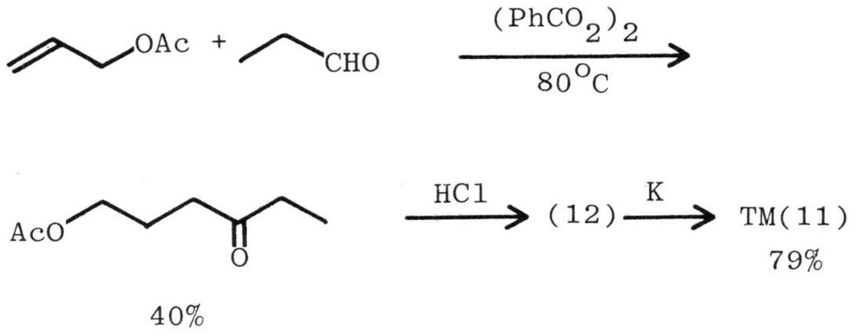

~~OAc + ~~CHO →[(PhCO₂)₂ / 80°C]

AcO~~~C(=O)~~ (40%) →[HCl] (12) →[K] TM(11) 79%

Addition Routes[3,4,5]

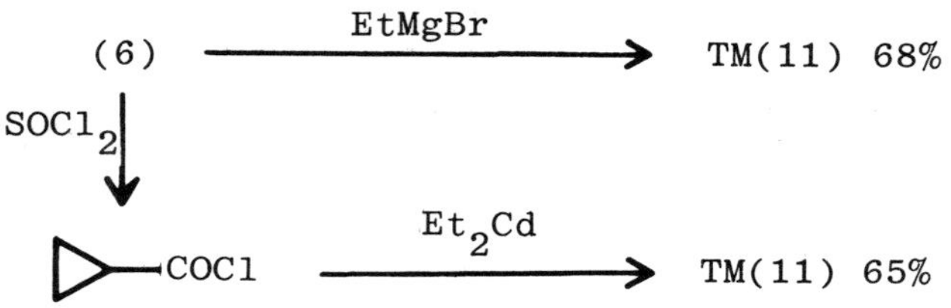

(6) —EtMgBr→ TM(11) 68%

SOCl$_2$ ↓

△—COCl —Et$_2$Cd→ TM(11) 65%

Problem :
Suggest a synthesis for (13).

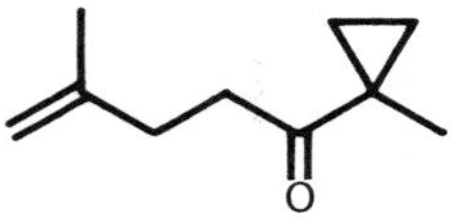

(13)

Answer :
Immediate disconnection (13a) of the three-membered ring gives a very complicated starting material (14) so it is better to simplify the left hand portion of the molecule first. Disconnection (13b) is good as it leaves simple ketone (16) and an allylic halide (15).

Analysis

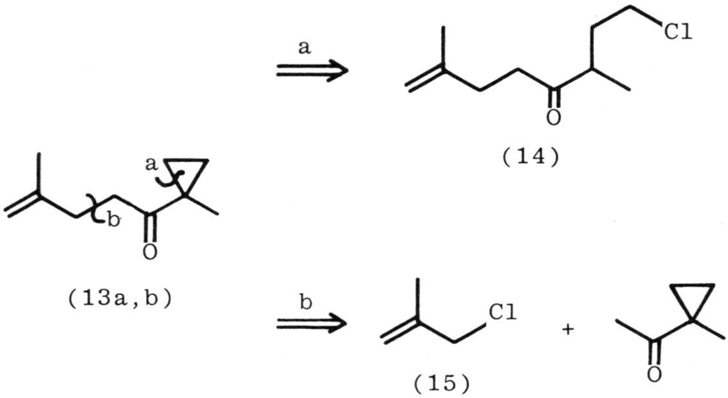

⇒ᵃ (14)

(13a,b)

⇒ᵇ (15) + (16)

Ketone (16) can be made by the cyclisation route[347] as the γ-chloroketone (17) comes from lactone (18) (p T 251) which is made by alkylation of lactone (19).

Analysis 2

(16a) ⇒ (17) ⇒ [alcohol intermediate]

add activating group ⇒ (18) ⇒ (19) (Page T 251)

The alkylation of ketone (16) will need control – the published method[347] uses an activating group added directly to (16) as one side of the carbonyl group is blocked.

Synthesis[347]

(19) $\xrightarrow[\text{MeI}]{\text{Na,xylene}}$ (18) $\xrightarrow{\text{HCl}}$ (17) $\xrightarrow{\text{KOH}}$ (16)
                                  74%   90%   83%

$\xrightarrow[\text{CO(OEt)}_2]{\text{NaH}}$ (20) $\xrightarrow[\text{2.(15)}]{\text{1.NaH,THF}}$ [intermediate]

$\xrightarrow[\text{2.H}^+\text{,heat}]{\text{1.Ba(OH)}_2\text{,H}_2\text{O}}$ TM(13) 88% from (20)

(b) *Insertion Reactions*
(1) *Epoxides*

Problem :
Suggest a synthesis for the perfumery compound 'strawberry aldehyde' (21) - not of course an aldehyde at all.

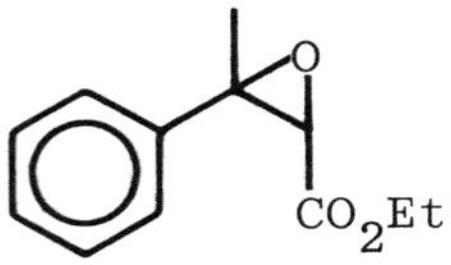

(21)

Answer :
Either strategy from p T 253 will work : the Darzens disconnection gives (21a) greater simplification.

Analysis

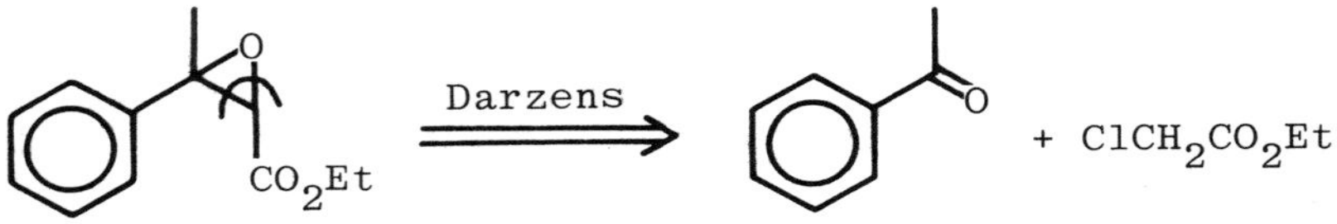

(21a) ⟹ Darzens ⟹ PhCOMe + ClCH$_2$CO$_2$Et

Synthesis[348]

PhCOMe + ClCH$_2$CO$_2$Et $\xrightarrow[\text{xylene}]{\text{NaNH}_2}$ TM(21) 64%

Problem :
Suggest a synthesis for epoxide (22).

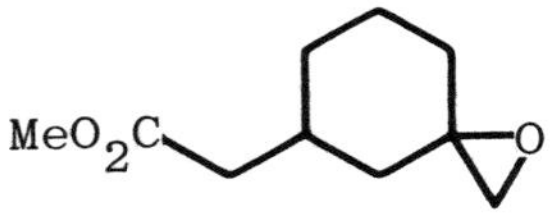

(22)

Answer :

Epoxidation of (23) would no doubt give (22) and we could make (23) by a Wittig reaction from ketone (24). It it better to disconnect (22) directly to (24) and use a sulphur ylid for the reaction. Ketone (24) is a 1,5-dicarbonyl compound with an obvious ring-chain disconnection.

Analysis

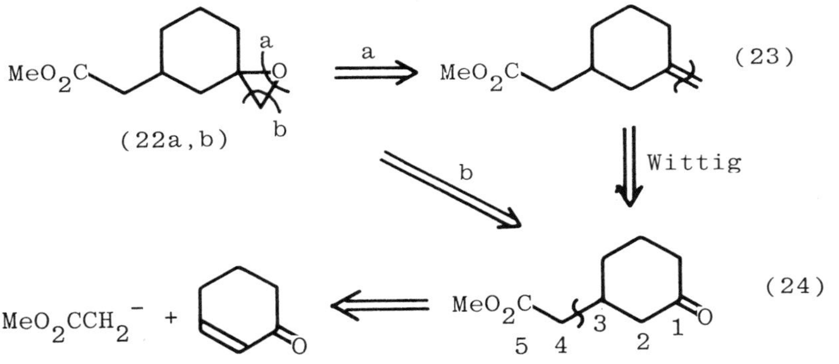

The 'ylid' from dimethyl sulphoxide, Me$_2$SO, DMSO, was used in this synthesis.

Synthesis[3,4,9]

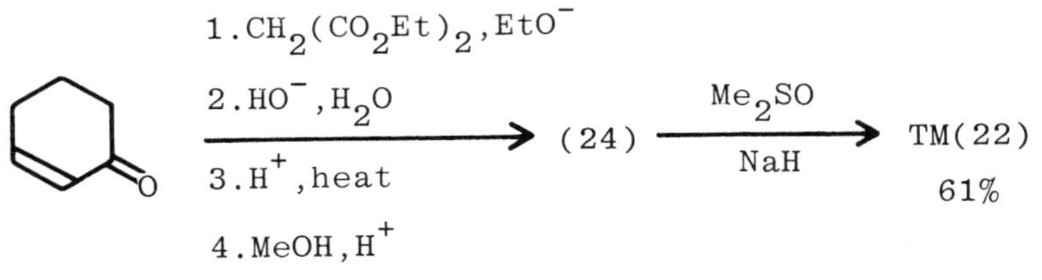

$$\text{cyclohexenone} \xrightarrow[\substack{3.\text{H}^+,\text{heat} \\ 4.\text{MeOH},\text{H}^+}]{\substack{1.\text{CH}_2(\text{CO}_2\text{Et})_2,\text{EtO}^- \\ 2.\text{HO}^-,\text{H}_2\text{O}}} (24) \xrightarrow[\text{NaH}]{\text{Me}_2\text{SO}} \text{TM}(22) \quad 61\%$$

(iii) *Cyclopropyl Ketones*

Example :

The alternative disconnection (25a) requires the Michael addition to a nucleophilic carbene equivalent to an enon

(25a)

The obvious choice for a reagent is again a sulphur ylid, but how are we to control the regioselectivity of the addition? The more reactive sulphur ylids, notably (26) and (27), add directly to the carbonyl group (kinetic control, cf p T 117) giving epoxides (29) while the more stable ylid (28), which combines the anion-stabilisations of (26) and (27), adds reversibly and gives the thermodynamic product (25).

$$^-CH_2SMe \quad\quad ^-CH_2\overset{+}{S}Me_2 \quad\quad ^-CH_2-\overset{+}{\overset{\displaystyle O}{\underset{\displaystyle \|}{S}}}Me_2$$

(26)   (27)   (28)

This reaction involves Michael addition of (28) to the enone followed by cyclisation (29) with displacement of DMSO. Sulphur ylids react in these ways rather than removing oxygen, as phosphorus ylids do, because the SO bond is weaker than the PO bond and the lower valency states of sulphur more stable than those of phosphorus.

Problem :
Suggest syntheses of (30) and (31).

(30)   (31)

Answer :
Carbene disconnection reveals easily made enones (32) and (33). The sulphur ylids must be correctly chosen.
Analysis

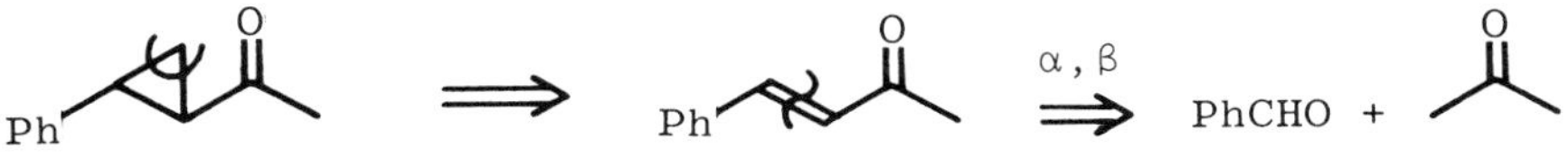

A stable sulphur ylid, such as (28) will be needed for (30) and a reactive one, such as (26), for (31).
Synthesis[350]

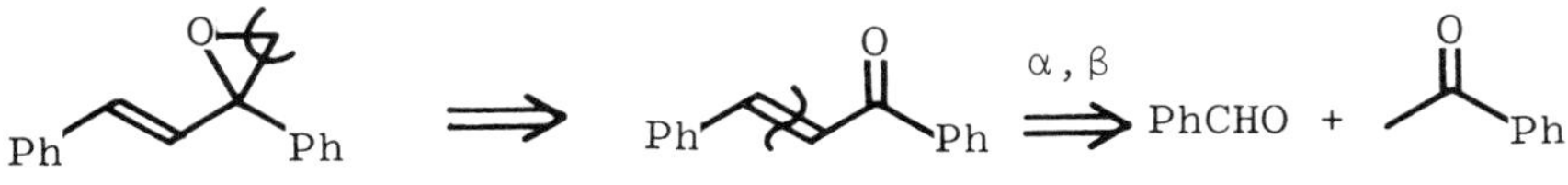

$$\text{PhCHO} + \text{Me}_2\text{CO} \xrightarrow{\text{NaOH}} (32) \xrightarrow{(28)} \text{TM}(30)\ 70\%$$

Synthesis[351]

$$\text{PhCHO} + \text{PhCOMe} \xrightarrow{\text{NaOH}} (33) \xrightarrow[\substack{\text{NaH}\\\text{THF}}]{\text{Me}_2\text{SO}} \text{TM}(31)\ 87\%$$

Problem :

The third component of the elm bark beetle pheromone (p T 4) is α-cubebene (34) which the beetle takes from the tree. Simple disconnections lead to ketone (35). What disconnection would you suggest next?

α-cubebene : *Analysis 1*

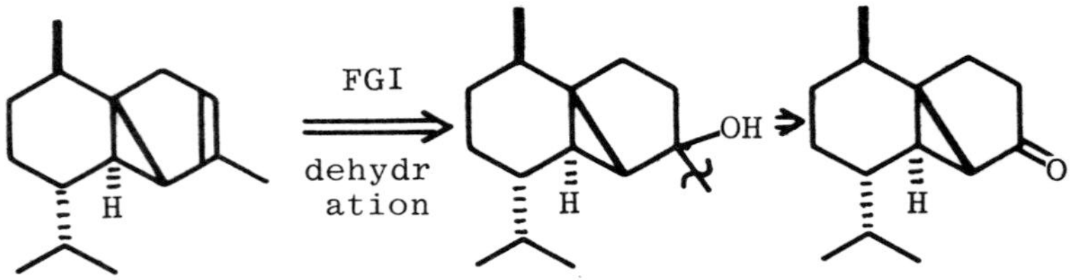

(34) $\xrightarrow{\text{FGI dehydration}}$ (OH) $\Rightarrow$ (35)

Answer :

Disconnection of the three-membered ring next to the carbonyl group (35a) to give diazoketone (36) achieves maximum simplification. Acid (37) becomes the key intermediate.

Analysis 2

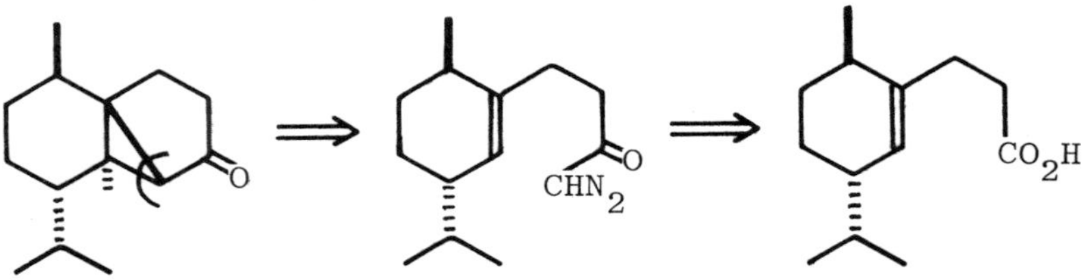

(35a) $\Rightarrow$ (36, CHN$_2$) $\Rightarrow$ (37, CO$_2$H)

This approach has been used in two published syntheses of cubebene, though there is a problem in that control over the stereochemistry has not yet been perfected.

Synthesis[352]

(37) $\xrightarrow[\text{2. CH}_2\text{N}_2]{\text{1. (COCl)}_2}$ (36) $\xrightarrow{\text{Cu}}$ (35)

When the right isomer of (35) is separated, conversion to (34) can be achieved by a variety of methods.

Synthesis

$$(35) \xrightarrow[\text{2. SOCl}_2, \text{pyr, heat}]{\text{1. MeMgI}} TM(34)$$

(iii) *Cyclopropane Acids*

Problem :

Suggest a synthesis for (38), needed for trial as an anti-histamine.

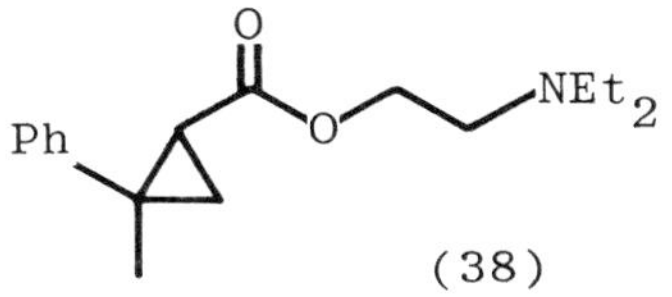

[Structure of (38): Ph-substituted dimethylcyclopropane with CO-O-CH$_2$CH$_2$-NEt$_2$ ester group]

Answer :

Ester disconnection gives easily made alcohol (39) and cyclopropane acid (40). Disconnection of diazoacetic ester leaves simple olefin (41).

Analysis

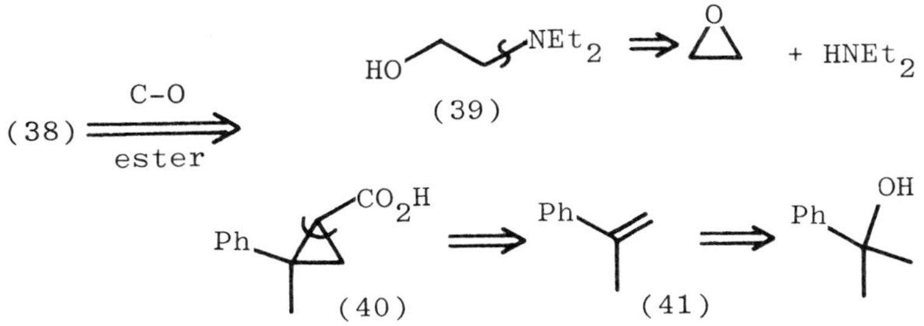

$$(38) \underset{\text{ester}}{\overset{\text{C-O}}{\Longrightarrow}} \quad HO\text{-}CH_2CH_2\text{-}NEt_2 \quad (39) \quad \Rightarrow \quad \triangle\text{=O} \;+\; HNEt_2$$

[Further analysis: cyclopropane acid (40) Ph-substituted with CO$_2$H ⇒ Ph-C(=CH$_2$)-CH$_3$ (41) ⇒ Ph-C(OH)(CH$_3$)$_2$]

It is not necessary to isolate acid (40) : (38) can be made directly from ester (42) by ester exchange.

Synthesis[353]

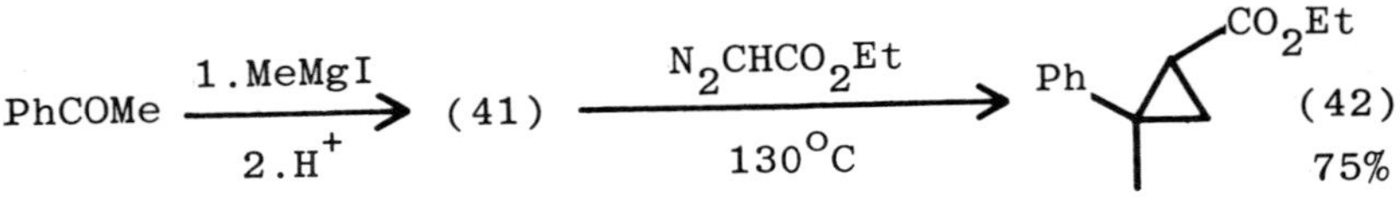

PhCOMe $\xrightarrow[\text{2. H}^+]{\text{1. MeMgI}}$ (41) $\xrightarrow[\text{130°C}]{\text{N}_2\text{CHCO}_2\text{Et}}$ Ph⟨CO₂Et (42) 75%

Et₂NH + ⟨O⟩ ⟶ (39) $\xrightarrow[\text{2. (42)}]{\text{1. Na, toluene}}$ TM(38) 97%

(iv) *Cyclopropanes*

Example :

When Winstein[354] was studying the possibility of generating non-classical cation (44) from *cis* tosylate (43), he needed both *cis* and *trans* alcohols to show there was a substantial rate difference in the reaction of the tosylates.

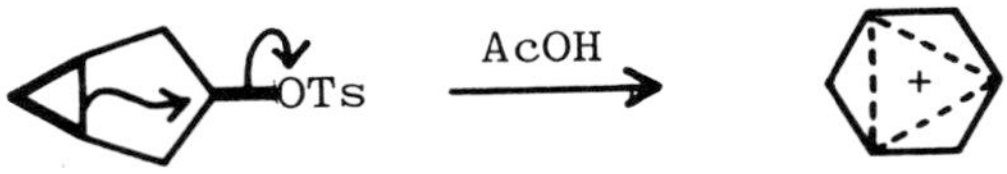

(43) $\xrightarrow{\text{AcOH}}$ (44)

The *cis* alcohol (45) is an obvious Simmons-Smith product (p T 257) since we require the CH₂ group to be added from the same side as the hydroxyl group. The starting alcohol (46) was available from cyclopentadiene.

Analysis

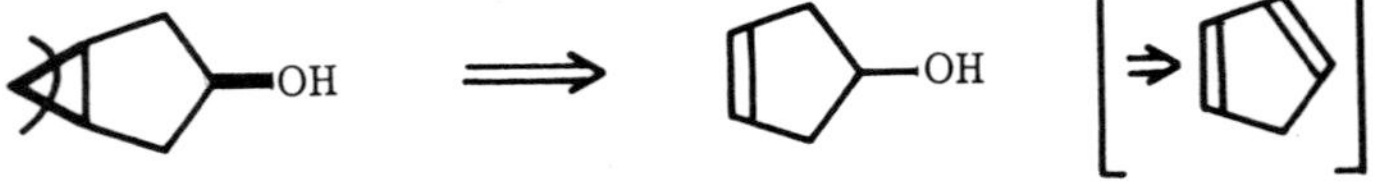

(45) ⟹ (46) [⇒ ⟨⟩]

Synthesis[354]

(46) $\xrightarrow[\text{Zn/Cu}]{\text{CH}_2\text{I}_2}$ TM(45) 75%

The *trans* alcohol (47) might be made by reduction of ketone (48). Oxidation of (45) would give (48), but an alternative is to add an activating group and disconnect as a 1,3-dicarbonyl compound - standard strategy for a symmetrical ketone.

Analysis 1

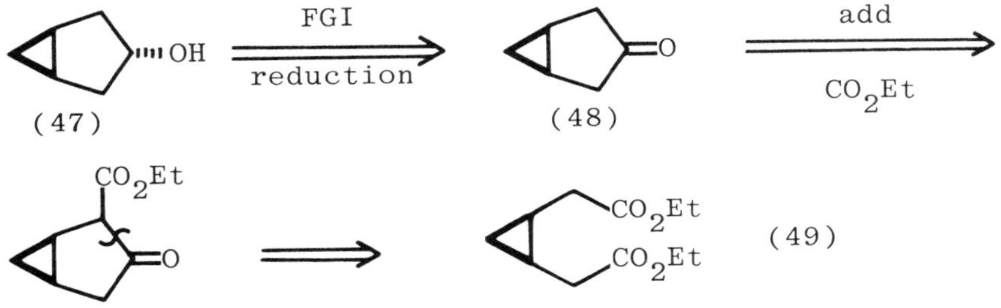

(47) ⟶ [FGI reduction] ⟶ (48) ⟶ [add CO_2Et] ⟶

⟶ ⟹ (49)

Diester (49) is a 1,6-dicarbonyl compound and reconnection gives (50) which might be made by CH_2 additio to diene (51).

Analysis 2

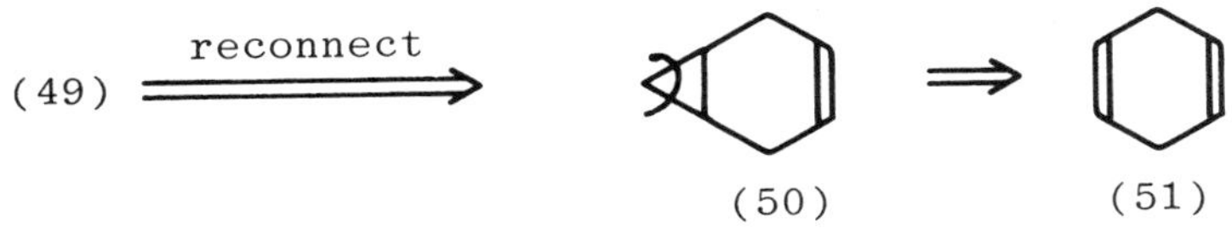

(49) ⟹ [reconnect] (50) ⟹ (51)

Workers[355] investigating the synthesis of (49) tried various carbenes with (51) but achieved success only with CBr_2. After cleavage of the remaining double bond, the bromine atoms were removed by hydrogenolysis.

Synthesis[355]

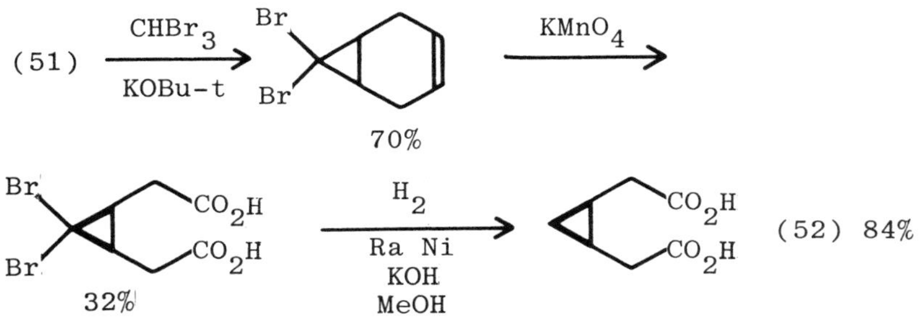

(51) ⟶ [$CHBr_3$, KOBu-t] ⟶ 70% ⟶ [$KMnO_4$] ⟶ 32% ⟶ [H_2, Ra Ni, KOH, MeOH] ⟶ (52) 84%

Cyclisation and decarboxylation gave ketone (48) which was reduced to a mixture of (47) and (45). The highest ratio of *trans/cis* was achieved with $(i\text{-PrO})_3\text{Al}$, which normally favours the thermodynamic product (p T 101).

Synthesis[354]

$$(52) \xrightarrow[\text{H}^+]{\text{EtOH}} (49) \xrightarrow[\substack{2.\text{KOH},\text{H}_2\text{O} \\ 3.\text{H}^+,\text{heat}}]{1.\text{EtO}^-} (48)$$

$$\xrightarrow{(i\text{-PrO})_3\text{Al}} \underset{60\%}{(47)} + \underset{40\%}{(45)}$$

Problem :
Suggest a synthesis for (53), needed for Grieco's synthesis[356] of the anti-tumour compound ivangulin.

(53)

The first disconnection should be obvious and you might find it helpful to try and recognise a familiar product in the residual structure.

Answer :
Cyclopropanation from the hydroxyl side of (54) should give (53). Removal of the acetal leaves ketone (55) in which the double bond has been returned to conjugation. The structure remaining is very like ketone (56), the classical product of a Robinson annelation (p T 175).

Analysis

$$(53) \xRightarrow{\text{carbene}} (54) \xRightarrow[\text{acetal}]{1,1\text{-diX}}$$

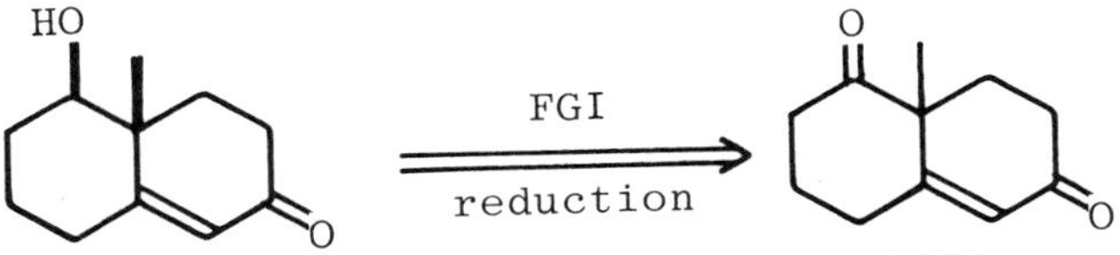

HO
O
FGI
reduction
O
O

(55) (56)

Chemo- and stereoselective reduction of (56) to (55) is achieved in highest yield by sodium borohydride in ethanol. The isolated ketone is reduced more rapidly than the enone and (55) is the equatorial alcohol. Protection moves the double bond out of conjugation and even the distant OH group in (54) successfully controls the stereochemistry of the Simmons-Smith reaction. No cyclopropanation occurred unless the OH group was there.

Synthesis[356,7]

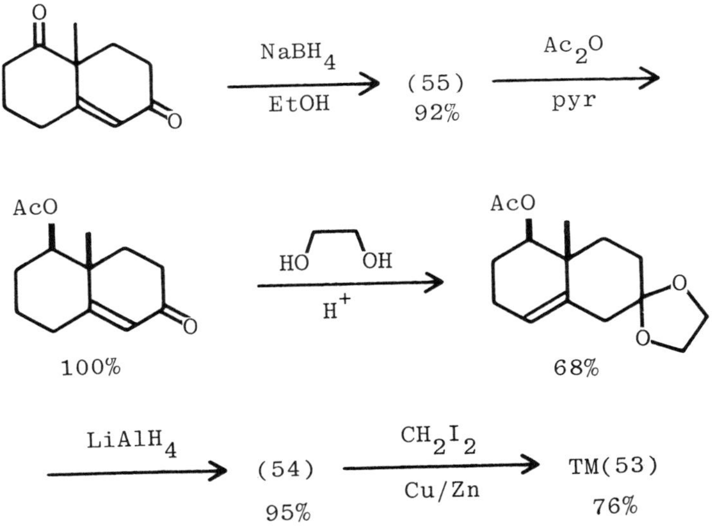

O
NaBH4
EtOH
(55)
92%
Ac2O
pyr
O
AcO
O
100%
HO OH
H+
AcO
O
O
68%
LiAlH4
(54)
95%
CH2I2
Cu/Zn
TM(53)
76%

CHAPTER 31

Strategy XIV: Rearrangements in Synthesis

(a) *Diazoalkanes*

Problem :

Suggest a synthesis for (1), needed for an alkaloid synthesis.

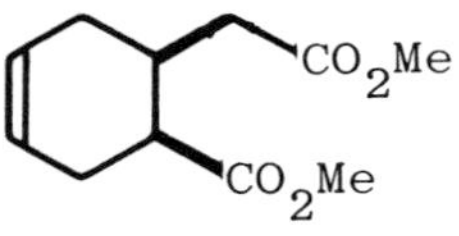

(1)

Answer :

This is nearly a Diels-Alder adduct and removal of the extra CH_2 group gives a Diels-Alder adduct (2). The reaction must be chemoselective addition of CH_2N_2 to one carbonyl group only of (2). This is most easily achieved from the anhydride (3).

Analysis

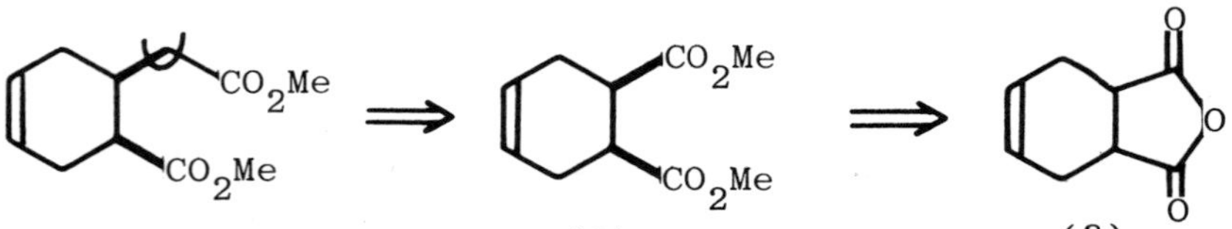

(1a) ⇒ (2) ⇒ (3)

Synthesis[358]

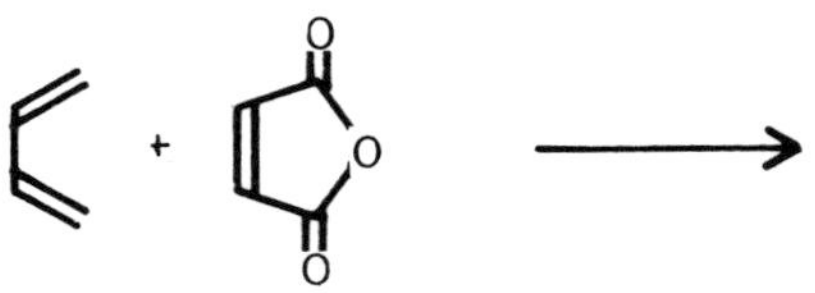

+ ⟶ (3)

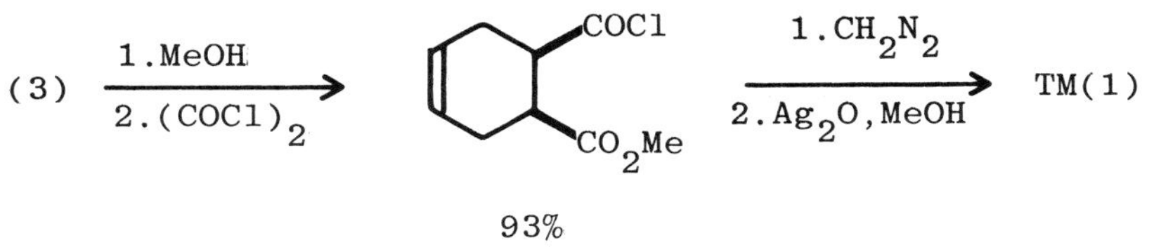

(3) $\xrightarrow[\text{2.(COCl)}_2]{\text{1.MeOH}}$ [cyclohexene with COCl and CO$_2$Me substituents] 93% $\xrightarrow[\text{2.Ag}_2\text{O,MeOH}]{\text{1.CH}_2\text{N}_2}$ TM(1)

Example :

Ketone (4) looks very difficult to disconnect. With a chain extension in mind, disconnection to acid (5) and removal of CH_2 gives an easily made cyclopropane acid (6).

Analysis

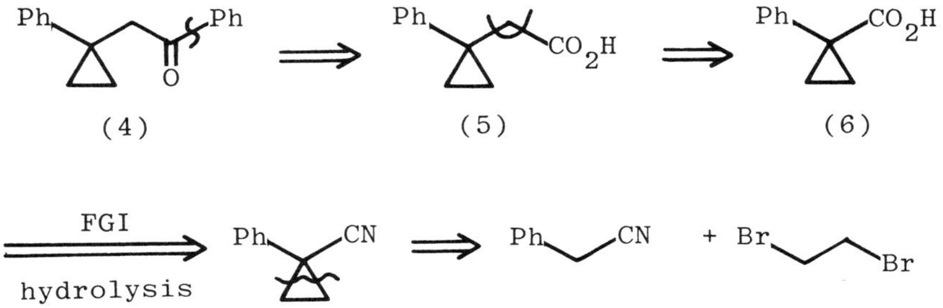

Ph—[cyclopropane]—CH$_2$—C(=O)—Ph $\Longrightarrow$ Ph—[cyclopropane(Ph)]—CO$_2$H $\Longrightarrow$ Ph—[cyclopropane]—CO$_2$H

(4)　　　　　　　　　(5)　　　　　　　(6)

$\xrightarrow[\text{hydrolysis}]{\text{FGI}}$ Ph—[cyclopropane(Ph)]—CN $\Longrightarrow$ Ph—CH$_2$—CN + Br—CH$_2$CH$_2$—Br

The formation of (6) gives only a poor yield[359] but the rest of the synthesis[360] is short and gives good yields with an organocadmium reagent used for the last step.

Synthesis[359,360]

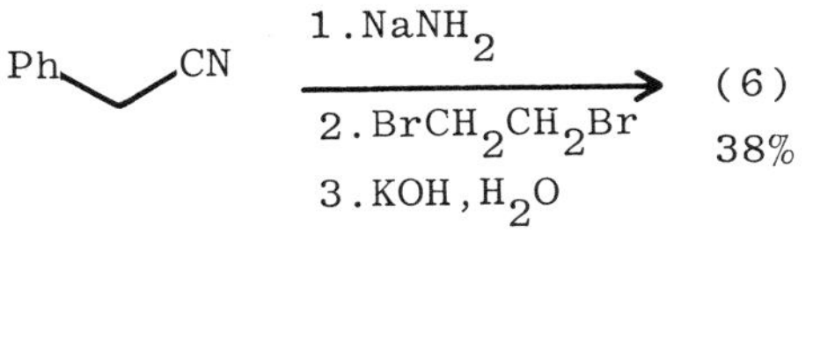

Ph—CH$_2$—CN $\xrightarrow[\text{3.KOH,H}_2\text{O}]{\text{1.NaNH}_2 \atop \text{2.BrCH}_2\text{CH}_2\text{Br}}$ (6) 38%

$$\xrightarrow[\substack{2.\,CH_2N_2 \\ 3.\,Cu,heat}]{1.\,(COCl)_2} \quad (5) \quad \xrightarrow[2.\,Ph_2Cd]{1.\,(COCl)_2} \quad TM(4)$$

Problem :

The Arndt-Eistert method is useful for making arylacetic acids from benzoic acids. How would you carry out this conversion?

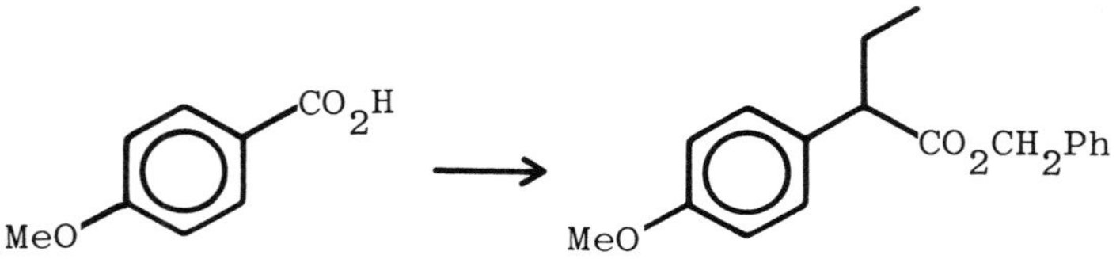

MeO-C6H4-CO2H (7) → MeO-C6H4-CH(Et)-CO2CH2Ph (8)

Answer :

Disconnection of the extra carbon atom reveals that we need diazoalkane (9) as the reagent.

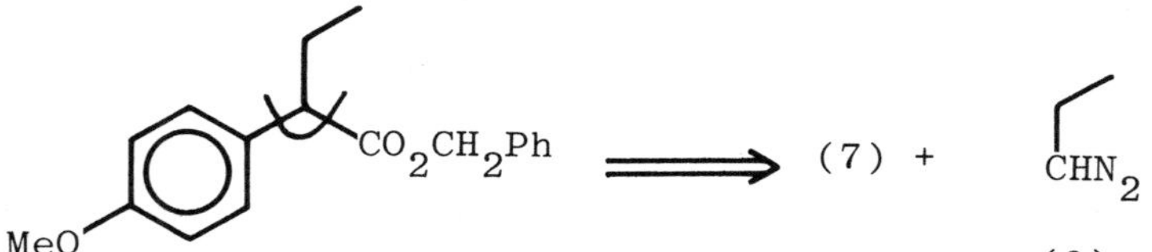

MeO-C6H4-CH(Et)-CO2CH2Ph (8a) ⟹ (7) + EtCHN2 (9)

The ketone produced in the rearrangement can be captured by benzyl alcohol in base to give (8) directly. *Synthesis*[361]

$$(7) \xrightarrow[\substack{2.\,EtCHN_2 \\ 3.\,PhCH_2OH,\,PhNEt_2}]{1.\,SOCl_2} TM(8)$$

(ii) *Diazoalkanes and Ketones*
Problem :
Suggest a synthesis for (10).

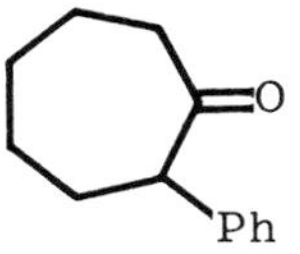

(10)

Answer :
Disconnection of the substituted carbon atom achieves maximum simplification and requires diazoalkane (11).
Analysis

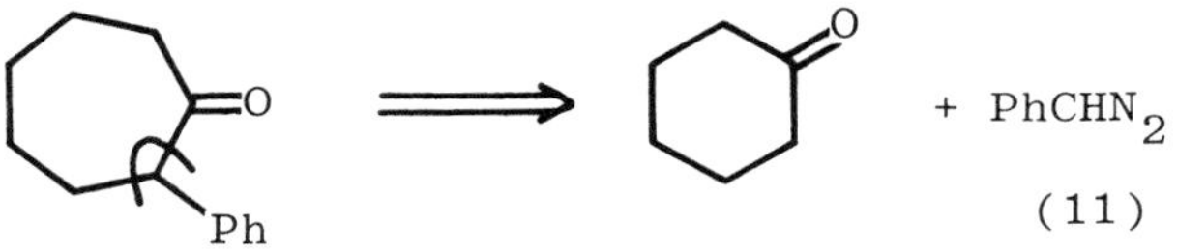

⟹ + PhCHN$_2$

(10a)   (11)

Synthesis[362]

$\xrightarrow[\text{MeOH}]{(11)}$ TM(10)

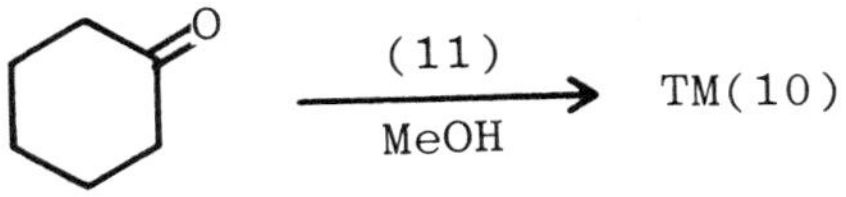

Example :
An alternative strategy, avoiding the danger of over-reaction with diazomethane, is to make the diazonium salt, by diazotisation of a hydroxy amine (14), available from the original ketone (12) via epoxide (13).
Synthesis[363]

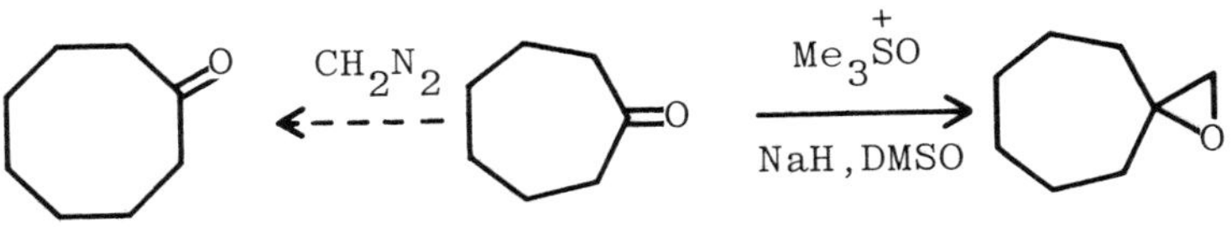

$\xleftarrow{CH_2N_2}$ $\xrightarrow[\text{NaH, DMSO}]{Me_3\overset{+}{S}O}$

(12)   (13)

(13) 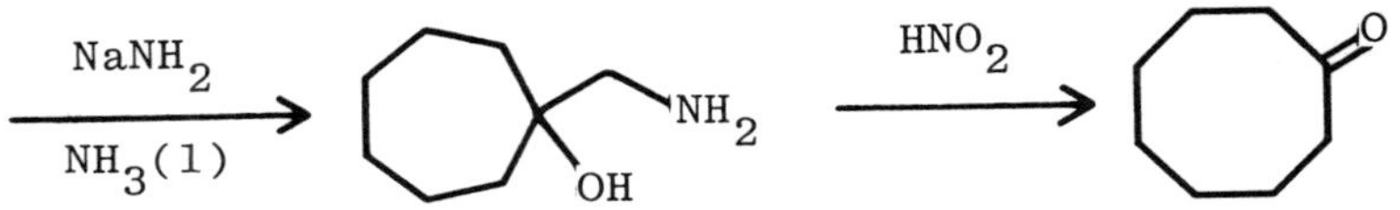 $\xrightarrow[NH_3(l)]{NaNH_2}$ [cycloheptane with CH$_2$NH$_2$ and OH] (14) $\xrightarrow{HNO_2}$ [cyclooctanone]

(b) *The Pinacol Rearrangement*
    Epoxide Rearrangements

Problem:
Suggest a synthesis for ketone (15) using a rearrangement step.

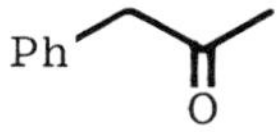

Ph–C(=O)–CH(CH$_3$)– (15)

Answer:
Either epoxide (17) or diol (18) will rearrange to (15).
Analysis

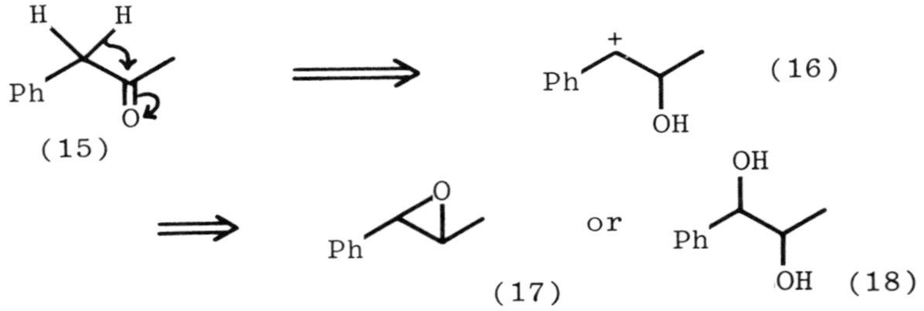

(15) $\Longrightarrow$ Ph–CH$^+$–CH(OH)–CH$_3$ (16)

$\Longrightarrow$ Ph–epoxide (17)   or   Ph–CH(OH)–CH(OH)–CH$_3$ (18)

There is no ambiguity in the rearrangement as the benzylic cation (16) is more stable than the alternative, and H shifts occur more readily than Me shifts.

Synthesis[364]

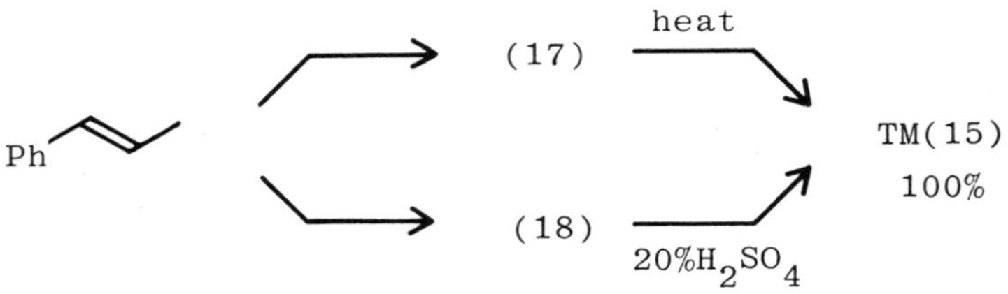

Ph–CH=CH–CH$_3$ → (17) $\xrightarrow{heat}$ TM(15) 100%
            → (18) $\xrightarrow{20\% H_2SO_4}$

Example :
Unsymmetrical pinacols, e.g. (19) can be made by reductive coupling of the mixed carbonyl precursors, but the yields are poor as the two dimers are also formed.

Synthesis[365]

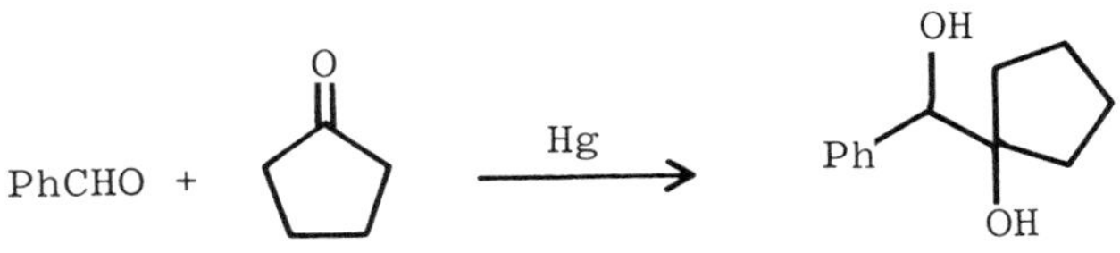

PhCHO + cyclopentanone —Hg→ (19)

In general, it is better to make such compounds by hydroxylation of the alkene. There is no problem with the alkene rearrangement here, the ring expanded ketone (20) being formed in good yield. This product can also be made by another rearrangement route (see p 120).

Synthesis

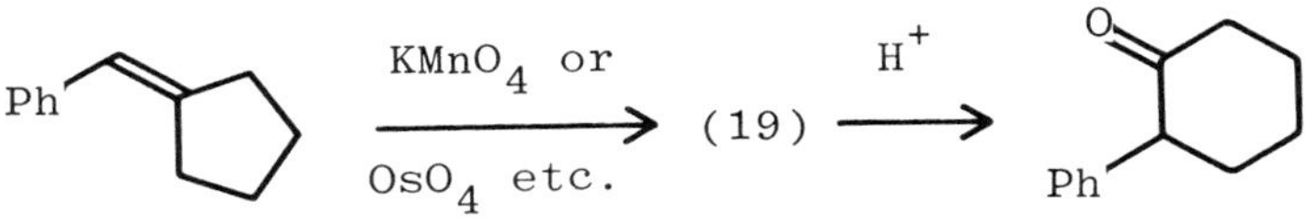

Ph=cyclopentane —KMnO$_4$ or OsO$_4$ etc.→ (19) —H$^+$→ (20)

Problem :
Suggest a synthesis of dienone (21).

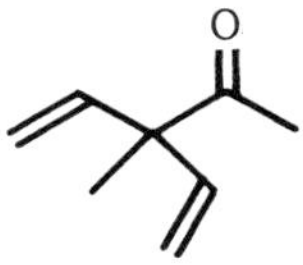

(21)

Answer :
Reversing the rearrangement and keeping the symmetry gives a pinacol (22) which can be made by reductive dimerisation of (23) but not from selective hydroxylation of (24).

Analysis

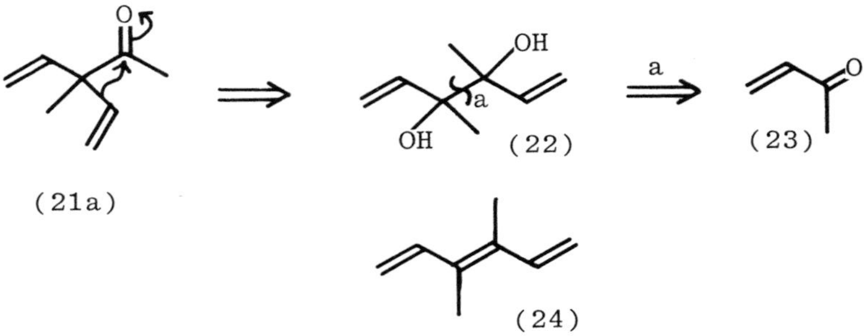

[structures (21a), (22), (23), (24)]

*Synthesis*³⁶⁶

(23) —Hg→ (22) —H⁺→ TM(21)
                              52%

Example :
Epoxides are of course as easy to make from unsymmetrical, e.g. (29) as from symmetrical, e.g. (26),
alkenes. The more stable carbonium ion is again formed.
Hydride shift is preferred to alkyl shift except in the
favourable 6 → 5 ring contraction.

*Synthesis*³⁶⁷

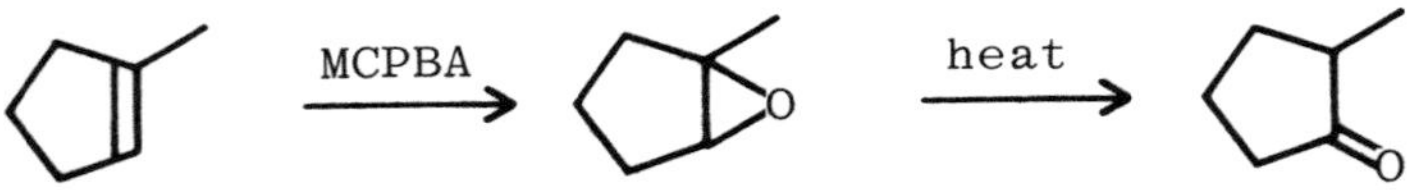

(25) —MCPBA→ [epoxide] —heat→ [2-methylcyclopentanone] 100%

*Synthesis*³⁶⁷

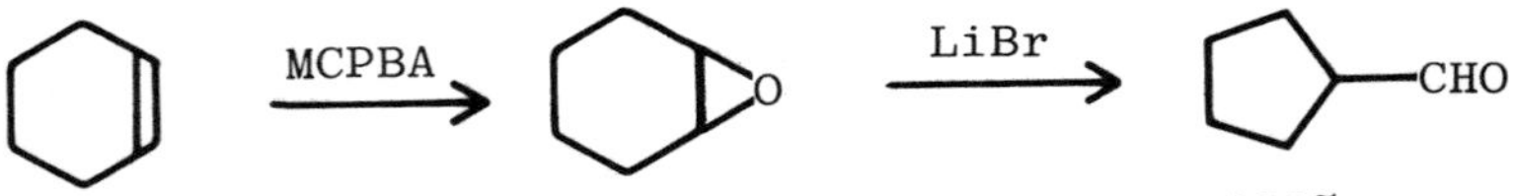

(26) —MCPBA→ [epoxide] —LiBr→ [cyclopentyl-CHO] 100%

Favorskii

Problem :

Both α-chloroketones (27) and (28) give the same product on treatment with MeO⁻. What is it?

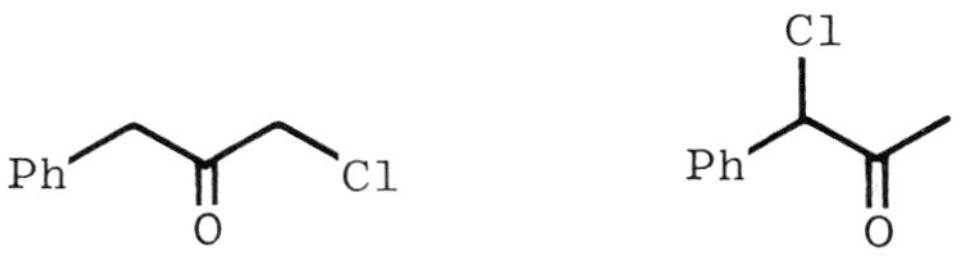

(27)

(28)

Answer :

The same cyclopropane (29) is formed from both, which opens to eliminate the more stable benzylic carbanion.[368] The product is (30).

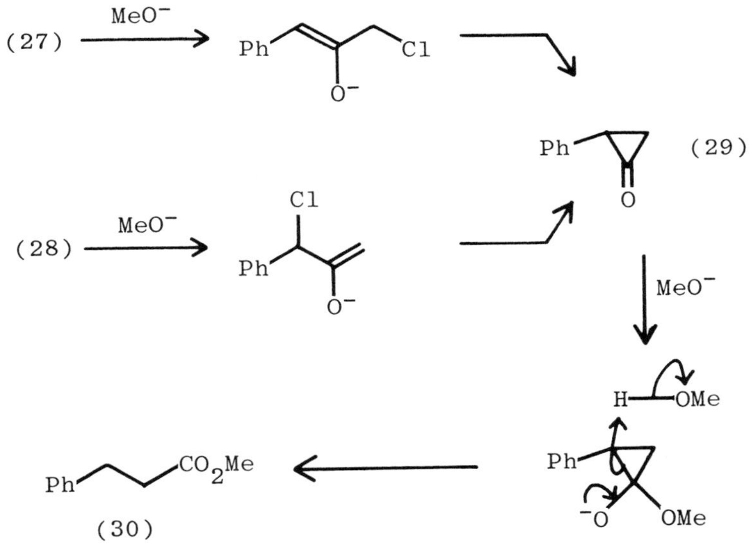

(27) →[MeO⁻] → (29) →[MeO⁻] → Ph~CH₂~CH₂~CO₂Me (30)

(28) →[MeO⁻] →

CHAPTER 32

Four-Membered Rings: Photochemistry in Synthesis

(a) *Photochemical 2 + 2 Cycloadditions*

Problem :

Suggest a synthesis for (1), used in a study on the mechanism of cycloadditions.

(1)

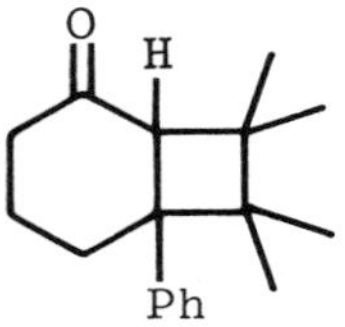

Answer :

Of the two possible 2 + 2 disconnections, (1b) is much the better as it gives two simple starting materials, one of which (3) is symmetrical so that there is no ambiguity in the reaction.

Analysis

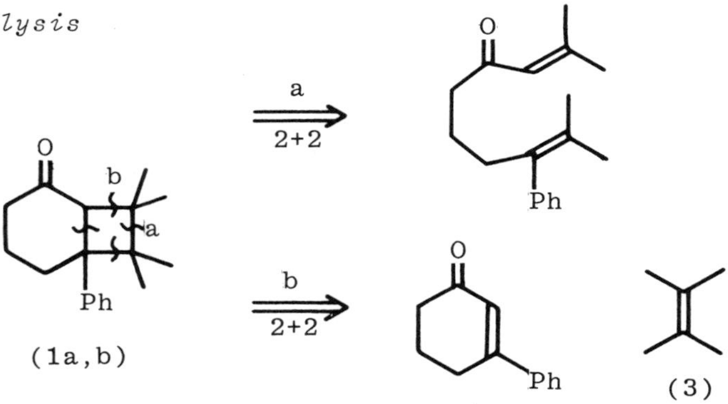

(1a,b)
(2)
(3)

Synthesis[369]

$$(2) + (3) \xrightarrow[\text{t-BuOH}]{h\nu} \text{TM}(1)$$

Example :
Synthesis of the propellane (4) requires FGA. With two four-membered rings probably to be made by photochemical reactions, a carbonyl group on the six-membered ring is best, allowing a 2 + 2 disconnection.
Analysis 1

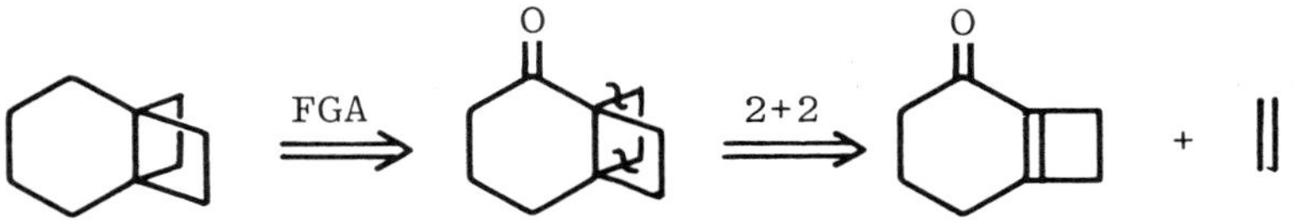

(4) (5) (6)

Further 2 + 2 is impossible, but (6) could be made by dehydration of (7) or (8), and 2 + 2 disconnections are possible on both of these.
Analysis 2

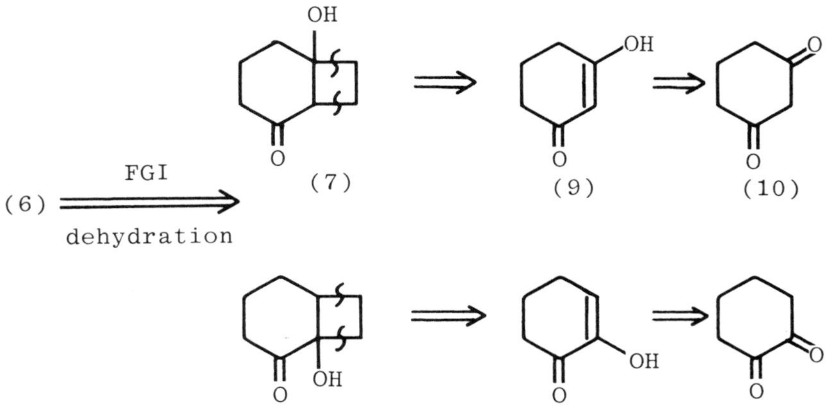

(7) (9) (10)

(8)

Enol (9) comes from available dione (10) and the synthesis was performed on the enol acetate (11) rather than enol (9). Elimination with strong base gave (6) and the synthesis was completed by Wolf-Kishner removal of the carbonyl group.

Synthesis[370]

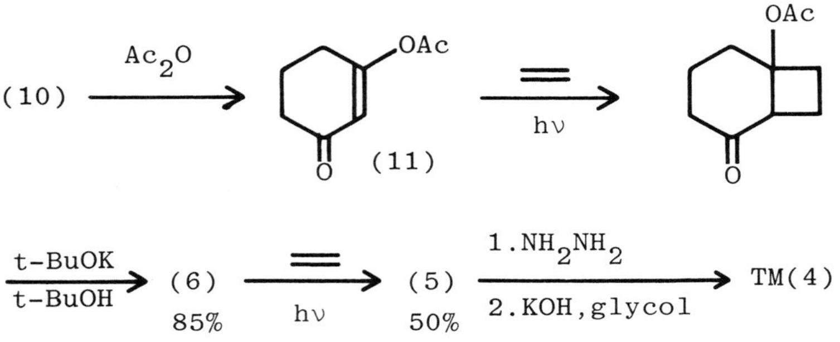

(10) $\xrightarrow{Ac_2O}$ (11) $\xrightarrow{h\nu}$ [OAc bicyclic ketone]

$\xrightarrow[t-BuOH]{t-BuOK}$ (6) 85% $\xrightarrow{h\nu}$ (5) 50% $\xrightarrow[2.KOH,glycol]{1.NH_2NH_2}$ TM(4)

Stereo- and Regioselectivity
Problem :
Suggest a synthesis for (12).

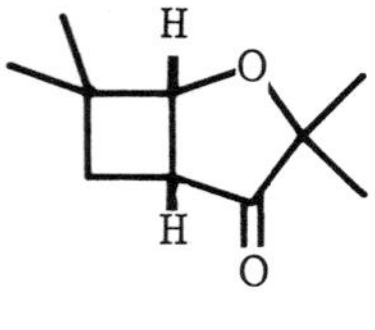

(12)

Answer :
The intermolecular reaction (disconnection 12b) has the correct orientation (p T 271) and starting material (13) is easy to make (p T 196).
Analysis

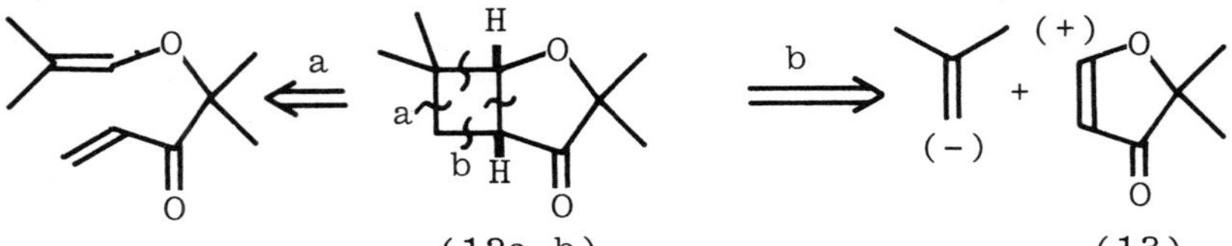

(12a,b) $\Longleftarrow$ (13)

Synthesis[371]

$$\text{(isobutylene)} + (13) \xrightarrow{h\nu} TM(12)\ 95\%$$

The product was used as an intermediate in the synthesis of cyclohexenones (16).

$$(12) \xrightarrow{MCPBA} (14) \xrightarrow[2.\ HO^-,\ H_2O]{HO\ \ OH,\ H^+} $$

(15)

$$\xrightarrow[\substack{1.\ MeLi \\ 2.\ H^+,\ H_2O \\ 3.\ base}]{} (16)\ 65\%$$

Revision Problem:
What is the structure of (14) and why is it formed? What are the reactions leading from (15) to (16)? What alternative synthesis of (16) can you suggest?

Answer:
The Baeyer-Villiger oxidation (p T 226) of (12) will occur with migration of the more substituted atom. The structure of (14) must therefore be:

$$(12) \xrightarrow{MCPBA} (14)$$

This is an acetal, so in acid solution it is in equilibrium with (17) which fragments to aldehyde (18) and hence (15).

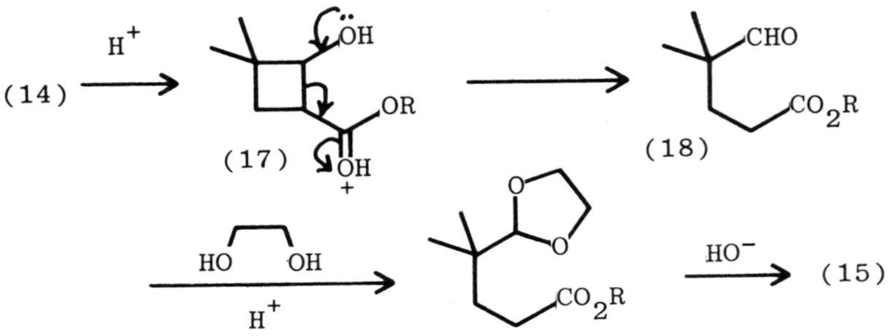

MeLi gives ketone (19), acid hydrolysis frees the aldehyde (20) which cyclises to (16) since only the ketone can enolise.

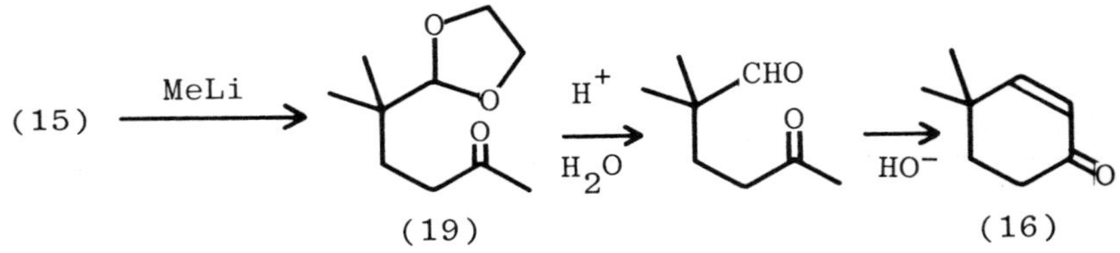

An alternative synthesis of (16) is found by disconnecting the 1,5-dicarbonyl relationship in (15).

Analysis

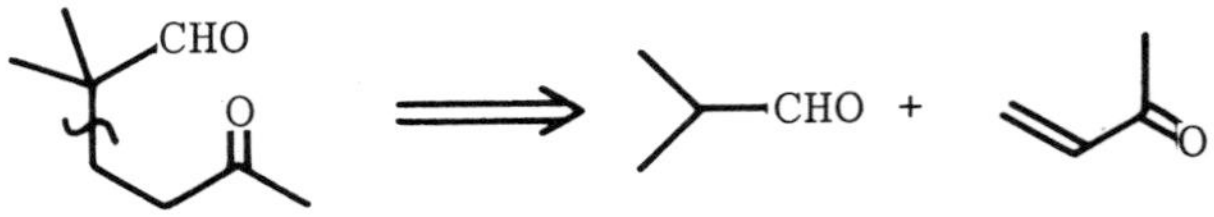

Control is best exercised by forming the enamine of the aldehyde. The reaction goes to (16) in one step and is a Robinson annelation (p T 175).

Synthesis[372]

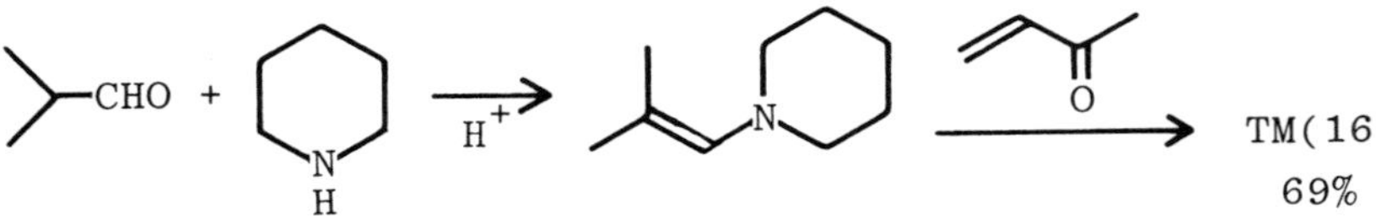

TM(16) 69%

Example :

The insect pheromone lineatin (21) contains a four-membered ring : it also contains an acetal and disconnection of this reveals the carbon skeleton (22).

Lineatin : *analysis*

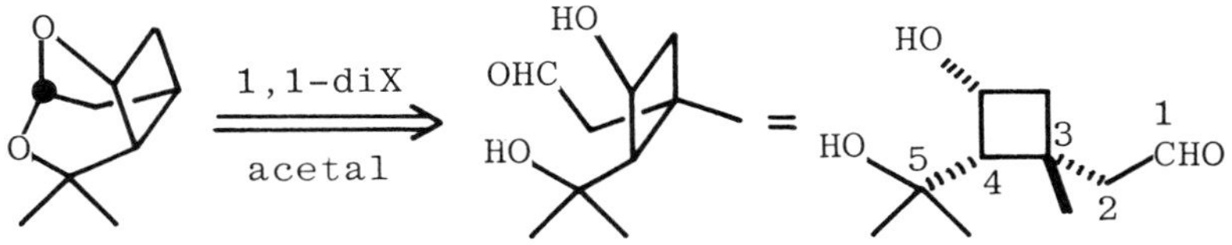

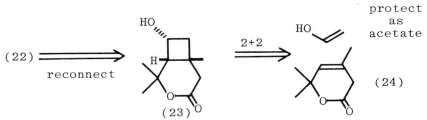

(21) (22)

Early syntheses of lineatin reconnected the 1,5-difunctionalised part to give structures like (23) so that syntheses by 2 + 2 cycloaddition would ensure the correct stereochemistry.

Analysis 2

Unfortunately the double bond in (24) is not conjugated with the carbonyl group so that the 2 + 2 cycloaddition does not give a good yield. It is also unsymmetrical so that a mixture of adducts is formed. Attempts to solve the first of these problems by using (25) or (26) made the second problem worse as the wrong adduct, e.g. (27) is the major isomer.

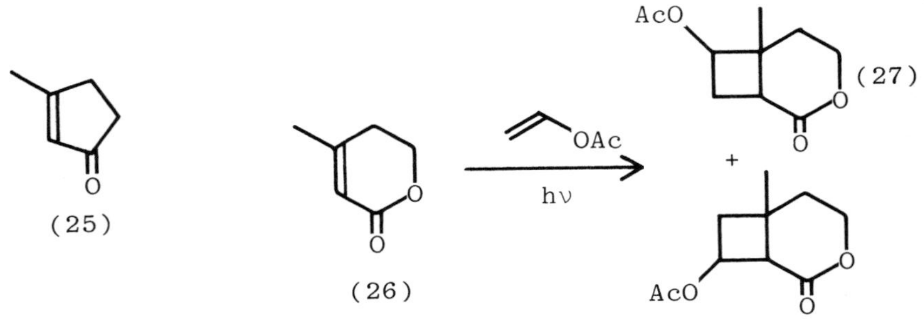

One solution is to leave out a CH_2 group in (24) so that it becomes conjugated (28). Cycloaddition now gives the right isomer (29) in reasonable yield but extensive transformations are needed to convert (29) into lineatin and less than 0.2% yield resulted.
Synthesis[373]

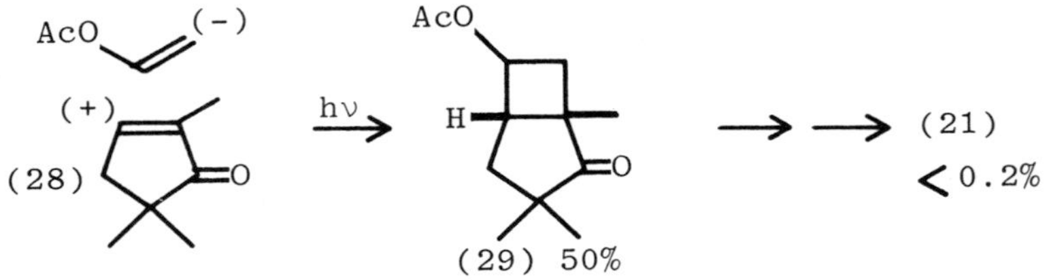

Another solution is to abandon the 2 + 2 cycloaddition and use the right compound (24) and a route involving three-membered ring formation and ring expansion. Reduction of (24) gave acetal (30) which added dichlorocarbene in good yield. Further transformations gave the epoxide (31); the same type of compound was used on page T 273. Rearrangement gave ketone (32), and hence lineatin. The overall yield from (24) was 2.8%.
Synthesis[374]

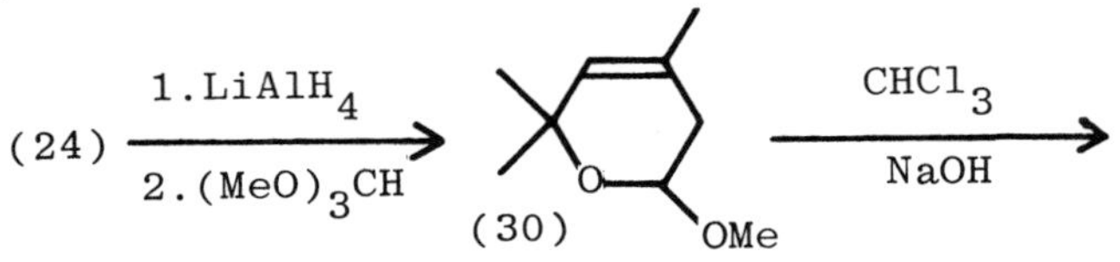

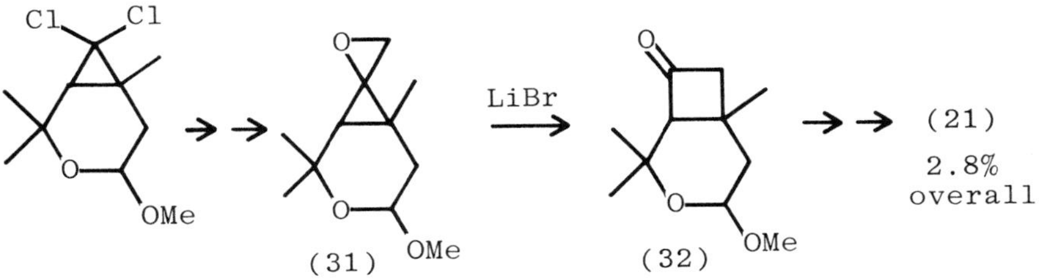

Cl Cl → → (31) → LiBr → (32) → → (21) 2.8% overall

(structures with OMe groups shown)

Problem :

How might (33) be synthesised so that its solvolysis reactions can be studied?

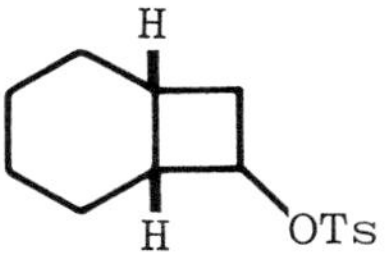

(33) — bicyclic structure with H, H, OTs

This is not a trivial problem: cyclohexene does not give good yields of 2 + 2 photo adducts with $CH_2=CHOR$.

Answer :

Alcohol (34) is needed and we shall want to use a photocycloaddition to make some derivative of it (ether or ester). One partner in the cycloaddition will be nucleophilic ($CH_2=CHOR$) so we need a carbonyl group conjugated with the cyclohexene. FGA is required.

Analysis

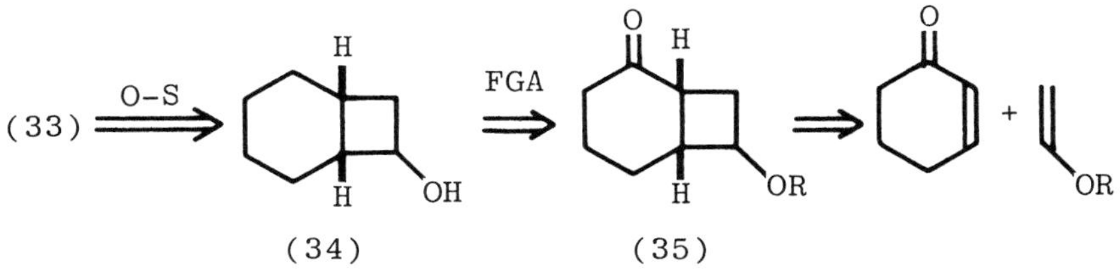

(33) $\overset{O-S}{\Longrightarrow}$ (34) $\overset{FGA}{\Longrightarrow}$ (35) $\Longrightarrow$ cyclohexenone + $CH_2=CHOR$

The benzyl ether is a good protecting group as it can be removed under reducing conditions. It turns out that the carbonyl group has another role to play as a mixture of *cis* and *trans* photoadducts is formed, equilibrated to (35, R=CH$_2$Ph) in base *via* enolisation. *Synthesis*[375]

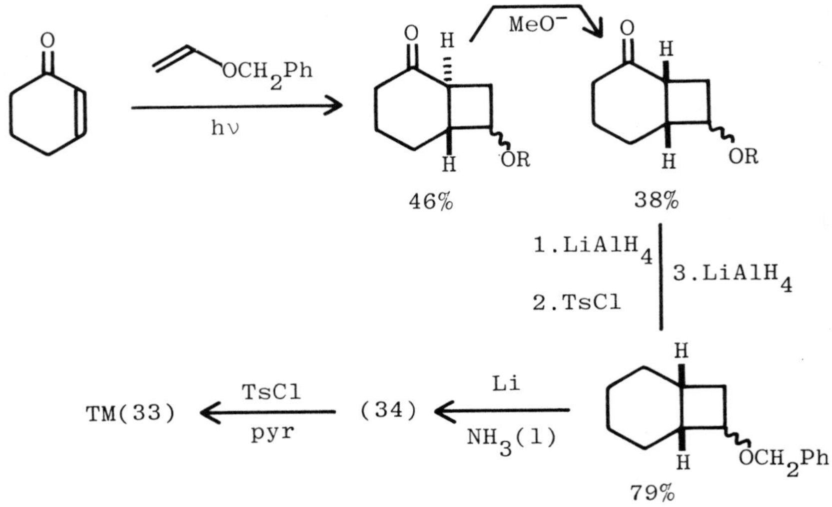

[Reaction scheme showing cyclohexenone + CH$_2$=CHOCH$_2$Ph with hν giving bicyclic ketone (46%) and epimer (38%) via MeO⁻; then 1. LiAlH$_4$, 2. TsCl; 3. LiAlH$_4$ giving bicyclic OCH$_2$Ph compound (79%); then Li/NH$_3$(l) to (34); then TsCl/pyr to TM(33).]

(b) *Four-Membered Rings by Ionic and Radical Reactions*

Example :
The acyloin reaction in the presence of Me$_3$SiCl (p T 202) is a way of making four-membered rings, e.g. (36). *Synthesis*[376]

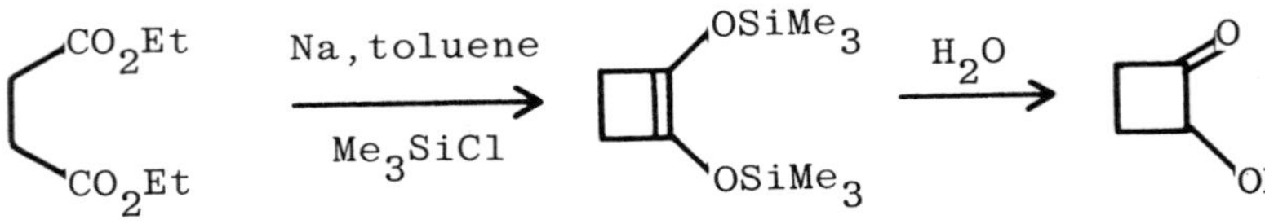

[Scheme: diethyl adipate + Na, toluene, Me$_3$SiCl → bis(trimethylsilyloxy)cyclobutene (36); H$_2$O → 2-hydroxycyclobutanone (37).]

The first product (36) is itself an important source of other four-membered rings, but it can be hydrolysed to the acyloin (37) if required.[377]

Example :

Cyclisation of malonates is one of the few reliable ionic reactions giving four-membered rings. Hence when the *cis* and *trans* ketones (38) were wanted for a photochemical study, acids (39) were the obvious starting materials as these could be made by cyclisation of (40).

Analysis

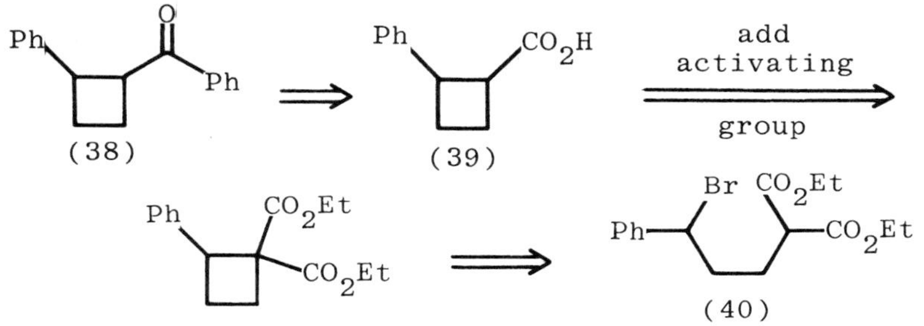

Direct disconnection of (40) to dibromide (41) is a possibility, but it is better to make (40) from alkene (42) as disconnection now gives an allylic bromide (43) available from cinnamic acid.

Analysis

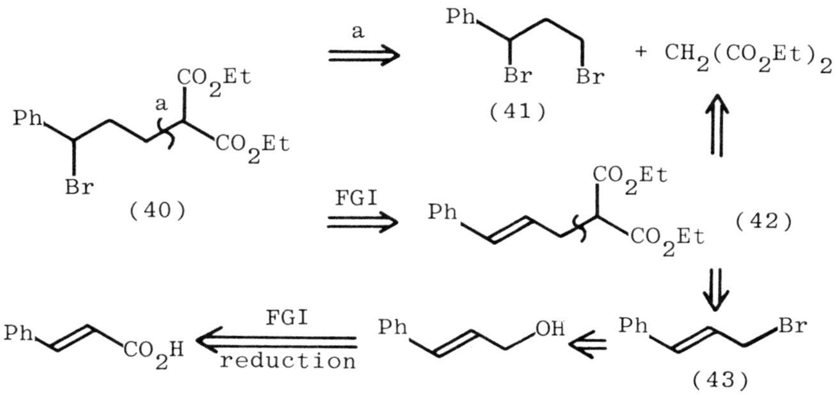

Synthesis[378]

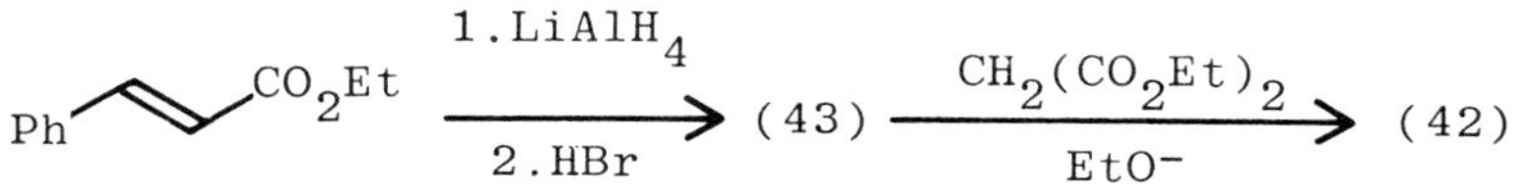

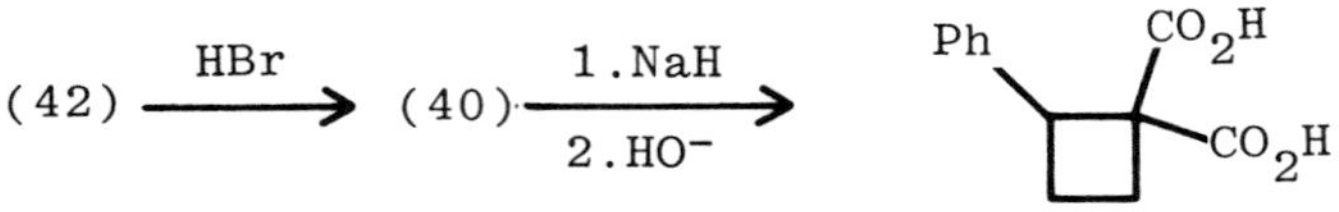

(42) —HBr→ (40) —1.NaH / 2.HO⁻→ [cyclobutane with Ph, CO$_2$H, CO$_2$H] (44)

79% from (42)

Decarboxylation of (44) on heating gave a mixture of *cis* and *trans* acids (39) which were separated by chromatography. Each was converted to the phenyl ketone (38).

Synthesis[379]

(44) —1.heat / 2.separate→

cis-(39) —PhLi→ *cis*-(38)

trans-(39) —PhLi→ *trans*-(38)

Problem :

Suggest a strategy for synthesising spiro compounds (45).

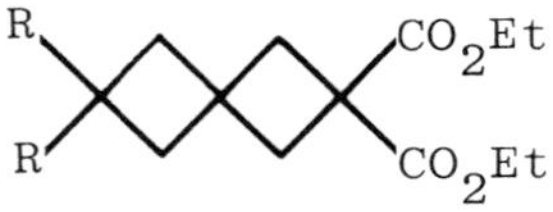

[structure showing two fused rings with R, R' and CO$_2$Et, CO$_2$Et substituents] (45)

Answer :

Malonate disconnection requires dibromide (46) available in turn from malonate (47) which requires dibromide (48) and so on.

Analysis

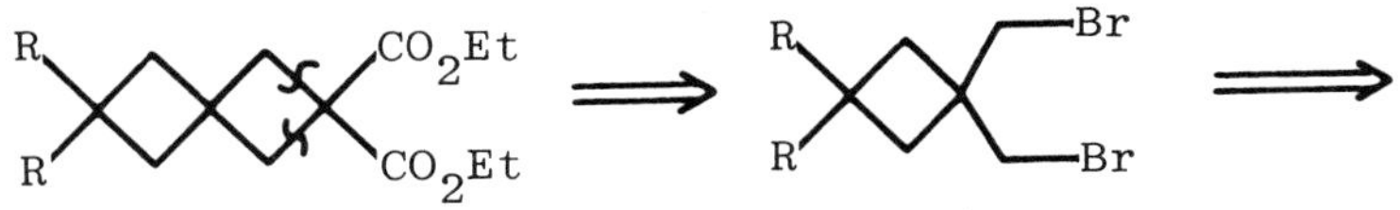

(45a) ⟹ (46) ⟹

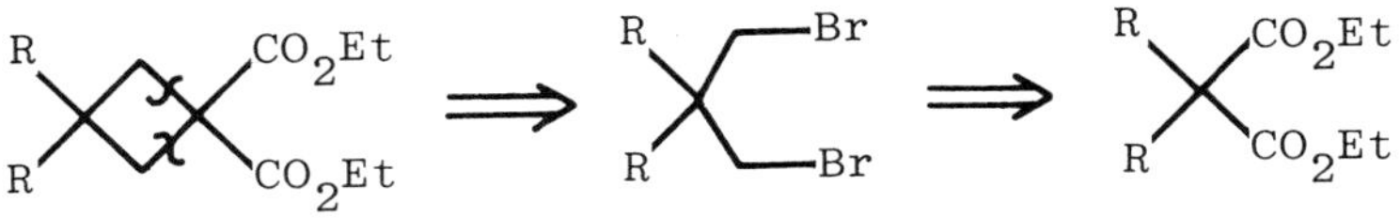

(47) ⟹ (48) ⟹ (49)

The synthesis consists of the same steps repeated and repeated, and can be extended to longer chains of spiro rings.

Synthesis[380]

$$(49) \xrightarrow[\text{2.PBr}_3]{\text{1.LiAlH}_4} (48) \xrightarrow[\text{EtO}^-]{\text{CH}_2(\text{CO}_2\text{Et})_2} (47)$$

$$\xrightarrow[\text{2.PBr}_3]{\text{1.LiAlH}_4} (46) \xrightarrow[\text{EtO}^-]{\text{CH}_2(\text{CO}_2\text{Et})_2} \text{TM}(45)$$

Problem:

What starting ketone would be necessary for a Trost synthesis of (50) (p T 273)?

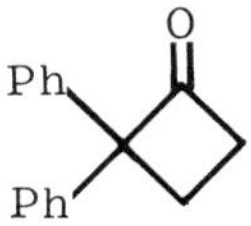

(50)

Answer:

Reversing the rearrangement gives epoxide (51) (cf 31) and hence benzophenone.

Analysis

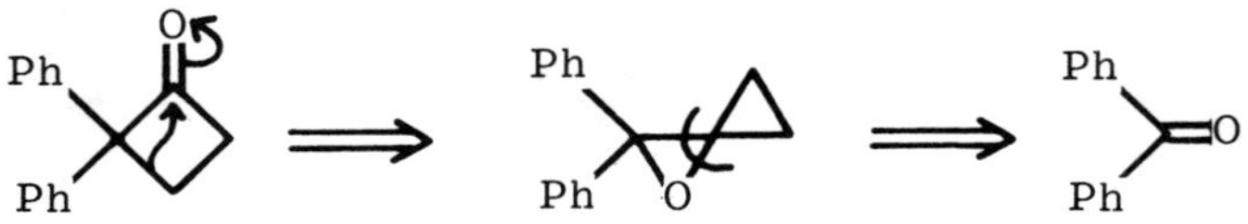

$$\text{(50a)} \Longrightarrow \text{(51)} \Longrightarrow$$

Synthesis[381]

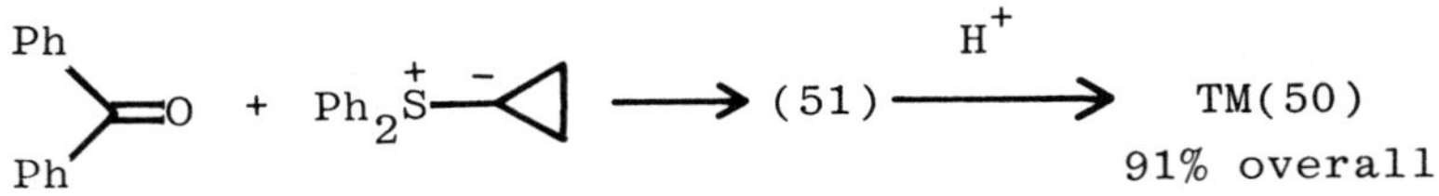

$$\underset{Ph}{\overset{Ph}{>}}\!\!=\!\!O \; + \; Ph_2\overset{+}{S}\!\!-\!\!\triangleleft^{-} \longrightarrow (51) \xrightarrow{H^+} TM(50)$$
91% overall

CHAPTER 33

Strategy XV: Use of Ketenes in Synthesis

Problem :
What would be the starting materials for the synthesis of (1) by thermal 2 + 2 cycloaddition?

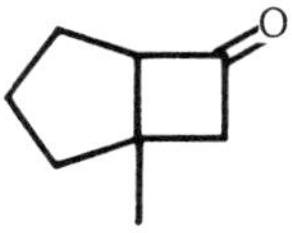

(1)

Answer :
There are two possible disconnections (1a) and (1b) each requiring a different ketene.
Analysis

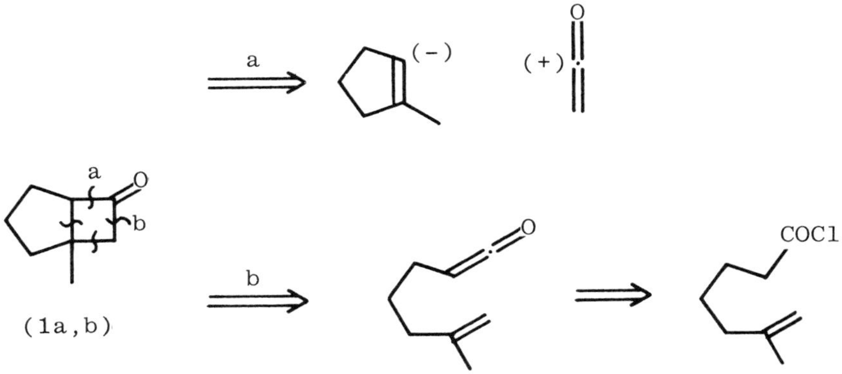

(1a,b)

(2)   (3)

Route (a) has the correct regiochemistry but cycloadditions of ketene itself are rare and usually work best

with dienes.   Route (b) is successful.
Synthesis[382]

$$(3) \xrightarrow{Et_3N} TM(1)\ 35\%$$

Revision Problem :
Suggest a synthesis for (3).
Answer :
The parent acid (4) could be made by a Wittig reaction from 1,6-dicarbonyl compound (5) which reconnects to (6). Alternatively a repeated malonate disconnection *via* bromide (7) is quite attractive.
Analysis

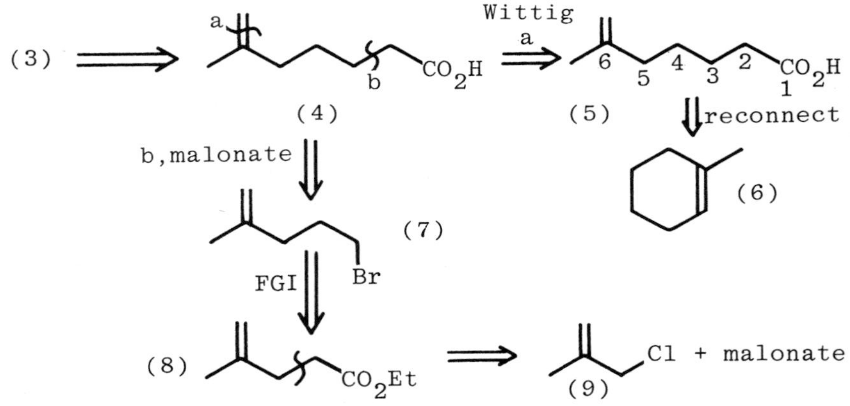

No synthesis of (3) appears in the original paper[382] so we may speculate on the success of these or other ideas. Route (a) is much shorter as no carboxyl protection is necessary if we are ready to waste the first mole of ylid.
Synthesis

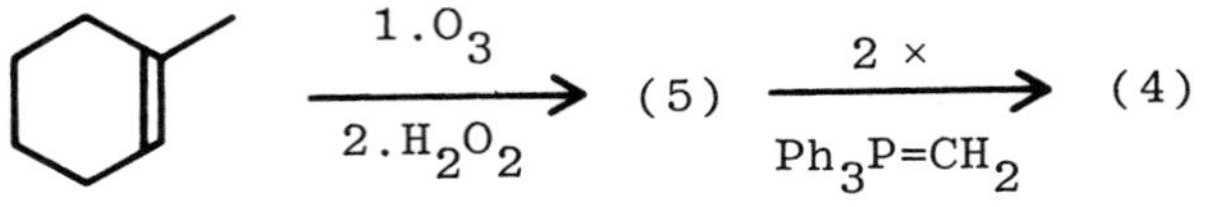

$CH_2(CO_2Et)_2$ →[1. EtO⁻ / 2.(9)] [structure with CO_2Et groups] →[1. KOH, H_2O / 2. H⁺, heat / 3. EtOH, H⁺] (8)

→[1. LiAlH₄ / 2. PBr₃] (7) →[$CH_2(CO_2Et)_2$ / EtO⁻] [structure with CO_2Et groups] →[1. KOH, H_2O / 2. H⁺, heat] TM(4)

Example:
Cycloadditions between chloroketenes and electron-rich double bonds occur very easily. The microbial toxin moniliformin (10) has been made using this reaction. *Synthesis*[383]

EtO—≡ →[$Cl_2CHCOCl$ / Et_3N] [cyclobutenone with Cl, Cl, OEt] →[HCl / H_2O] [structure (10)]

Note that the regiochemistry of the cycloaddition (11) combines the most electrophilic atom of the ketene with the most nucleophilic atom of the acetylene.

(11)

Problem:
Predict the structures of the adducts of (12) and (13) with dichloroketene $Cl_2C=C=O$.

(12)   (13)

Answer:
The more nucleophilic atom in (12) is easy to identify,

and the product is a mixture of stereo isomers of a single regioisomer (14).
Synthesis[384]

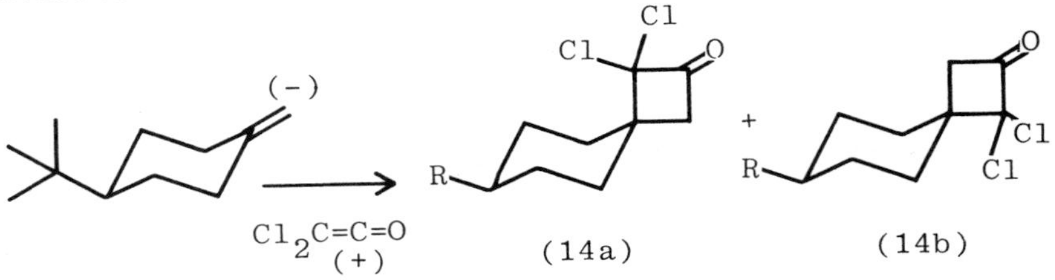

(14a)   (14b)

The difference in the reactivity of the two ends of the double bond in (13) is more subtle : adduct (15) is formed so presumably the electron-donating effects of the methyl groups outweigh their bulk.
Synthesis[385]

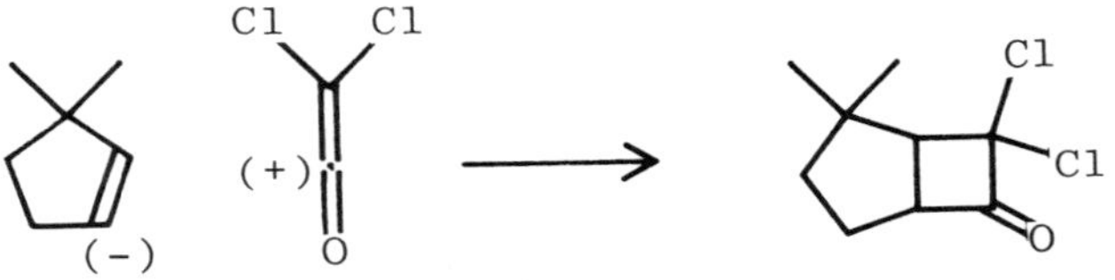

(15)

Example :
The intermediate (19), used on p 382-3 in a synthesis of lineatin, is a more surprising ketene product. It can be made on a large scale by the BF_3 catalysed addition of ketene to enone (16). This is presumably not a 2 + 2 cycloaddition to (17), but an ionic reaction via (18).
Synthesis[374]

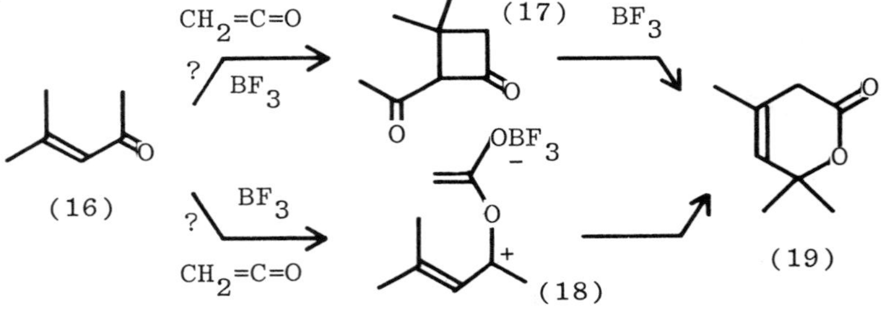

Example :

Ketenes also add to imines in 2 + 2 cycloadditions giving the important β-lactams (20). The regiochemistry is predictable as the nitrogen is the nucleophilic atom in the imine. This is true even in conjugated imine (21) which gives (22) with azidoketene. The azido group can be reduced to an imine for antibiotic synthesis.

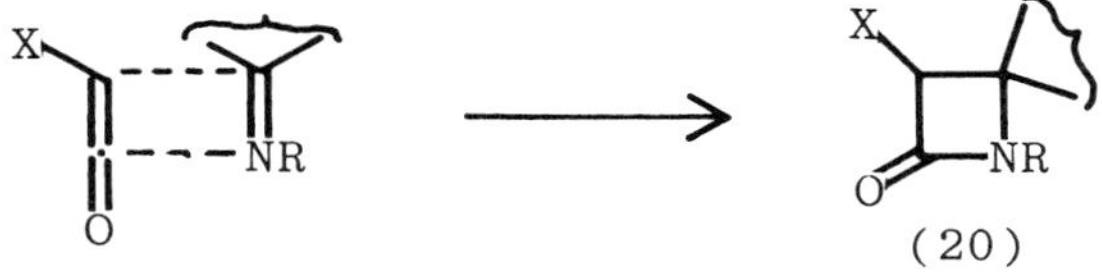

Synthesis[387]

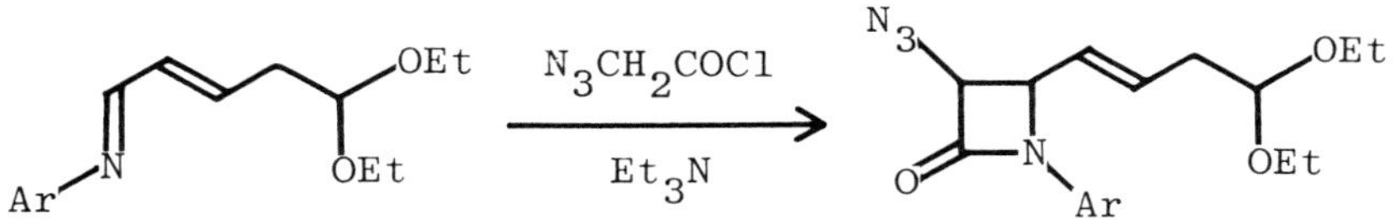

An alternative 2 + 2 route to β-lactams is to use ketene-like (23) on alkenes and reduce away the sulphonyl group.[387]

Synthesis

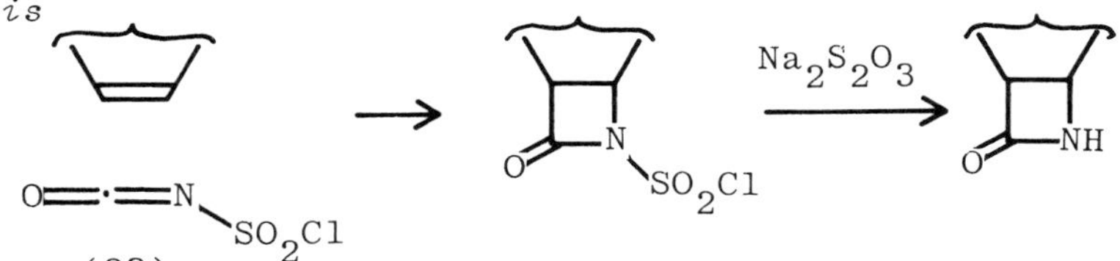

Problem :

The halogen atom in haloketene adducts such as (24) is strongly electron-withdrawing. Suggest a synthesis for (24) and predict the product of Baeyer-Villiger oxidation.

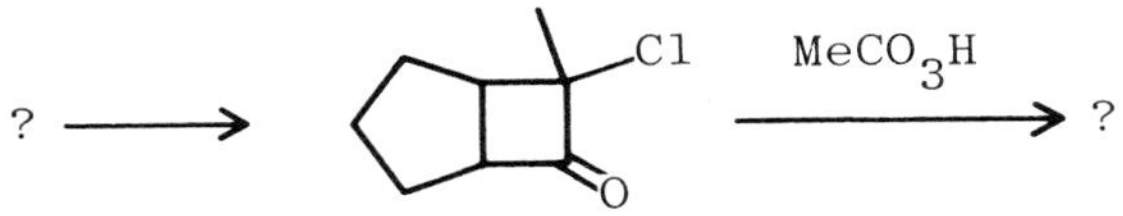

Answer :
Disconnection is to cyclopentene and a chloroketene generated from acid chloride (25) and base.
Analysis

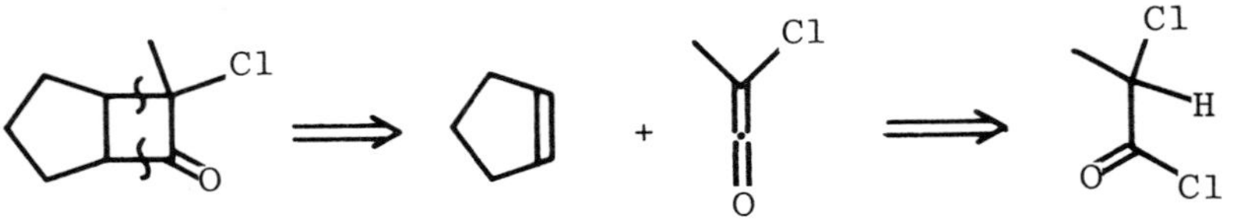

(24a) ⟹ cyclopentene + chloroketene ⟹ (25)

Synthesis[389]

cyclopentene —(25), Et$_3$N→ TM(24)

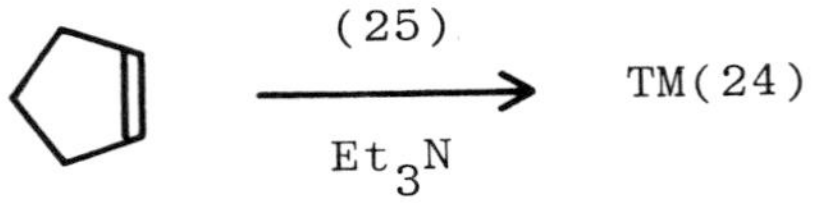

The group which migrates in a Baeyer-Villiger reaction (p T 226) is the more electron-donating one, so it will be the carbon *not* bearing the halogen. The product (26) was transformed into biologically active α-methylene lactone (27) with DBN (p 150).

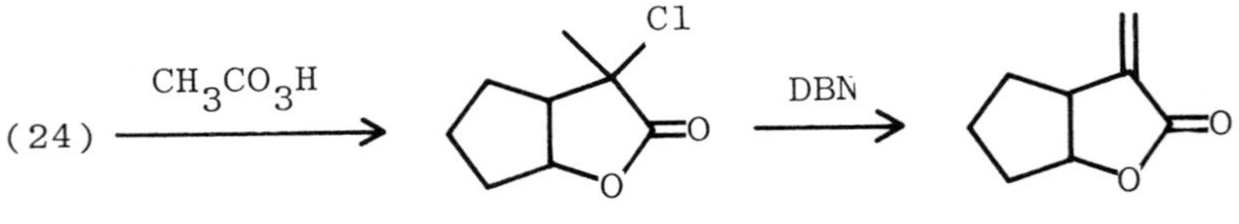

(24) —CH$_3$CO$_3$H→ (26) —DBN→ (27)

(c) *Ketene Dimers*
Example :
Ketene dimer (28), made from [^{14}C]-labelled acetic acid, has been used to make doubly labelled mevalonic lactone (29) for studies on the biosynthesis of terpenes. Note

the use of unlabelled ketene in a later step

Synthesis [389]

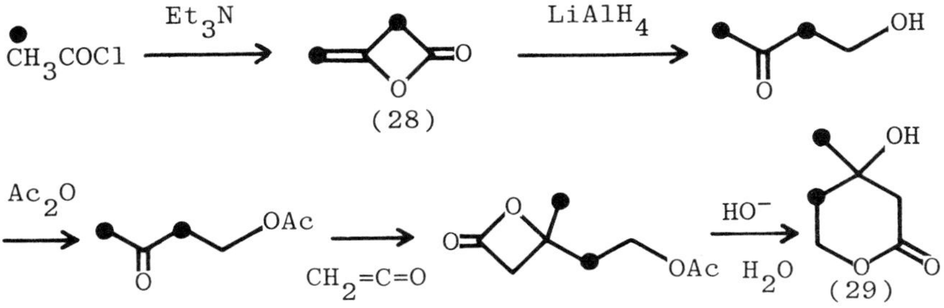

CH$_3$COCl $\xrightarrow{Et_3N}$ (28) $\xrightarrow{LiAlH_4}$ ●—●—OH (with C=O)

$\xrightarrow{Ac_2O}$ ●—C(=O)—●—●—OAc $\xrightarrow{CH_2=C=O}$ [β-lactone with OAc side chain] $\xrightarrow{HO^-\ H_2O}$ (29)

Problem :

Compound (30) was needed for a synthesis of an insecticide. How might it be made?

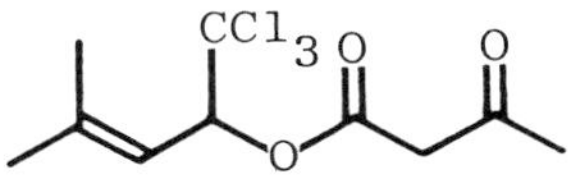

(30)

Answer :

The acetoacetyl group may be supplied as ketene dimer (p T277), leaving alcohol (31). You may have considered chlorinating ketone (32) or adding CCl_3^- to aldehyde (33) to make (31) but you probably did not see the disconnection to chloral and isobutylene (31b).

Analysis

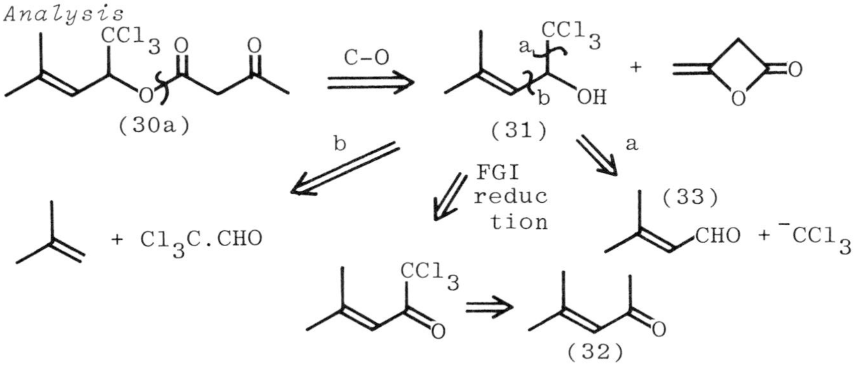

(30a) $\xrightarrow{C-O}$ (31) + ketene dimer

(31) $\xrightarrow{b}$ isobutylene + $Cl_3C.CHO$

(31) $\xrightarrow{a,\ FGI\ reduction}$ (33) [CHO + $^-CCl_3$]

$\begin{array}{c} CCl_3 \\ \diagup\!\!\!\diagdown \end{array}$ (enone with CCl$_3$) $\Rightarrow$ (32)

Route (b), based on an aliphatic Friedel-Crafts reaction, gives a good yield of (31) which is duly acylated with ketene dimer.

Synthesis[389]

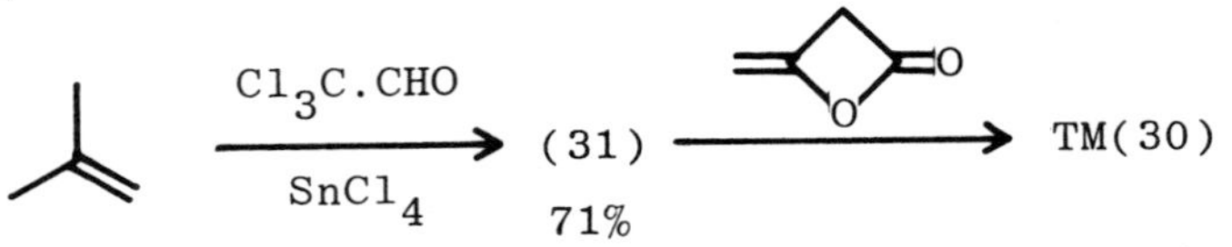

$$\text{Me}_2\text{C=CH}_2 \xrightarrow[\text{SnCl}_4]{\text{Cl}_3\text{C.CHO}} (31) \quad 71\% \xrightarrow{\text{ketene dimer}} \text{TM}(30)$$

Example :
An interesting use of the diketone type of ketene dimer (34) is reduction and cleavage[390] to β,γ-unsaturated aldehyde (35).

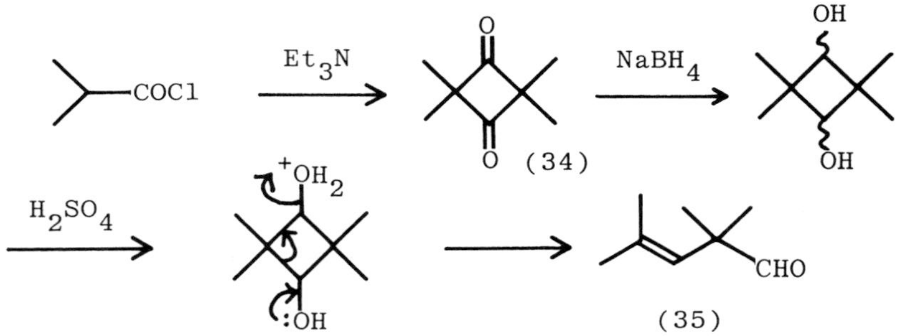

$$\text{Me}_2\text{CH-COCl} \xrightarrow{\text{Et}_3\text{N}} (34) \xrightarrow{\text{NaBH}_4} \text{diol} \xrightarrow{\text{H}_2\text{SO}_4} \text{intermediate} \longrightarrow (35)$$

This aldehyde (35), also made by other routes (p 225), is used in the synthesis of insecticidal derivatives.

CHAPTER 34

Five-Membered Rings

This is partly a revision chapter so it is not divided into sections as in Chapter T 31 : any method may appear at any stage.

Problem :

Suggest a synthesis of cyclopentenone (1).

Answer :

Enone disconnection gives ketoaldehyde (2) as the key intermediate. Cyclisation of (2) to (1) will be unambiguous as no other stable ring can be formed. Disconnection of (2) is more of a problem. The branchpoint suggests two possibilities (a) and (b).

Analysis

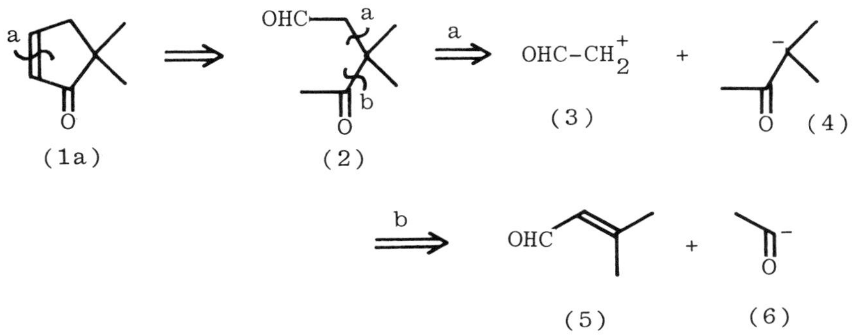

The only acyl anion (6) equivalents we have met which will carry out Michael additions are nitro compounds (p T183 and 213) but the scales are weighted against such a Michael reaction as the most hindered atom must be attacked.

Possible Synthesis 1

$$MeCH_2NO_2 \xrightarrow{\text{base} \atop (5)} OHC\underset{NO_2}{\overset{}{\diagdown\!\!\diagup}} \xrightarrow{TiCl_3} (2) \xrightarrow{\text{base}} (1)$$

In approach (a) both synthons present problems : the regioselectivity of (4) is a serious difficulty and we do not like to use α-bromo aldehydes for (3). Allyl cleavage (Chapter T 26) provides an answer to the last problem.

Analysis 2

$$(2) \xRightarrow{\text{reconnect}} \underset{O}{\diagdown\!\!\diagup} \Rightarrow \diagdown\!\!\diagup Br + (4)$$

It might be possible to add the allyl group regiospecifically, using an enamine, but usually the less substituted enamine is formed preferentially.

A solution to all these problems can be found by an alternative disconnection. Enone (1) could be made by an aliphatic Friedel-Crafts reaction from acid (7). Disconnection of the allyl group now gives synthon (8) and the regioselectivity problem disappears.

Analysis 3

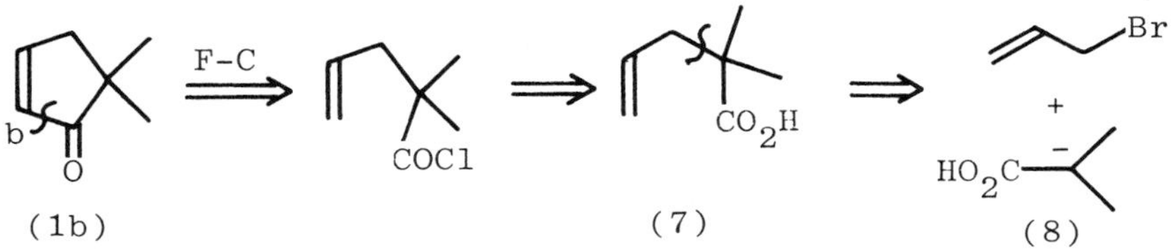

[Scheme: (1b) ⇒ (F-C) ⇒ structure with COCl ⇒ (7) structure with CO₂H ⇒ allyl bromide (8) + HO₂C-C(CH₃)₂⁻]

(1b)   (7)   (8)

There is no room for an activating group on (8) so we can either use the nitrile or simply the acid with the strong hindered base LDA.

Synthesis[391]

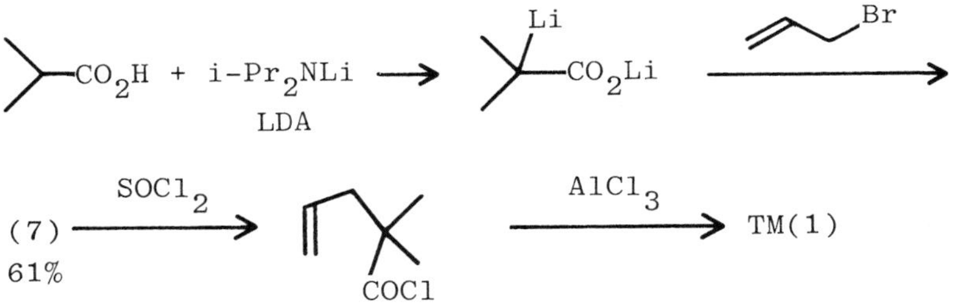

$\rightarrow$CO$_2$H + i-Pr$_2$NLi $\rightarrow$ $\rightarrow$CO$_2$Li   allyl-Br $\rightarrow$
                LDA

(7) $\xrightarrow{SOCl_2}$ [COCl intermediate] $\xrightarrow{AlCl_3}$ TM(1)
61%

Example :

For his synthesis of the anti-tumour compound diketo-coriolin, Danishefsky[392] needed (9) with its two fused five-membered rings. Enone disconnection suggests two possible precursors.

Analysis

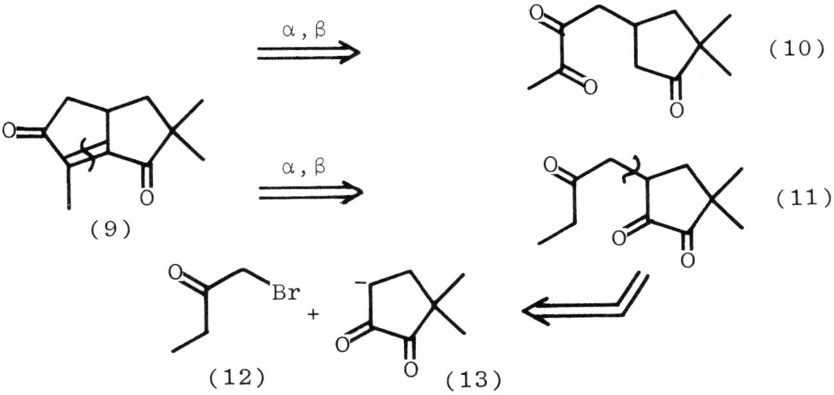

(9) $\xRightarrow{\alpha,\beta}$ (10)

(9) $\xRightarrow{\alpha,\beta}$ (11) $\Leftarrow$ (12) + (13)

Triketone (11) has an obvious 1,4-dicarbonyl disconnection without ambiguity, providing we can make (12), as (13) can enolise on one side only. Ring-chain (1,5-dicarbonyl) disconnection on (10) is more attractive as it gives symmetrical (14) and the enone (1) we have just made.

Analysis 2

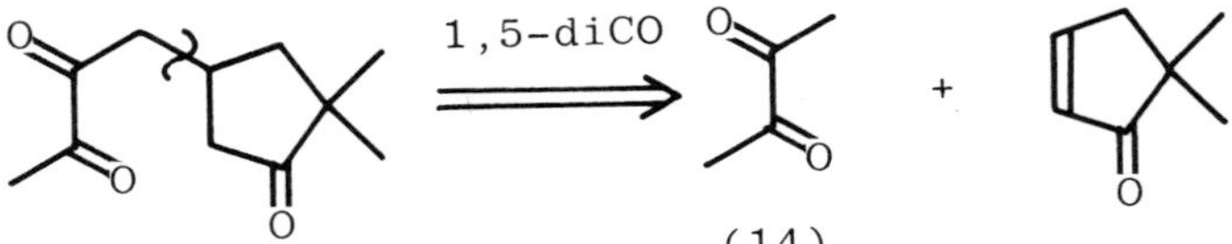

Control will be needed for the Michael addition, and it proved necessary to protect one carbonyl of (14) as an acetal and add an activating group to the other to give (16). There is no ambiguity in either of these steps as protection of one carbonyl also deactivates the other (Chapter T5) and (15) can enolise on one side only. Removal of the acetal from (17), cyclisation, and decarboxylation can all be accomplished in one step.

Synthesis[392]

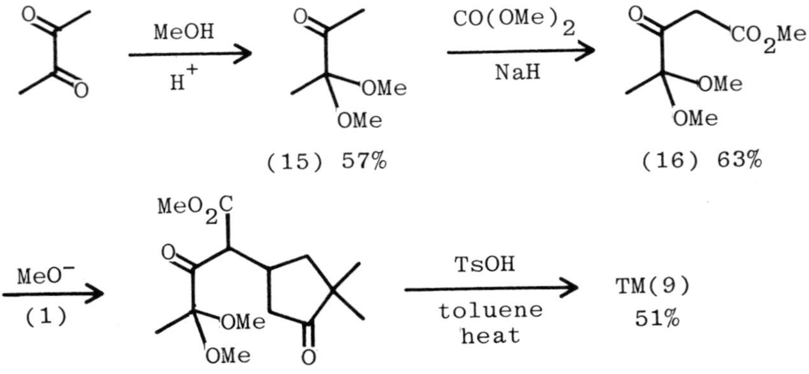

Danishefsky's analysis[392] of this problem in general and of his attempts to add the next five-membered ring to (9) are worth reading.

Example :

Spiro compound (18), also containing two five-membered rings, can be made by oxidation of the acyloin (19) (Chapter T24).

Analysis 1

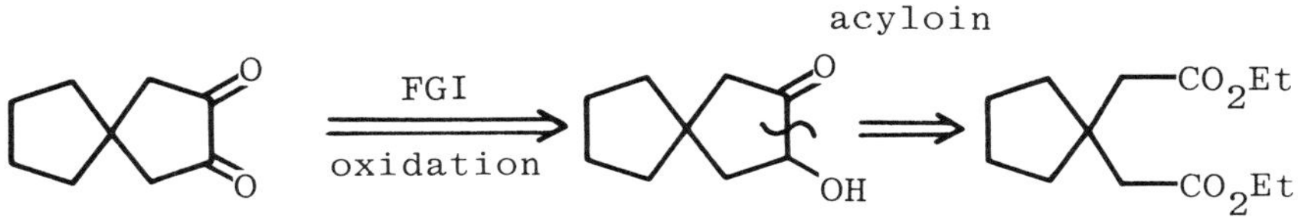

(18) $\xRightarrow{\text{FGI oxidation}}$ (19) acyloin $\Rightarrow$ (20)

Diester (20) could be made by the usual 1,5-dicarbonyl route involving Michael addition to (21), but we can use its symmetry to devise an alternative route from malonate adduct (22).

Analysis 2

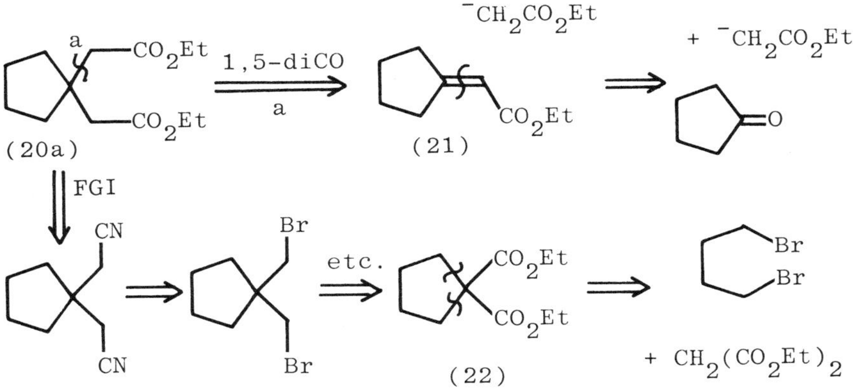

(20a) $\xRightarrow[a]{\text{1,5-diCO}}$ (21) $\Rightarrow$ $^-CH_2CO_2Et$ + $^-CH_2CO_2Et$ + cyclopentanone

$\Downarrow$ FGI

(CN,CN) $\Rightarrow$ (Br,Br) $\xRightarrow{\text{etc.}}$ (22) $\Rightarrow$ (Br,Br) + $CH_2(CO_2Et)_2$

No doubt either route could be used for (20), or one of the two routes discussed for TM (27) below. The acyloin reaction needs the assistance of silicon, and the product (23) can be oxidised directly to (18) with bromine.

Synthesis[393]

(20) 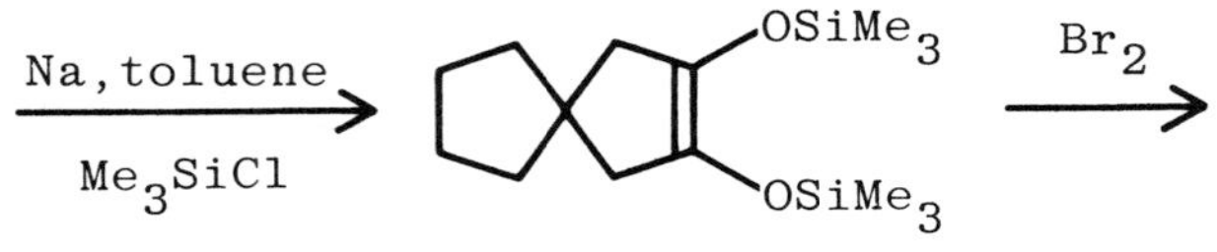 $\xrightarrow{\text{Na, toluene}}_{\text{Me}_3\text{SiCl}}$ [spirocyclic bis(OSiMe$_3$) enediol ether] (23) 59% $\xrightarrow{\text{Br}_2}$ TM(18)

Problem :
Suggest a synthesis for (24), needed for enzyme model studies.

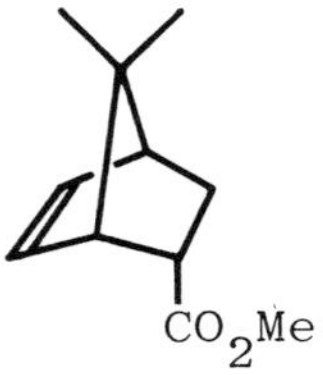

(24) [norbornene with gem-dimethyl bridge and CO$_2$Me]

Answer :
This compound contains two five-membered rings, but it also contains a six-membered ring with all the hallmarks of a Diels-Alder adduct. The real problem is how to make (25).

Analysis 1

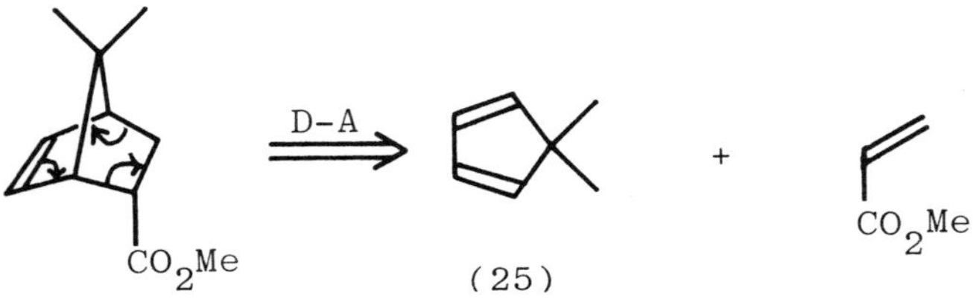

(24a) $\xRightarrow{\text{D-A}}$ (25) + CH$_2$=CH-CO$_2$Me

One route uses the acyloin reaction to make (26) which can be converted into (25) by FGI.

Analysis 2

(25) $\xrightarrow{FGI}$ (26) $\xrightarrow{acyloin}$ (27)

The problems involved in making (27) are similar to those for (20). Neither of the routes discussed above for (20) provides convenient routes to (27) : instead a short route involves the strategy of re-connection to readily available dimedone (28),(Chapter T 21). The haloform reaction provides the cleavage.

Synthesis 1[394]

(28) $\xrightarrow{NaOCl}$ (29) 98% $\xrightarrow[H^+]{EtOH}$ (27)

An alternative is the strategy outlined above for (20) *via* (21). Since the two enolates needed are the same, a malonate derivative supplies them both and in practice cyanoester (30) is the most convenient. The version with ammonia to trap the intermediate as (31) is the Guaresci reaction.

Analysis

(27a) $\xrightarrow{1,5-diCO}$ $EtO_2C.CH_2^-$ + EtO_2C $\Longrightarrow$ + $(EtO_2C)CH^-$

Synthesis[395]

(30) acetone + NC-CH₂-CO₂Et → NH₃ / EtOH → [(31) cyclic imide with two CN groups] → H₂SO₄ / H₂O → (29)

The latter part of the synthesis is straightforward, though silicon must again be used to control the acyloin.

Synthesis[396]

(27) → Na, toluene, Me₃SiCl → bis(trimethylsilyloxy) dimethylcyclopentene (87%) → MeOH → α-hydroxy cyclopentanone (90%) → P₂O₅ →

dimethyl cyclopentenone (80%) → LiAlH₄ → hydroxy cyclopentene (90%) → Al₂O₃, 200°C, 70% → (25) → CH₂=CH-CO₂Me, 73% → TM(24)

Problem :
Suggest a synthesis for bicyclic ketone (32).

(32) bicyclic ketone   (34) bicyclic alkene

Answer :
The symmetry of (32) suggests addition of an activating group followed by 1,3-dicarbonyl disconnection to symmetrical (33). This is a 1,6-dicarbonyl compound but reconnection gives (34) with no obvious disconnections. An alternative is to work back to readily available (37) by FGA and chain extension.

Analysis

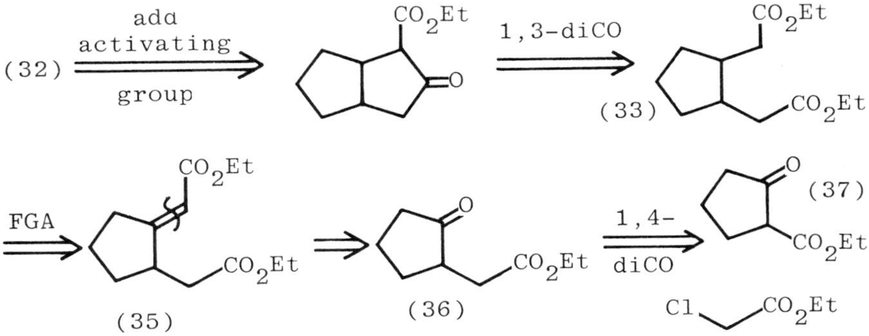

(32) $\xrightarrow{\text{add activating group}}$ [bicyclic CO$_2$Et ketone] $\xrightarrow{1,3\text{-diCO}}$ [cyclopentane with CO$_2$Et and CH$_2$CO$_2$Et] (33)

$\xrightarrow{\text{FGA}}$ (35) [cyclopentanone enoate with CH$_2$CO$_2$Et] $\Rightarrow$ (36) [cyclopentanone with CH$_2$CO$_2$Et] $\xrightarrow{1,4\text{-diCO}}$ (37) [cyclopentanone with CO$_2$Et] + Cl–CH$_2$CO$_2$Et

The Knoevenagel method (Chapter T 20) was used to control the condensation of (36) to (35) and aluminium amalgam to remove the double bond from (35).

Synthesis[397]

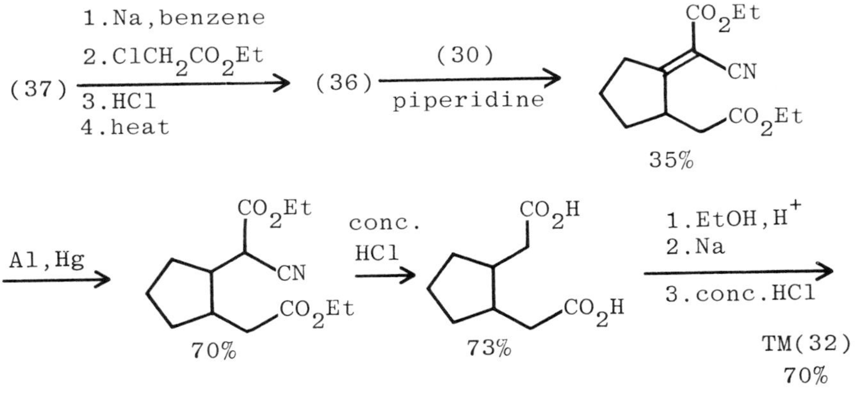

(37) $\xrightarrow[\substack{\text{1. Na, benzene}\\ \text{2. ClCH}_2\text{CO}_2\text{Et}\\ \text{3. HCl}\\ \text{4. heat}}]{}$ (36) $\xrightarrow[\text{piperidine}]{(30)}$ [product, 35%]

$\xrightarrow{\text{Al, Hg}}$ [cyclopentane with CH(CO$_2$Et)CN and CH$_2$CO$_2$Et, 70%] $\xrightarrow{\text{conc. HCl}}$ [dicarboxylic acid, 73%] $\xrightarrow[\substack{\text{1. EtOH, H}^+\\ \text{2. Na}\\ \text{3. conc. HCl}}]{}$ TM(32) 70%

Examples:

Ring contractions, such as the Favorskii reaction on (38), or the cyclopentane aldehyde synthesis on p 374 can be used to make five membered rings.[398]

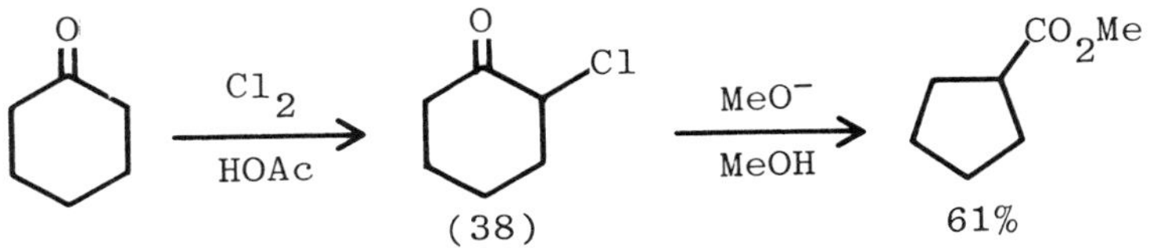

cyclohexanone $\xrightarrow[\text{HOAc}]{\text{Cl}_2}$ (38) 2-chlorocyclohexanone $\xrightarrow[\text{MeOH}]{\text{MeO}^-}$ methyl cyclopentanecarboxylate, 61%

The ready availability of cyclobutanones (34) by ketene cycloaddition (Chapter T 33) makes them useful intermediates for cyclopentanone synthesis by ring expansion. Dichloro ketene gives higher yields of adducts (39) and ring expansion with a diazoalkane than goes with migration of the carbon atom *not* carrying the chlorines (cf p T 276). The chlorines atoms can be removed by reduction to give cyclopentanones (40). Synthesis[399]

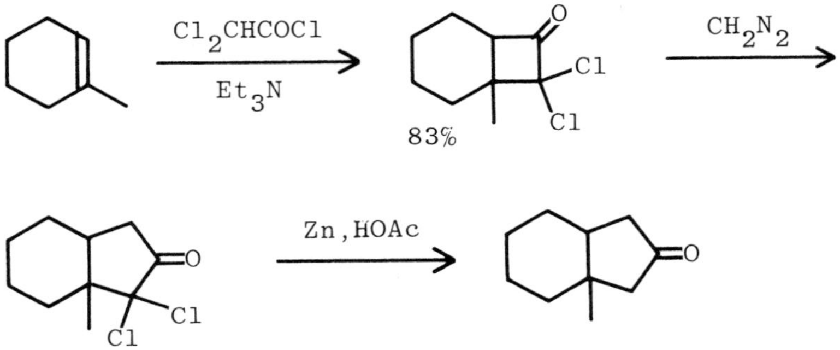

CHAPTER 35

Strategy XVI: Pericyclic Rearrangements in Synthesis: Special Methods for Five-Membered Rings

a) *Electrocyclic Reactions*
Problem:
Suggest a synthesis for (1).

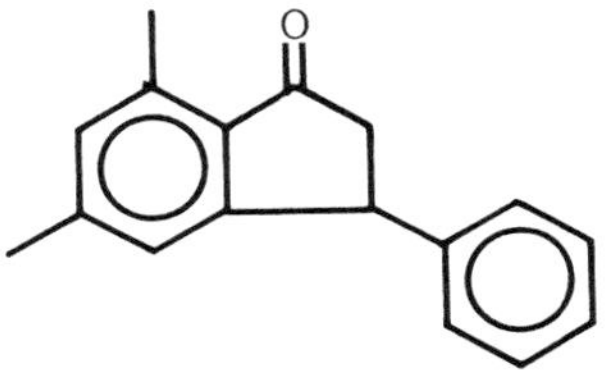

(1)

Answer:
With the dienone to cyclopentenone cyclisation in mind, we should disconnect the bond in the five-membered ring opposite the carbonyl group to reveal enone (2) simply made by a condensation.

Analysis

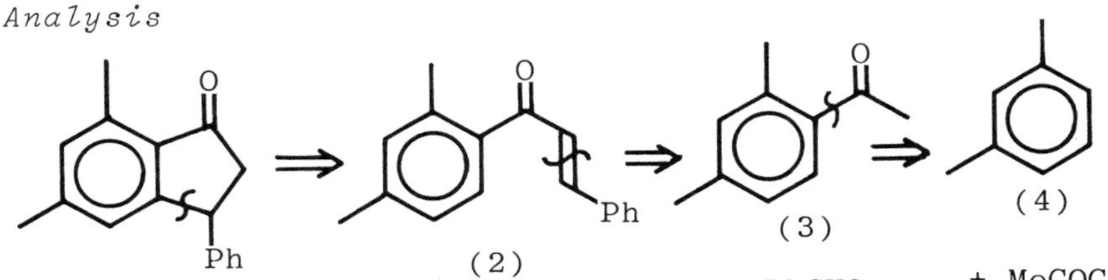

(1a) ⇒ (2) ⇒ (3) + PhCHO ⇒ (4) + MeCOCl

Synthesis[400]

(4) —MeCOCl / AlCl$_3$→ (3) 90% —PhCHO / NaOH→ (2) —PPA→ TM(1) 87%

Example :

The alternative analysis via conventional condensations and FGA on (5) reveals just how much of a short cut this cyclisation is.

Analysis

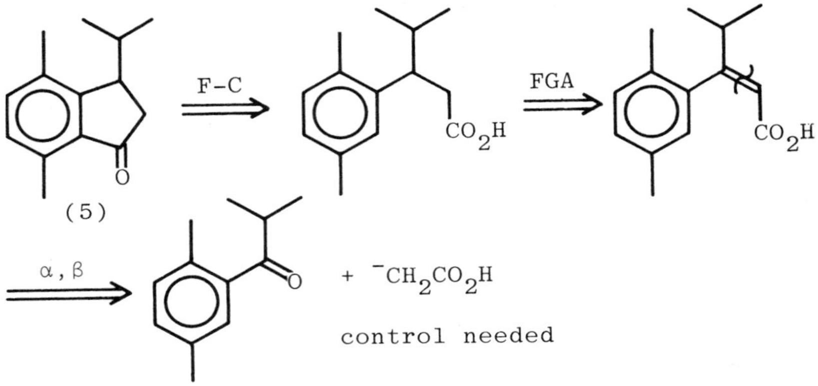

[Scheme: indanone (5) ⇒ (F-C) arylpropanoic acid with isopropyl ⇒ (FGA) aryl alkene with CO₂H ⇒ (α,β) aryl isobutyl ketone + ⁻CH₂CO₂H, control needed]

The four step conventional sequence gives about 6 per cent yield : the one step pericyclic reaction gives 66 per cent yield - an order of magnitude better.

Synthesis[401]

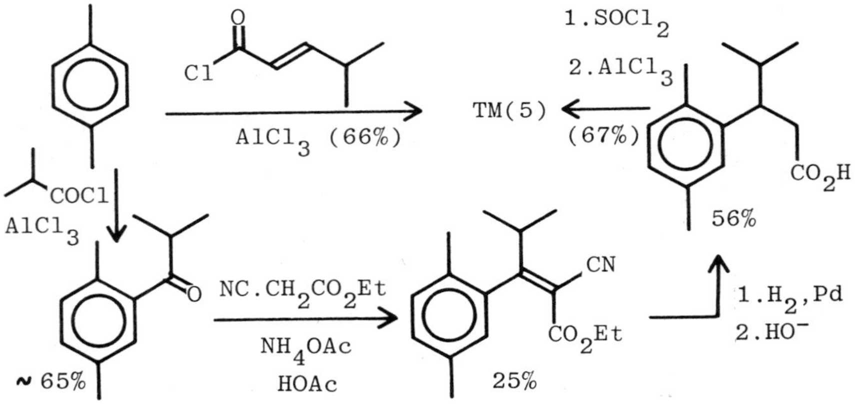

[Scheme showing synthesis: toluene + ClCO-CH=CH-CH(CH₃)₂ / AlCl₃ (66%) → TM(5) ← (67%) 1. SOCl₂ 2. AlCl₃ from arylpropanoic acid (56%); toluene + iPr-COCl / AlCl₃ (~65%) → aryl isobutyl ketone + NC·CH₂CO₂Et / NH₄OAc, HOAc → alkene with CN, CO₂Et (25%) → 1. H₂, Pd 2. HO⁻]

Example :

Two similar strategies are available for tricyclic compounds such as (6). The key intermediate (7) may be disconnected by reverse aromatic (a) or aliphatic (b) Friedel-Crafts reactions.

Analysis

(6) ⇒ (7) ⇒ (a) benzene + cyclohexenoyl chloride

(7) ⇒ (b) PhCOCl + cyclohexene

Strategy (b) has been used to make (6) and strategy (a) to make the corresponding five-membered compound (8).

Synthesis[402]

cyclohexene $\xrightarrow[\text{AlCl}_3]{\text{PhCOCl}}$ (7) $\xrightarrow{\text{PPA}}$ TM(6)
                          40%   65%

Synthesis[403]

cyclopentanone $\xrightarrow[\substack{2.\,\text{dehydrate}\\ 3.\,\text{hydrolyse}}]{1.\,\text{CN}^-}$ cyclopentene-CO$_2$H $\xrightarrow[\substack{2.\,\text{PhH,}\\ \text{AlCl}_3}]{1.\,\text{SOCl}_2}$ (8) 58%

Sigmatropic Rearrangements
Vinyl Cyclopropane to Cyclopentenone Rearrangement
Problem :
What vinyl cyclopropane(s) should rearrange to (9) on heating?

(9) with CO$_2$Et

Answer :

Reversing the rearrangement as described in Chapter T 35 gives two possible starting materials (10) and (11).

(9a) 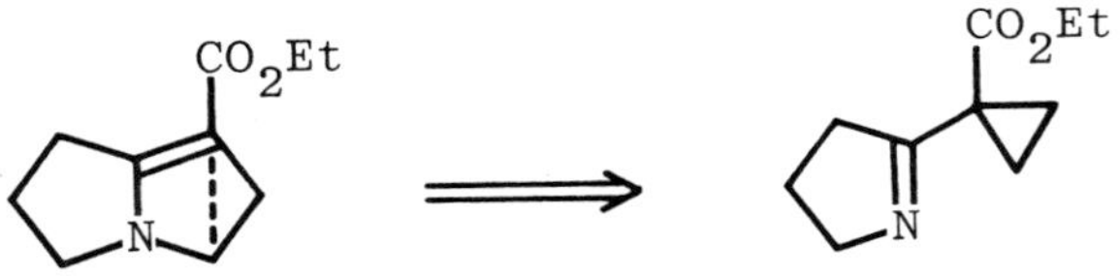 ⟹ (10)

(9b) 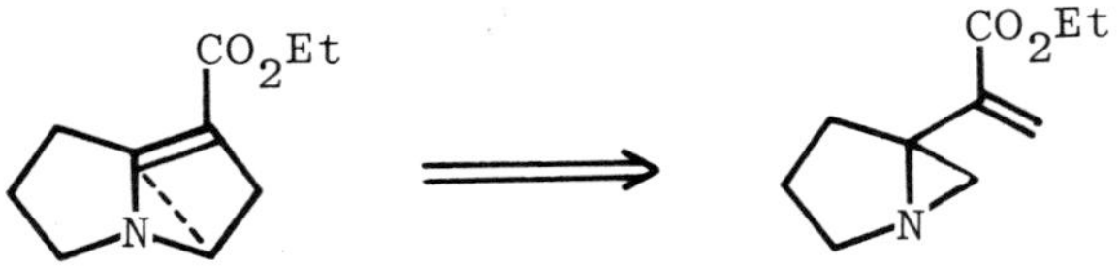 ⟹ (11)

No doubt both will give (9) on heating, though the synthesis has been carried out only with (10). The synthesis of (10) proved rather difficult as the obvious disconnection (a) failed to give a good route. The more laborious route via imine disconnection (b) had to be used.

Analysis

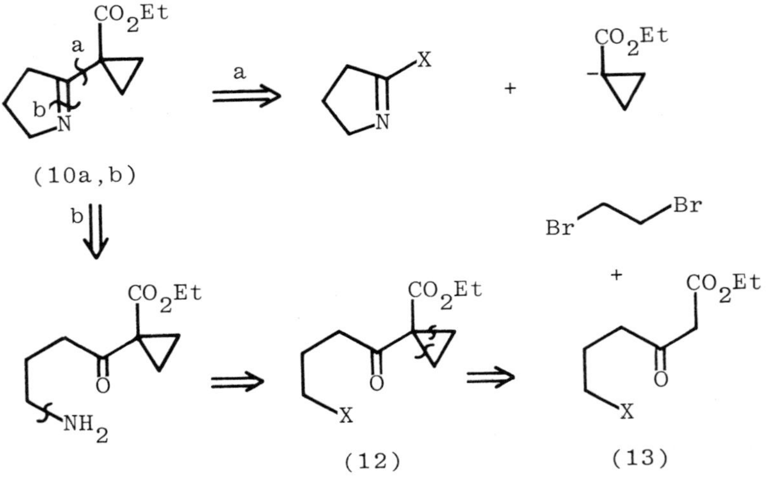

(10a,b) ⟹ +

⇓ b

⟹ (12) + BrCH₂CH₂Br ⟹ (13)

The problem of putting X in a 1,4 relationship to the ketone in (13) was solved by using unsaturated compound (14) and cleaving with ozone after three-membered ring formation and protection of the ketone.

The phthalimide method (Chapter T8) was used to make the amine and cyclisation occurred spontaneously on removal of the protecting group.

Synthesis[404]

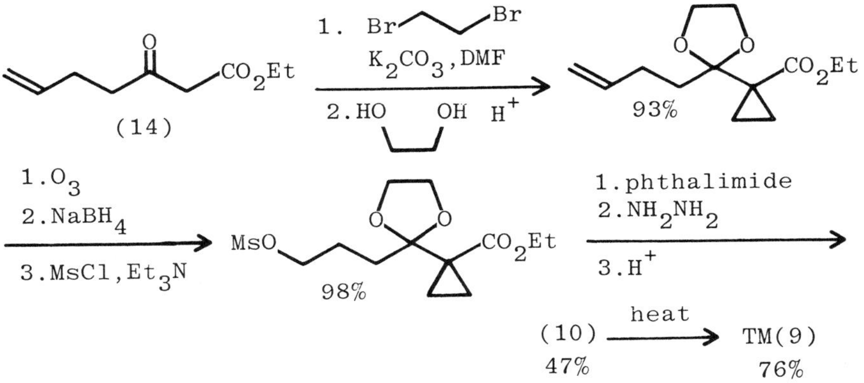

[Scheme showing (14) → with 1. Br-CH₂CH₂-Br, K_2CO_3, DMF; 2. HO-CH₂CH₂-OH, H^+ → 93% intermediate with dioxolane and cyclopropane bearing CO_2Et]

1. O_3
2. $NaBH_4$
3. $MsCl, Et_3N$

→ MsO-intermediate, 98%

1. phthalimide
2. NH_2NH_2
3. H^+

(10) 47% —heat→ TM(9) 76%

Example and Problem:

The anions of vinyl cyclopropanols (16), conveniently released from ethers (15) with BuLi, rearrange rapidly to cyclopentenes (17).

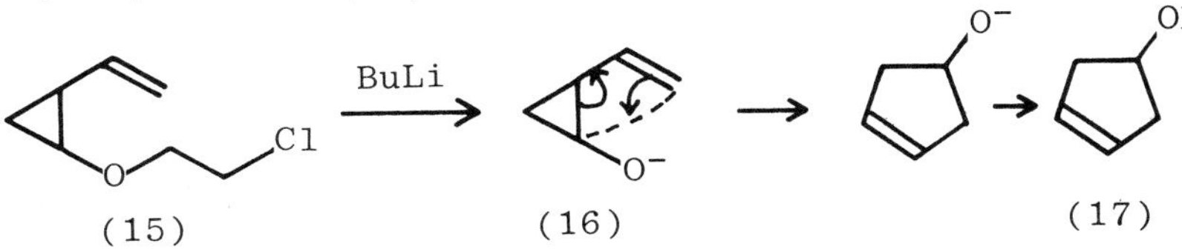

(15) —BuLi→ (16) → → (17)

What will be the product from forming adducts such as (15) from (18) and carbene (19) and rearranging them in the same way?

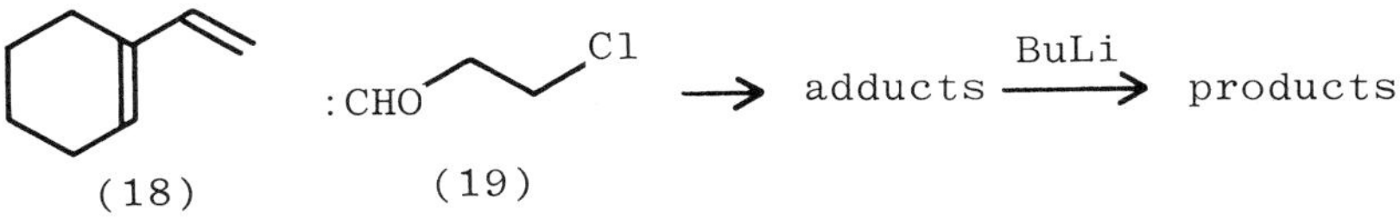

(18)   (19) :CHO-CH₂CH₂-Cl → adducts —BuLi→ products

Answer :

Two adducts can be formed, (20) and (21), from the addition of (19) to each of the double bonds in (18). Each rearranges to the same cyclopentenol (22).

Synthesis[405]

(18) $\xrightarrow{(19)}$ (20) + (21) $\xrightarrow{\text{BuLi}}$ (22)

In practice it is even more complicated as diastereoisomers of (20) and (21) mean that four adducts are formed. All rearrange to the same diastereoisomer of (22) with the OH group on the outside of the molecule.

[3,3] *Sigmatropic Rearrangements*

Example :

Carbofuran (23), a systemic insecticide, can be made from phenol (24) by reaction with MeNCO. Further disconnection of the ether reveals triol (25) as a key intermediate.

Analysis 1

(23) $\xRightarrow{\text{C-O ester}}$ (24) $\Rightarrow$ (25)

The problem of assembling three adjacent substituents on the benzene ring of (25) suggests using the Claisen rearrangement and hence making (25) from (26).

Analysis 2

(25) $\xrightarrow{\text{FGI}}_{\text{hydration}}$ [structure with OH, OH, and isobutenyl group] $\xrightarrow{[3,3]}$ [phenol with O-allyl ether] (27) $\Rightarrow$ [methallyl chloride] Cl + [catechol] (28)

This is the strategy of the manufacture of carbofuran, though the mono alkylation of (28) is avoided by using available (29) and converting the NO_2 to OH by reduction and substitution. The acidic nitrophenol (30) cyclises on heating.

Synthesis[406]

[o-nitrophenol] (29) $\xrightarrow[K_2CO_3]{\text{Cl-methallyl}}$ [nitro aryl methallyl ether] $\xrightarrow{\text{heat}}$ [nitrophenol with allyl] (30)

$\xrightarrow{275°C}$ [nitro-dihydrobenzofuran] $\xrightarrow[\text{cat}]{H_2}$ [amino-dihydrobenzofuran] (24) $\xrightarrow{\text{MeNCO}}$ TM(23)

$\xrightarrow{\substack{1.\text{HONO} \\ 2.\text{H}_2\text{O}}}$

Problem :

What is the starting material required for a Claisen-Cope route to (31)?

[structure: (CH₃)₂CH-CH=CH-C(CH₃)₂-CH₂-CO₂Et] (31)

Answer :

We might have recognised this as a Claisen-Cope product since it is a γ,δ-unsaturated ester. Disconnecting and inverting the allylic group from the ester enolate gives the starting material (32). Ortho esters are usually used as the reagents.

Analysis

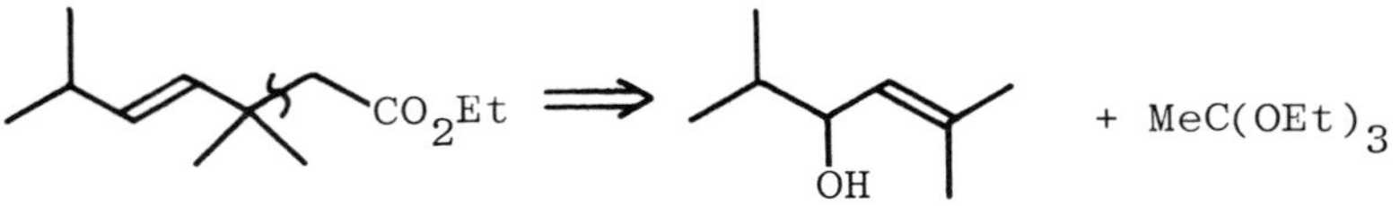

(31a) ⟹ (32) + MeC(OEt)$_3$

Synthesis[407]

(32) $\xrightarrow{\text{MeC(OEt)}_3, \text{H}^+}$ [intermediate] ⟶ TM(31)

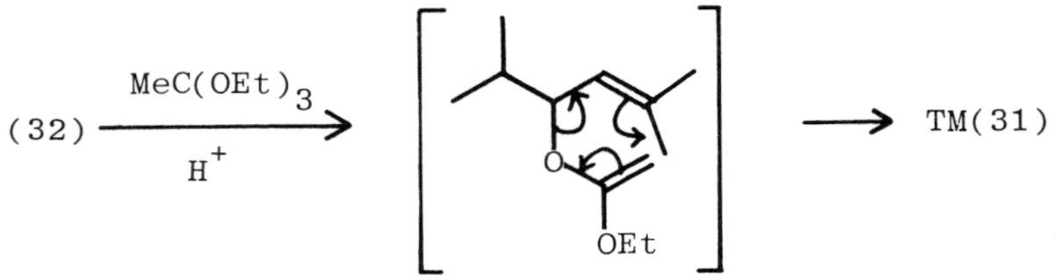

Example :

Bicyclic ketone (33) was needed for a chrysanthemic acid synthesis.[408] Carbene disconnection next to the ketone group (Chapter T30) reveals γ,δ-unsaturated acid (35) as an intermediate, available by a Claisen-Cope rearrangement.

Analysis

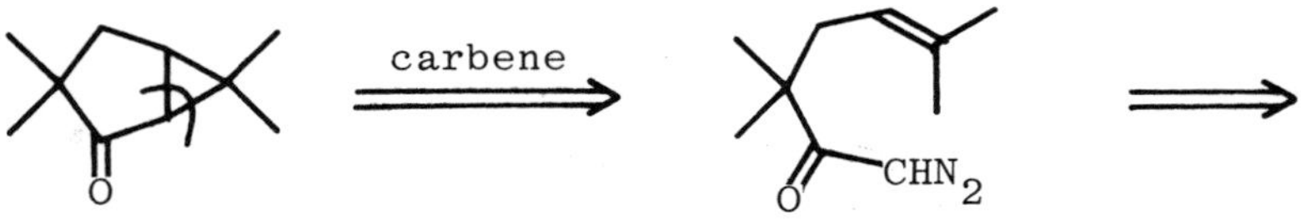

(33) $\xrightarrow{\text{carbene}}$ (34) ⟹

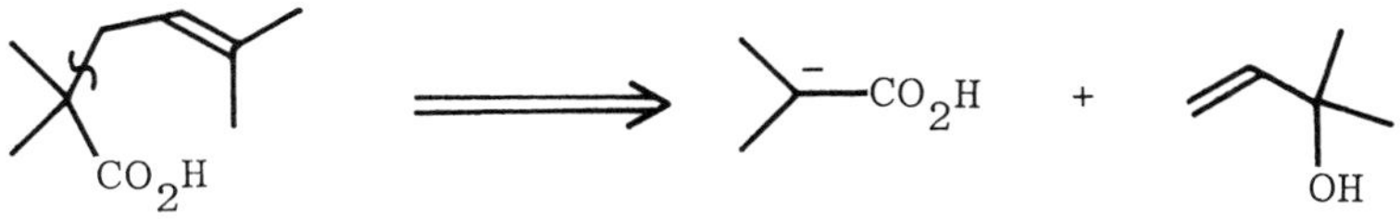

This time conventional ester (38) was treated with base to give enolate (37). No doubt an ortho ester would do the same job.

Synthesis

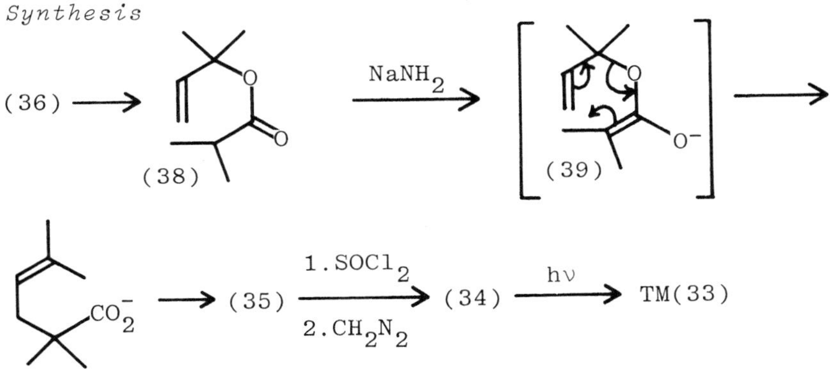

(36) ⟶ (38) —NaNH$_2$→ [(39)] ⟶

⟶ (35) —1. SOCl$_2$ / 2. CH$_2$N$_2$→ (34) —hν→ TM(33)

Problem:
Devise a route to (40) using the Carroll rearrangement.

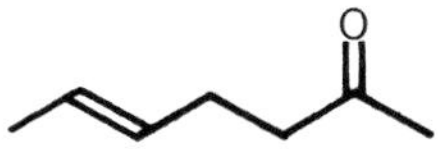

(40)

Answer:
The disconnection is the same (40a), we must again remember to invert the allyl group, and the only difference is the use of the acetoacetic ester (41) in the synthesis.

Analysis

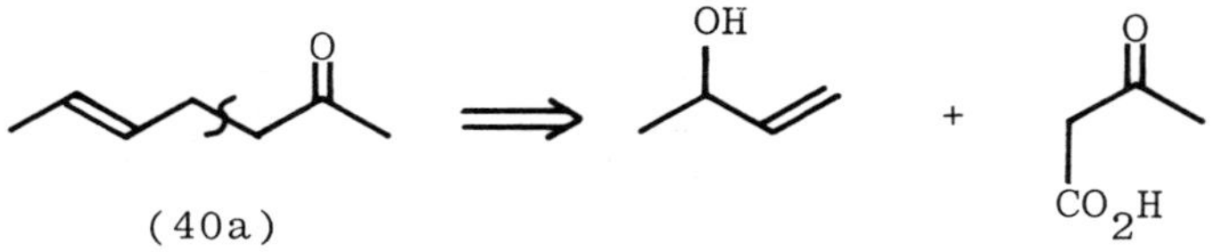

(40a)

Synthesis[409]

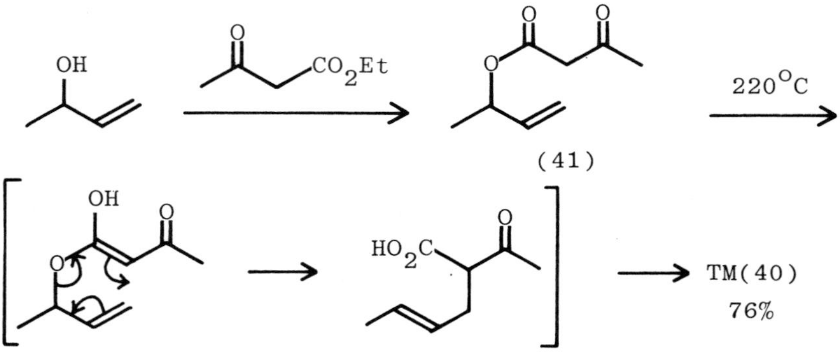

(41)

220°C

⟶ TM(40)
76%

CHAPTER 36

Six-Membered Rings

(a) *Carbonyl Condensations*

Problem :
Suggest a synthesis for triene (1) needed for photo-chemical experiments.

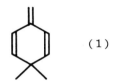

Answer :
With the Robinson annelation in mind, we can get back to a cyclohexadienone (2) by a Wittig disconnection. One of the double bonds could be put in by quinone oxidation leaving cyclohexenone (3), which we have already made by a Robinson annelation and by another route (p 380).

Analysis

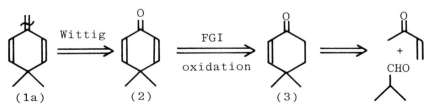

The oxidation of (3) to (2) can be carried out by SeO_2 or DDQ (Chapter T 36).

Synthesis[372,410]

[CHO (isobutyraldehyde) + CH$_2$=CH-CO-CH$_3$, HO$^-$] → (3) → [DDQ or SeO$_2$] → (2) → [Ph$_3$P=CH$_2$] → TM(1)

(b) *Diels-Alder Reaction*

Bicyclic compounds of structure (4) are needed for the synthesis of the alkaloid dendrobine. What is the starting material for a Diels-Alder synthesis of (4)? Stereochemistry is obviously vital here.

(4)

Answer :
The cyclohexene is easy to see so that the Diels-Alder disconnection follows. The stereochemistry of the double bonds comes from two separate arguments; the dienophile (a in 5) must be *trans* as the two substituents it produces in (4) are also *trans*. The diene must be all *cis* or all *trans* since the two substituents it produces in (4) are *cis* (both down). The all *trans* is needed because endo approach (6) is preferred. (See Chapter T 17).

Analysis 1

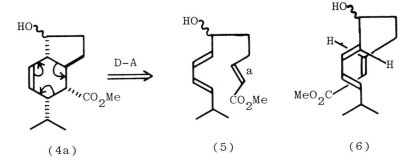

The synthesis of (5) was approached conventionally by Wittig and Grignard disconnections to simple starting materials. The 1,4-disconnection (7a) is unusual.

Analysis 2

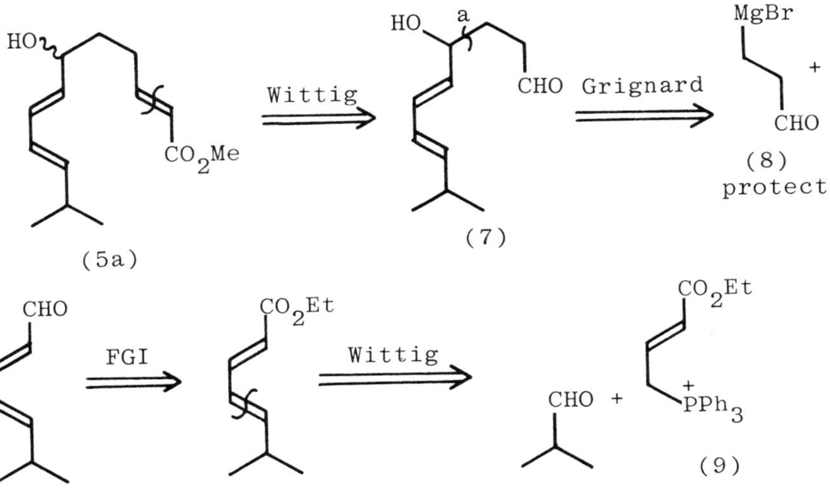

Both Wittigs use stabilised ylids and so will give *trans* double bonds as required. Protection of

(8) was as an acetal. Ylid (9) was made on p 270.

Synthesis[411]

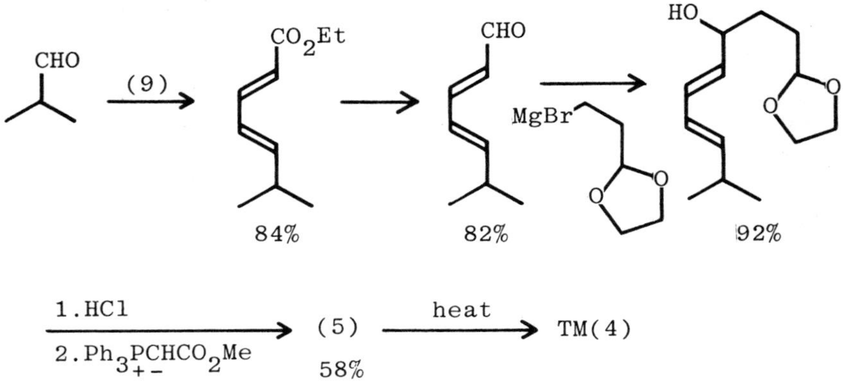

CHO
(9)
CO_2Et
84%
CHO
82%
MgBr
O
O
HO
O
O
92%
1.HCl
2.$Ph_3\overset{+}{P}\overset{-}{C}HCO_2Me$
(5)
58%
heat
TM(4)

Roush[411] gives an analysis of how he came to choose (4) as an intermediate and how it was developed into dendrobine. Another synthesis of dendrobine uses Diels-Alder reaction (10) → (11). What is odd about the stereochemistry of (11)?

Synthesis[412]

(10)

O
O
CO_2Me
180°C
O
O
H
H
H
H
CO_2Me

(11)

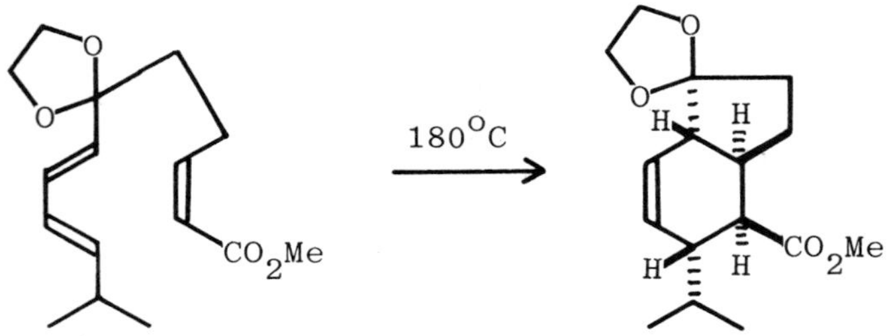

Answer :

The stereospecificity is of course correct : all *trans* diene producing *cis* substituents and *cis* dienophile producing *cis* substituents. It is the stereoselectivity which is odd: (11) is the exo product evidently preferred because the strain in the endo approach (12) is too much. The exo approach is much easier (13).

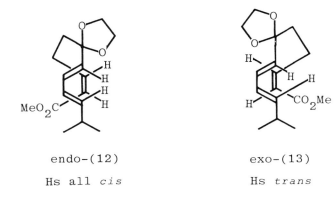

endo-(12) exo-(13)
Hs all *cis* Hs *trans*

(c) *Reduction of Aromatic Compounds*

Problem :

Suggest a synthesis of the nasal decongestant Propylhexedrine (14).

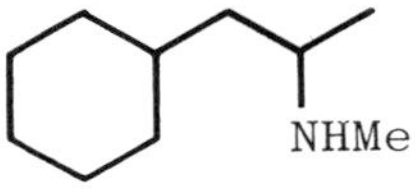

Answer :

Reduction of the aromatic amine (15) is the usual source of (14), and reductive amination of (16) gives (15). There are many published routes to (15) of which addition of an activating group (17) is probably easiest on a large scale. You may also have considered using nitro compound (18) or epoxide (19).

Analysis

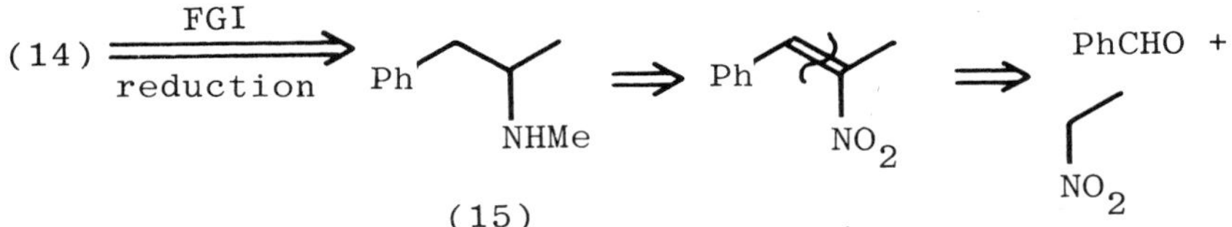

$$\underset{(15)}{\underset{\text{NHMe}}{\text{Ph}\diagdown}} \Longrightarrow \underset{(16)}{\overset{b}{\text{Ph}\diagdown\overset{\text{O}}{\underset{}{\diagup}}}} \Longrightarrow \underset{1,3\text{-diCO}}{\overset{\overset{\text{CO}_2\text{Et}}{a}}{\text{Ph}\diagdown\overset{\text{O}}{\underset{}{\diagup}}}} \quad (17)$$

$$\Downarrow 1.\text{FGI} \quad 2.\text{b} \qquad \Downarrow$$

$$\text{PhLi} + \underset{(19)}{\triangle\text{O}} \qquad \underset{}{\text{Ph}\diagdown\overset{\text{CO}_2\text{Et}}{}} + \text{MeCO}_2\text{Et}$$

One synthesis uses cyanide as the activating group and carries out both reductions catalytically.

Synthesis[413]

$$\underset{}{\text{Ph}\diagdown\text{CN}} \xrightarrow[\text{EtO}^-]{\text{MeCO}_2\text{Et}} \underset{64\%}{\overset{\overset{\text{CN}}{|}}{\text{Ph}\diagdown\overset{\text{O}}{\underset{}{\diagup}}}} \xrightarrow{\text{H}_2\text{SO}_4} (16)$$

$$\xrightarrow[\text{H}_2,\text{PtO}_2]{\text{MeNH}_2} (15) \xrightarrow[\text{RuO}_2,\text{Al}_2\text{O}_3]{\text{H}_2} \underset{78\%}{\text{TM}(14)}$$

Example :

Electron-rich rings are reduced more easily so that β-naphthol (20) can be reduced to ketone (21) in base.[414]

Problem :

Suggest a synthesis for cyclic ether (22).

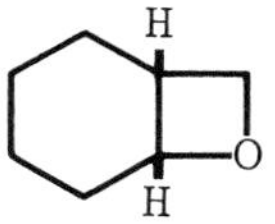

(22)

Answer :
Diol (23) is clearly needed and can be made by the reduction of a suitable aromatic precursor. Since salicyclic acid (24) is very cheap - it is used in the manufacture of aspirin - a salicylate ester is ideal.

Analysis

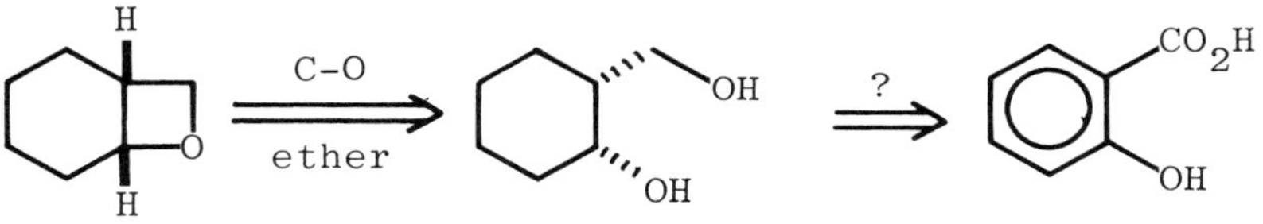

(22a) ⟹ (C-O ether) (23) ⟹ ? (24)

Hydrogenation of methyl salicylate gave mostly *cis* (25) implying that the last double bond to be reduced is often between the two functional groups. Selective esterification of diol (23) with p-bromobenzenesulphonylchloride (BsCl) gave (26) which cyclised in strong base.

Synthesis[4][1][5]

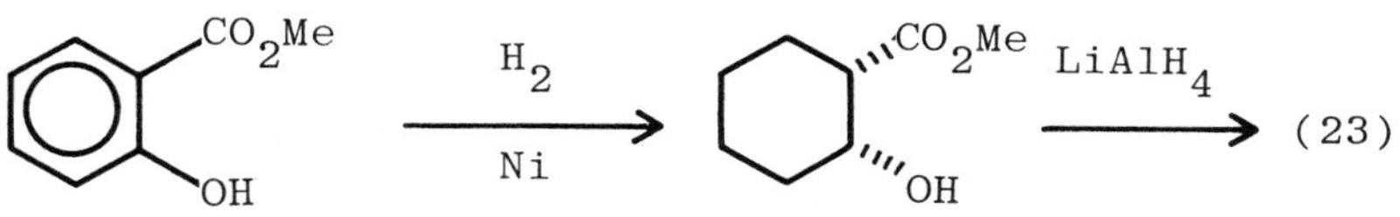

methyl salicylate —H₂/Ni→ (25) 63% (+22% trans) —LiAlH₄→ (23)

—BsCl/pyr→ (26) 72% —KOBu-t/HOBu-t→ TM(22) 67%

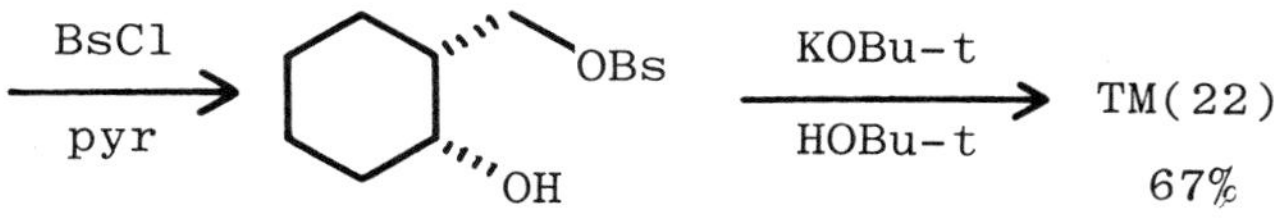

Birch Reduction

Problem :

Ketone (27) was needed for a synthesis of the boll weevil pheromone (28). Suggest a synthesis of (27) using Birch reduction.

(27) 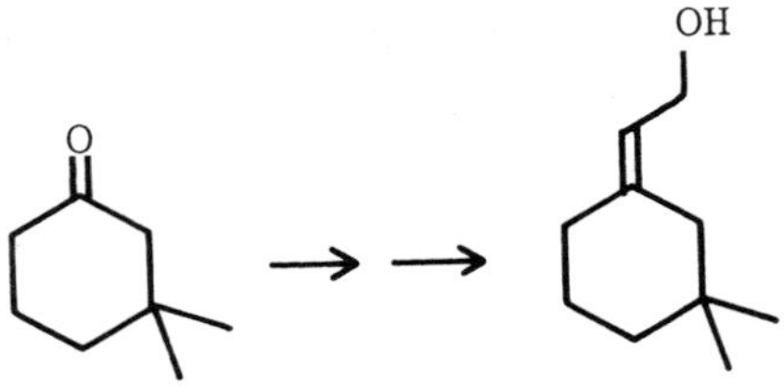 (28)
O
OH

Answer :

The quaternary centre must be removed before an aromatic ring can be seen so a reverse Michael disconnection gives enone (29), available from (30) by acid hydrolysis and hence from *m*-cresol by Birch reduction.

Analysis

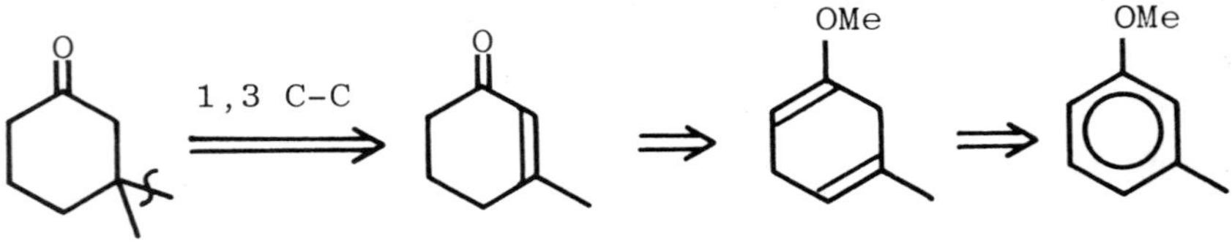
O
1,3 C-C
O
OMe
OMe

(27a) (29) (30)

Note that (30) is the only possible Birch reduction product in which both electron-donating groups are on double bonds. Vigorous acid hydrolysis moves the double bond into conjugation.

Synthesis[416]

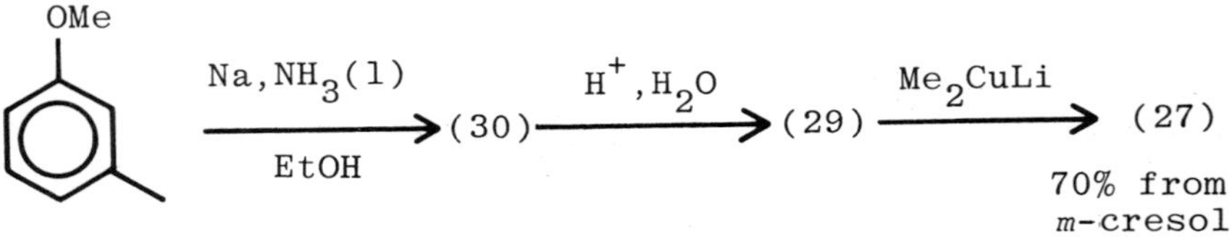
OMe
Na, NH3(l)
EtOH
(30)
H+, H2O
(29)
Me2CuLi
(27)

70% from *m*-cresol

Example :
Ketone (31) was used in a synthesis of juvabione (32) an insect hormone mimic produced by conifers.

(31) 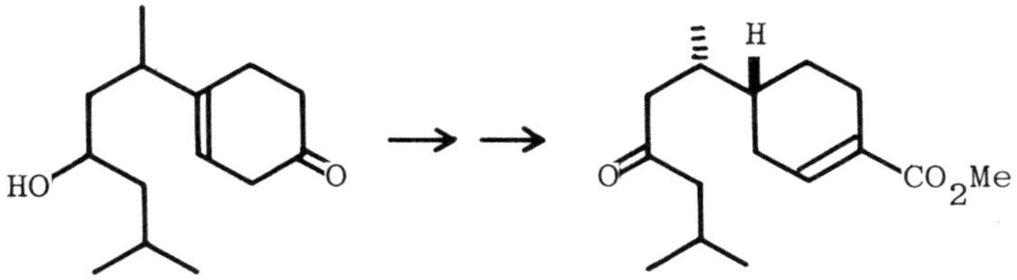 (32)

Non-conjugated enone (31) is clearly a Birch reduction product from ether (33). Grignard disconnection leaves aldehyde (34), and FGA reveals a condensation product from (35).

Analysis

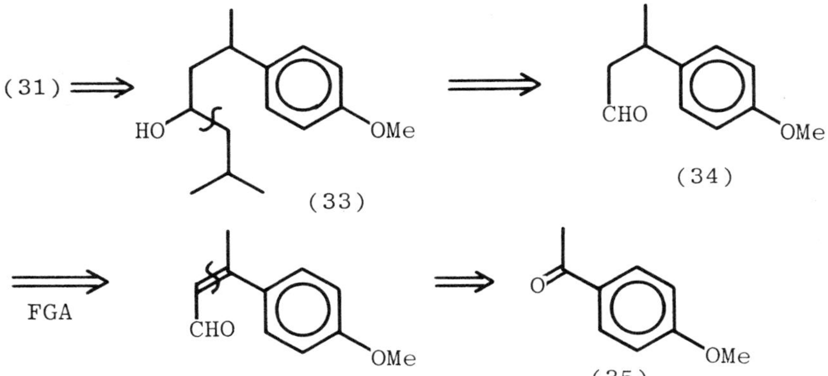

(33) (34) (35)

The initial condensation is best carried out at the ester oxidation level to take advantage of the Reformatsky method of control. Oxalic acid hydrolyses (36) to (31) without moving the double bond into conjugation.

Synthesis[417]

(35) 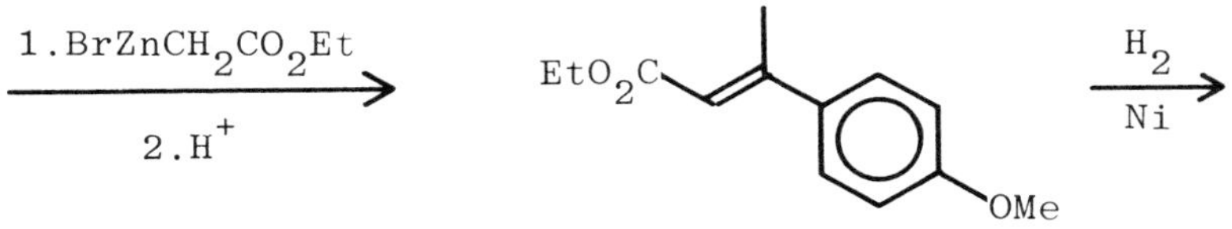

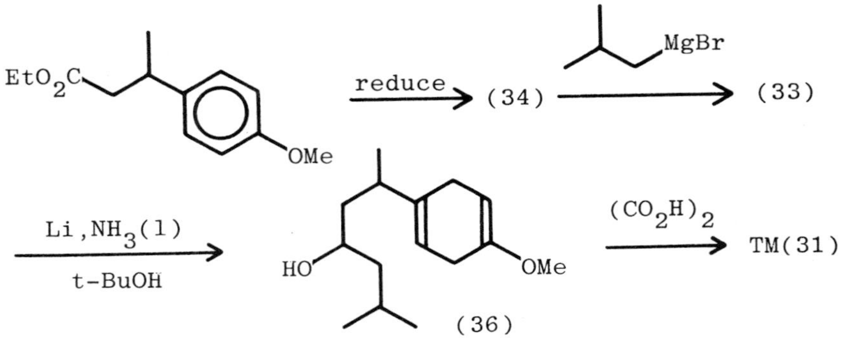

General Problem:

Select one compound from (37) - (40) for synthesis by each of the methods of this chapter : Robinson annelation, Diels-Alder reaction, total reduction of aromatics, and Birch reduction.

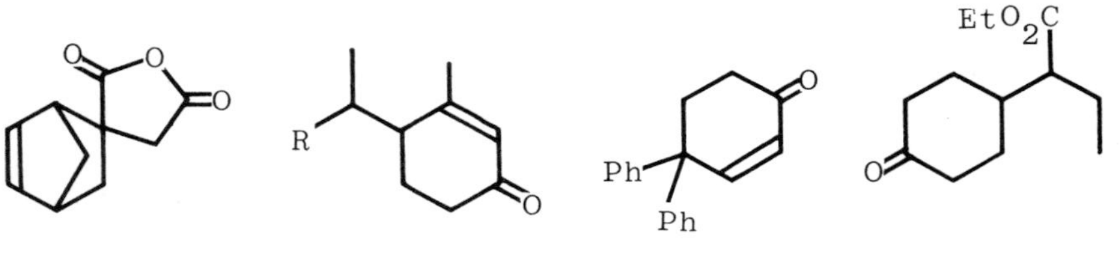

(37)   (38)   (39)   (40)

Answer:

The first compound (37) is clearly a Diels-Alder adduct and was made this way in a synthesis of ipsenol.[418]

Analysis

(37a)

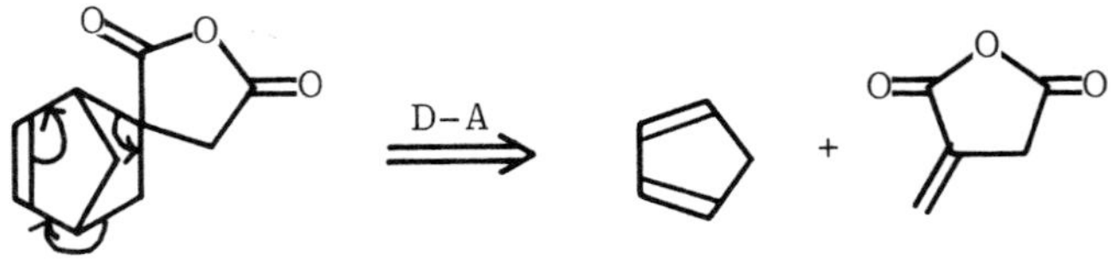

The rest are more difficult to classify, but (39) looks like the Robinson annelation product as the two phenyl groups prevent synthesis from an aromatic compound.

Analysis

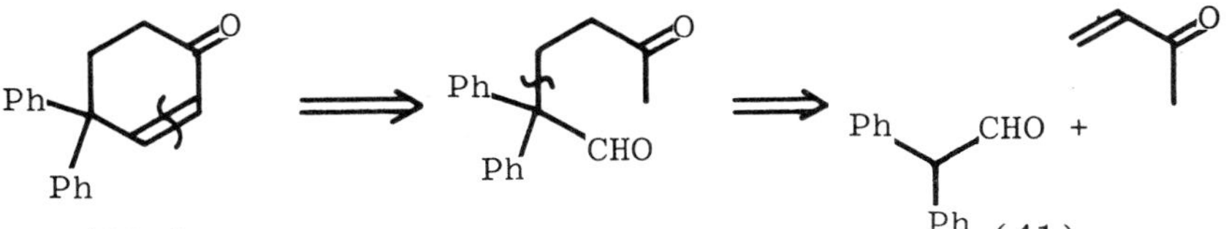

[Scheme: (39a) ⇒ Ph,Ph-substituted ketoaldehyde ⇒ Ph-CH(Ph)-CHO (41) + methyl vinyl ketone]

Aldehyde (41) can be made by rearrangement[419] of epoxide (42) and the Robinson annelation gives (39) in one step.

Synthesis[419,420]

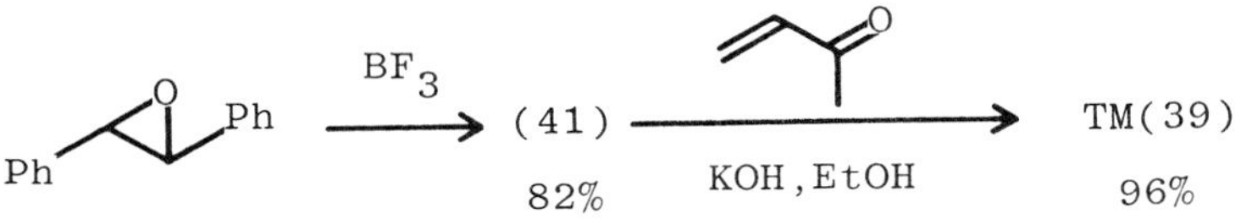

[Scheme: epoxide (42) —BF$_3$→ (41) 82% —methyl vinyl ketone, KOH, EtOH→ TM(39) 96%]

The compound without a double bond (40) must then be made by total reduction of (43) - oxidation is necessary here. Note the basic conditions, cf TM (21).

Synthesis[421]

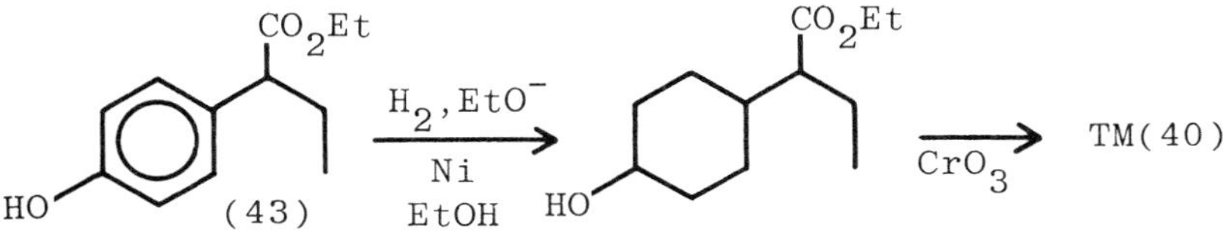

[Scheme: (43) —H$_2$, EtO$^-$, Ni, EtOH→ cyclohexanol intermediate —CrO$_3$→ TM(40)]

So, finally, (38) must be the Birch reduction product, derived from (44) via (45). This time base was used to move the double bond into conjugation.

Synthesis[422]

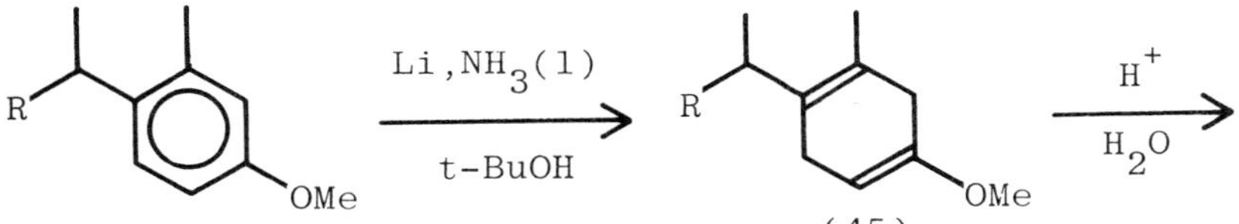

[Scheme: (44) —Li, NH$_3$(l), t-BuOH→ (45) —H$^+$, H$_2$O→]

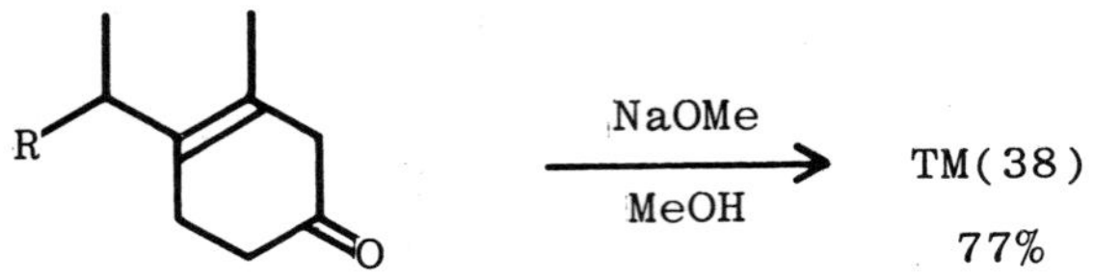

[structure: cyclohexenone with methyl at 3-position and CH(CH3)R at 4-position] →(NaOMe / MeOH) TM(38) 77%

85% from (44)

You may have come to other choices and these may well provide good syntheses too.

CHAPTER 37

General Strategy C: Strategy of Ring Synthesis

This Chapter does not follow the organisation of Chapter T 37 : all the same points are covered but several apply to each problem. This is a revision Chapter for ring synthesis.

Problem :

Suggest a synthesis for 5-MeO tetralone (1) used for a tetracycline synthesis (see p 431).

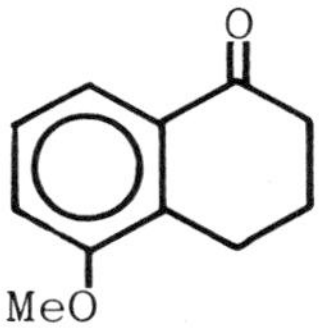

Answer :

The difficult disconnection is (a) so we make it the first disconnection as it will then be a cyclisation in the synthesis. Further disconnections can follow the pattern of the synthesis of 7-MeO tetralone in Chapter T 24.

Analysis

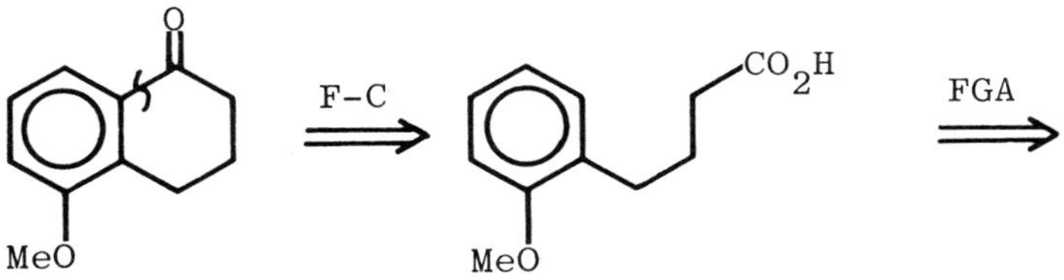

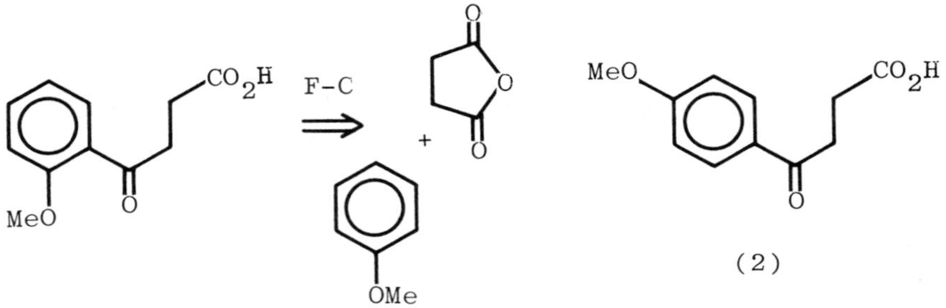

The problem with this synthesis is that essentially only *para* product (2) is formed in the initial Friedel-Crafts reaction. A solution is to block the *para* position with a chlorine atom which can be removed by hydrogenolysis. This tetralone (1) is now a commercial product.

Synthesis[4,2,3]

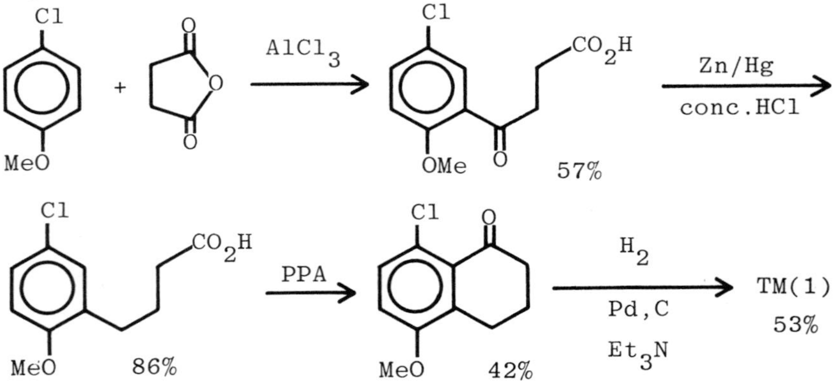

Example:
Compound (3) is, appropriately enough, a useful intermediate for the synthesis of tetracycline antibiotics. Disconnection (a) gives considerable simplification.

Analysis 1

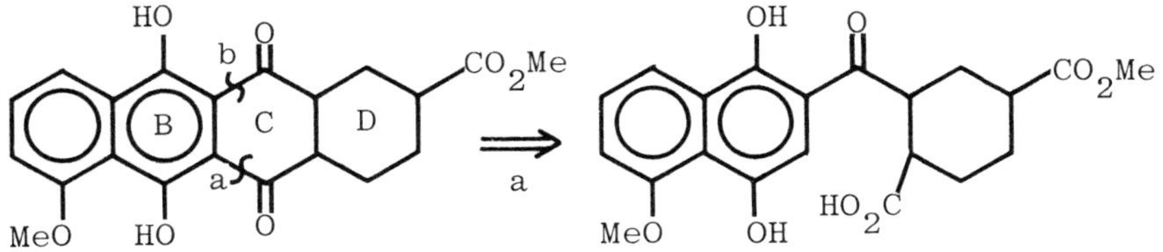

(3) ⇒ (4)

We could have disconnected (b) in the same way, but we want eventually to get back to (1) so we aim to make ring B by oxidative aromatisation of (5); and bond (b) can then be disconnected because of the remaining 1,3-dicarbonyl relationship.

Analysis 2

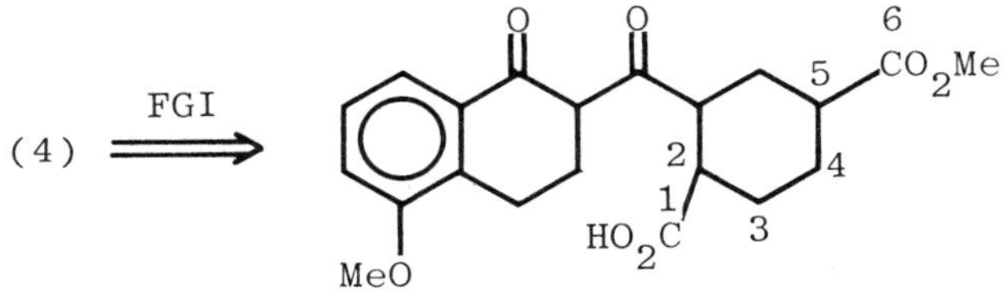

(4) ⇒ (5)   FGI

First we can reconnect the 1,6-dicarbonyl relationship in (5) to reveal a Diels-Alder adduct (6). Either disconnection (1,3-diCO or Diels-Alder) may be made first; the 1,3-dicarbonyl is nearer the centre of the molecule so is preferred.

Analysis 3

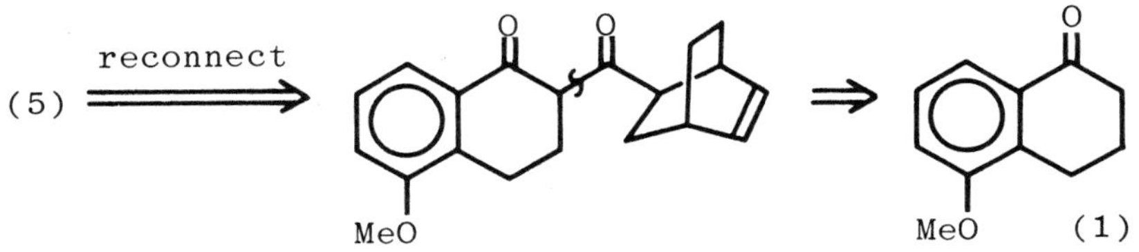

(5) ⇒ reconnect ⇒ (6) ⇒ (1) + (7)

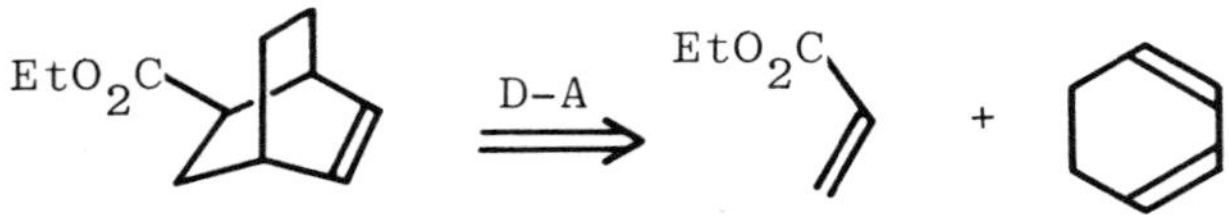

(7) [structure] ⇒[D-A] EtO₂C-CH=CH₂ + [cyclohexadiene]

In the synthesis, aromatisation of ring B and oxidation to quinone were two separate steps.

Synthesis[424]

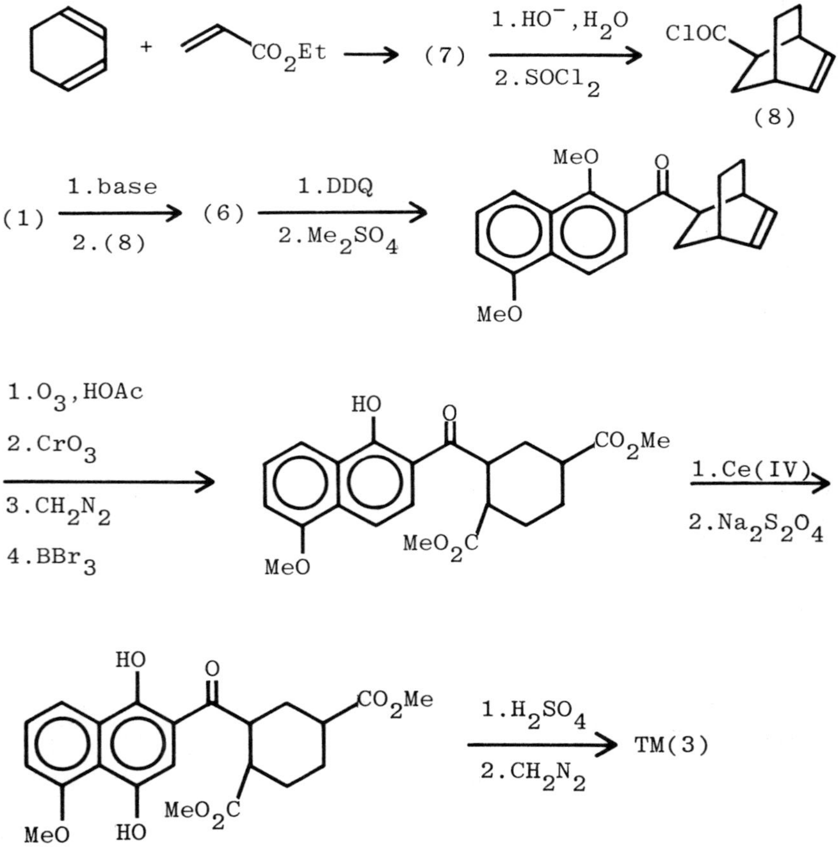

[cyclohexadiene] + CH₂=CH-CO₂Et → (7) →[1. HO⁻, H₂O; 2. SOCl₂] ClOC-[bicyclic] (8)

(1) →[1. base; 2. (8)] (6) →[1. DDQ; 2. Me₂SO₄] [naphthalene with MeO groups and ketone to bicyclic]

→[1. O₃, HOAc; 2. CrO₃; 3. CH₂N₂; 4. BBr₃] [naphthalene-OH, MeO, ketone, cyclohexane with CO₂Me and MeO₂C] →[1. Ce(IV); 2. Na₂S₂O₄]

[naphthalene with HO, MeO, HO, ketone, cyclohexane with CO₂Me, MeO₂C] →[1. H₂SO₄; 2. CH₂N₂] TM(3)

Problem :

Compound (9) was a target molecule in Chapter T 37. Stork needed it for the following sequence of reactions.

Synthesis[425]

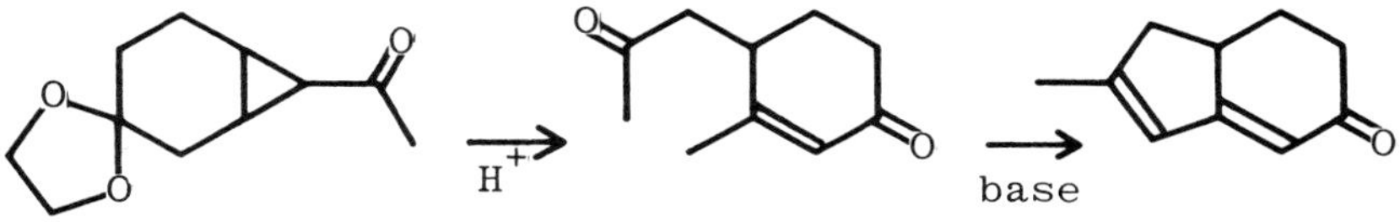

<chemical structures: (9) →H+ (10) →base (11)>

What would be a more conventional approach to (11) and why is it unsatisfactory?

Answer :

The Robinson annelation looks good.

Analysis

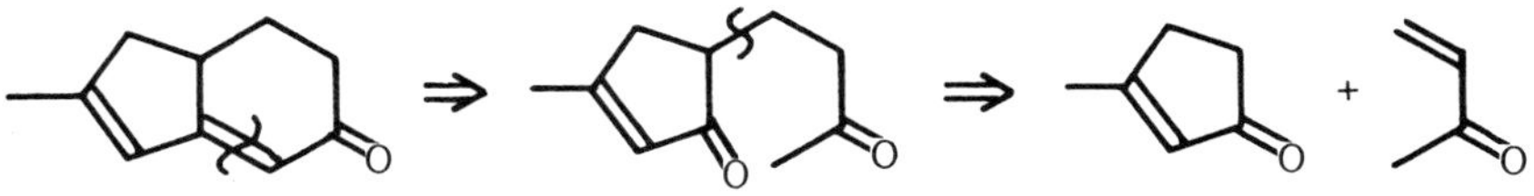

<chemical structures: (11a) ⇒ ⇒ (12) +>

The problem arises in controlling the enolisation of (12) as all saturated CHs in the molecule could conceivably be removed by base. The obvious disconnection (10a) leads to similar problems.

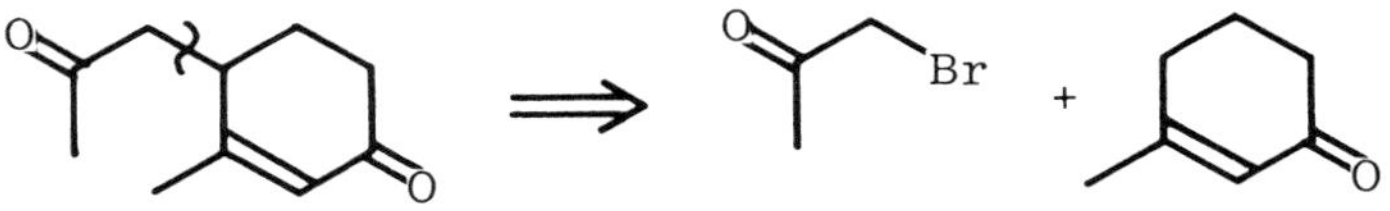

<chemical structures: (10a) ⇒ Br + (13)>

Ketone (13) can again enolise in a variety of ways and these are difficult to control. How is the cyclisation of (10) to (11) controlled?

Answer :

Enolisation probably occurs at all the marked atoms in (10b) but only enol (14) can cyclise to give a stable ring.

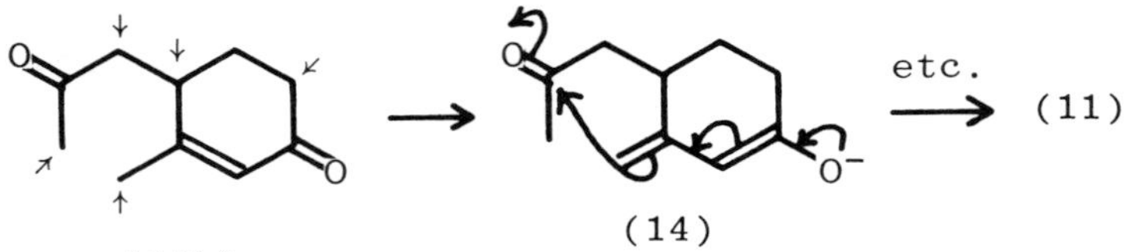

(10b) → (14) —etc.→ (11)

Hagemann's ester (15) is ketone (13) with an activating group. It forms one enolate only (16, the anion being stabilised by both carbonyl groups) and it reacts at only one position with alkylating agents.

Synthesis

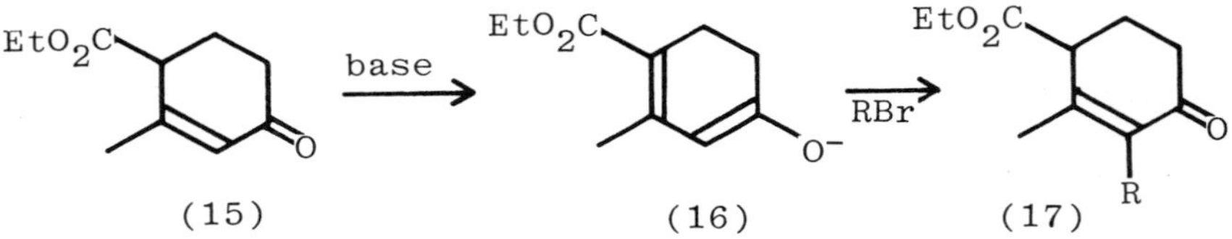

(15) —base→ (16) —RBr→ (17)

Problem :

Suggest a synthesis of (18).

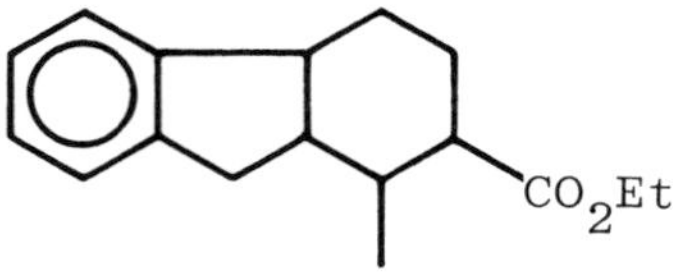

(18)

Answer :

Disconnection of the central ring is strategically best and disconnection (a) gives a starting material (19) resembling (17).

Analysis

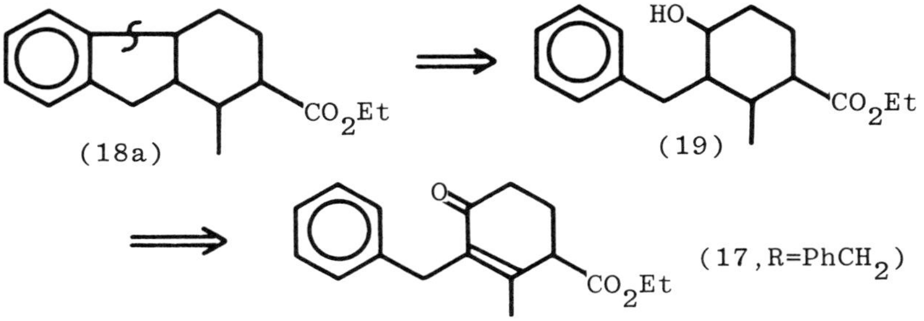

(18a) ⟹ (19) ⟹ (17, R=PhCH$_2$)

It is not necessary to reduce the ketone in (17): PPA catalyses cyclisation of (20), and the double bond in (21) can be removed by hydrogenation.

Synthesis[426]

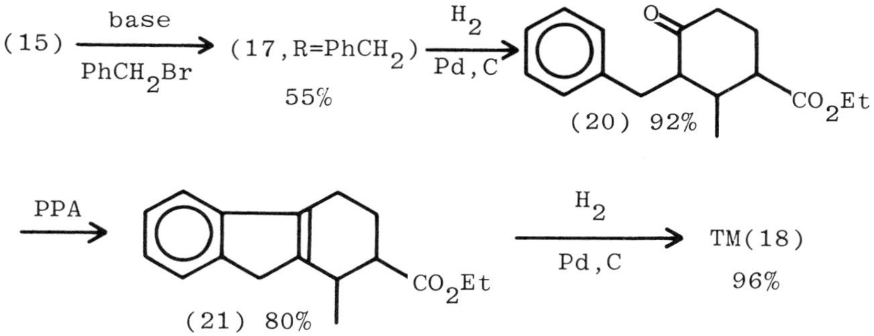

(15) —base, PhCH$_2$Br→ (17, R=PhCH$_2$) 55% —H$_2$, Pd,C→ (20) 92%

—PPA→ (21) 80% —H$_2$, Pd,C→ TM(18) 96%

Example:
Bicyclic keto ester (22) was needed for conformational studies. The common atoms are marked (●) and the obvious disconnections of this symmetrical molecule require double alkylation of cyclohexanone with a reagent such as (23). Double 1,5-diCO disconnection of (22) is impossible as you will discover if you attempt it.

Analysis

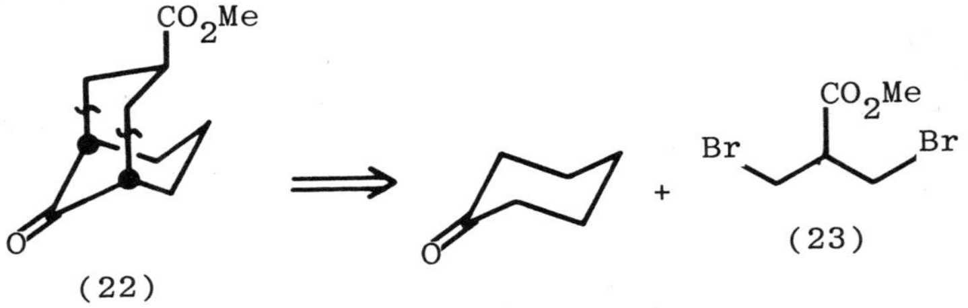

(22) ⟹ [cyclohexanone] + Br-CH₂-CH(CO₂Me)-CH₂-Br (23)

Dibromide (23) comes from diol (24) a double formaldehyde adduct of methyl acetate. This route is practical if we use malonate as there is room for only two molecules of formaldehyde.

Analysis 2

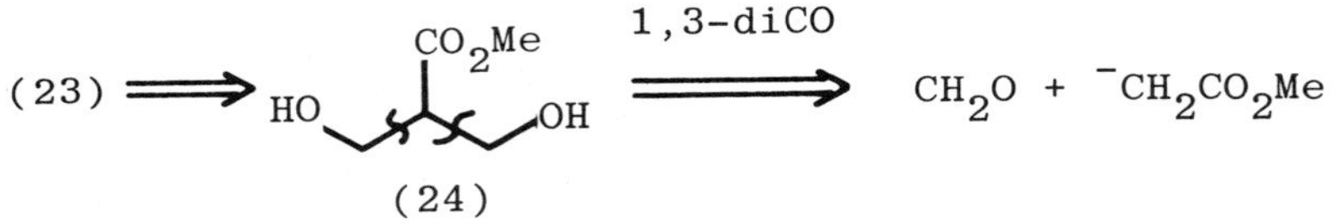

(23) ⟹ HO-CH₂-C(CO₂Me)(H)-CH₂-OH (24) $\xrightarrow{1,3-diCO}$ CH_2O + $^-CH_2CO_2Me$

Synthesis[427]

$CH_2(CO_2Et)_2$ $\xrightarrow[\text{pH 8.5}]{CH_2O}$ HO-CH₂-C(CO₂Et)₂-CH₂-OH $\xrightarrow{HBr}$ Br-CH₂-CH(CO₂H)-CH₂-Br 66%

$\xrightarrow[H^+]{MeOH}$ (23)

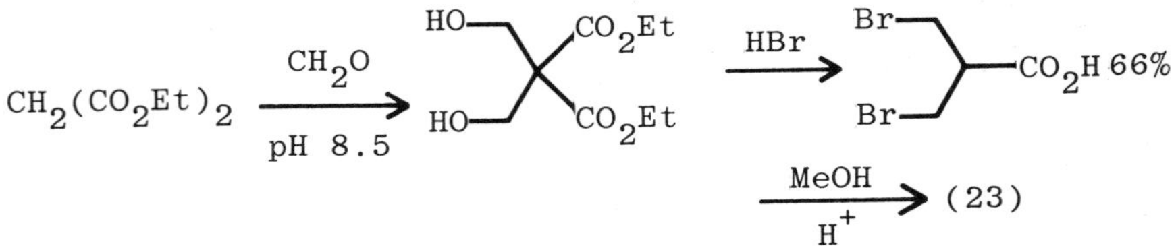

The double alkylation of cyclohexanone with (23) is easily carried out in one step *via* an enamine. The first alkylation is unambiguous and the second is a cyclisation so occurs very easily.

Synthesis[428]

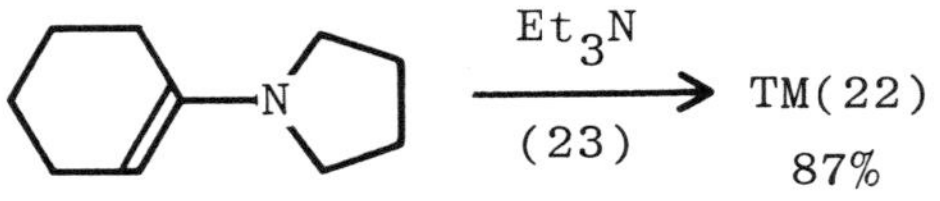

[cyclohexene-pyrrolidine enamine] $\xrightarrow[(23)]{Et_3N}$ TM(22) 87%

Problem :
Identify the common atoms in fenchone (25) and suggest possible disconnections.

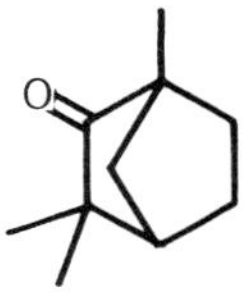

(25)

Answer :
The common atoms are marked ● in (25a) and the best disconnections correspond to the intramolecular alkylation of ketone (26) or (27).

Analysis

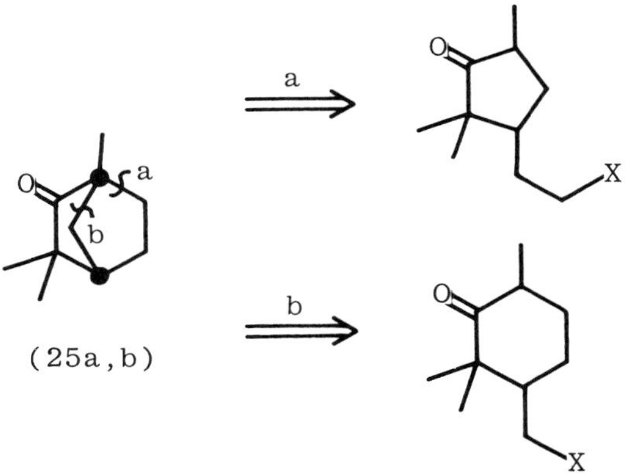

(25a,b) ⟹ᵃ (26)

(25a,b) ⟹ᵇ (27)

The cyclisation of (26, X=Cl) is one route to fenchone.[429]

Problem :
Mark the common atoms in tricyclic ketone (28) and suggest disconnections.

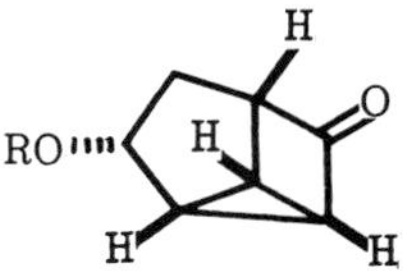

(28)

Answer :

There are four common atoms here (28a) and two sensible disconnections of common bonds.

Analysis

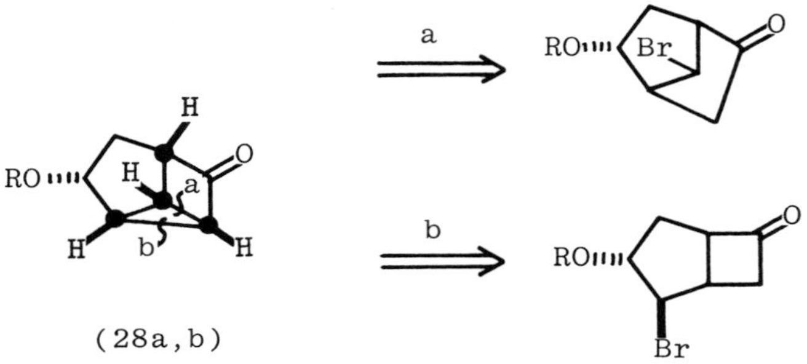

[Scheme: (28a,b) ⇒ a: (29) RO···Br with bicyclic ketone; ⇒ b: (30) RO···/Br bicyclic ketone]

The basic 5 + 4 ring structure in (30) is that of a ketene cycloadduct to a cyclopentene, or, if the 1,2-difunctionality (Br and OR) come from a double bond, of adduct (31) (Chapter T 33).

Analysis

(30) 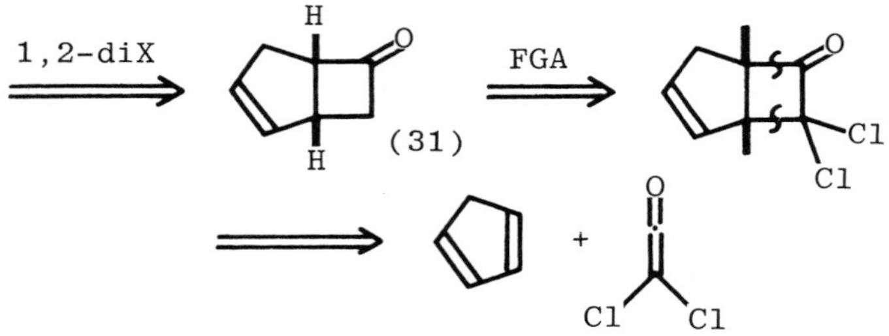 $\xRightarrow{1,2-diX}$ (31) $\xRightarrow{FGA}$ [bicyclic with Cl, Cl] ⇒ [cyclopentadiene] + $Cl_2C=C=O$ (Cl-CO-Cl shown)

Compound (28, R= $SiMe_2Bu\text{-}t$) is an important intermediate in prostaglandin synthesis and can indeed be made by this strategy.

This is a very strained ketone but three-membered ring formation is kinetically favourable and the hindered base offers no cleavage reactions. Notice that the rings are added in order of size.
Synthesis[430]

cyclopentene $\xrightarrow[\text{2.Zn}]{\text{1.Cl}_2\text{CHCOCl,Et}_3\text{N}}$ (31) $\xrightarrow[\text{2.t-BuMe}_2\text{SiCl(75\%)}]{\text{1.MeCONHBr(75\%)}}$

(30) $\xrightarrow{\text{t-BuOK}}$ TM(28)
100%

CHAPTER 38

Strategy XVII: Stereoselectivity B

This Chapter contains problems and examples related to Chapter T 38, using the same principles and underlining the same points without following the same organisation.

Problem :

Ester (1) is a perfumery compound from civet cats. What factor favours a stereoselective synthesis?

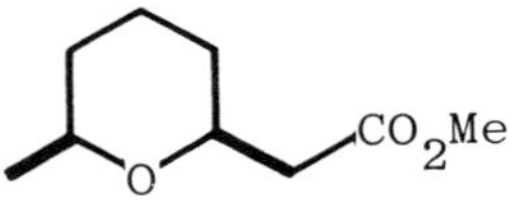

<img of structure (1): tetrahydropyran with methyl and CH2CO2Me substituents>

Answer :

A conformational drawing (1a) reveals that the two substituents are equatorial and that (1) is therefore the more stable of the two diasteroisomers.

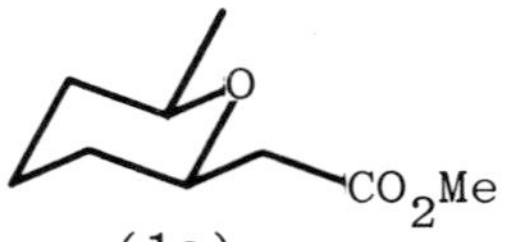

<img of chair conformation (1a)>

Further development :

Neither chiral centre in (1) can be epimerised by acid or base, but a reverse Michael reaction, which should happen in base, provides both a suitable disconnection and a way to epimerise one chiral centre.

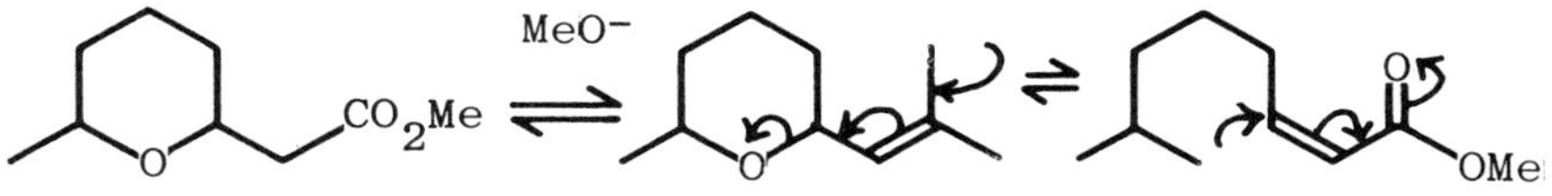

One published stereoselective synthesis uses the base-catalysed cyclisation of optically active enone (2) with a prolonged reaction time to get *cis*-(3) which is converted into (1) by degradation of the oxime. *Synthesis*[431]

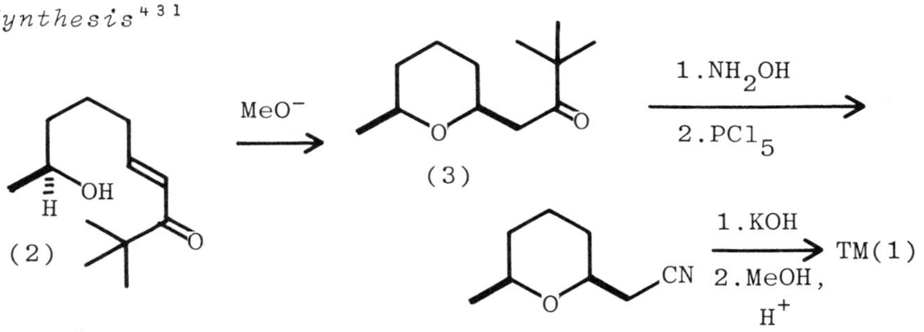

Examples :

An alternative strategy for (1) would have been to re-connect both substituents into a ring to ensure their *cis* arrangement. This strategy was used in syntheses of the antibiotic methylenomycin (4) and the trail pheromone of pharaoh's ant, faranal (5).

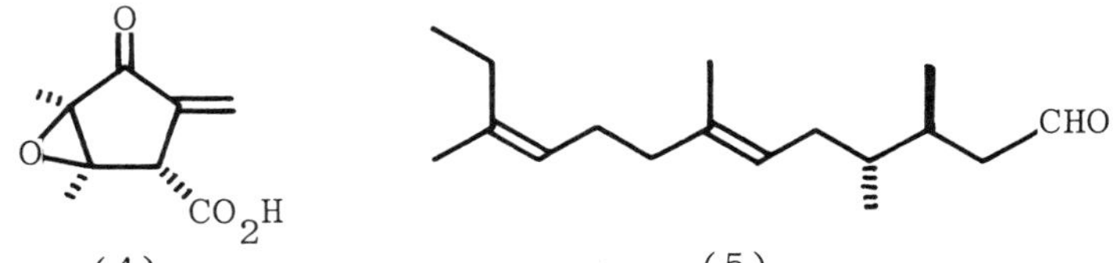

Methylenomycin (4) contains three adjacent chiral centres but two are related as the epoxide. This epoxide could be formed stereoselectively from a suitable olefin as the oxygen has been added on the opposite side from the only substituent. The simplest candidate (6) is most unsuitable as it would be impossible to distinguish between the two double bonds. Reconnection of the acid with the highly reactive α-methylene group gives (7), a much better candidate for stereoselective epoxidation.

Analysis 1

(6) ⇐ (4) reconnect ⟹ (7)

(7) ⟹ (8)

Bicyclic (7) is bowl-shaped and epoxidation ought to take place on the outside of the bowl. Further disconnection of the enone in (7) reveals a 1,4-diketone, which must be *cis* if cyclisation to (7) is to occur. Reconnection to (9) fixes the stereochemistry.

Analysis 2

(8a) —α,β⟹ — reconnect ⟹ (9)

The lactone ring in (9) is now a serious problem. If it were a symmetrical ring - say an anhydride (10) - we could make it easily by a 2 + 2 cycloaddition, as it would be symmetrical.

Analysis 3

(9) ⟹ [FGA] (10) ⟹ [2+2] alkyne + maleic anhydride

This is such a simple strategy that it is worth solving the problem of removing one carbonyl group. The best way turns out to be to reduce (10) all the way to the symmetrical ether (11) and then to *add* one carbonyl group after condensation to (12) has made the molecule sufficiently unsymmetrical.

Synthesis[4,3,2]

alkyne + maleic anhydride →[hv, MeCN, Ph$_2$CO] (10) 79% →[LiAlH$_4$] diol 96% →[TsCl, pyridine, heat]

(11) 86% →[1. O$_3$, -78°C, MeOH; 2. Ph$_3$P] dialdehyde 75% →[NaOH, MeOH] (12) 85%

Epoxidation of (12) with a nucleophilic agent occurred only from the outside of the molecule and the necessary oxidation to (7) could be carried out with RuO$_4$ and NaIO$_4$. The remaining steps required new reactions based on selenium chemistry and gave methylenomycin in 12% overall yield from maleic anhydride.

Synthesis[432]

(12) $\xrightarrow[\text{HO}^-]{\text{H}_2\text{O}_2}$ [epoxy bicyclic ketone] $\xrightarrow[\text{NaIO}_4 \cdot 47\%]{\text{RuO}_4}$ (7) $\rightarrow$ $\rightarrow$ TM(4)

90%

Faranal (5) is also an important compound as pharaoh's ant has invaded many hospitals, carrying infections by its trails along heating pipes. The trail pheromone might be useful to lure it away. Wittig disconnection of the central bond reveals the vital fragment (13) containing both chiral centres. Unfortunately this is a dialdehyde so a better starting material is (14), derived by a chain lengthening strategy. The two chiral centres in (14) are still difficult as they are remote from the functional groups.

Analysis

(5a) $\xrightarrow{\text{Wittig}}$ (13)

$\Longrightarrow$ (14)

Problem :
Devise a reconnection strategy to synthesise (14).
Answer :
The most obvious source of (14) is lactone (15) in which the two methyl groups are *cis* as the chain must be turned over in 'cyclisation'.

Analysis 2

(14) $\xrightarrow{\text{reconnect}}$ (15)

Lactone (15) can be derived by Baeyer-Villiger oxidation of symmetrical ketone (16) whose further analysis follows the usual strategy for symmetrical ketones.

Analysis 3

(15) $\xRightarrow{B-V}$ (16) $\xRightarrow[\text{activating group}]{\text{add}}$ [cyclopentanone with CO_2Et]

$\xRightarrow{1,3-diCO}$ [chain: 1-CO_2Et, 2, 3, 4, 5, 6-CO_2Et] $\xRightarrow{\text{reconnect}}$ (17)

Cyclohexene (17) looks remarkably like a Diels-Alder adduct and the addition of some carbonyl groups makes it one. The best starting material is again maleic anhydride.

Analysis 4

(17) $\xRightarrow{FGA}$ [cyclohexene with XOC and COX] $\Longrightarrow$ [maleic anhydride] + [butadiene]

The synthesis is long (and not given in full detail) but it ensures the correct diastereoisomer of (14) and an early resolution - actually of (18) - ensures optically active faranal. This synthesis confirmed the configuration of the natural product.

Synthesis[433]

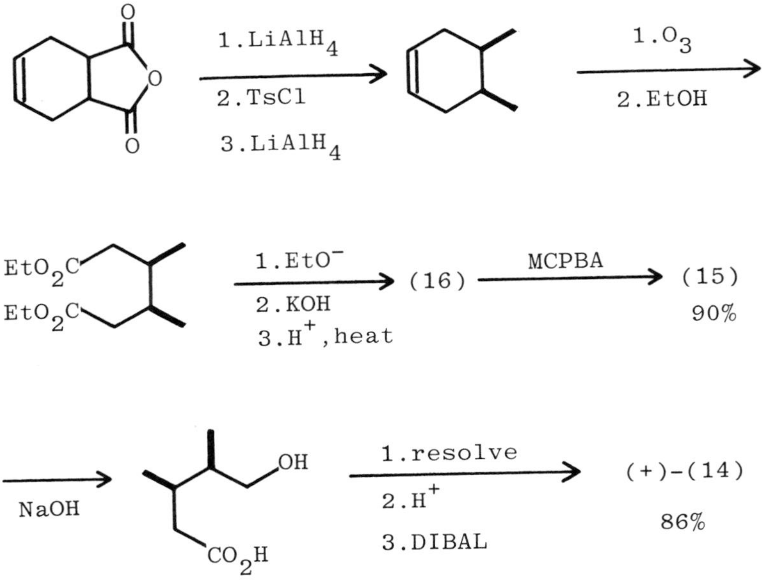

[Scheme: cyclohexene-dicarboxylic anhydride → (1. LiAlH₄, 2. TsCl, 3. LiAlH₄) → 4,5-dimethylcyclohexene → (1. O₃, 2. EtOH) → ...]

[Scheme: EtO₂C-CH₂-CH(Me)-CH(Me)-CH₂-CO₂Et → (1. EtO⁻, 2. KOH, 3. H⁺, heat) → (16) → MCPBA → (15) 90%]

[Scheme: → NaOH → (18) HOCH₂-CH(Me)-CH(Me)-CH₂-CO₂H → (1. resolve, 2. H⁺, 3. DIBAL) → (+)-(14) 86%]

Example :

The strategy of cleaving cyclic compounds can be taken much further. All four chiral centres in (20) are fixed by its formation from cage compound (19).

Synthesis[434]

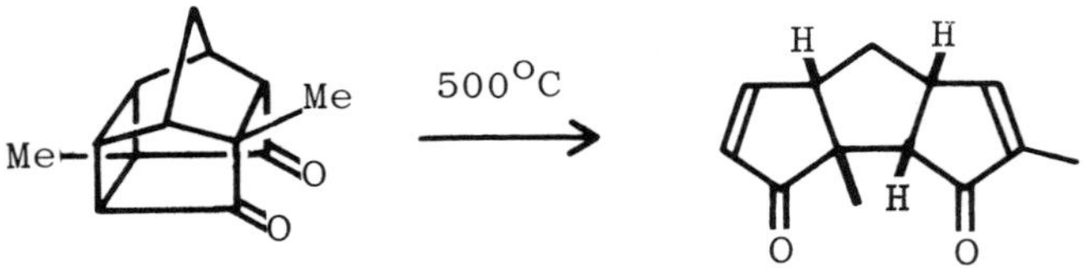

[Scheme: cage compound (19) with two Me groups and two C=O → 500°C → tricyclic diketone (20) 100%]

Problem :

Suggest a synthesis of the starting material (19).

Answer :

This polycyclic compound contains four five-membered

rings and many common atoms. Disconnection of two
common bonds in the four-membered ring leaves (21), a
Diels-Alder adduct from cyclopentadiene and quinone (22).
Analysis

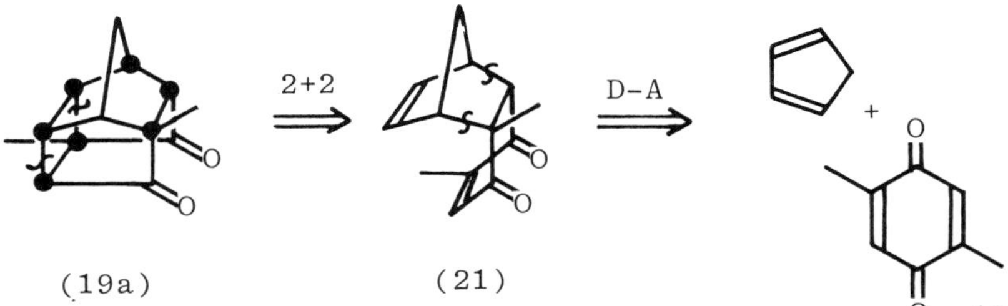

(19a)   2+2 ⟹   (21)   D-A ⟹   + (22)

Notice that the Diels-Alder disconnection also
cleaves common bonds. The synthesis is short and high
yielding.
Synthesis[434]

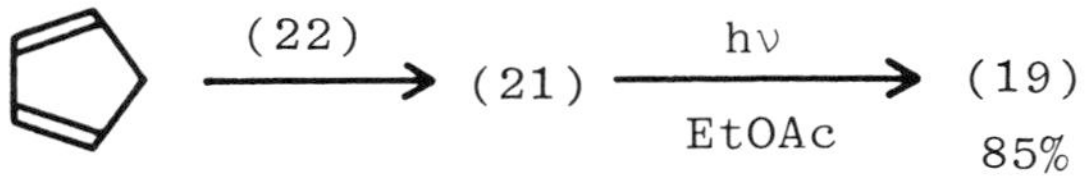

$$\bigcirc \xrightarrow{(22)} (21) \xrightarrow[\text{EtOAc}]{h\nu} (19) \quad 85\%$$

Example :
The synthesis of (23) illustrates how a six-membered ring
may be used to control even more remote chiral centres.
Reverse Michael disconnection leaves enone (24), an
oxidation product from allylic alcohol (25). The double
bond can come from elimination on bromohydrin (26) and
hence from (27).
Analysis

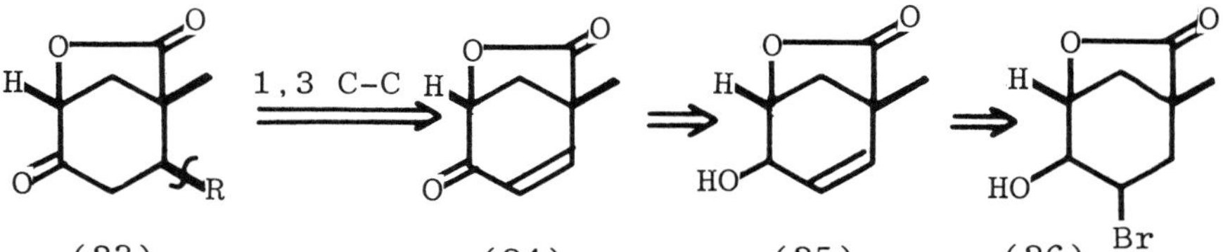

(23) ⟹ 1,3 C-C (24) ⟹ (25) ⟹ (26)

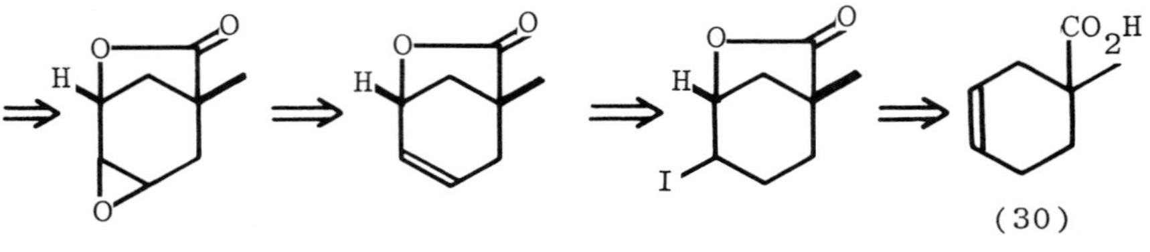

(27) ⇒ (28) ⇒ (29) ⇒ (30)

The double bond in (28) can come from iodolactone (29) by elimination and hence Diels-Alder adduct (30) is the starting material.

Problem :

The one chiral centre in (30) is used to set up four other chiral centres stereoselectively. Draw out the synthesis, giving the correct stereochemistry for all compounds.

Answer :

Attack by iodine and CO_2H on the double bond of (30) must be *trans*, hence I is axial in (29) and elimination to (28) is easy. Epoxidation occurs on the less hindered side of the double bond to give (27) which is opened by bromide to *trans* diaxial (26). This compound was protected as a silyl ether.

Synthesis[435]

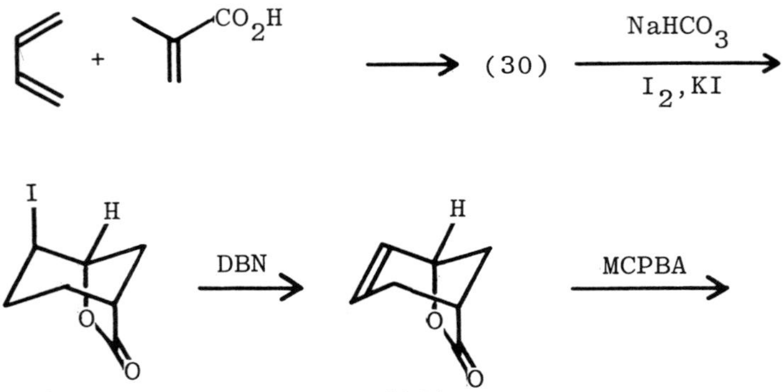

butadiene + methacrylic acid ⟶ (30) $\xrightarrow[I_2, KI]{NaHCO_3}$ (29) $\xrightarrow{DBN}$ (28) $\xrightarrow{MCPBA}$

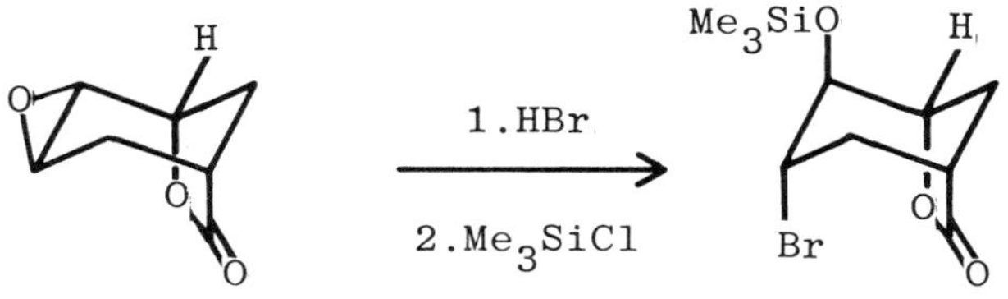

(27) (26)

Elimination on (26) must remove an axial proton so can give only (25), oxidation gives (24) and axial addition of 'R$^-$' gives (23).

Synthesis 2[435]

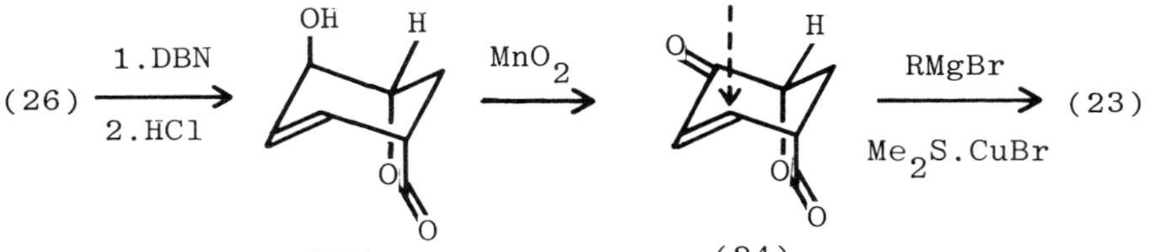

(25) (24)

Problem :

The Diels-Alder reaction is a favourite way of introducing many chiral centres at once, and we saw examples of this in Chapter 36. Compound (31) was used in a synthesis of quassin.[436] It has seven adjacent chiral centres. How many can be introduced in one step by a Diels-Alder reaction and what should the starting materials be?

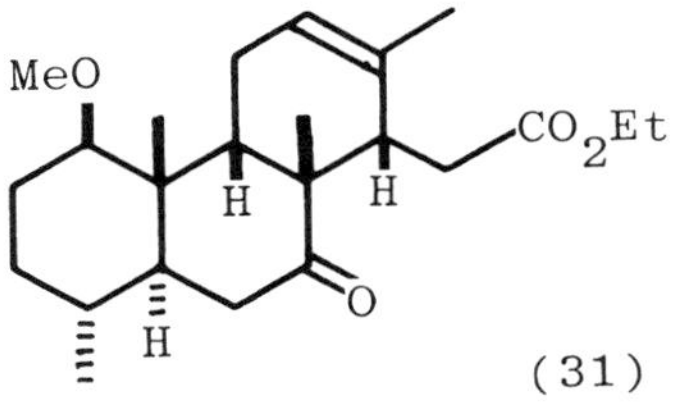

(31)

Answer :

The cyclohexene (ring C) can be disconnected by a Diels-Alder reaction to reveal enone (32) and diene (33). Four chiral centres remain in (32) so three are introduced in the Diels-Alder reaction.

Analysis

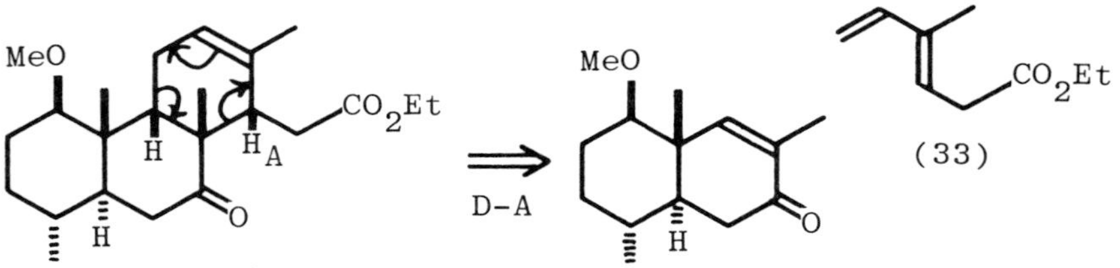

(31a) ⇒ D-A ⇒ (32) + (33)

The diene must attack from the underside of enone (32) to avoid the axial methyl group and the position of H^A in (31) comes from the endo transition state if E-(33) is used.

Further development :

Enone (32) is synthesised from (34) which we discussed on page 366. After methylation, acid-catalysed opening of the three-membered ring gives (35) in which the new methyl group (Me^A) has equilibrated to the equatorial position by enolisation.

Synthesis[437]

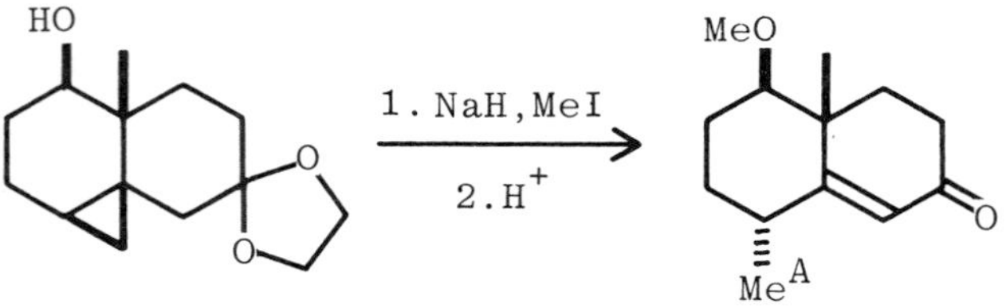

(34) → 1. NaH, MeI 2. H^+ → (35)

Methylation of the ketone occurs from the less crowded underside to give enone (36) which is reduced to saturated ketone (37) by lithium in liquid ammonia.

The more stable *trans* ring junction is formed. Finally the required double bond is introduced by axial bromination and *trans*-diaxial elimination.

Synthesis[437]

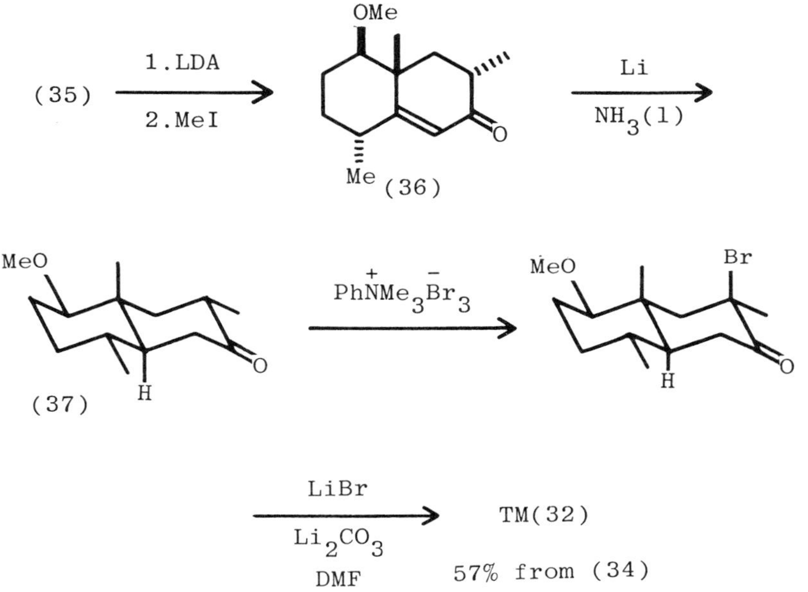

(35) →(1. LDA, 2. MeI)→ (36) →(Li, NH$_3$(l))→

(37) →(PhN$^+$Me$_3$ Br$_3^-$)→

→(LiBr, Li$_2$CO$_3$, DMF)→ TM(32)

57% from (34)

CHAPTER 39

Aromatic Heterocycles

Five-Membered Heterocycles
Example :
Thiophene saccharine (1) is a synthetic sweetening agent with a distinct resemblance to saccharine itself. Imide disconnection shows that we need some derivative of diacid (2).
Analysis 1

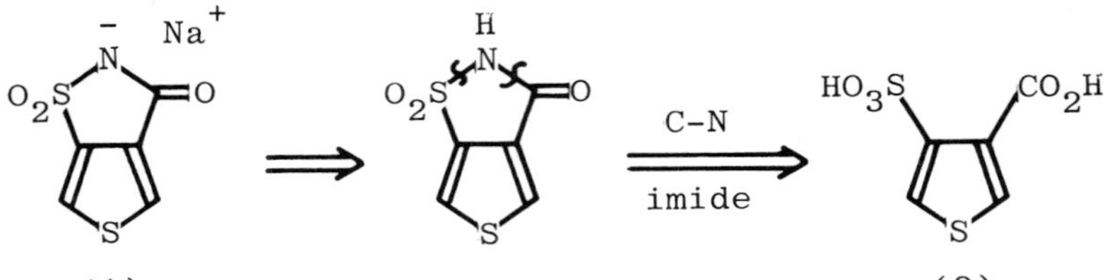

(1) ⇒ ⇒ (2)
          C-N
          imide

Derivatives of any of pyrrole, thiophene, or furan substituted in the 3 and 4 positions are difficult to make so the simple synthesis of (3) (Chapter T 29) might be put to good use.
Synthesis[438]

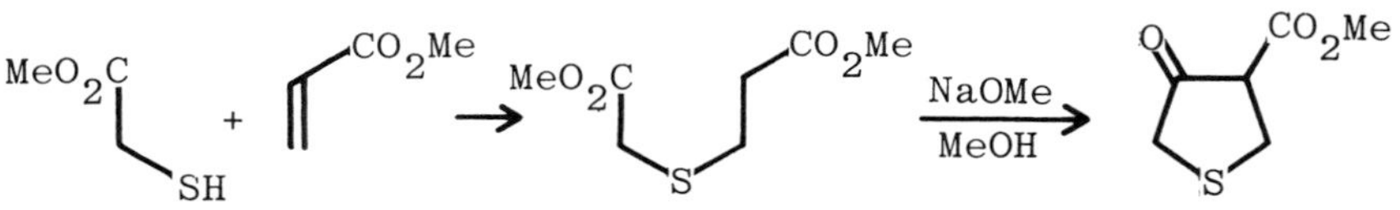

(3) 50%

Replacement of carbonyl by sulphur is achieved via enol tosylate (4) and the rest is simply an adjustment of oxidation levels. The thiophene ring is much more stable than pyrrole or furan so that treatment with acids and oxidising agents do not destroy it.

Synthesis[438]

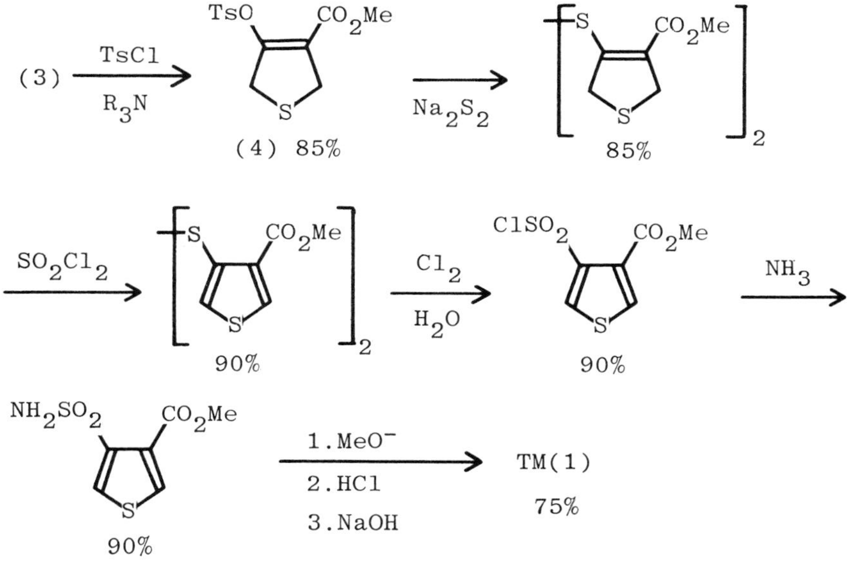

$$(3) \xrightarrow{\text{TsCl}, R_3N} \underset{(4)\ 85\%}{\text{TsO-thiophene-CO}_2\text{Me}} \xrightarrow{\text{Na}_2\text{S}_2} \left[\underset{85\%}{\text{S-thiophene-CO}_2\text{Me}}\right]_2$$

$$\xrightarrow{\text{SO}_2\text{Cl}_2} \left[\underset{90\%}{\text{S-thiophene-CO}_2\text{Me}}\right]_2 \xrightarrow[\text{H}_2\text{O}]{\text{Cl}_2} \underset{90\%}{\text{ClSO}_2\text{-thiophene-CO}_2\text{Me}} \xrightarrow{\text{NH}_3}$$

$$\underset{90\%}{\text{NH}_2\text{SO}_2\text{-thiophene-CO}_2\text{Me}} \xrightarrow[\text{2. HCl} \atop \text{3. NaOH}]{\text{1. MeO}^-} \underset{75\%}{\text{TM}(1)}$$

Problem :

Suggest a synthesis for amino pyrrole (5), given that Me_2NNH_2 is available from nitrosation and reduction of Me_2NH.

(5) 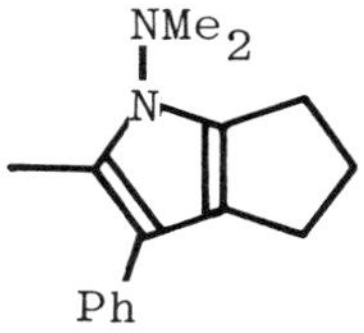 [pyrrole fused to cyclopentane ring, with NMe$_2$ on N, Me at 2-position, Ph at 3-position]

$$Me_2NH \xrightarrow[\text{2. reduce}]{\text{1. HONO}} Me_2N.NH_2$$

Answer :

This is an elementary exercise in 1,4-dicarbonyl chemistry. Bromoketone (7) was discussed in Chapter T 7.

Analysis

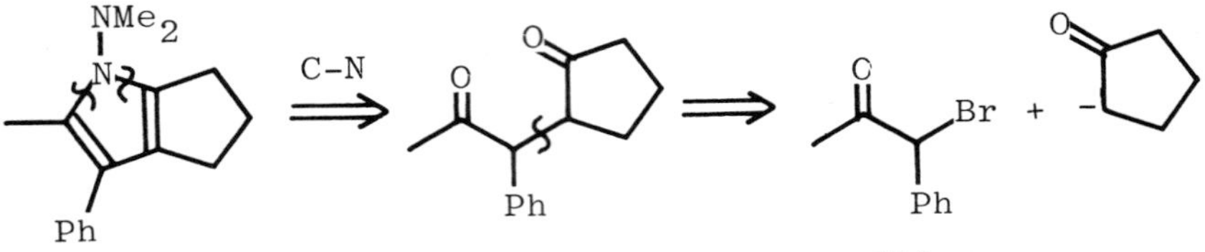

[Structures: (5a) pyrrole with NMe2, Ph; ⟹ C-N ⟹ (6) 1,4-diketone with Ph; ⟹ (7) α-bromoketone with Ph + cyclopentanone enolate]

Synthesis[439]

Cyclopentanone —1. pyrrolidine/H⁺; 2.(7)→ (6) —Me₂N.NH₂→ TM(5)

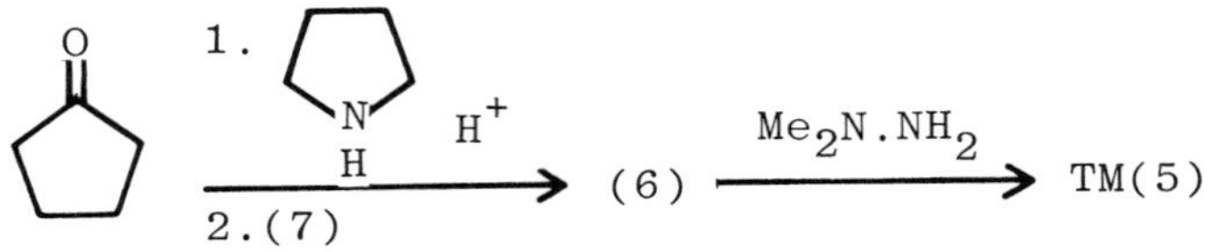

Problem :

How could the Hantsch pyrrole synthesis be used for the synthesis of TM (8)?

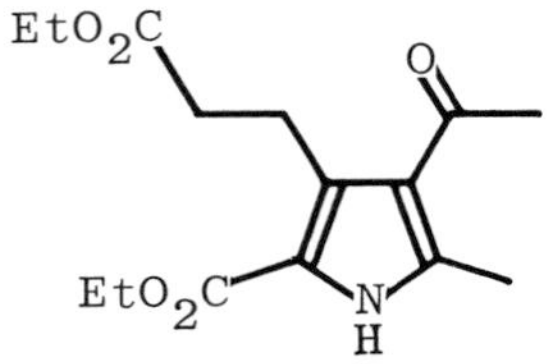

[Structure (8): pyrrole with EtO₂C-CH₂CH₂- , acetyl, EtO₂C, and methyl substituents]

Answer :

The amino group must be left attached to a carbon atom next to a carbonyl group so we must disconnect bond (a) first. The next disconnection follows from the symmetry of (10).

Analysis

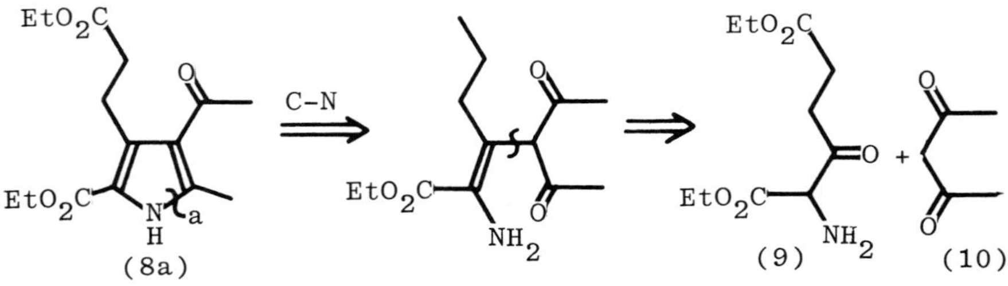

[Structures showing (8a) → C-N disconnection → enamine intermediate → (9) α-amino ketoester + (10) pentane-2,4-dione]

The amino group in (9) can be inserted by nitrosation of (11) available from laevulinic acid (Table T 25.1).

As usual, the amino ketone (9) is trapped by (10) as it is formed. The product (8) was used in a synthesis of porphobilinogen.

Synthesis[440]

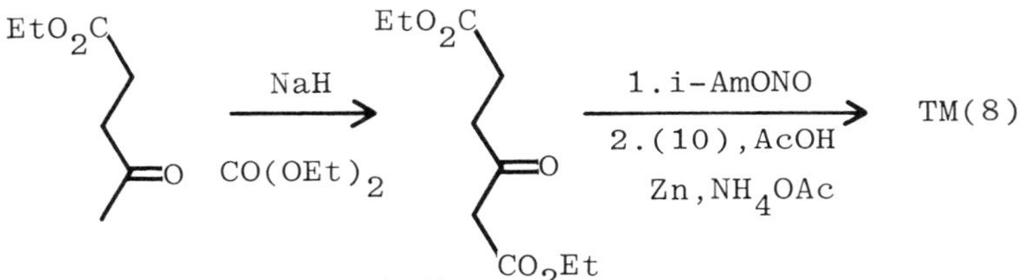

EtO$_2$C-CH$_2$CH$_2$-CO-CH$_3$ $\xrightarrow{\text{NaH, CO(OEt)}_2}$ EtO$_2$C-CH$_2$CH$_2$-CO-CH$_2$-CO$_2$Et (11) $\xrightarrow[\text{Zn, NH}_4\text{OAc}]{\text{1. i-AmONO}\atop\text{2. (10), AcOH}}$ TM(8)

Problem :

A new pain-killer (12), a possible rival to aspirin, has recently been introduced. Speculate on possible syntheses for this compound.

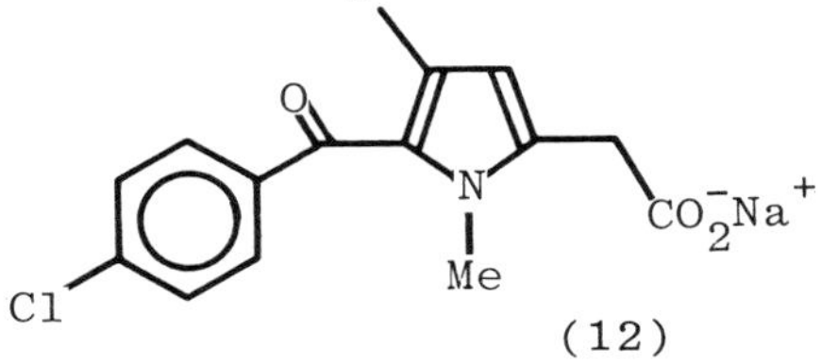

(12)

Answer :

Simple C-N disconnections (12a + b) give a triketo acid (13) which does not look like a very stable compound, and could react with MeNH$_2$ to give a number of heterocycles. Hantsch disconnection (12b) looks more promising as addition of a necessary activating group to the right hand part of the molecule suggests symmetrical (15) as starting material. There is still an ambiguity in the reaction of (15) with (14) as either carbonyl group could be attached.

Analysis

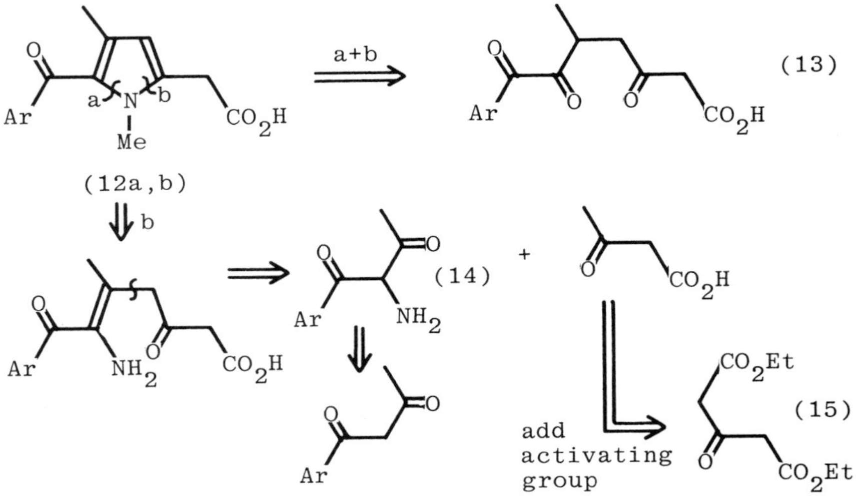

[Structural diagrams showing retrosynthesis of (12a,b) to (13) via disconnections a+b, with Ar-CO group, methyl-pyrrole N-Me, CO₂H; leading to intermediate (14) Ar-CO-CH(NH₂)-C(=O)-Me plus CH₃-CO-CH₂-CO₂H; and via "add activating group" to (15) with CO₂Et groups.]

Electrophilic substitution provides a third and best synthesis. The *p*-chlorobenzoyl group can be added in a Friedel-Crafts reaction, leaving (16) which can be disconnected to (15) and chloroacetone.

Analysis 2

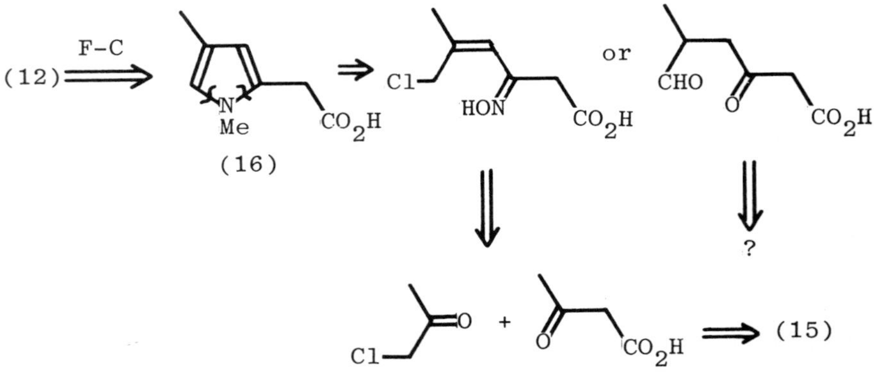

[Retrosynthetic scheme: (12) ⟹ (F-C) (16) pyrrole with N-Me and CO₂H; ⟹ Cl-C(=CHMe)-C(=NOH)-CH₂-CO₂H or CHO-CH(Me)-CO-CH₂-CO₂H; ⟹ Cl-CH₂-CO-CH₃ + CH₃-CO-CH₂-CO₂H ⟹ (15)]

Hydrolysis of both ester groups and chemoselective re-esterification of the more reactive alkyl acid distinguishes the two ester groups. The pyrrole acid decarboxylates on heating.

Synthesis[441]

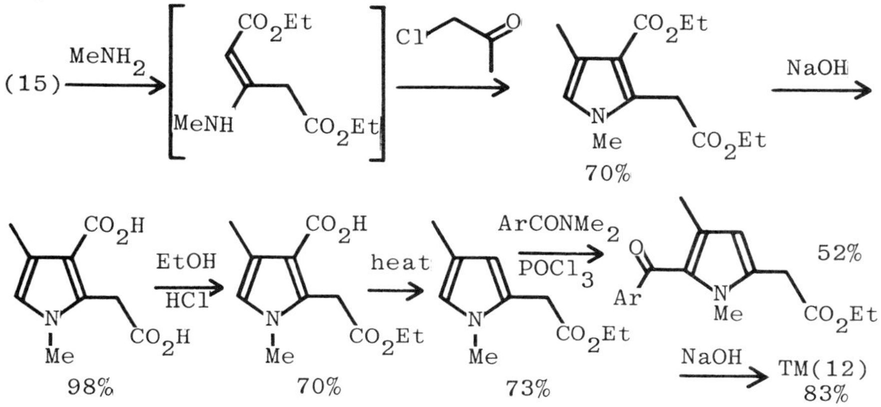

(15) $\xrightarrow{\text{MeNH}_2}$ [intermediate with CO$_2$Et, MeNH, CO$_2$Et] $\xrightarrow{\text{Cl}\diagdown\text{CO}}$ pyrrole (CO$_2$Et, Me, CH$_2$CO$_2$Et) 70% $\xrightarrow{\text{NaOH}}$

pyrrole(CO$_2$H, Me, CH$_2$CO$_2$H) 98% $\xrightarrow[\text{HCl}]{\text{EtOH}}$ pyrrole(CO$_2$H, Me, CH$_2$CO$_2$Et) 70% $\xrightarrow{\text{heat}}$ pyrrole(Me, CH$_2$CO$_2$Et) 73% $\xrightarrow[\text{POCl}_3]{\text{ArCONMe}_2}$ Ar-CO-pyrrole(Me, CH$_2$CO$_2$Et) 52% $\xrightarrow{\text{NaOH}}$ TM(12) 83%

Benzofurans and Indoles

Example:

Indoles with no substituent on the 2-position, e.g. (17) cannot be made by the simple Fischer indole method as acetaldehyde gives a very poor yield in this reaction. An alternative is to use pyruvic acid as the α-CO$_2$H group in (18) can easily be decarboxylated to give the free indole.

Analysis

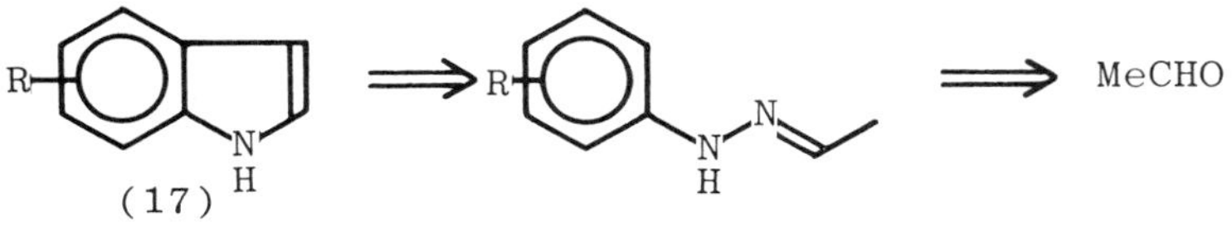

R—indole (17) $\Longrightarrow$ R—C$_6$H$_4$—NH—N=CH— $\Longrightarrow$ MeCHO

Synthesis[442]

R—C$_6$H$_4$—NH—NH$_2$ $\xrightarrow{\text{MeCOCO}_2\text{H}}$ R—C$_6$H$_4$—NH—N=C(Me)(CO$_2$H)

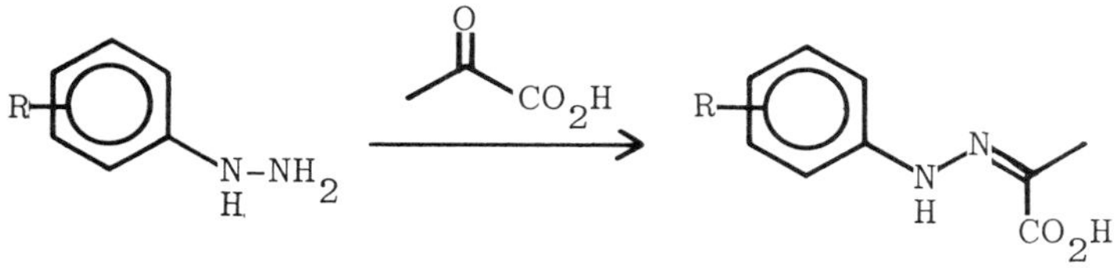

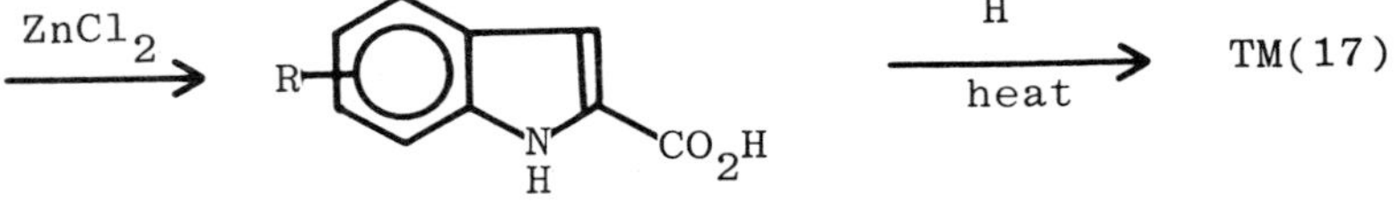

$$\xrightarrow{ZnCl_2}$$ (18) $$\xrightarrow[heat]{H^+}$$ TM(17)

Problem :

Suggest a synthesis for (19), needed in the search for new anti-histamines.

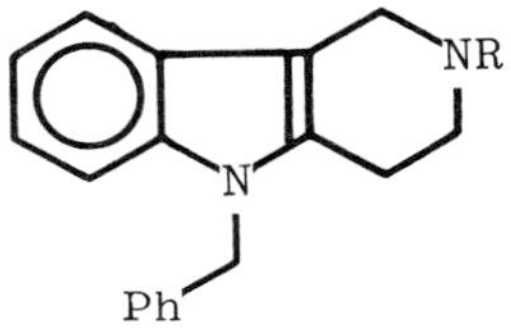

(19)

Answer :

Fischer indole disconnection reveals hydrazine (20) and symmetrical ketone (21) (Chapter T 30) as starting materials. The hydrazine can be made from (22) by nitrosation and reduction.

Analysis

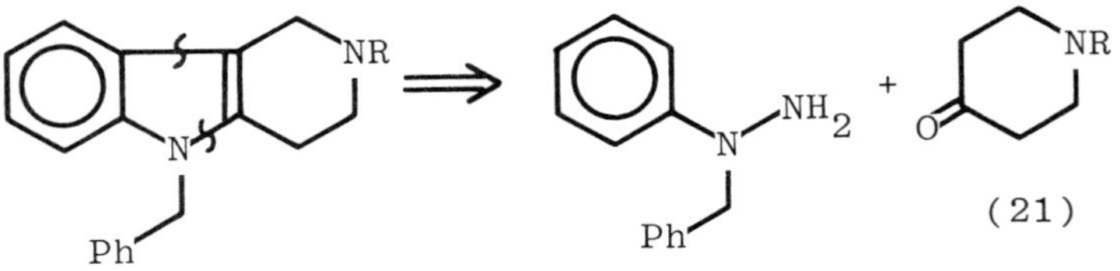

(19a) ⟹ (20) + (21)

Synthesis[443]

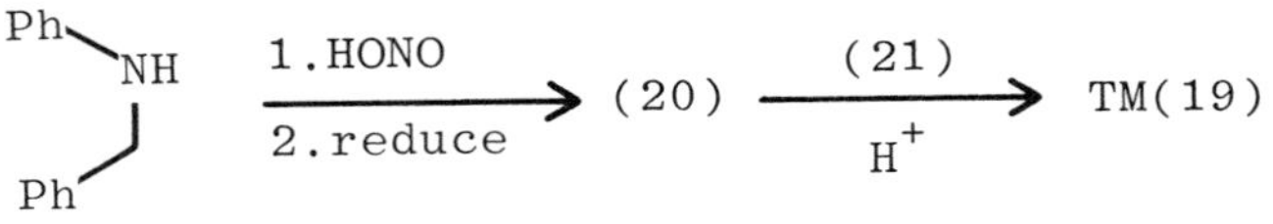

Ph-NH-CH-Ph $\xrightarrow[\text{2.reduce}]{\text{1.HONO}}$ (20) $\xrightarrow[H^+]{(21)}$ TM(19)

Pyridines

Example :

The principles expounded in Chapter T 39 apply to di-hydropyridines, e.g., (23), wanted in their own right in this case for the synthesis of morphine analogues. The 1,5-dicarbonyl compound needed for (23) is (24), available by a Michael reaction.

Analysis

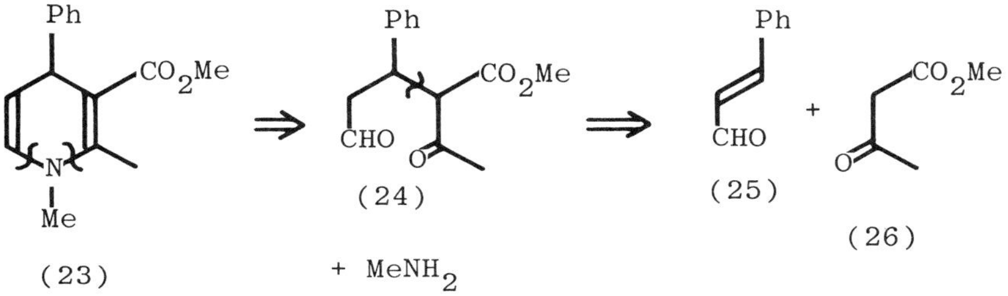

(23) ⇒ (24) ⇒ (25) + (26)

+ MeNH$_2$

In practice it is simpler to use MeNH$_2$ to make an enamine from (26) and to encourage the Michael reaction whereupon cyclisation occurs rapidly.

Synthesis[444]

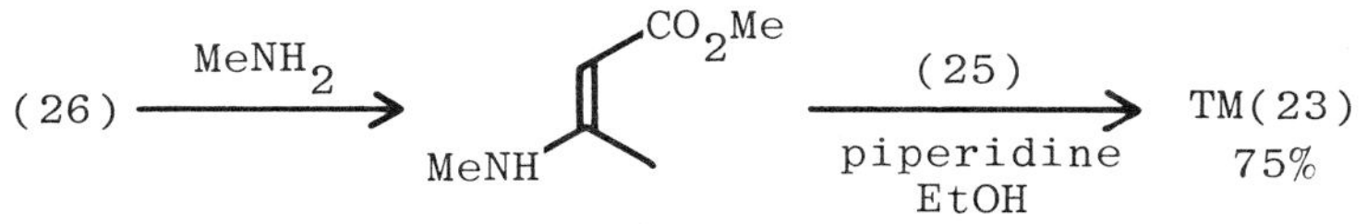

(26) →[MeNH$_2$] MeNH-C(CO$_2$Me)=... →[(25), piperidine, EtOH] TM(23) 75%

Benzoderivatives

Problem :

Identify the possible initial disconnections on the alkaloid (27).

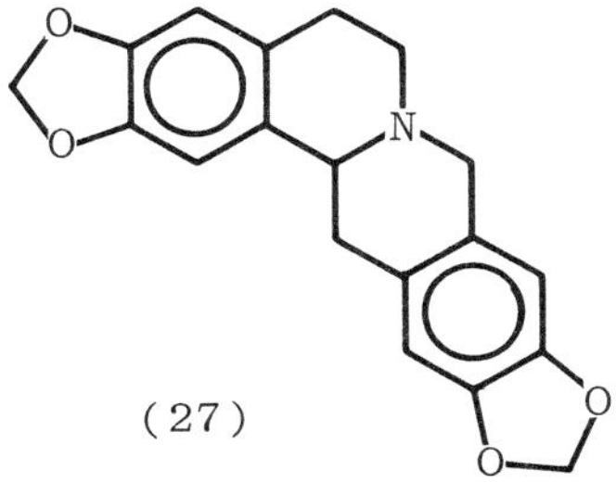

(27)

Answer :

Two atoms (● in 27a) occupy positions between a benzene ring and the nitrogen atom. Disconnection might be to an aldehyde, *via* (28), or an amide (e.g. 30).

Analysis

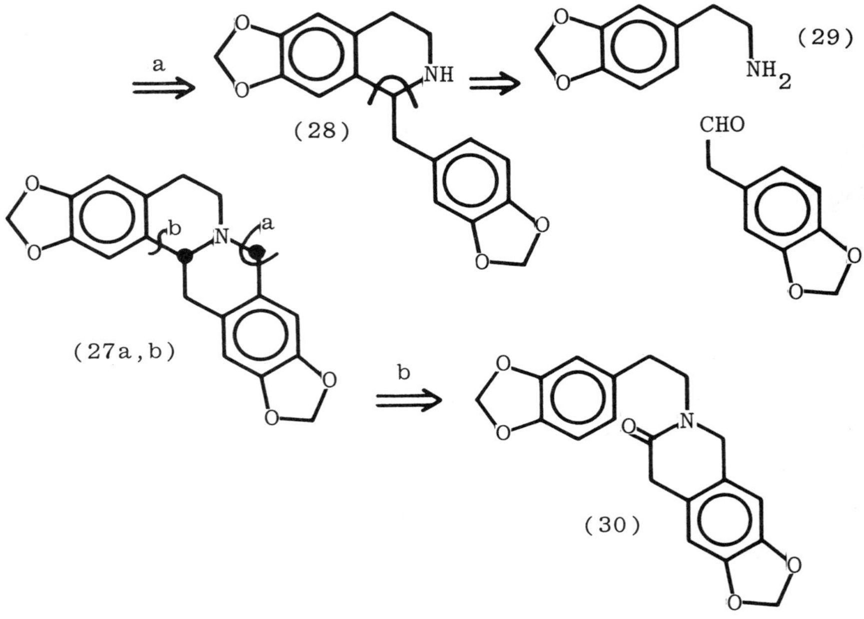

Further Development

Disconnection of (28) is by the methods described in the chapter so we shall explore the alternative (30). Disconnection of amine (29) from (30) leaves (31) which is again a formaldehyde product.

Analysis 2

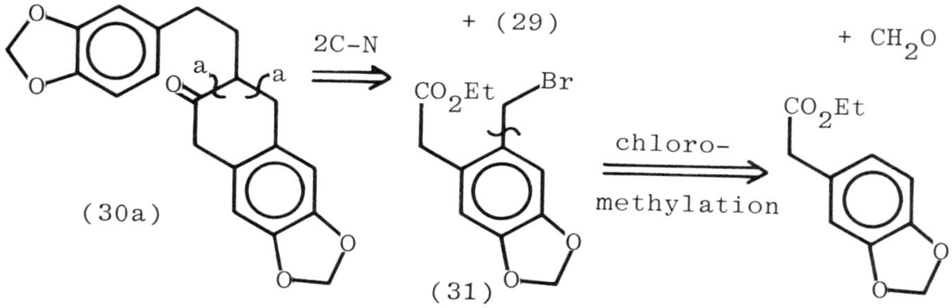

[Scheme: (30a) ⇒ (2C-N) (31) + (29); (31) ← chloromethylation ← CO₂Et-CH₂-(methylenedioxybenzene) + CH₂O]

In practice it is more convenient to use free acid (32) and isolate the formaldehyde adduct as (33) than to use conventional chloromethylation. The rest of the synthesis is as expected.

Synthesis[4,4,5]

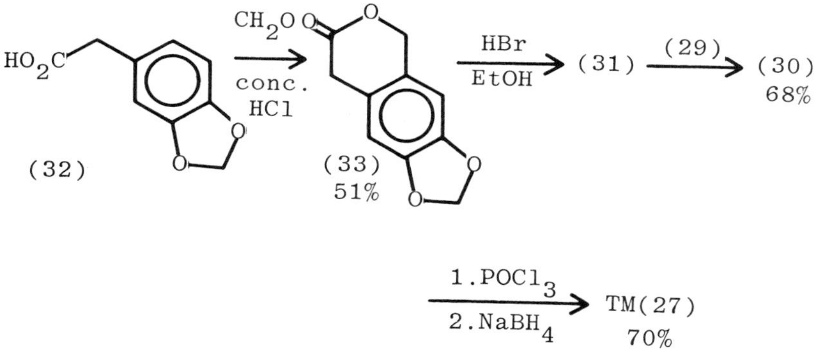

[Scheme: (32) HO₂C-CH₂-(methylenedioxybenzene) → (conc. HCl, CH₂O) → (33) 51% → (HBr/EtOH) → (31) → (29) → (30) 68% → 1. POCl₃, 2. NaBH₄ → TM(27) 70%]

Problem :

The Distillers Company have patented a mehod to make warfarin (36) as follows:-

Synthesis[446]

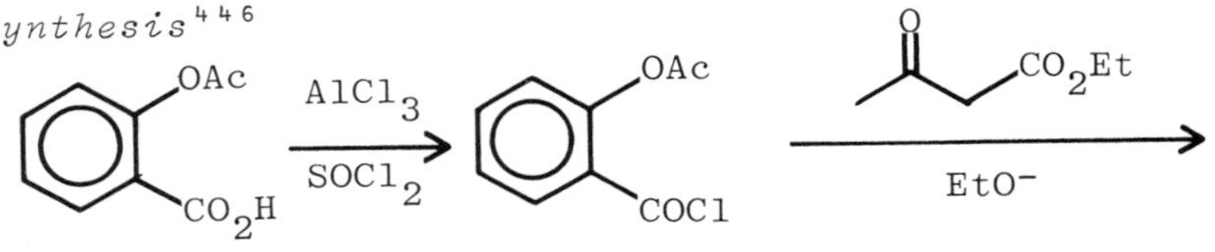

[Scheme: (34) salicylic acid derivative with OAc and CO$_2$H → AlCl$_3$, SOCl$_2$ → acid chloride with OAc and COCl → with CH$_3$COCH$_2$CO$_2$Et / EtO$^-$ →]

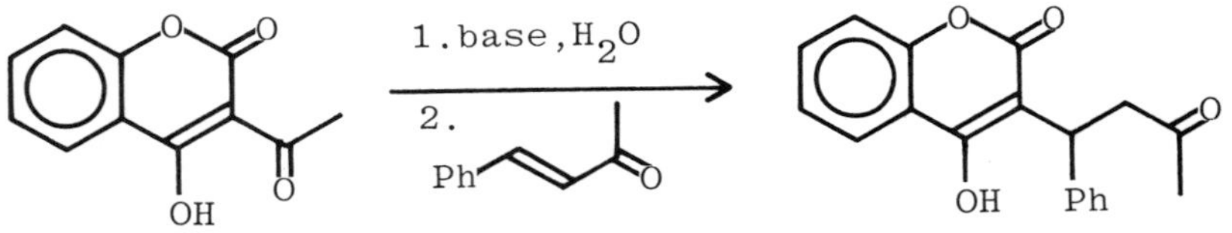

[Scheme: (35) 3-acetyl-4-hydroxycoumarin — 1. base, H$_2$O; 2. PhCH=CHCOCH$_3$ → (36) coumarin with CH(Ph)CH$_2$COCH$_3$ side chain]

What is the starting material? Analyse the strategy of the method.

Answer :

The starting material is aspirin. The acetyl group in (35) is an activating group, present to allow the synthesis of (35) and removed by 1,3-dicarbonyl cleavage. The strategy is therefore a 1,5-dicarbonyl disconnection to (37) and a 1,3-dicarbonyl disconnection to (38).

Analysis

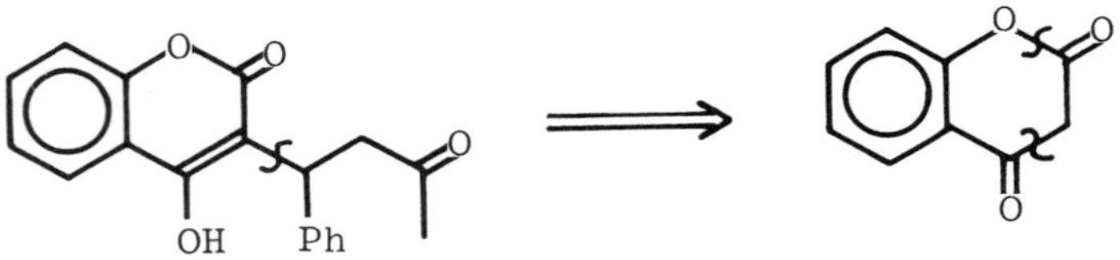

[Scheme: (36a) ⇒ (37)]

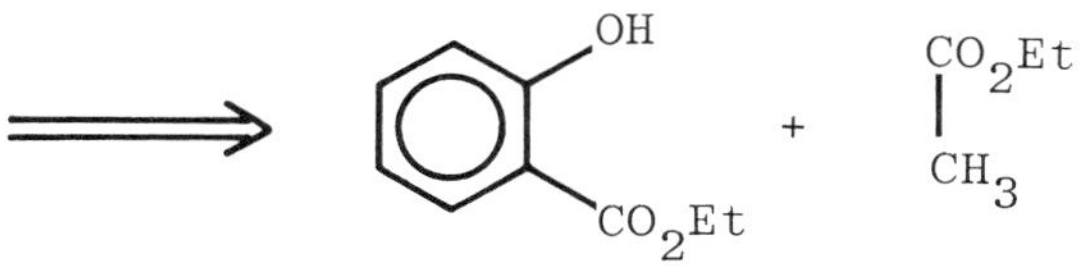

$$\Longrightarrow$$ [structure with OH and CO₂Et on benzene] (38) + CO₂Et—CH₃

Aspirin (34) is an available protected form of (38) and aceto acetate is an activated form of ethyl acetate, instead of the more usual malonate.

Example :

The precocenes, e.g. (39), are the opposite of juvenile hormones in that they stimulate a juvenile insect to become an adult. There is some hope that premature application of precocenes might lead to sterile adults and hence control of that species.

The double bond in (39) can be derived from ketone (40) and hence by simple disconnections from (41). The synthesis is straightforward.

Analysis

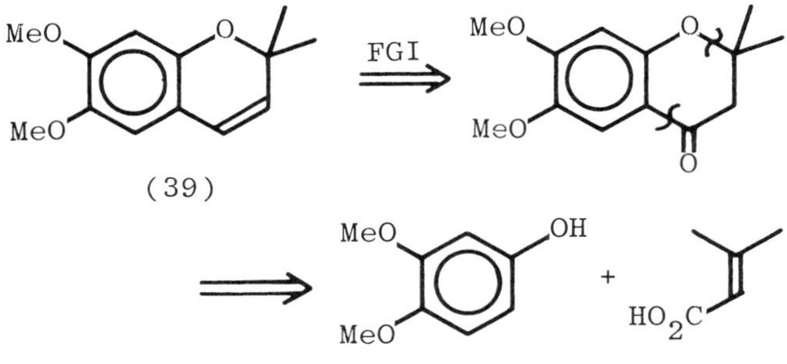

(39) $\xRightarrow{FGI}$ (40)

$\Longrightarrow$ (41) + (42)

Synthesis[447]

(41) $\xrightarrow[PPA]{(42)}$ (40) $\xrightarrow[\text{2.HCl}]{\text{1.LiAlH}_4}$ TM(39)

Six-Membered Heterocycles with two hetero-atoms
Problem :
Choose suitable starting materials for syntheses of the insecticide base for diazinone (43) and the ant alarm pheromone (44).

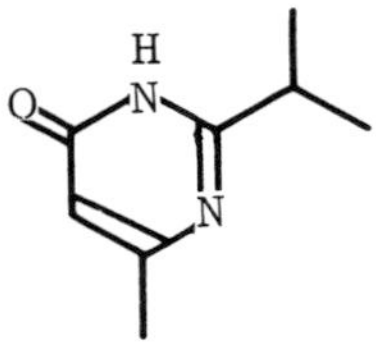

(43)

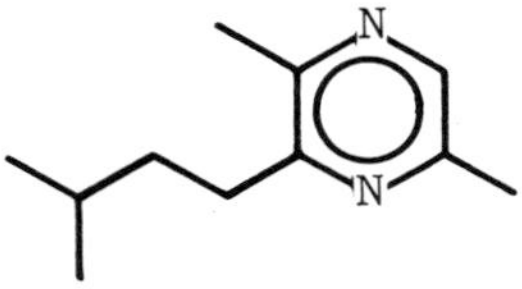

(44)

Answer :
Removing the piece with the two heteroatoms from (43) leaves acetoacetate as one starting material; the other is an amidine, easily made from nitrile (45).

Analysis

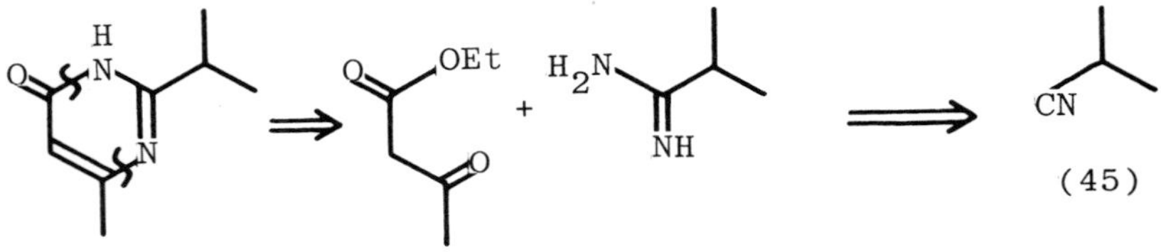

(43a)

(45)

Synthesis[448]

$$(45) \xrightarrow[\text{HCl}]{\text{MeOH}} \text{MeO-C(NH}_2^+\text{)-iPr} \xrightarrow{\text{NH}_3} \text{H}_2\text{N-C(NH)-iPr} \xrightarrow{\text{MeCOCH}_2\text{CO}_2\text{Et}} \text{TM(43)}$$

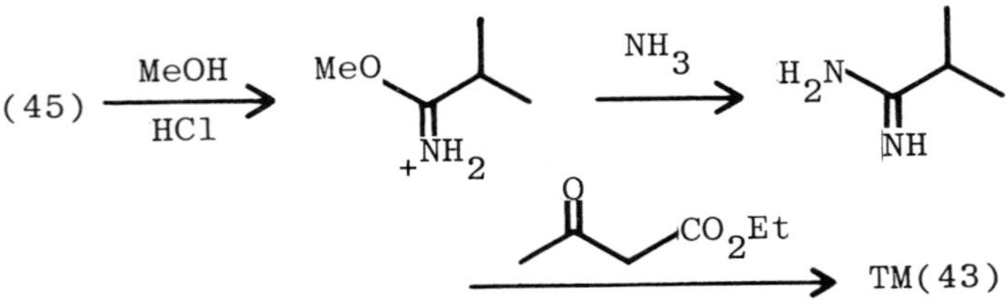

Pheromone (44) is more difficult. Disconnections such as (a) or (b) lead to problems because of the lack of symmetry. Hence reaction of diamine (46) and diketone (47) gives a 50:50 mixture of the two isomers.[449]

Analysis

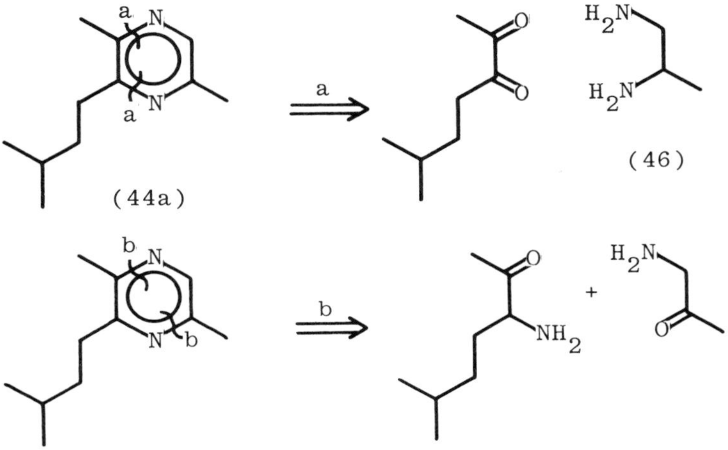

(44a) (46)

Synthesis[449] (no good)

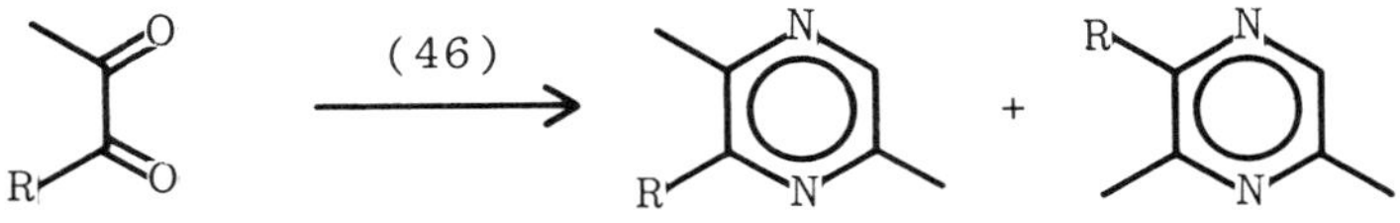

(47)

Removal of the C_5 side chain gives a symmetrical pyrazine (48), best made from a single precursor according to disconnection (a). The side chain can be added by treatment of (48) with the alkyl lithium at the electrophilic centre next to nitrogen (LiH is displaced).

Analysis

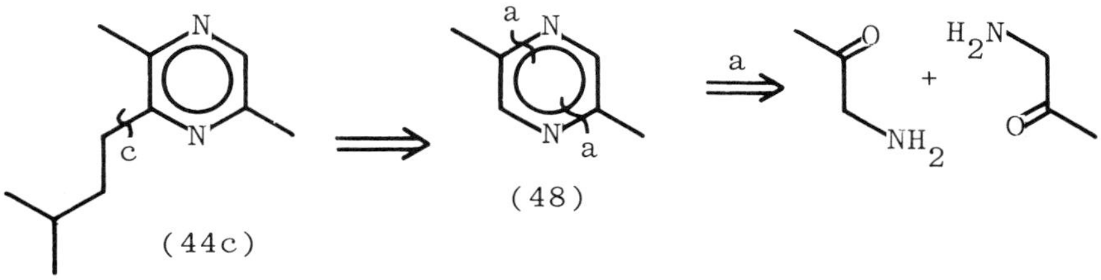

(48) (44c)

Amino acetone is not a stable compound. It can be made from acetone by nitrosation and reduction whereupon it dimerises to give (48).

Synthesis[449]

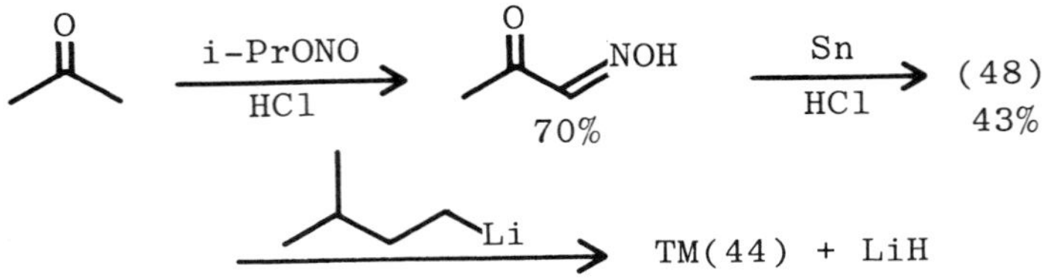

$$\underset{}{\text{Me-CO-Me}} \xrightarrow[\text{HCl}]{\text{i-PrONO}} \underset{70\%}{\text{Me-CO-CH=NOH}} \xrightarrow[\text{HCl}]{\text{Sn}} (48) \quad 43\%$$

$$\xrightarrow{\text{i-Bu-CH}_2\text{-Li}} \text{TM}(44) + \text{LiH}$$

Imidazoles

Example :

Compounds such as (49) are of interest in studies on aromaticity : this compound has 10 π electrons (including the lone pair on the bridgehead nitrogen atom) and is an analogue of naphthalene.

Disconnecting the five-membered ring from the six is a sensible start as it leaves a stable pyridine (50) after some adjustment of the double bonds.

Analysis 1

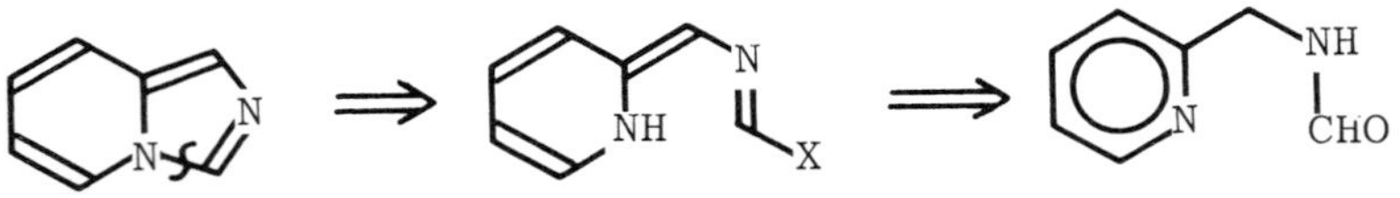

(49) ⇒ [pyridine-NH-CH=N-C(X)] ⇒ [pyridine-CH₂-NH-CHO] (50)

Amide disconnection leaves simple amine (51), available by reductive amination from aldehyde (52).

Analysis 2

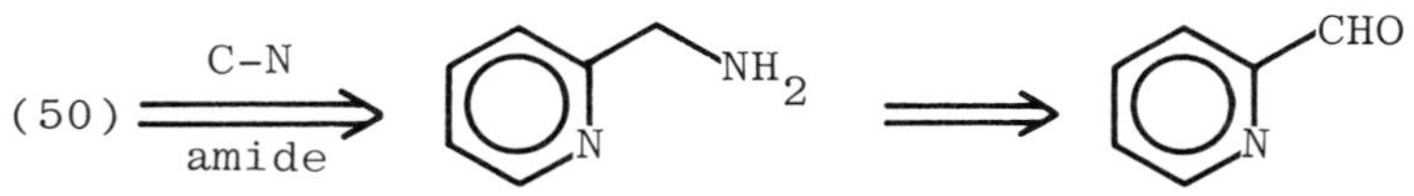

(50) $\underset{\text{amide}}{\overset{\text{C-N}}{\Longrightarrow}}$ [pyridine-CH₂-NH₂] (51) ⇒ [pyridine-CHO] (52)

Formic acid itself gives amide (50) which is cyclised to (49) in good yield with $POCl_3$.

Synthesis[450]

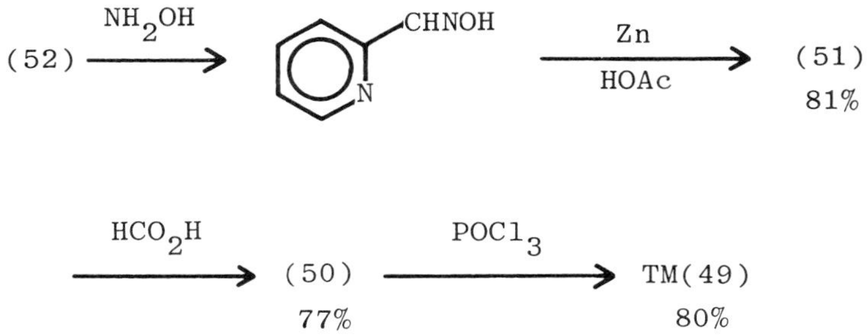

(52) $\xrightarrow{NH_2OH}$ [pyridine-2-CHNOH] $\xrightarrow[HOAc]{Zn}$ (51) 81%

$\xrightarrow{HCO_2H}$ (50) $\xrightarrow{POCl_3}$ TM(49)
77%   80%

Problem :

Lilly have recently introduced an anti-inflammatory drug opren (53) which may repair damage as well as reduce inflammation. Suggest a synthesis for (53).

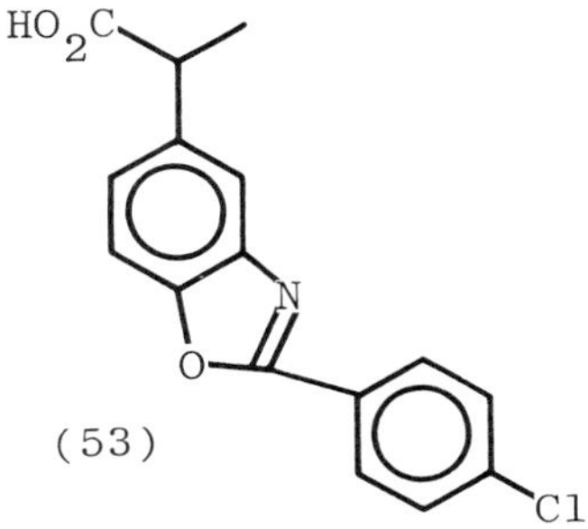

Answer :

This compound is a benzoxazole with a disconnection to an *ortho* amino phenol (54) and an acid derivative. Further disconnections on (54) are straightforward.

Analysis

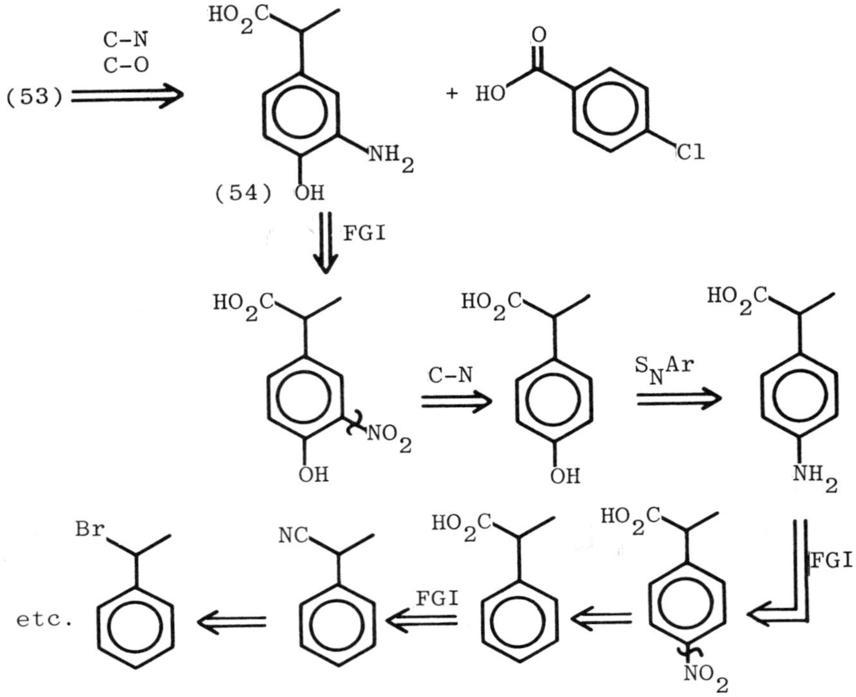

[Retrosynthetic scheme: (53) ⇒ (C-N, C-O disconnection) ⇒ (54) 3-amino-4-hydroxy-α-methylphenylacetic acid + 4-chlorobenzoic acid; (54) ⇒ (FGI) ⇒ nitro-hydroxy compound ⇒ (C-N) ⇒ 4-hydroxy-α-methylphenylacetic acid ⇒ (S_NAr) ⇒ 4-amino-α-methylphenylacetic acid ⇒ (FGI) ⇒ 4-nitro-α-methylphenylacetic acid ⇒ ⇒ (FGI) ⇒ 2-phenylpropanenitrile ⇒ α-bromoethylbenzene ⇒ etc.]

The order of events in this synthesis can be varied — one possibility is to keep the nitrile until after the oxazole has been made. The acid chloride is used for the cyclisation.

Synthesis[451]

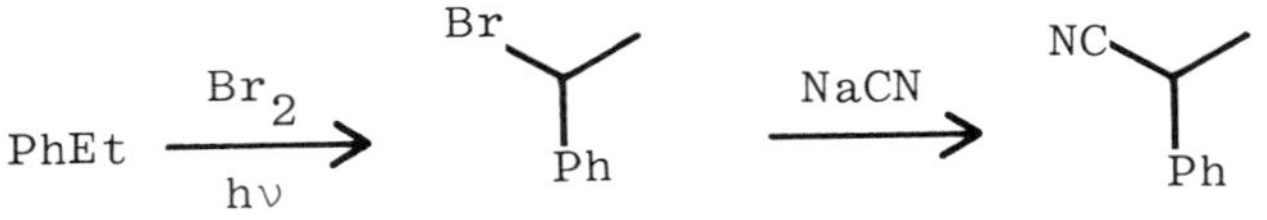

PhEt $\xrightarrow{\text{Br}_2, h\nu}$ PhCHBrMe $\xrightarrow{\text{NaCN}}$ PhCH(CN)Me

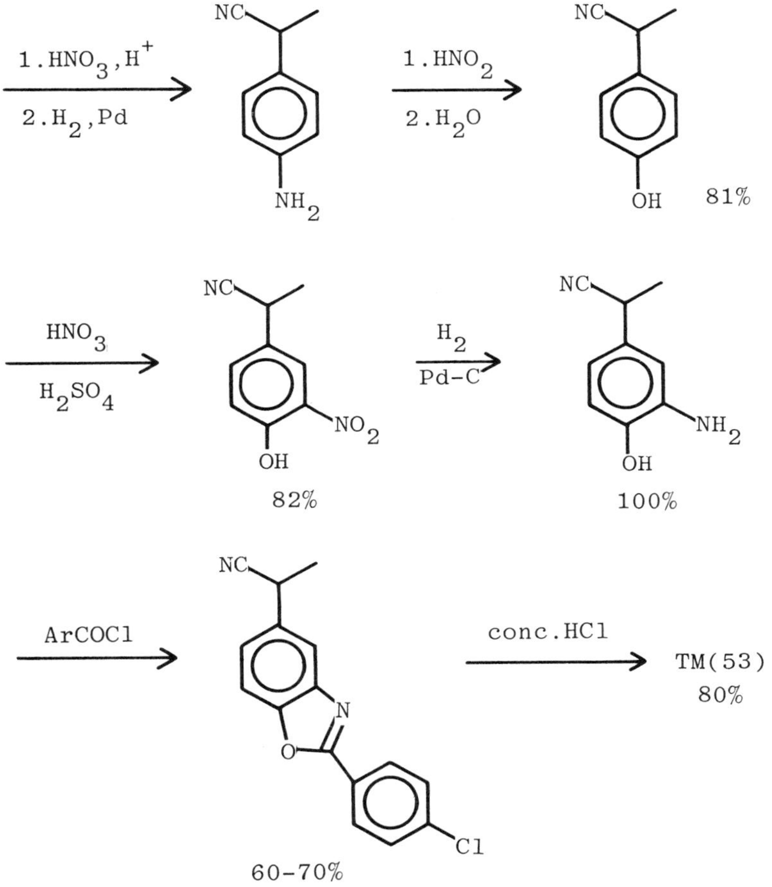

$\xrightarrow[\text{2.H}_2,\text{Pd}]{\text{1.HNO}_3,\text{H}^+}$ [4-aminophenyl CH(CN)CH$_3$] $\xrightarrow[\text{2.H}_2\text{O}]{\text{1.HNO}_2}$ [4-hydroxyphenyl CH(CN)CH$_3$] 81%

$\xrightarrow[\text{H}_2\text{SO}_4]{\text{HNO}_3}$ [4-hydroxy-3-nitrophenyl CH(CN)CH$_3$] 82% $\xrightarrow[\text{Pd-C}]{\text{H}_2}$ [4-hydroxy-3-aminophenyl CH(CN)CH$_3$] 100%

$\xrightarrow{\text{ArCOCl}}$ [2-(4-chlorophenyl)benzoxazole with CH(CN)CH$_3$] 60-70% $\xrightarrow{\text{conc.HCl}}$ TM(53) 80%

CHAPTER 40

General Strategy D: Advanced Strategy

This chapter roughly follows the organisation of Chapter T 40 but also contains revision examples and problems.

Convergence
Problem :
Examine the tetracycline synthesis on page 432
Does it follow a linear or convergent strategy?
Answer :
The first disconnections are made on ring C (1) separating the molecule into two roughly equal fragments (2) and (3) each of which has to be synthesised. This is a convergent plan. A linear analysis would disconnect ring A or D first.

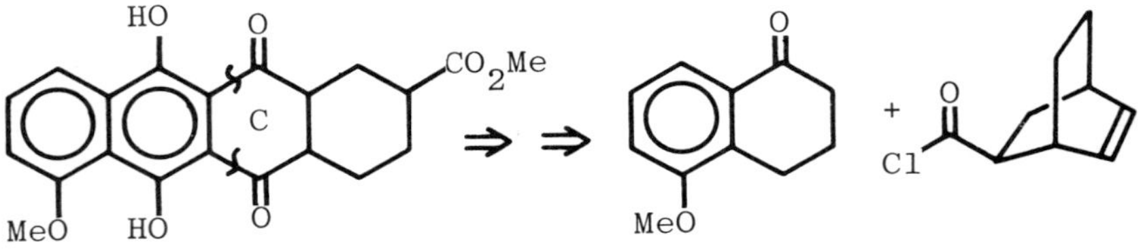

Example :
Thorough analyses of linear and convergent strategies for tetraene (3) are given by Corey.[452] His approach is for computer-based synthetic design but the logic is helpful to humans. This would be a good example to look

up if you are interested in computer syntheses.

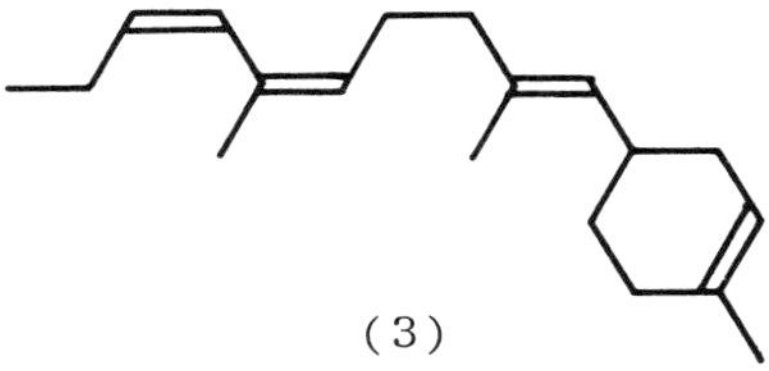

(3)

Problem :

Choose the first disconnection on (4) so that a convergent plan should emerge from further analysis.

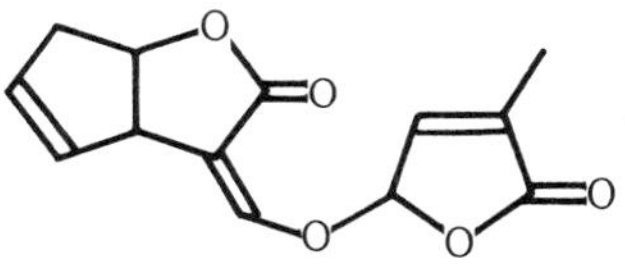

(4)

Answer :

Disconnection of the ether linkage (4a) between the two ring systems breaks the molecule into two sizeable fragments.

Analysis

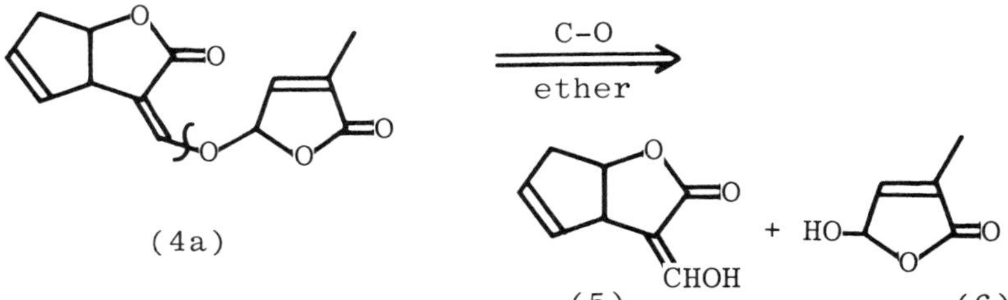

(4a) $\xRightarrow{\text{C-O ether}}$ (5) + HO—(6)

How could (5) be synthesised?

Answer :

Enone disconnection gives bicyclic lactone (7) which

comes from ketene cycloadduct (8) by a route we discussed in Chapter T 33. Lactone (7) qualifies as an advanced available starting material (Table T 40.1).

Analysis

[structure (5a) with CHOH substituent] $\xrightarrow{\alpha,\beta}$ [structure (7)] $\xrightarrow{B-V}$ [cyclobutanone structure]

$\xrightarrow{FGA}$ [structure (8) with CCl$_2$] $\xrightarrow{2+2}$ [cyclopentadiene] + [Cl-CO-Cl]

The coupling of the two parts of TM (4) is achieved by using the mesylate of (6) with enol (5).

Synthesis[4,5,3]

(7) $\xrightarrow[\text{base}]{HCO_2Et}$ (5) + MsO-[butenolide structure] $\xrightarrow{\text{base}}$ TM(4)

The product (4) is a synthetic germination stimulant for parasitic plants. Normally these plants germinate only when the host is growing. Premature germination induced by (4) increases yields of sorghum (an African food crop) planted later.

Revision Problem :

Suggest a synthesis of the uterus relaxant vetraburine (9).

[structure (9): Ph-CH$_2$CH$_2$CH$_2$-CH(NMe$_2$)-Ar where Ar = 3,4-dimethoxyphenyl]

Answer :

One route is by reductive amination of ketone (10) available by Friedel-Crafts reaction from acid (11) which we can make by an FGA strategy (Chapter T 24).
Analysis 1

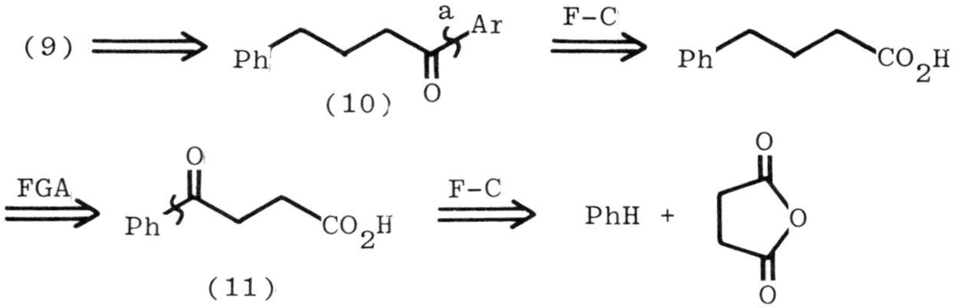

$$(9) \Longrightarrow \underset{(10)}{Ph\frown\frown\underset{O}{\overset{a}{\underset{|}{C}}}Ar} \overset{F-C}{\Longrightarrow} Ph\frown\frown\frown CO_2H$$

$$\overset{FGA}{\Longrightarrow} \underset{(11)}{Ph\overset{O}{\overset{\|}{C}}\frown\frown CO_2H} \overset{F-C}{\Longrightarrow} PhH + \text{(succinic anhydride)}$$

This is clearly a linear approach. Disconnection on the other side of the carbonyl group leads to a convergent synthesis.
Analysis 2

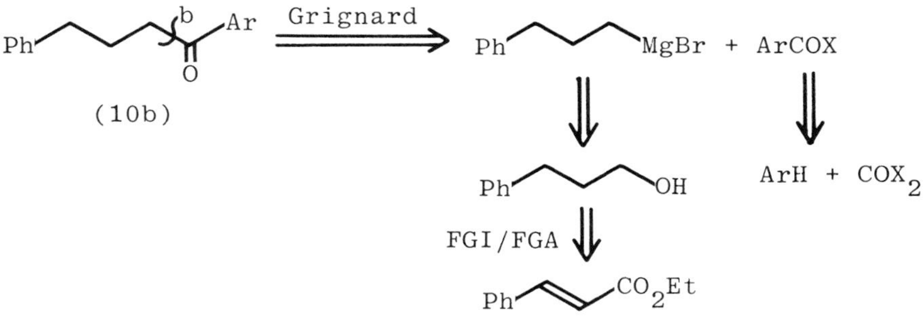

$$Ph\frown\frown\frown\underset{(10b)}{\overset{O}{\overset{\|}{C}}}^{b}Ar \overset{Grignard}{\Longrightarrow} Ph\frown\frown\frown MgBr + ArCOX$$

$$\Downarrow \qquad\qquad \Downarrow$$

$$Ph\frown\frown\frown OH \qquad ArH + COX_2$$

$$FGI/FGA \Downarrow$$

$$Ph\frown CO_2Et$$

There is a short cut in the synthesis. The cyanoamine (13), available from aldehyde (12), reacts with a Grignard reagent with loss of cyanide to give (9) directly.

Convergent Synthesis[454]

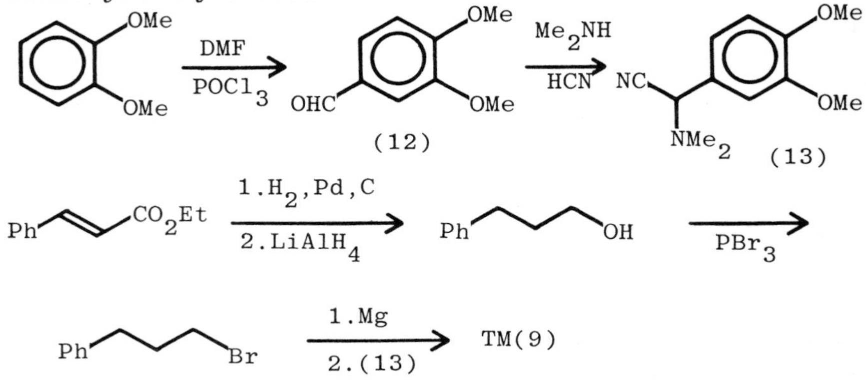

[Scheme showing: 1,2-dimethoxybenzene + DMF/POCl$_3$ → (12) aldehyde with OMe, OMe; then Me$_2$NH, HCN → (13) with NC, NMe$_2$ groups]

[Scheme: Ph-CH=CH-CO$_2$Et → (1. H$_2$, Pd,C; 2. LiAlH$_4$) → Ph-CH$_2$CH$_2$CH$_2$-OH → PBr$_3$ →]

[Ph-CH$_2$CH$_2$CH$_2$-Br → (1. Mg; 2. (13)) → TM(9)]

Problem :

Where would you make the first disconnection to devise a convergent synthesis for (14), whose epoxide is the cecropia juvenile hormone? What reaction and what starting materials might you use?

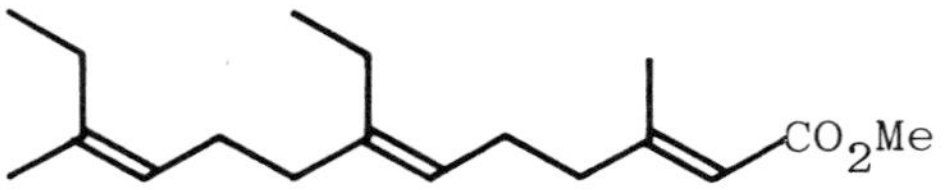

[Structure of (14): methyl ester of a triene with ethyl substituents]

(14)

Answer :

A Wittig disconnection of the central of the three double bonds is the obvious place to start, though any disconnection in this region of the molecule makes sense if you have found a reaction to go with it. Two sets of starting materials could be used.

Analysis

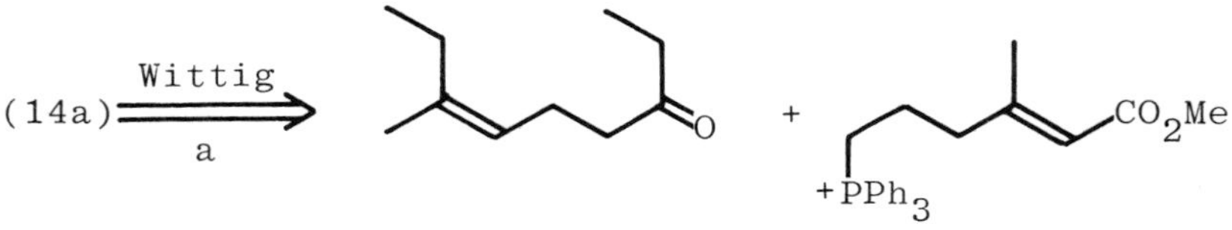

(14a) ⇒ (Wittig a) [ketone fragment] + [phosphorane fragment with CO$_2$Me] + PPh$_3$

(14b) $\xrightarrow{\text{Wittig}}_{b}$ (15) [alkyl-PPh$_3^+$] + OHC–...–CO$_2$Me (16)

Further development

Of the ten published syntheses of juvenile hormone,[455] only one uses this disconnection first, though many use it later in a more linear approach. These workers used route (b), made (15) by alkylation of another phosphonium salt, and used a reconnection strategy for (16).

Analysis of (15)

[structure with PPh$_3^+$] $\Rightarrow$ [structure-Br] + [structure-PPh$_3^+$]

Analysis of (16)

(16) $\xrightarrow{\text{reconnect}}$ (17) [structure-CO$_2$Me] $\xrightarrow{\text{Wittig}}$

(18) [ketone structure] + Br–CH$_2$–CO$_2$Me

Ketone (18) is a familiar starting material (Chapter T 1). The first Wittig reaction gave mostly E-(17) but the second, to give (14), is not very stereoselective so that a mixture of isomers was obtained.

Synthesis[4,5,5]

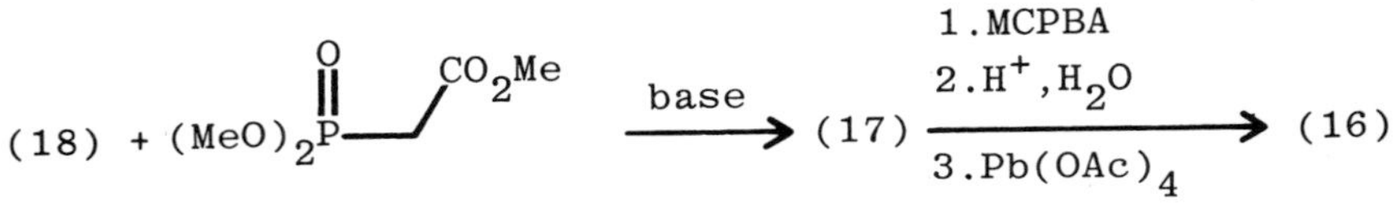

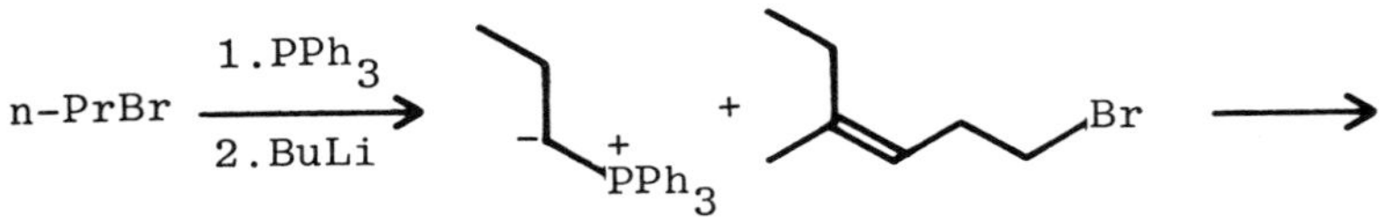

(18) + (MeO)$_2$P(O)CH$_2$CO$_2$Me $\xrightarrow{\text{base}}$ (17) $\xrightarrow[\text{3.Pb(OAc)}_4]{\text{1.MCPBA; 2.H}^+,\text{H}_2\text{O}}$ (16)

n-PrBr $\xrightarrow[\text{2.BuLi}]{\text{1.PPh}_3}$ [ylide] + [allylic bromide] $\longrightarrow$

(15) $\xrightarrow[\text{2.(16)}]{\text{1.BuLi}}$ TM(14)

Example :

Compound (19) is an advanced intermediate in the synthesis of the ring system (20) of fusidic acid, a steroidal antibiotic.

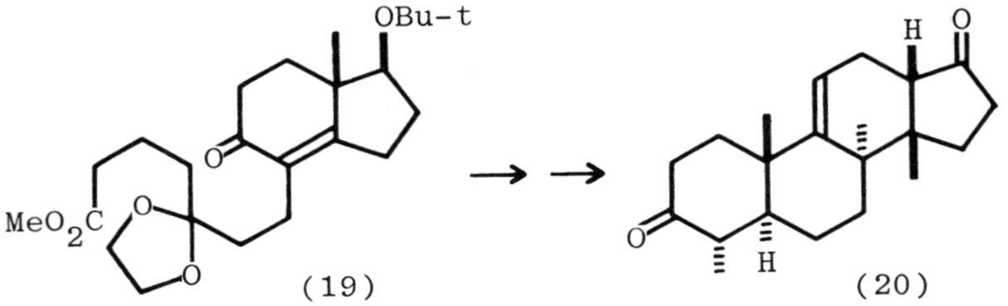

(19) $\rightarrow \rightarrow$ (20)

The ideal disconnection strategically is (19a) which separates the ring from the chain. This corresponds to the α-alkylation of enone (22) (as discussed on p 434) with alkyl halide (21).

Analysis

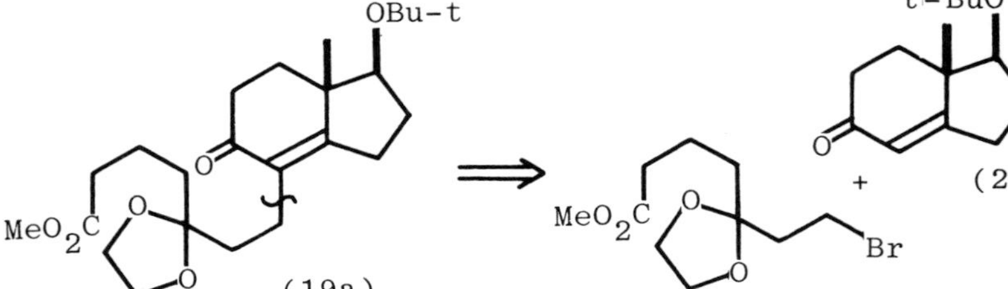

(19a) $\Rightarrow$ (21) + (22)

Disconnection of (21), after removal of the ketal, can lead back by aliphatic Friedel;Crafts reaction to anhydride (24). Bicyclic (22) is clearly derived from Robinson annelation product (25).

Analysis 2

(21) $\overset{FGI}{\Longrightarrow}$ MeO$_2$C—CO—⋯—Br $\Longrightarrow$ MeO$_2$C—⋯—COCl (23)   anhydride (24)

(22) $\Longrightarrow$ $\Longrightarrow$ (25) $\overset{Robinson}{\Longrightarrow}$ methyl vinyl ketone + (26)

Reduction of (25) with sterically demanding reducing agent (27) gives alcohol (28) in good yield. The convergent step gives only moderate yields of (19), but this is acceptable as it accomplishes so much.

Synthesis[4,5,6]

(26) $\xrightarrow{\text{base}}$ (25) $\xrightarrow[73\%]{\text{LiAlH(OBu-t)}_3 \text{ (27)}}$ (28) $\xrightarrow[H^+]{\text{CH}_2=\text{C(CH}_3\text{)}_2}$ (22)

(24) $\xrightarrow[\text{2. SOCl}_2]{\text{1. MeOH}}$ (23) $\xrightarrow[\text{2. HO⋯OH, H}^+]{\text{1. CH}_2=\text{CH}_2, \text{AlBr}_3}$ (21) $\xrightarrow[\text{DMSO}]{(22)\;\text{NaH}}$ TM(19) 50%

Key Reaction Strategy

Problem :

Carry out the first two standard disconnections on compound (28).

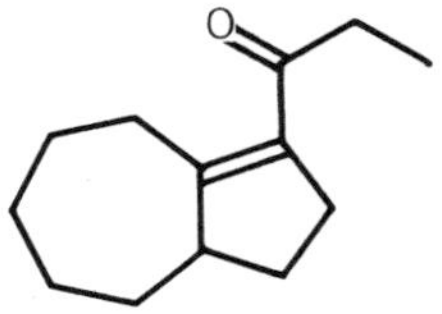

(28)

Answer :
Enone disconnection (28a) and reconnection of 1,6-diketone (29) gives (30).
Analysis

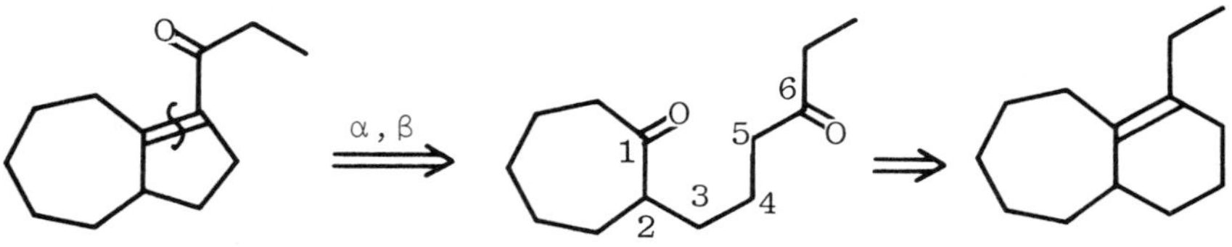

(28a) (29) (30)

Consider an FGA strategy for (30) in which you add a carbonyl group somewhere in the molecule so that it can be made by a key reaction from simple starting materials.

Answer :
Addition of, say, a CO_2Et group would allow Diels-Alder disconnection leading eventually to cycloheptanone. A better strategy is to add a carbonyl group to make an enone (31) with Robinson annelation as the key reaction.
Analysis

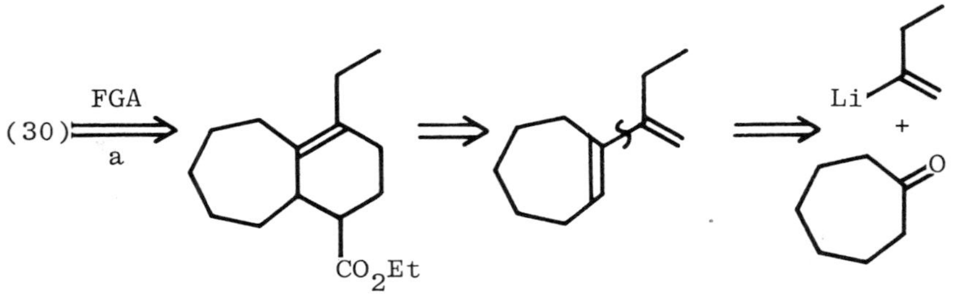

(30) $\overset{FGA}{\underset{a}{\Longrightarrow}}$

(30) 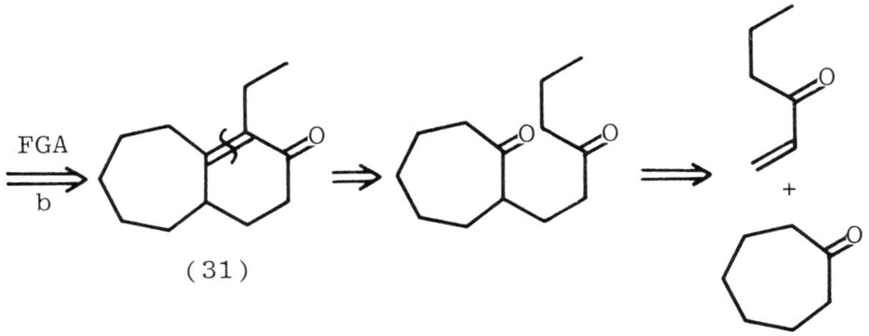 ⇒ᵇ^(FGA) [structure 31] ⇒ [structure] ⇒ [vinyl ketone] + [cycloheptanone]

(31)

One published synthesis follows strategy (b) but adds the ethyl group later by alkylation of enone (32) in the manner of TM (19).

Synthesis[457]

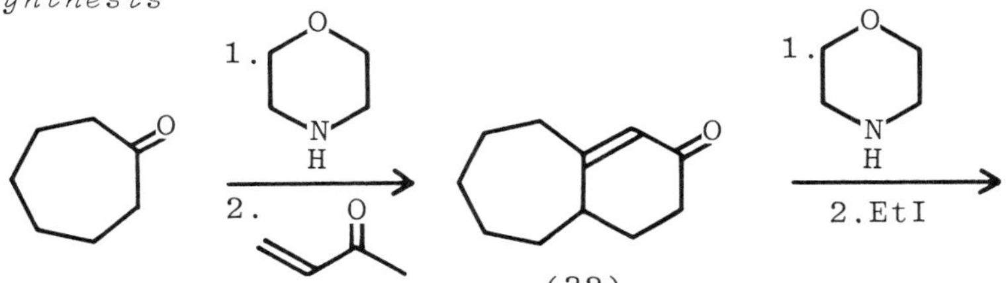

[cycloheptanone] →(1. morpholine; 2. methyl vinyl ketone) [structure 32] →(1. morpholine; 2. EtI)

(31) →(1. LiAlH$_4$; 2. Ac$_2$O, pyr; 3. Li, NH$_3$) (30) →(1. O$_3$; 2. MeO⁻) TM(28)

Example :

The synthesis of lycorane (33) is discussed in Chapter T 40. A recent synthesis of lycorane also uses the Diels-Alder as a key reaction but approaches it

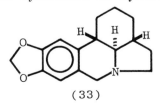

(33)

in a different way. There is no preliminary disconnection of the other rings : a double FGA to give (34) reveals the key reaction from (35).

Analysis

(33) $\xrightarrow{FGA}$ (34) $\xRightarrow{D-A}$ (35)

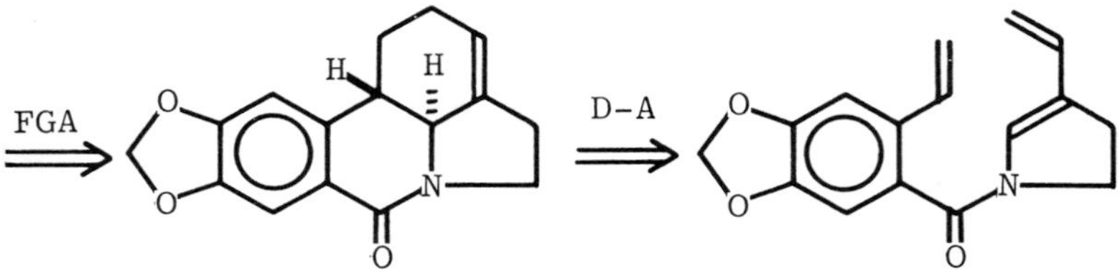

This simple strategy was not used before because (34) is the *exo* adduct from (35), and it was supposed that the *endo* adduct would be preferred. However, Stork[458] examined the reaction carefully and realised that the *endo* transition state would be impossibly crowded (cf p 420). Heating (35) to 140°C does indeed give (34) which is hydrogenated from above to give (36).

Synthesis[458]

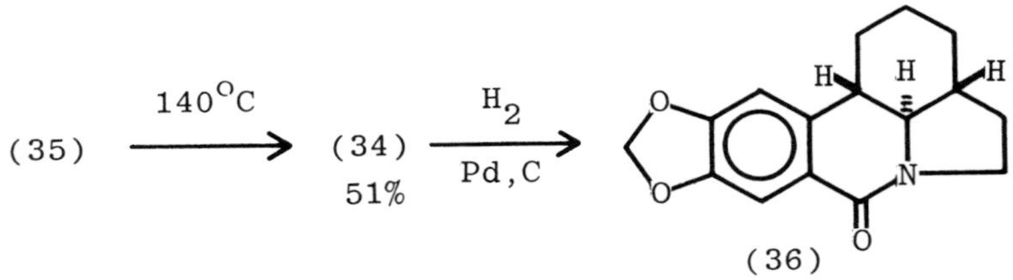

(35) $\xrightarrow{140°C}$ (34) 51% $\xrightarrow{H_2, Pd,C}$ (36)

Revision Problem :
Amide (35) is made from acid (37) which is derived by elimination from (38). How would you make (38)?

Analysis

(35) 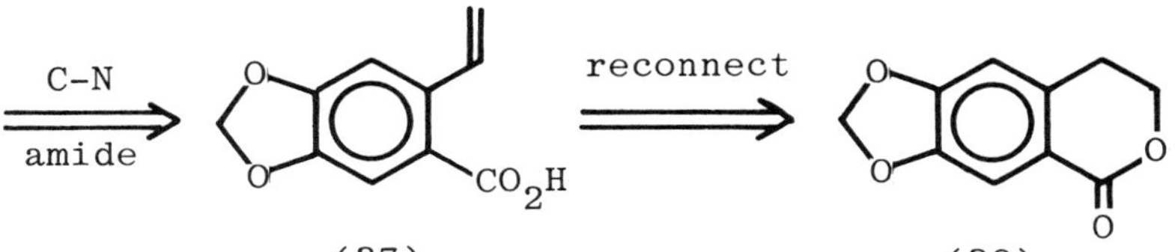 $\xrightarrow[\text{amide}]{\text{C-N}}$ (37) $\xrightarrow{\text{reconnect}}$ (38)

Answer :

This very simple-looking problem is not easy at all. Ester disconnection gives (39) which we might analyse by keeping the two aryl oxygens and removing the rest.

Analysis 1

(38) 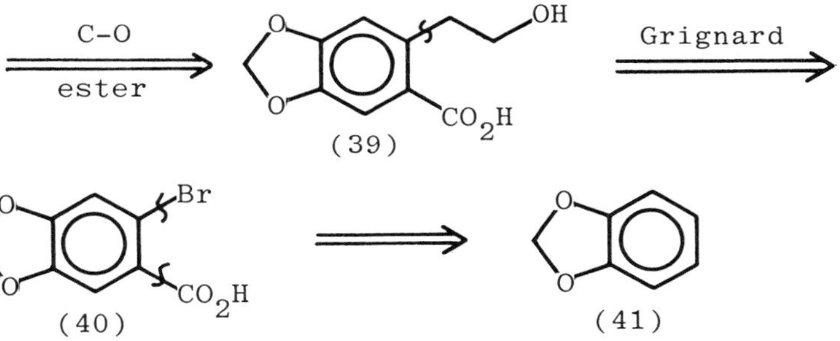 $\xrightarrow[\text{ester}]{\text{C-O}}$ (39) $\xrightarrow{\text{Grignard}}$

(40) $\Longrightarrow$ (41)

But how do we convert (41) to (40)? Is there an order of events which will ensure the right isomer? Adding a one-carbon electrophile, e.g. to give (42), is easy, but bromination could now occur anywhere.

Synthesis

(41) 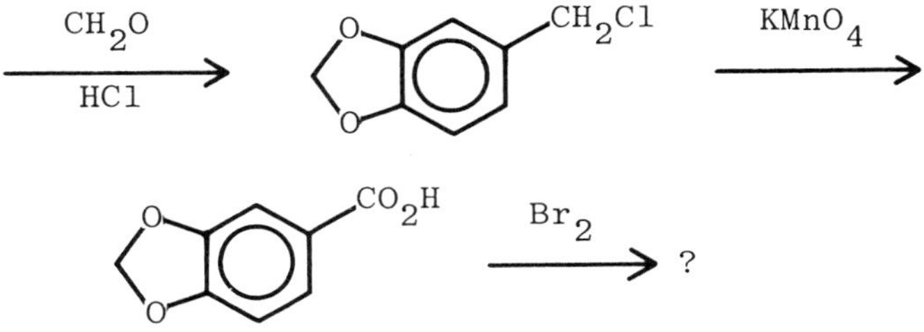 $\xrightarrow[\text{HCl}]{\text{CH}_2\text{O}}$ → $\xrightarrow{\text{KMnO}_4}$

(42) $\xrightarrow{\text{Br}_2}$?

Any carefully thought out synthesis on these lines might well work, but an alternative is to condense (42) with chloral to give (43) and hence (44), in which the new groups must be *ortho*, because they form a five-membered ring.

Synthesis[459]

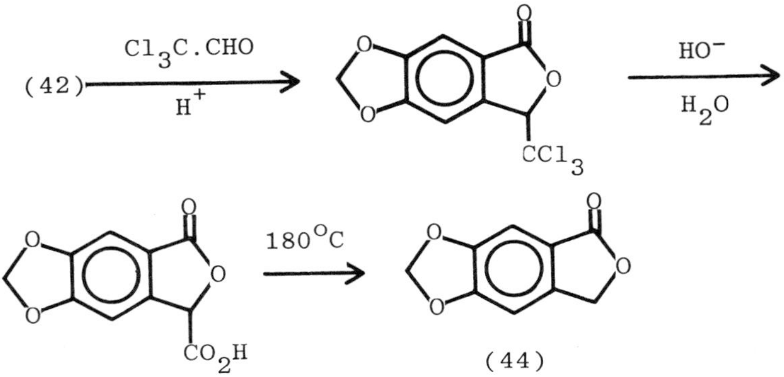

An alternative is to brominate (41) first, and add the two carbon side chains next, to give (45) which will add formaldehyde to give (46).

Synthesis[460,461]

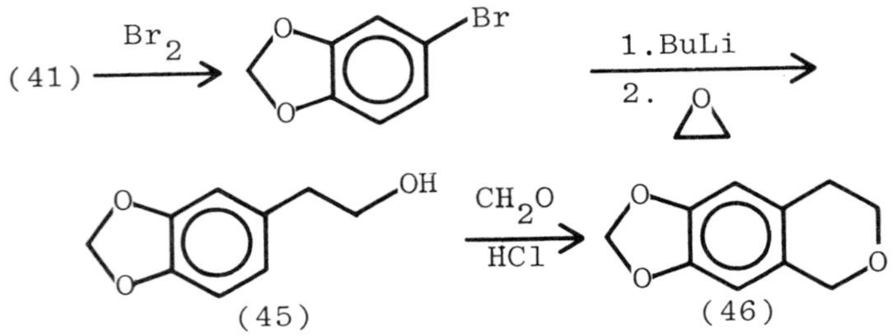

Problem :

Using the formation of (44) or (46) as a key reaction, develop another synthesis of (38).

Answer :

Development of (46) is easier because it needs only to be functionalised at the unique carbon atom between the

oxygen atom and the benzene ring. Experiments showed that NBS (Chapter T 24) is the right reagent.
Synthesis[461]

$$(46) \xrightarrow[\substack{(PhCO_2)_2 \\ CCl_4}]{NBS} TM(38) \quad 74\%$$

The simplest route from (44) is to react with cyanide ion to give (47) and convert this to (46) before using the same reaction.
Synthesis[461,462]

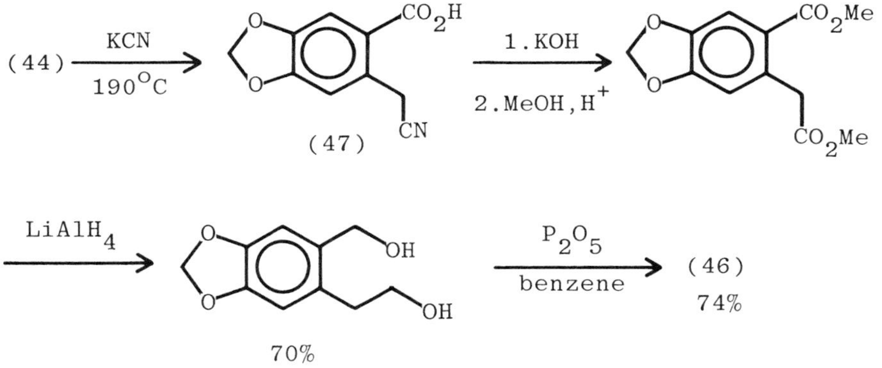

(44) $\xrightarrow[190°C]{KCN}$ (47) [methylenedioxybenzene with CO$_2$H and CH$_2$CN groups] $\xrightarrow[\text{2.MeOH,H}^+]{1.KOH}$ [methylenedioxybenzene with CO$_2$Me and CH$_2$CO$_2$Me groups]

$\xrightarrow{LiAlH_4}$ [methylenedioxybenzene with CH$_2$OH and CH$_2$CH$_2$OH groups] 70% $\xrightarrow[\text{benzene}]{P_2O_5}$ (46) 74%

Problem :

Identify the most strategically important aspect of TM (48) and disconnect it to an intermediate which could be made by the key reaction most appropriate for that key feature. You may be able to suggest a synthesis for (48) without such elaborate thinking.

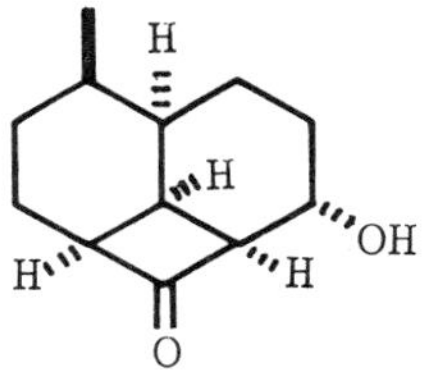

(48)

Answer :

The cyclobutanone is the important aspect. We shall want to make it by ketene cycloaddition (the key reaction) (Chapter T 33) and this we can do after an aldol disconnection.

Analysis

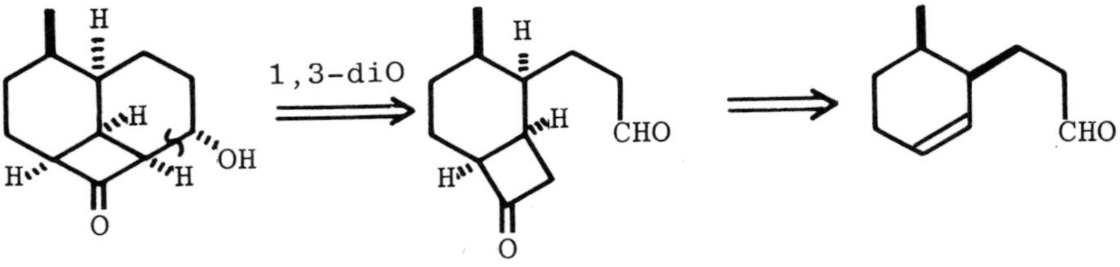
1,3-diO
CHO
CHO
OH

(48a)

In practice simple alkenes give poor cycloaddition so diene (49) was the chosen starting material and the aldol was protected and hydrogenated to give (48). Note that (50) is emphatically bowl shaped (Chapter T 34) so that substituents (OR) or reagents (H_2) prefer to be on the outside of the bowl. The regioselectivity of the addition of $Cl_2C{=}C{=}O$ to the diene (49) is correct.

Synthesis[463]

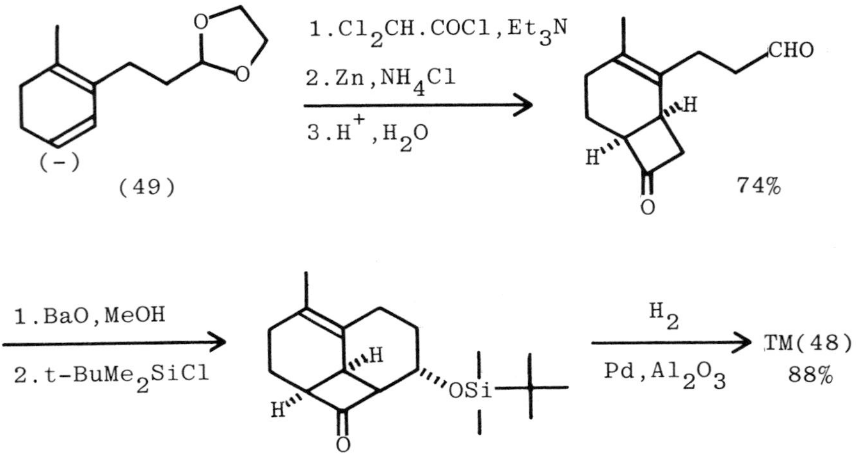
1.$Cl_2CH.COCl, Et_3N$
2.Zn, NH_4Cl
3.H^+, H_2O
(-)
(49)
CHO
74%
1.BaO,MeOH
2.t-$BuMe_2SiCl$
(50) 70%
H_2
Pd, Al_2O_3
TM(48)
88%

Strategy of Available Starting Materials
Problem :
Propanolol (51) is a very important heart drug (beta blocker). Suggest a synthesis.

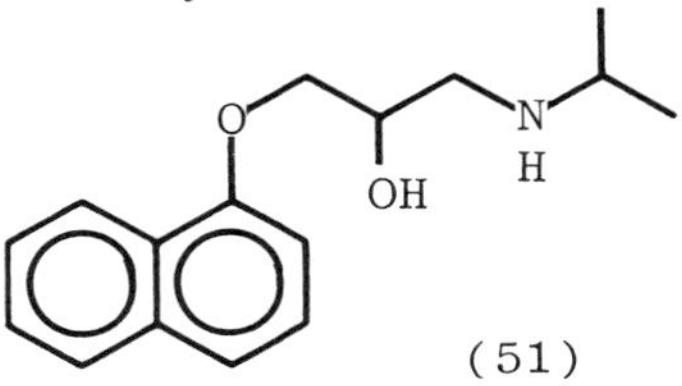

(51)

Answer :
The skeleton of epichlorhydrin (52) is clear (● in 51a). In one published ICI synthesis the phenol is added before the amine.

Analysis

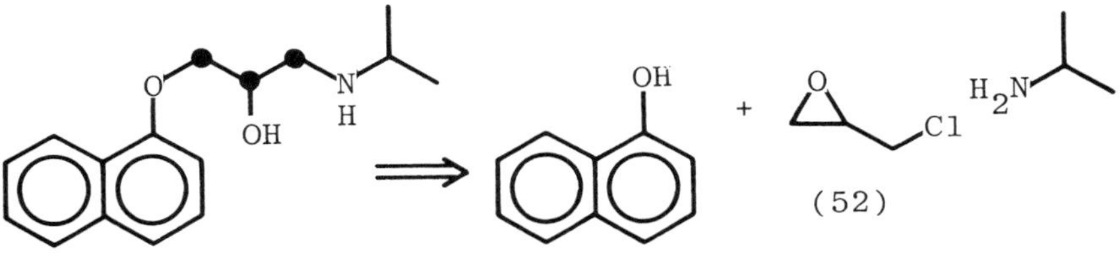

(52)

Synthesis[464]

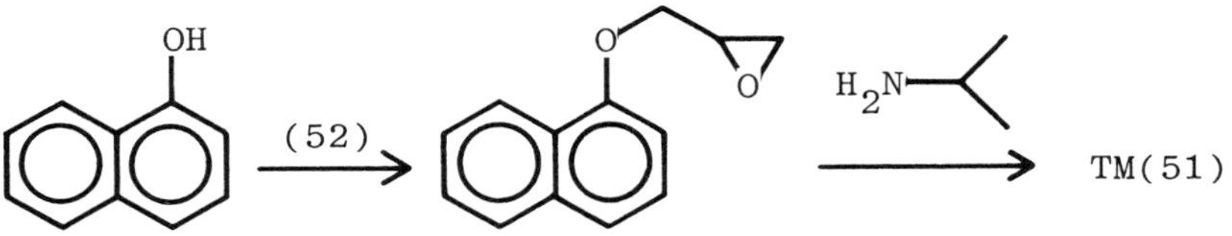

(52) → → TM(51)

Example :
Some starting materials can be unrecognisable even in quite small target molecules. The anti-spasmodic piperidolate (53) is clearly made from (54) which could come from cyclisation of (55). Reconnection of (55) shows that it can be made from furfuraldehyde (56).

Analysis

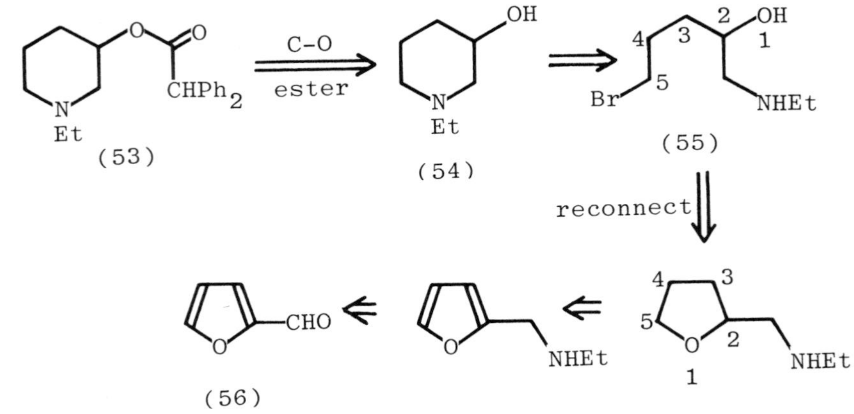

[Structures showing retrosynthetic analysis of (53) → (54) → (55) → tetrahydrofuran intermediate → (56) furfural]

(53), (54), (55), (56)

Synthesis[465]

(56) →[1. EtNH$_2$; 2. H$_2$, cat] (tetrahydrofurfuryl-NHEt) →[HBr] (54) →[Ph$_2$CH.COCl] TM(53)

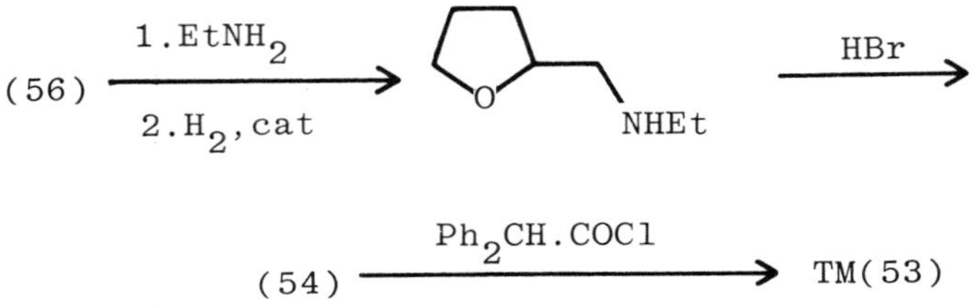

Problem :

Starting materials for TMs(57) and (58) are in Tables T 40.1-3. Attempt to recognise them.

(57) Intal, the Fison's asthma drug.

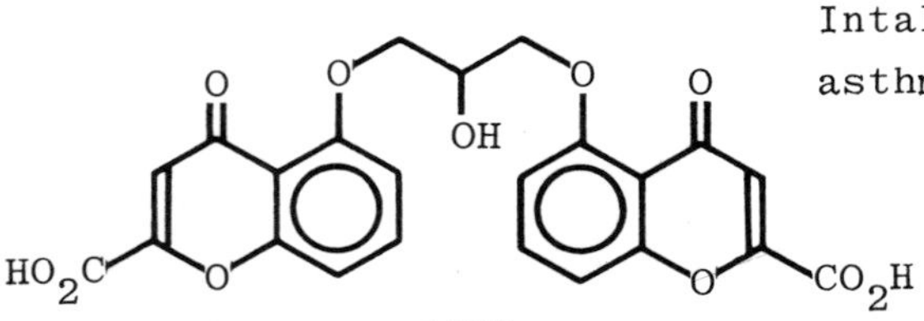

(57)

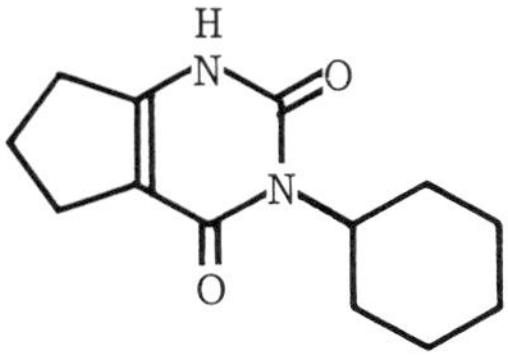

(58) Lenacil, a herbicide which inhibits photosynthesis.

Answer:

Epichlorohydrin is in (57) which is made from an oxalate ester and (59), and hence from (60).

Analysis

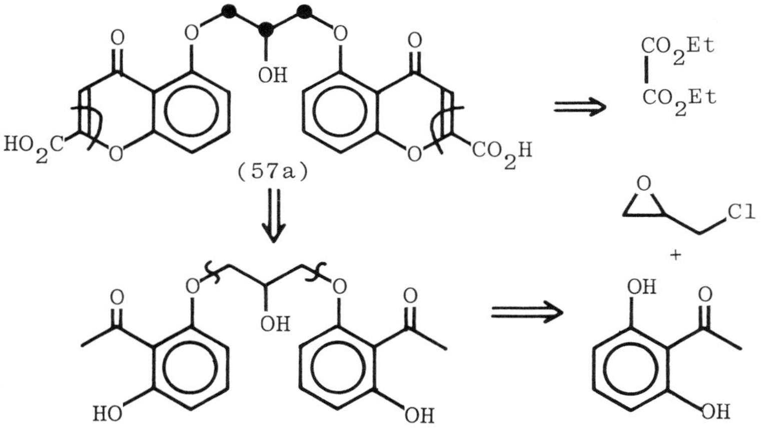

Synthesis[466]

$$(60) \xrightarrow{(52)} (59) \xrightarrow[\text{2. HO}^-, \text{H}_2\text{O}]{\text{1. (CO}_2\text{Et)}_2 \text{ base}} \text{TM(57)}$$

Removal of the two heteroatoms from (58) leaves keto ester (61) - also readily available.

Analysis

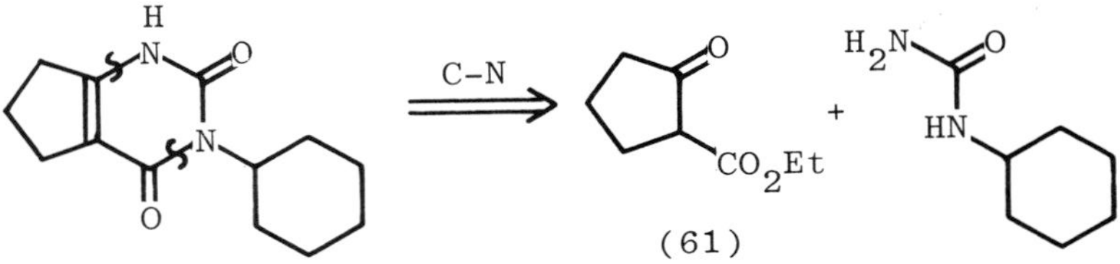

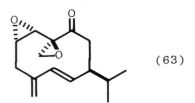

(58) ⇒ (61) + (62)

Synthesis[467]

(61) + (62) —H$_3$PO$_4$ / dioxan, benzene→ TM(58)

General Example :

One of the most remarkable recent achievements was the synthesis of periplanone-(B) (63) the sex excitant of the American cockroach.[468] At the time the synthesis was planned, the stereochemistry of periplanone was unknown and even the gross structure was in doubt. W.C. Still synthesised three diastereoisomers before he found the one identical with the natural product.

His approach was to remove both epoxides to leave alkene and ketone and to mask the exocyclic double bond and the original ketone-giving intermediate (64) - so that he could use a Cope rearrangement from (65) as a key reaction.

Analysis

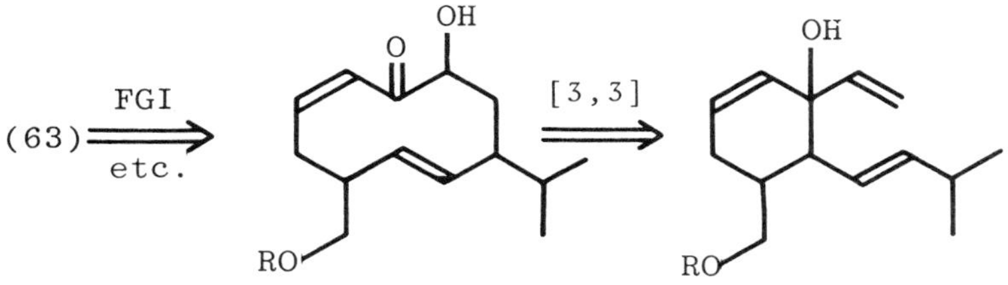

(63) $\xrightarrow{\text{FGI} \atop \text{etc.}}$ (64) $\xRightarrow{[3,3]}$ (65)

He was guided by the knowledge that (66), a suitable starting material for (65), had been made from available (67) in good yield.[469]

Analysis 2

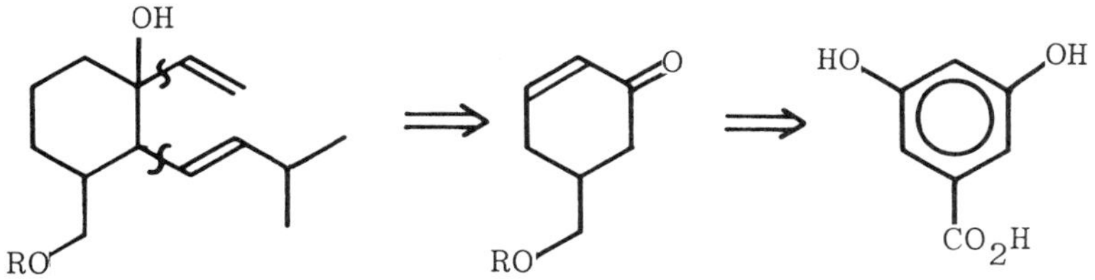

(65a) $\Rightarrow$ (66) $\Rightarrow$ (67)

The synthesis is too advanced to give here but it was remarkably economical and stereoselective, considering the complexity of (63). Still was able to carry out these three syntheses largely because of his confidence in his stereochemical and synthetic predictions.

The Final Problem :

Semburin and *iso*-semburin[470] are epimers of structure (68). They may be of pharmaceutical interest as the Japanese herb 'semburi' in which they occur is used in folk medicine for stomach complaints. Nearer home, γ-muurolene[471] (69) has been identified in gin and may even be necessary for the flavour of a good gin.

As far as I am aware, neither of these compounds has been synthesised. Can you devise routes? Stereochemical control is essential for (69) but the chiral centre to which the vinyl group is attached in (68) is unimportant as both epimers are natural products.

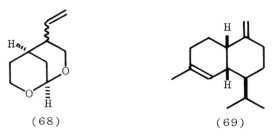

(68)   (69)

No answer is available at the time of writing so your proposals can be challenged only as a result of laboratory trial.

REFERENCES

General References are given in full in the main text.

1. W. J. Elliott and J. Fried, *J. Org. Chem.*, 1976, 41, 2475.
2. W. E. Gore, G. T. Pearce and R. M. Silverstein, *J. Org. Chem.*, 1975, 40, 1705.
3. R. L. Metcalf, I. P. Kapoor and A. S. Hirwe, *Bull. WHO*, 1971, 44, 363., *Chem. Abstr.*, 1971, 75, 97522.
4. E. A. Prill, A. Hartzell and J. M. Arthur, *Science*, 1945, 101, 464.
5. C. G. Brown and B. R. Sanderson, *Chem. Ind. (London)*, 1980, 68.
6. H. C. Brown and B. C. Subba Rao, *J. Am. Chem. Soc.*, 1956, 78, 2582.
7. E. P. Di Bella, *U.S. Pat.*, 3,000,975 (1959); *Chem. Abstr.*, 1962, 56, 5884 e.
8. *Vogel*, p. 621, 660, 698, 821.
9. *Perfumes*, p. 8.
10. N. Yoneda, Y. Takahashi and A. Suzuki, *Chem. Lett.*, 1978, 231.
11. W. S. Calcott, J. M. Tinker and V. Weinmayr, *J. Am. Chem. Soc.*, 1939, 61, 1010.
12. C. C. Price, *Org. React.*, 1946, 3, 38.
13. G. A. Olah, Friedel-Crafts and Related Reactions, III/1, p. 107, Interscience, New York, 1964.
14. G. G. I. Moore and A. R. Kirk, *J. Org. Chem.*, 1979, 44, 925.
15. K. Friedrich and H. Öster, *Chem. Ber.*, 1961, 94, 834.
16. R. Breslow, M. F. Czarniecki, J. Emert and H. Hamaguchi, *J. Am. Chem. Soc.*, 1980, 102, 762.
17. A. Higginbottom, P. Hill and W. F. Short, *J. Chem. Soc.*, 1937, 263.
18. C. S. Gibson and J. D. A. Johnson, *J. Chem. Soc.*, 1929, 1229.

19. J. H. Simons, S. Archer and H. J. Passino, *J. Am. Chem. Soc.*, 1938, **60**, 2956.
20. R. Kirchlechner and W. Rogalski, *Tetrahedron Lett.*, 1980, 247.
21. R. A. Benkeser and G. Schroll, *J. Am. Chem. Soc.*, 1953, **75**, 3196.
22. H. van Erp, *J. Prakt. Chem.*, 1930, **127**, 20.
23. H.-J. Teuber and K. Schnee, *Chem. Ber.*, 1958, **91**, 2089.
24. J. R. J. Pare and J. Berlanger, *Synth. Commun.*, 1980, **10**, 711.
25. E. Schraufstätter, W. Meiser and R. Gönnert, *Z. Naturforsch., Teil B*, 1961, **16**, 95; *Pesticide Manual*, p. 381.
26. L. Varnholt, *J. Prakt. Chem.*, 1887, **(2) 36**, 16.
27. *Vogel*, p. 661; C. R. Noller and P. Liang, *J. Am. Chem. Soc.*, 1932, **54**, 670.
28. N. V. Sidgwick and H. E. Rubie, *J. Chem. Soc.*, 1921 1013; C. W. N. Holmes and J. D. Loudon, *Ibid.*, 1940 1521.
29. *Pesticides*, p. 114, *Pesticide Manual*, p. 78.
30. M. P. Mertes, P. E. Hanna and A. A. Ramsey, *J. Med. Chem.*, 1970, **13**, 125.
31. R. E. Harmon, B. L. Jensen, S. K. Gupta and J. D. Nelson, *J. Org. Chem.*, 1970, **35**, 825.
32. *Pesticide Manual*, p. 32.
33. W. E. Conrad and S. M. Dec, *J. Org. Chem.*, 1958, **23**, 1700.
34. A. J. Ewins, *J. Chem. Soc.*, 1912, 544.
35. O. Kamm and J. B. Segur, *Org. Synth. Coll.*, 1941, **1**, 372, 391.
36. F. M. Beringer and S. Sands, *J. Am. Chem. Soc.*, 1953, **75**, 3319.
37. J. Büchi, E. Stünzi, M. Flury, R. Hirt, P. Labhart and L. Ragaz, *Helv. Chim. Acta*, 1951, **34**, 1002.
38. T. Jojima, H. Takeshiba and T. Kinoto, *Bull. Chem. Soc. Jpn.*, 1979, **52**, 2441.
39. R. B. Silverman and M. A. Levy, *J. Org. Chem.*, 1980, **45**, 816.
40. R. A. Micheli, Z. G. Hajos, N. Cohen, D. R. Parrish, L. A. Portland, W. Sciamanna, M. A. Scott and P. A. Wehrli, *J. Org. Chem.*, 1975, **40**, 675.
41. M. Haslanger and R. G. Lawton, *Synth. Commun.*, 1974, **4**, 155; J. A. Marshall and G. A. Flynn, *Ibid.*, 1979, **9**, 123.
42. P. Radlick and H. T. Crawford, *J. Org. Chem.*, 1972, **37**, 1669.
43. K. Adler and H. von Brachel, *Liebigs Ann. Chem.*, 1962, **651**, 141.

44. R. K. Boeckman, D. M. Blum and S. D. Arthur, *J. Am. Chem. Soc.*, 1979, 101, 5060.
45. K. Fujioka, K. Hashmoto and K. Honda, *Jap. Pat.*, 7,426,244 (1974); *Chem. Abstr.*, 1974, 81, 151806.
46. K. J. Divakar, P. P. Sane and A. S. Rao, *Tetrahedron Lett.*, 1974, 399.
47. D. Papa, E. Schwenk, F. Villani and E. Klingsperg *J. Am. Chem. Soc.*, 1948, 70, 3356.
48. G. W. Kinzer, A. F. Fentiman, T. F. Page, R. L. Foltz, J. P. Vite and G. P. Pitman, *Nature (London)*, 1969, 221, 477.
49. *Perfumes*, p. 322.
50. T. G. Clarke, N. A. Hampson, J. B. Lee, J. R. Morley and B. Scanlon, *Can. J. Chem.*, 1969, 47, 1649; J. B. Lee and T. G. Clarke, *Tetrahedron Lett.*, 1967, 415.
51. W. Francke, G. Hindorf and W. Reith, *Angew. Chem., Int. Ed. Engl.*, 1978, 17, 862.
52. *Perfumes*, p. 141; C. Feugeas, *Bull. Soc. Chim. Fr.*, 1964, 1892.
53. L. Claisen, *Ber.*, 1907, 40, 3903.
54. Röhm and Haas Co., *Fr. Pat.*, 822,326 (1937); *Chem. Abstr.*, 1938, 32, 4250.
55. R. A. Smith and C. J. Rodden, *J. Am. Chem. Soc.*, 1937, 59, 2353; O. Kamm and J. H. Waldo, *Ibid.*, 1921, 43, 2223.
56. *Pesticide Manual*, p. 154-5, 301.
57. C. C. Price, M. Hori, T. Parasaran and M. Polk, *J. Am. Chem. Soc.*, 1963, 85, 2278.
58. A. K. Kiang and F. G. Mann, *J. Chem. Soc.*, 1951, 1909.
59. A. S. Hay, J. W. Eustance and H. S. Blanchard, *J. Org. Chem.*, 1960, 25, 616.
60. Tedder and Nechvatal, Vol. 5, p. 419; E. Ador and A. A. Rilliet, *Ber.*, 1879, 12, 2298.
61. H. A. J. Carless and D. J. Haywood, *J. Chem. Soc., Chem. Commun.*, 1980, 657.
62. N. A. Milas, E. Sakal, J. T. Plati, J. T. Rivers, J. K. Gladding, F. X. Grossi, Z. Weiss, M. A. Campbell and H. F. Wright, *J. Am. Chem. Soc.*, 1948, 70, 1597.
63. E. Rouvier, J.-C. Giacomoni and A. Cambon, *Bull. Soc. Chim. Fr.*, 1971, 1717; R. B. Moffett, *J. Org. Chem.*, 1949, 14, 862.
64. S. M. McElvain and T. P. Carney, *J. Am. Chem. Soc.*, 1946, 68, 2592.
65. C. S. Marvel and H. O. Calvery, *Org. Synth. Coll.*, 1941, 1, 533.
66. N. J. Leonard, F. E. Fischer, E. Barthel, J. Figueras and W. C. Wildman, *J. Am. Chem. Soc.*, 1951 73, 2371.

67. Y. Izumi, I. Chibata and T. Itoh, *Angew. Chem., Int. Ed. Engl.*, 1978, **17**, 176.
68. R. Zurflüh, E. N. Wall, J. B. Siddall and J. A. Edwards, *J. Am. Chem. Soc.*, 1968, **90**, 6224; K. H. Dahm, B. M. Trost and H. Roller, *Ibid.*, 1967, **89**, 5292.
69. E. J. Corey, J. A. Katzenellenbogen, N. W. Gilman, S. A. Roman and B. W. Erickson, *J. Am. Chem. Soc.*, 1968, **90**, 5618.
70. E. E. van Tamelan and T. J. Curphey, *Tetrahedron Lett.*, 1962, 121.
71. K. Hiroi and S. Sato, *Chem. Lett.*, 1979, 923.
72. A. Bresson, G. Dauphin, J.-M. Geneste, A. Kergomard and A. Lacourt, *Bull. Soc. Chim. Fr.*, 1970, 2432.
73. *Pesticide Manual*, p. 360.
74. C. Vogel and R. Aebi, *Swiss Pat.*, 581,607 (1976); *Chem. Abstr.*, 1977, **86**, 72242.
75. W. Reeve and A. Sadle, *J. Am. Chem. Soc.*, 1950, **72**, 1251.
76. P. E. Peterson and F. J. Slama, *J. Org. Chem.*, 1970, **35**, 529.
77. P. L. Southwick, D. I. Sapper and L. A. Pursglove, *J. Am. Chem. Soc.*, 1950, **72**, 4940.
78. S. Terashima, M. Wagatsuma and S.-I. Yamada, *Chem. Pharm. Bull.*, 1970, **18**, 1137.
79. A. W. Ingersoll, *Org. Synth. Coll.*, 1943, **2**, 506; J. C. Robinson and H. R. Snyder, *Ibid.*, 1953, **3**, 71
80. *Pesticide Manual*, p. 61.
81. A. R. Surrey and M. K. Rukwid, *J. Am. Chem. Soc.*, 1955, **77**, 3798.
82. P. D. Kennewell, S. S. Matharu, J. B. Taylor and P. G. Sammes, *J.C.S. Perkin I*, 1980, 2542.
83. F.-H. Marquardt and S. Edwards, *J. Org. Chem.*, 1972 **37**, 1861; L. I. Krimen and D. J. Cota, *Org. React.*, 1969, **17**, 213.
84. R. T. Coutts, A. Benderly, A. L. C. Mak and W. G. Taylor, *Can. J. Chem.*, 1978, **56**, 3054.
85. H. Spanig and C. Schuster, *U.S. Pat.*, 2,623,880 (1952); *Chem. Abstr.*, 1953, **47**, 11203i; K. A. Schellenberg, *J. Org. Chem.*, 1963, **28**, 3259.
86. H. A. Bruson, *Org. React.*, 1949, **5**, 79.
87. R. Baltzly, W. S. Ide and J. S. Buck, *J. Am. Chem. Soc.*, 1942, **64**, 2231.
88. A. Kleeman, W. Leuchtenberger, J. Martens and H. Weigel, *Angew. Chem., Int. Ed. Engl.*, 1980, **19**, 627.
89. M. J. S. A. Amaral, *J. Chem. Soc.* (C), 1969, 2495.
90. J. Grimshaw, *J. Chem. Soc.*, 1965, 7136.
91. C. Harries and O. Schauwecker, *Ber.*, 1901, **34**, 2981.

92. B. Glatz, G. Helmchen, H. Muxfeldt, H. Porcher, R. Prewo, J. Senn, J. J. Stezowski, R. J. Stojda and D. R. White, *J. Am. Chem. Soc.*, 1979, 101, 2171.
93. J. A. Marshall and J. J. Partridge, *J. Am. Chem. Soc.*, 1968, 90, 1090; *Tetrahedron*, 1969, 25, 2159.
94. P. G. Katsoyannis, A. Tometsko and K. Fukuda, *J. Am. Chem. Soc.*, 1963, 85, 2863.
95. *Perfumes*, p. 69.
96. J. W. Howard, *J. Am. Chem. Soc.*, 1925, 47, 455.
97. *Drug Synthesis*, p. 42.
98. O. M. Halse, *J. Prakt. Chem.*, 1914, (2) 89, 451.
99. A. F. Plate, R. N. Shafran and M. I. Batuev, *Zh. Obshch. Khim.*, 1950, 20, 472; *Chem. Abstr.*, 1950, 44, 7785c.
100. A. S. Hussey, *J. Am. Chem. Soc.*, 1951, 73, 1364.
101. T. G. Clarke, N. A. Hampson, J. B. Lee, J. R. Morley and B. Scanlon, *Tetrahedron Lett.*, 1968, 5685.
102. R. C. Huston and A. H. Agett, *J. Org. Chem.*, 1941, 6, 123.
103. C. T. Lester and J. R. Proffitt, *J. Am. Chem. Soc.*, 1949, 71, 1877.
104. G. Edgar, G. Calingaert and R. E. Marker, *J. Am. Chem. Soc.*, 1929, 51, 1483.
105. *Perfumes*, p. 26. Taub, Wingler and Schulemann, *Ger. Pat.*, 423,544 (1924).
106. C. Weizmann and F. Bergmann, *J. Am. Chem. Soc.*, 1938, 60, 2647.
107. F. Ullmann and R. von Wurstemberger, *Ber.*, 1905, 38, 4105.
108. M. Tuot and M. Guyard, *Bull. Soc. Chim. Fr.*, 1947, 1086; *Chem. Abstr.*, 1948, 42, 5833f.
109. R. C. Huston and H. E. Tiefenthal, *J. Org. Chem.*, 1951, 16, 673; T.-T. Chu and C. S. Marvel, *J. Am. Chem. Soc.*, 1931, 53, 4449.
110. J. Werner and M. T. Bogert, *J. Org. Chem.*, 1939, 3, 578.
111. R. C. Elderfield, B. M. Pitt and I. Wempen, *J. Am. Chem. Soc.*, 1950, 72, 1334.
112. R. L. Letsinger and A. W. Schnizer, *J. Org. Chem.*, 1951, 16, 704.
113. W. H. Pirkle and P. E. Adams, *J. Org. Chem.*, 1979, 44, 2169.
114. K. Q. Do, P. Thanei, M. Caviezel and R. Schwyzer, *Helv. Chim. Acta*, 1979, 62, 956.
115. P. N. Confalone, G. Pizzolato, E. G. Baggiolini, D. Lollar and M. R. Uskovic, *J. Am. Chem. Soc.*, 1975, 97, 5936.

116. B. D. Johnston and K. N. Slessor, *Can. J. Chem.*, 1979, **57**, 233.
117. B. Seuring and D. Seebach, *Helv. Chim. Acta*, 1977, **60**, 1175; R. G. Ghirardelli, *J. Am. Chem. Soc.*, 1973, **95**, 4987.
118. K. D. Paranjape, N. L. Phalnikar, B. V. Bhide and K. S. Nargund, *Nature (London)*, 1944, **153**, 141.
119. J. W. Cornforth, R. H. Cornforth and M. J. S. Dewar, *Nature (London)*, 1944, **153**, 317; J. M. O'Gorman, *J. Am. Chem. Soc.*, 1944, **66**, 1041; G. R. Clemo, W. Cocker and S. Hornsby, *J. Chem. Soc.*, 1946, 616.
120. R. C. Dougherty, *J. Am. Chem. Soc.*, 1980, **102**, 380.
121. C. A. Mead and A. Moscowitz, *J. Am. Chem. Soc.*, 1980, **102**, 7301; A. Peres, *Ibid.*, 7389.
122. G. D. Malpass, R. A. Palmer and R. G. Ghirardelli, *Tetrahedron Lett.*, 1980, 1489.
123. G. Stork and E. W. Logusch, *Tetrahedron Lett.*, 1979, 3361.
124. B. M. Trost, J. M. Timko and J. L. Stanton, *J. Chem. Soc., Chem. Commun.*, 1978, 436.
125. A. S. Hussey and R. R. Herr, *J. Org. Chem.*, 1959, **24**, 843.
126. C. W. Alexander, M. S. Hamdam and W. R. Jackson, *J. Chem. Soc., Chem. Commun.*, 1972, 94.
127. *Perfumes*, p. 149.
128. C. R. Hauser and W. J. Humphlett, *J. Org. Chem.*, 1950, **15**, 359.
129. M. Muskatirovic and V. Krstic, *Liebigs Ann. Chem.*, 1976, 1964.
130. B. E. Hudson and C. R. Hauser, *J. Am. Chem. Soc.*, 1940, **62**, 2457.
131. Y. Yamase, *Bull. Chem. Soc. Jpn.*, 1961, **34**, 480; L. L. Abell, W. F. Bruce and J. Seifter, *U.S. Pat.*, (1952), 2,590,079; *Chem. Abstr.*, 1952, **46**, 10200e.
132. C. S. Marvel and F. D. Hager, *Org. Synth. Coll.*, 1932, **1**, 248.
133. J. R. Johnson and F. D. Hager, *Org. Synth. Coll.*, 1932, **1**, 351.
134. R. Rossi and P. A. Salvadori, *Synthesis*, 1979, 209.
135. D. St. C. Black and J. E. Doyle, *Aust. J. Chem.*, 1978, **31**, 2247.
136. W. Kimel and A. C. Cope, *J. Am. Chem. Soc.*, 1943, **65**, 1992.
137. B. K. Wasson, C. H. Gleason, I. Levi, J. M. Parker, L. M. Thompson and C. H. Yates, *Can. J. Chem.*, 1961, **39**, 923.
138. A. J. Birch and Sir R. Robinson, *J. Chem. Soc.*, 1943, 501.

139. E. R. Alexander and G. R. Coraor, *J. Am. Chem. Soc.*, 1951, $\underline{73}$, 2721.
140. A. C. Cope, H. L. Holmes and H. O. House, *Org. React.*, 1957, $\underline{9}$, 107 (see p. 301).
141. J. Levy and P. Jullien, *Bull. Soc. Chim. Fr.*, 1929, $\underline{45}$, 941; *Chem. Abstr.*, 1930, $\underline{24}$, 1104.
142. M.-J. Brienne, C. Ouannes and J. Jacques, *Bull. Soc. Chim. Fr.*, 1967, 613.
143. Ref. 140, p. 297.
144. H. Gilman and J. F. Nelson, *Recl. Trav. Chim. Pays-Bas*, 1936, $\underline{55}$, 518.
145. H. O. House and W. F. Fischer, *J. Org. Chem.*, 1969, $\underline{34}$, 3615.
146. M. Antennis, *Ind. Chim. Belge*, 1960, 485.
147. M. S. Kharasch and O. Reinmuth, 'Grignard Reactions of Nonmetallic Substances', Constable, London, 1954, p. 609.
148. H. O. House and H. W. Thompson, *J. Org. Chem.*, 1963, $\underline{28}$, 360.
149. H. Riviere and J. Tostain, *Bull. Soc. Chim. Fr.*, 1969, 568.
150. H. O. House, W. L. Respess and G. M. Whitesides, *J. Org. Chem.*, 1966, $\underline{31}$, 3128.
151. S. O. Winthrop and L. G. Humber, *J. Org. Chem.*, 1961, $\underline{26}$, 2834.
152. S. Danishefsky and B. L. Migdalof, *J. Chem. Soc., Chem. Commun.*, 1969, 1107.
153. T. Nakajima and S. Suga, *Tetrahedron*, 1969, $\underline{25}$, 1807.
154. D. Ginsberg and R. Pappo, *J. Chem. Soc.*, 1951, 516.
155. H. J. Lucas, *J. Am. Chem. Soc.*, 1929, $\underline{51}$, 248.
156. R. A. Bartsch and J. F. Bunnett, *J. Amer. Chem. Soc.*, 1969, $\underline{91}$, 1382; H. van Risseghem, *Bull. Soc. Chim. Belge*, 1938, $\underline{47}$, 47; *Chem. Abstr.*, 1938, $\underline{32}$, 3753.
157. C. H. DePuy, D. L. Storm, J. T. Frey and C. G. Naylor, *J. Org. Chem.*, 1970, $\underline{35}$, 2746.
158. M. Orchin, *J. Amer. Chem. Soc.*, 1946, $\underline{68}$, 571.
159. A. Klages and S. Heilmann, *Ber.*, 1904, $\underline{37}$, 1447.
160. H. Oediger, F. Möller and K. Eiter, *Synthesis*, 1972, 591.
161. S. Cabiddu, A. Maccioni and M. Secci, *Ann. Chim. (Rome)*, 1962, $\underline{52}$, 1261; *Chem. Abstr.*, 1963, $\underline{59}$, 489g.
162. D. A. Carlson, M. S. Mayer, D. L. Silhacek, J. D. James, M. Beroza and B. A. Bierl, *Science*, 1971, $\underline{174}$, 76.
163. W. Roelops, A. Comeau, A. Hill and G. Milicevik, *Science*, 1971, $\underline{174}$, 297.

164. W. Hoffmann and M. Baumann, *Ger. Pat.*, 2,534,859, (1977); *Chem. Abstr.*, 1977, 86, 189229.
165. H. J. Bestmann, O. Vostrowsky, H. Paulus, W. Billmann and W. Stransky, *Tetrahedron Lett.*, 1977, 121.
166. M. Barbier, E. Lederer and T. Nomura *Compt. Rend.*, 1960, 251, 1133; H. J. Bestmann, R. Kunstmann and H. Schulz, *Liebigs Ann. Chem.*, 1966, 699, 33.
167. E. J. Corey and D. J. Beames, *J. Am. Chem. Soc.*, 1972, 94, 7210.
168. T. L. Jacobs, *Org. React.*, 1949, 5, 1; (see p 55).
169. K. Junkman and H. Pfeiffer, *U.S. Pat.*, 2,816,910 (1957); *Drug Synthesis*, p. 38.
170. N. Gilman and B. Holland, *Synth. Commun.*, 1974, 4, 199.
171. K. E. Schulte and K. P. Reiss, *Angew. Chem., Int. Ed. Eng.*, 1955, 67, 516.
172. A. Butenandt, E. Hecker, M. Hopp and W. Koch, *Liebigs Ann. Chem.*, 1962, 658, 39.
173. H. Disselnkötter and K. Eiter, *Tetrahedron*, 1976, 32, 1591.
174. M. A. Adams, A. J. Duggan, J. Smolanoff and J. Meinwald, *J. Am. Chem. Soc.*, 1979, 101, 5364.
175. I. M. Gverdtsiteli and N. G. Pataraya, *Zh. Obshch. Khim.*, 1949, 19, 1479; *Chem. Abstr.*, 1950, 44, 1038e.
176. A. M. Islam and R. A. Raphael, *J. Chem. Soc.*, 1952, 4086.
177. C. Schorlemmer, *Liebigs Ann. Chem.*, 1872, 161, 263; H. Masson, *Compt. Rend.*, 1909, 149, 630.
178. J. D. Billimoria and N. F. Maclagan, *J. Chem. Soc.*, 1954, 3257.
179. E. de B. Barnett and C. A. Lawrence, *J. Chem. Soc.*, 1935, 1104.
180. N. Galloway and B. Halton, *Aust. J. Chem.*, 1979, 32, 1743.
181. W. J. Greenlee and R. B. Woodward, *J. Am. Chem. Soc.*, 1976, 98, 6075.
182. N. A. Natsinskaya and A. A. Petrov, *J. Gen. Chem. USSR*, 1941, 11, 665; *Chem. Abstr.*, 1941, 35, 6934[1].
183. F. Fringuelli, F. Pizzo and A. Taticchi, *Synth. Commun.*, 1979, 9, 391.
184. O. Diels and K. Alder, *Liebigs Ann. Chem.*, 1929, 470, 62.
185. E. Dane, J. Schmitt and C. Rautenstrauch, *Liebigs Ann. Chem.*, 1937, 532, 29.
186. S. W. Baldwin and R. J. Doll, *Tetrahedron Lett.*, 1979, 3275.
187. F. V. Brutcher and D. D. Rosenfeld, *J. Org. Chem.*, 1964, 29, 3154.

188. D. F. Taber and B. P. Gunn, *J. Am. Chem. Soc.*, 1979, 101, 3992.
189. V. F. Kucherov and E. P. Serebryakov, *Izv. Akad. Nauk SSSR, Ser. Khim.*, 1960, 1057; *Chem. Abstr.*, 1961, 55, 475h.
190. O. N. Jitkow and M. T. Bogert, *J. Am. Chem. Soc.*, 1941, 63, 1979.
191. S. P. Tanis and K. Nakanishi, *J. Am. Chem. Soc.*, 1979, 101, 4398.
192. S. C. Howell, S. V. Ley, M. Mahon and P. A. Worthington, *J. Chem. Soc., Chem. Commun.*, 1981, 507.
193. Ref. 168, p. 57; C. Moureau and R. Delange, *Bull. Soc. Chim. Fr.*, 1903 (3) 29, 648.
194. C. E. Wood and F. Scarf, *J. Soc. Chem. Ind.*, 1923, 42, 13T.
195. B. Wojcik and H. Adkins, *J. Am. Chem. Soc.*, 1934, 56, 2419.
196. J. Cason and F. S. Prout, *J. Am. Chem. Soc.*, 1944, 66, 46.
197. A. E. Vanstone and J. S. Whitehurst, *J. Chem. Soc. (C)*, 1966, 1972.
198. K. Mori, S. Kobayashi and M. Matsui, *Agric. Biol. Chem.*, 1975, 39, 1889.
199. R. Kirchlechner and W. Rogalsk, *Tetrahedron Lett.*, 1980, 247.
200. P. J. Wagner, M. J. Lindstrom, J. H. Sedon and D. R. Ward, *J. Am. Chem. Soc.*, 1981, 103, 3842.
201. C. R. Hauser and B. E. Hudson, *Org. React.*, 1942, 1, 291.
202. R. D. Desai, *J. Chem. Soc.*, 1932, 1079.
203. J. Colonge, R. Falcotet and R. Gaumont, *Bull. Soc. Chim. Fr.*, 1958, 211.
204. A. T. Nielson and W. J. Houlihan, *Org. React.*, 1968, 16, 1; see p. 88.
205. P. L. Southwick, E. P. Previc, J. Casanova and E. H. Carlson, *J. Org. Chem.*, 1956, 21, 1087.
206. J. A. Barltrop and N. A. J. Rogers, *J. Chem. Soc.*, 1958, 2566.
207. Ref. 204, p. 95.
208. K. Bruno and J. Conrad, *Tetrahedron*, 1979, 35, 2523.
209. J. R. Pfister, *Tetrahedron Lett.*, 1981, 1281.
210. Ref. 204, p. 199.
211. V. Dave and J. S. Whitehurst, *J. Chem. Soc., Perkin Trans. 1*, 1973, 393.
212. G. Stork and R. N. Guthikonda, *Tetrahedron Lett.*, 1972, 2755.
213. W. Borsche and W. Menz, *Ber.*, 1908, 41, 190.
214. Ref. 204, p. 115.
215. P. Schuster, O. E. Polansky and F. Wessely, *Monatsh. Chem.*, 1964, 95, 53.

216. Ref. 204, p. 112.
217. V. V. Chelintsev and A. V. Pataraya, *Zh. Obshch. Khim.*, 1941, **11**, 461; *Chem. Abstr.*, 1941, **35**, 6571³.
218. L. N. Stukanova, N. V. Zhdanova, V. I. Epishev and A. A. Petrov, *Neftekhimiya*, 1964, **4**, 521; *Chem. Abstr.*, 1964, **61**, 13204g.
219. G. L. Buchanan and G. A. R. Young, *J. Chem. Soc., Chem. Commun.*, 1971, 643.
220. M. W. Goldberg and P. Müller, *Helv. Chim. Acta*, 1938, **21**, 1699.
221. G. A. Kraus and M. J. Taschner, *J. Am. Chem. Soc.*, 1980, **102**, 1974.
222. Ref. 204 p. 103; M. Lipp and F. Dallacker, *Chem. Ber.*, 1957, **90**, 1730.
223. W. C. Lumma and O. H. Ma, *J. Org. Chem.*, 1970, **35**, 2391.
224. I. T. Harrison, B. Lythgoe and S. Trippett, *J. Chem. Soc.*, 1955, 4016.
225. A. Pohland and H. R. Sullivan, *J. Am. Chem. Soc.*, 1953, **75**, 4458; 1955, **77**, 3400.
226. C. A. Grob, K. Seckinger, S. W. Tam and R. Traber, *Tetrahedron Lett.*, 1973, 3051.
227. C. R. Hauser and B. E. Hudson, *Org. React.*, 1942, **1**, 266; see p. 293; T. Reffstrup and P. M. Boll, *Acta Chem. Scand. Ser. B.*, 1977, **31**, 727.
228. M. J. Bullivant and G. Pattenden, *J. Chem. Soc., Perkin Trans. 1*, 1975, 256.
229. W. B. Trapp and D. E. Pletcher, *U.S. Pat.*, 2,834,800 (1958); *Chem. Abstr.*, 1958, **52**, 15575i.
230. C. R. Hauser, F. W. Swamer and J. T. Adams, *Org. React.*, 1954, **8**, 59.
231. S. Hünig, E. Benzing and E. Lücke, *Chem. Ber.*, 1957, **90**, 2833.
232. R. F. B. Cox and S. M. McElvain, *J. Am. Chem. Soc.*, 1934, **56**, 2459.
233. J. H. Babler, B. J. Invergo and S. J. Sarussi, *J. Org. Chem.*, 1980, **45**, 4241.
234. G. Stork, A. Brizzolara, H. Landesman, J. Szmuszkovicz and R. Terrell, *J. Am. Chem. Soc.*, 1963, **85**, 207.
235. J. E. D. Barton and J. Harley-Mason, *J. Chem. Soc., Chem. Commun.*, 1965, 197.
236. M. E. Kuehne, D. M. Roland and R. Hafter, *J. Org. Chem.*, 1978, **43**, 3705.
237. J. V. Silverton, M. Ziffer and H. Ziffer, *J. Org. Chem.*, 1979, **44**, 3959.
238. c.f. G. Stork and R. N. Guthikonda, *J. Am. Chem. Soc.*, 1972, **94**, 5109.

239. P. A. Grieco, J. A. Noguez, and Y. Masaki, *J. Org. Chem.*, 1977, 42, 495.
240. A. S. Hussey, H. P. Liao, and R. H. Baker, *J. Am. Chem. Soc.*, 1953, 75, 4727.
241. L. S. Minckler, A. S. Hussey and R. H. Baker, *J. Am. Chem. Soc.*, 1956, 78, 1009.
242. A. W. Crossley and W. R. Pratt, *J. Chem. Soc.*, 1915, 171; Ref. 204, p. 178.
243. A. W. Crossley and N. Renouf, *J. Chem. Soc.*, 1915, 602.
244. W. J. Gensler and C. M. Samour, *J. Am. Chem. Soc.*, 1951, 73, 5555.
245. G. Jones, *Org. React.*, 1967, 15, 493-509.
246. J. R. Butterick and A. M. Unrau, *J. Chem. Soc., Chem. Commun.*, 1974, 307.
247. J. Controulis, M. C. Rebstock and H. M. Crooks, *J. Am. Chem. Soc.*, 1949, 71, 2463.
248. H. G. Johnson, *J. Am. Chem. Soc.*, 1946, 68, 12, 14.
249. W. D. Emmons and A. S. Pagand, *J. Am. Chem. Soc.*, 1955, 77, 4557.
250. R. K. Hill, *J. Org. Chem.*, 1957, 22, 830.
251. E. McDonald and R. T. Martin, *Tetrahedron Lett.*, 1977, 1317.
252. Ref. 245 p. 499.
253. W. C. Wildman and R. B. Wildman, *J. Org. Chem.*, 1952, 17, 581.
254. R. Locquin and L. Leers, *Bull. Soc. Chim. Fr.*, 1925, 37, 1113. See A. W. Johnson, 'The Chemistry of the Acetylenic Compounds', Arnold, London, 1946, Vol. 1, p 103.
255. D. Guillerm-Dron, M. L. Capmau and W. Chodkiewicz, *Bull. Soc. Chim. Fr.*, 1973, 1417.
256. H. A. Bruson, F. W. Grant and E. Bobko, *J. Am. Chem. Soc.*, 1958, 80, 3633; Ref. 208 p. 377-378.
257. C. A. Heathcock, M. C. Pirrung, C. T. Buse, J. P. Hagen, S. D. Young and J. E. Sohn, *J. Am. Chem. Soc.*, 1979, 101, 7077.
258. T. E. Bellas, R. G. Brownlee and R. M. Silverstein, *Tetrahedron*, 1969, 25, 5149.
259. E. Ghera, *J. Org. Chem.*, 1970, 35, 660.
260. E. D. Venus-Danilova and E. P. Brichko, *J. Gen. Chem. USSR*, 1947, 17, 1849; *Chem. Abstr.*, 1948, 42, 4160c.
261. *Drug Synthesis*, p. 68.
262. J. Kapfhammer and A. Matthes, *Hoppe-Seyler's Z. Physiol. Chem.*, 1933, 223, 43 ; *Chem. Abstr.*, 1934, 28, 2353[3].
263. H. Scheibler and A. Fischer, *Ber.*, 1922, 55, 2903.
264. D. J. Abbott, S. Colonna and C. J. M. Stirling, *J. Chem. Soc., Chem. Commun.*, 1971, 471.

265. D. R. Gedge and G. Pattenden, *J. Chem. Soc., Perkin Trans. 1*, 1979, 89.
266. J. W. Wilson, C. L. Zirkle, E. L. Anderson, J. J. Stehle and G. E. Ullyot, *J. Org. Chem.*, 1951, 16, 792.
267. P. S. Skell, D. L. Tuleen and P. D. Readio, *J. Am. Chem. Soc.*, 1963, 85, 2580; C. Walling, A. L. Rieger and D. D. Tanner, *ibid.*, 3129; R. E. Pearson and J. C. Martin, *ibid.*, 3142.
268. F. L. Greenwood, M. D. Kellert and J. Sedlack, *Org. Synth. Coll.*, 1963, 4, 108.
269. E. Buchta and F. Andree, *Chem. Ber.*, 1959, 92, 3111.
270. *Vogel*, 3rd Edn. p. 926.
271. Ref. 245 p. 274
272. C. Barkenbus and J. B. Holtzclaw, *J. Am. Chem. Soc.*, 1925, 47, 2189.
273. J. M. Snell and A. Weissberger, *Org. Synth. Coll.*, 1955, 3, 788.
274. Polymer-Supported Reagents in Organic Synthesis, eds. P. Hodge and D. C. Sherrington, Wiley, Chichester, 1980, p. 28.
275. D. L. Tuleen and B. A. Hess, *J. Chem. Educ.*, 1971, 48, 476; R. Broos, D. Tavernier and M. Antennis, *ibid.*, 1978, 55, 813.
276. H. Barbier, *Helv. Chim. Acta*, 1940, 23, 524, 1477.
277. M. M. Baizer, *Chem. Ind. (London)*, 1979, 435; D. E. Danly, *ibid.*, 439.
278. D. D. Coffman, E. L. Jenner and R. D. Lipscomb, *J. Am. Chem. Soc.*, 1958, 80, 2864.
279. N. Katsaros, E. Vrachnov-Astra, J. Konstantatos and C. I. Stassinopoulou, *Tetrahedron Lett.*, 1979, 4319.
280. H.-D. Scharf, H. Plum, J. Fleischhauer and W. Schleker, *Chem. Ber.*, 1979, 112, 862.
281. A. C. Cope, J. W. Barthel and R. D. Smith, *Org. Synth. Coll.*, 1963, 4, 218; W. Reusch and R. LeMahieu, *J. Am. Chem. Soc.*, 1964, 86, 3068.
282. P. E. Verkade, K. S. de Vries and B. M. Wepster, *Recl. Trav. Chim. Pays-Bas*, 1964, 83, 367.
283. A. H. Stuart, A. J. Shukis, R. C. Tallman, C. McCann and G. R. Treves, *J. Am. Chem. Soc.*, 1946, 68, 729.
284. K. J. Shea and R. B. Phillips, *J. Am. Chem. Soc.*, 1980, 102, 3156.
285. S. F. Martin and T. Chou, *J. Org. Chem.*, 1978, 43, 1027.
286. J. Attenburrow, J. Elks, B. A. Hems and K. N. Speyer, *J. Chem. Soc.*, 1949, 510.

287. M. Larchevêque and A. Debal, *Synth. Commun.*, 1980, <u>10</u>, 49.
288. B. Rothstein, *Bull. Soc. Chim. Fr.*, 1935, [5] 2, 80, 1936; *Chem. Abstr.*, 1935, <u>29</u>, 3308^2; *ibid.*, 1936, <u>30</u>, 1742^7.
289. J. Cason and F. S. Prout, *J. Am. Chem. Soc.*, 1944, <u>66</u>, 46.
290. D. St. C. Black, *Tetrahedron Lett.*, 1972, 1331.
291. J. W. Loder, R. Eibl, M. J. Falkiner, R. H. Nearn and R. W. Parr, *Aust. J. Chem.*, 1978, <u>31</u>, 1011.
292. P. H. Carter, J. C. Craig, R. E. Lack and M. Moyle, *Org. Synth. Coll.*, 1973, <u>5</u>, 339.
293. S. Torii, K. Uneyama and K. Okamoto, *Bull. Chem. Soc. Jpn.*, 1978, <u>51</u>, 3590.
294. R. D. Desai and M. A. Wali, *Proc. Indian Acad. Sci., Sect. A*, 1937, <u>6</u>, 144; *Chem. Abstr.*, 1938, <u>32</u>, 5096.
295. R. H. Leonard, *Ind. Eng. Chem.*, 1956, <u>48</u>, 1331.
296. J. A. Edwards, V. Schwarz, J. Fajkos, M. L. Maddox and J. H. Fried, *J. Chem. Soc., Chem. Commun.*, 1971, 292.
297. W. Francke, *Naturwissenschaften*, 1977, <u>64</u>, 590.
298. L. R. Smith, H. J. Williams and R. M. Silverstein, *Tetrahedron Lett.*, 1978, 3231.
299. B. Helferich and W. Dommer, *Ber.*, 1920, <u>53</u>, 2004.
300. O. A. Moe and D. T. Warner, *J. Am. Chem. Soc.*, 1952, <u>74</u>, 2690.
301. A. J. Zambito and E. E. Howe, *Org. Synth. Coll.*, 1973, <u>5</u>, 373.
302. D. Yang and S. W. Pelletier, *J. Chem. Soc., Chem. Commun.*, 1968, 1055.
303. P. L. Stotter and J. B. Eppner, *Tetrahedron Lett.*, 1973, 2417.
304. A. M. Islam and R. A. Raphael, *J. Chem. Soc.*, 1952, 4086.
305. N. R. Easton, J. H. Gardner and J. R. Stevens, *J. Am. Chem. Soc.*, 1947, <u>69</u>, 2941.
306. W. S. Johnson, J. W. Petersen and W. P. Schneider, *J. Am. Chem. Soc.*, 1947, <u>69</u>, 74.
307. L. F. Fieser, *Organic Experiments*, Heath, Lexington, 1968, 2nd Ed., p. 85.
308. C. R. Noller and R. Adams, *J. Am. Chem. Soc.*, 1926, <u>48</u>, 2444; J. v. Braun and W. Münch, *Liebigs Ann. Chem.*, 1928, <u>465</u>, 52.
309. J. Meinwald and H. C. Hwang, *J. Am. Chem. Soc.*, 1957, <u>79</u>, 2910.
310. L. F. Fieser and J. Szmuszkovicz, *J. Am. Chem. Soc.*, 1948, <u>70</u>, 3352.
311. A. W. Schmidt and C. Hartmann, *Ber.*, 1941, <u>74</u>, 1325.

312. R. E. Doolittle, M. S. Blum and R. Boch, *Ann. Entomol. Soc. Am.*, 1970, **63**, 1180; Ap Simon, Vol. IV, p. 103.
313. O. Ceder and H. G. Nilsson, *Synth. Commun.*, 1976, **6**, 381.
314. J. Brugidou, B. H. Chiche-Trinh, H. Christol and J. Poncet, *Tetrahedron Lett.*, 1979, 1223.
315. F. Effenberger and K. Drauz, *Angew. Chem., Int. Ed. Engl.*, 1979, **18**, 474.
316. P. E. Eaton, R. H. Mueller, G. R. Carlson, D. A. Cullison, G. F. Cooper, T.-C. Chou and E.-P. Krebs, *J. Am. Chem. Soc.*, 1977, **99**, 2751.
317. H. Stetter, I. Krüger-Hansen and M. Rizk, *Chem. Ber.*, 1961, **94**, 2702.
318. C. Owens and R. A. Raphael, unpublished observations.
319. R. K. Boeckman, P. C. Naegely and S. D. Arthur, *J. Org. Chem.*, 1980, **45**, 753.
320. S. Hünig and W. Lendle, *Chem. Ber.*, 1960, **93**, 913; D. G. M. Diaper, *Can. J. Chem.*, 1955, **33**, 1720.
321. V. M. Mićović, S. Stojčić, M. Bralović, S. Mladenović, D. Jeremić and M. Stefanović, *Tetrahedron*, 1969, **25**, 985.
322. H. Stetter and H. Rauhut, *Chem. Ber.*, 1958, **91**, 2543.
323. R. Baker, R. Herbert, P. E. Howse and O. T. Jones, *J. Chem. Soc., Chem. Commun.*, 1980, 52.
324. M. Elliott, A. W. Farnham, N. F. Janes, P. H. Needham and B. C. Pearson, *Nature (London)*, 1967, **213**, 493.
325. L. T. Burka, B. J. Wilson and T. M. Harris, *J. Org. Chem.*, 1974, **39**, 2212.
326. G. H. Coleman, G. Nichols and T. F. Martens, *Org. Synth.*, 1945, **25**, 14.
327. H. Iida, Y. Yuasa and C. Kibayashi, *Synthesis*, 1977, 879.
328. H. H. Wasserman, D. J. Hlasta, A. W. Tremper and J. S. Wu, *Tetrahedron Lett.*, 1979, 549.
329. G. Ohloff and W. Giersch, *Helv. Chim. Acta*, 1980, **63**, 76.
330. J. D. McChesney and A. F. Wycpalek, *J. Chem. Soc., Chem. Commun.*, 1971, 542.
331. M. S. Newman and B.J. Magerlein, *Org. React.*, 1949, **5**, 413.
332. W. E. Barnett and J. C. McKenna, *Tetrahedron Lett.*, 1971, 2595.
333. E. A. Brown, *J. Med. Chem.*, 1967, **10**, 546.
334. N. J. Leonard, F. E. Fischer, E. Barthel, J. Figueras and W. C. Wildman, *J. Am. Chem. Soc.*, 1951, **73**, 2371.

335. G. Nannini, P. N. Giraldi, G. Mologora, G. Biasoli, F. Spinelli, W. Logemann, E. Dradi, G. Zanni, A. Buttinoni and R. Tommasini, *Arz. Forsch.*, 1973, 23, 10; *Chem. Abstr.*, 1974, 80, 3335.
336. D. G. Farnum and M. Burr, *J. Org. Chem.*, 1963, 28, 1387; B. T. Gillis and P. E. Peck, *ibid.*, 1388; A. Lapworth and J. A. McRae, *J. Chem. Soc.*, 1922, 1699.
337. *Drug Synthesis*, p 234.
338. K. Hofman, *Imidazole*, Interscience, New York, 1953, Vol. I, p. 226.
339. J. F. Mulvaney and R. L. Evans, *Ind. Eng. Chem.*, 1948, 40, 393.
340. Tedder and Nechvatel, Vo. 5, p. 269.
341. J. Tufariello and R. C. Gatrone, *Tetrahedron Lett.*, 1978, 2753.
342. A. H. Beckett, R. G. Lingard and A. E. E. Theobald, *J. Med. Chem.*, 1969, 12, 563.
343. V. E. Marquez, P. S. Liu, J. A. Kelley and J. S. Driscoll, *J. Org. Chem.*, 1980, 45, 485.
344. O. E. Curtis and J. M. Sandri, *Org. Synth. Coll.*, 1963, 4, 278.
345. M. Julia, S. Julia and S.-Y. Tchen, *Bull. Soc. Chim. Fr.*, 1961, 1849.
346. C. M. McCloskey and G. H. Coleman, *Org. Synth. Coll.*, 1932, 1, 156; M. J. Schlatter, *ibid.*, 223; C. F. H. Allen, *ibid.*, 1932, 1, 156; C. S. Marvel and H. O. Calvery, *ibid.*, 533.
347. M. Julia, S. Julia and Y. Noël, *Bull. Soc. Chim. Fr.*, 1960, 1708; S. F. Brady, M. A. Ilton and W. S. Johnson, *J. Am. Chem. Soc.*, 1968, 90, 2882.
348. C. F. H. Allen and J. van Allen, *Org. Synth. Coll.*, 1955, 3, 727.
349. H. O. House, S. G. Boots and V. K. Jones, *J. Org. Chem.*, 1965, 30, 2519.
350. C. Agami and J. Aubouet, *Bull. Soc. Chim. Fr.*, 1967, 1391.
351. E. J. Corey and M. Chaykovski, *J. Am. Chem. Soc.*, 1965, 87, 1353.
352. E. Piers, R. W. Britton and W. de Waal, *Tetrahedron Lett.*, 1969, 1251; A. Tanaka, H. Uda and A. Yoshikoshi, *J. Chem. Soc., Chem. Commun.*, 1969, 308.
353. C. H. Tilford, M. C. van Campen and R. S. Shelton, *J. Am. Chem. Soc.*, 1947, 69, 2902.
354. S. Winstein, J. Sonnenberg and L. de Vries, *J. Am. Chem. Soc.*, 1959, 81, 6523.
355. K. Hofman, S. F. Orochena, S. M. Sax and G. A. Jeffrey, *J. Am. Chem. Soc.*, 1959, 81, 992.

356. P. A. Grieca, T. Oguri, 1977, C.-L. Wang and E. Williams, *J. Org. Chem.*, 42, 4113.
357. C. H. Heathcock and R. Ratcliffe, *J. Am. Chem. Soc.*, 1971, 93, 1746.
358. F. V. Brutcher and D. D. Rosenfeld, *J. Org. Chem.*, 1964, 29, 3154.
359. F. H. Case, *J. Am. Chem. Soc.*, 1934, 56, 715.
360. M. J. Perkins, N. B. Peynircioğlu and B. V. Smith, *J. Chem. Soc., Perkin Trans. 2*, 1978, 1025.
361. A. L. Wilds and A. L. Meader, *J. Org. Chem.*, 1948, 13, 763.
362. C. D. Gutsche, *Org. React.*, 1954, 8, 364.
363. E. J. Corey and M. Chaykovsky, *J. Am. Chem. Soc.*, 1965, 87, 1353
364. *Houben/Weyl 7/2a* p. 935.
365. D. G. Botterton and G. Wood, *J. Org. Chem.*, 1965, 30, 3871.
366. M. J. Wiemann and S.-L. Thuan, *Bull. Soc. Chim. Fr.*, 1959, 1537.
367. J. H. Kennedy and C. Buse, *J. Org. Chem.*, 1971, 36, 3135; B. Rickborn and R. M. Gerkin, *J. Am. Chem. Soc.*, 1971, 93, 1693.
368. F. G. Bordwell, R. G. Scamehorn and W. R. Springer, *J. Am. Chem. Soc.*, 1969, 91, 2087.
369. J. J. McCullough and B. R. Ramachandran, *J. Chem. Soc., Chem. Commun.*, 1971, 1180.
370. P. E. Eaton and K. Nyi, *J. Am. Chem. Soc.*, 1971, 93, 2786.
371. S. W. Baldwin and J. M. Wilkinson, *Tetrahedron Lett.*, 1979, 2657.
372. R. N. L. Harris, F. Komitsky and C. Djerassi, *J. Am. Chem. Soc.*, 1967, 89, 4765.
373. K. Mori and M. Sasaki, *Tetrahedron*, 1980, 36, 2197.
374. K. N. Slessor, A. C. Oehlschlager, B. D. Johnston, H. D. Pierce, S. K. Grewal and L. K. Wickremesinghe, *J. Org. Chem.*, 1980, 45, 2290.
375. K. B. Wiberg and J. G. Pfeiffer, *J. Am. Chem. Soc.*, 1970, 92, 553.
376. J. J. Bloomfield, *Tetrahedron Lett.*, 1968, 591; K. Rühlmann, *Synthesis*, 1971, 236.
377. J. M. Conia and J. P. Barnier, *Tetrahedron Lett.*, 1971, 4981.
378. C. Beard and A. Burger, *J. Org. Chem.*, 1961, 26, 2335.
379. A. Padwa, E. Alexander and M. Niemcyzk, *J. Am. Chem. Soc.*, 1969, 99, 456.
380. E. Buchta and K. Geibel, *Liebigs Ann. Chem.*, 1961, 648, 36.

381. B. M. Trost, M. J. Bogdanowicz and J. Kern, *J. Am. Chem. Soc.*, 1975, 97, 2218; B. M. Trost, M. Preckel and L. M. Leichter, *ibid.*, 2224.
382. R. H. Bisceglia and C. J. Cheer, *J. Chem. Soc., Chem. Commun.*, 1973, 165.
383. J. P. Springer, J. Clardy, R. J. Cole, J. W. Kirksey, R. K. Hill, R. M. Carlson and J. L. Isidor, *J. Am. Chem. Soc.*, 1974, 96, 2267.
384. E. Dunkelblum, *Tetrahedron*, 1976, 32, 975.
385. A. Hassner, R. M. Cory and N. Sartoris, *J. Am. Chem. Soc.*, 1976, 98, 7698.
386. T. Fukuyama, R. K. Frank and C. F. Jewell, *J. Am. Chem. Soc.*, 1980, 102, 2122.
387. J. K. Rasmussen and A. Hassner, *Chem. Rev.*, 1976, 76, 389.
388. S. M. Ali and S. M. Roberts, *J. Chem. Soc., Chem. Commun.*, 1975, 887.
389. J. W. Cornforth and R. H. Cornforth in *Natural Substances formed Biologically from Mevalonic Acid*, Ed. T. W. Goodwin, Biochemical Society Symposia No. 29, Academic Press, London and New York, 1970, p. 5.
390. R. H. Hasek, R. D. Clark and J. H. Chaudet, *J. Org. Chem.*, 1961, 26, 3130.
391. W. C. Agosta and A. B. Smith, *J. Am. Chem. Soc.*, 1971, 93, 5513.
392. S. Danishefsky, R. Zamboni, M. Kahn and S. J. Etheredge, *J. Am. Chem. Soc.*, 1981, 103, 3460.
393. J. Strating and H. Wynberg, *Synthesis*, 1971, 211.
394. G. Tschudi and H. Schinz, *Helv. Chim. Acta*, 1950, 33, 1865.
395. A. I. Vogel, *J. Chem. Soc.*, 1934, 1758.
396. D. E. Ryono and G. M. Loudon, *J. Am. Chem. Soc.*, 1976, 98, 1889; P. Eilbracht and P. Dahler, *Liebigs Ann. Chem.*, 1979, 1890.
397. R. P. Linstead and E. M. Meade, *J. Chem. Soc.*, 1934, 935.
398. A. S. Kende, *Org. React.*, 1960, 11, 261.
399. A. E. Greene and J.-P. Deprés, *J. Am. Chem. Soc.*, 1979, 101, 4003.
400. J. M. Allen, K. M. Johnston, J. F. Jones and R. G. Shotter, *Tetrahedron*, 1977, 33, 2083.
401. W. Herz, *J. Am. Chem. Soc.*, 1953, 75, 73.
402. S. Dev, *J. Indian Chem. Soc.*, 1957, 34, 169.
403. W. Baker and P. G. Jones, *J. Chem. Soc.*, 1951, 787.
404. H. W. Pinnick and Y.-H. Chang, *Tetrahedron Lett.*, 1979, 837.
405. R. K. Danheiser, C. Martinez-Davila, R. J. Auchus and J. T. Kadonaga, *J. Am. Chem. Soc.*, 1981, 103, 2443.

406. *Pesticide Manual*, p. 82.
407. J. Ficini and J. d'Angelo, *Tetrahedron Lett.*, 1976, 2441.
408. S. Julia, M. Julia and G. Linstrumelle, *Bull. Soc. Chim. Fr.*, 1966, 3499.
409. N. Wakabayashi, R. M. Waters and J. P. Church, *Tetrahedron Lett.*, 1969, 3253.
410. H. E. Zimmerman, P. Hackett; D. F. Juers, J. M. McCall and B. Schröder, *J. Am. Chem. Soc.*, 1971. 93, 3653.
411. W. R. Roush, *J. Am. Chem. Soc.*, 1980, 102, 1390.
412. W. R. Roush and H. R. Gillis, *J. Org. Chem.*, 1980, 45, 4283.
413. P. L. Julian, J. J. Oliver, R. H. Kimball, A. B. Pike and G. D. Jefferson, *Org. Synth. Coll.*, 1943, 2, 487; P. L. Julian and J. J. Oliver, *ibid.*, 391; M. Freifelder and G. R. Stone, *U.S. Pat.*, 3,014,966 (1961); *Chem. Abstr.*, 1962, 56, 12772.
414. G. Stork and E. L. Foreman, *J. Am. Chem. Soc.*, 1946, 68, 2172.
415. H. B. Henbest and B. B. Millward, *J. Chem. Soc.*, 1960, 3575.
416. J. H. Babler and T. R. Mortell, *Tetrahedron Lett.*, 1972, 669.
417. K. Mori and M. Matsui, *Tetrahedron*, 1968, 24, 3127.
418. J. Haslouin and F. Rouessac, *Bull. Soc. Chim. Fr.*, 1977, 1242.
419. D. J. Reif and H. O. House, *Org. Synth. Coll.*, 1963, 4, 375.
420. H. E. Zimmerman, K. G. Hancock and G. C. Licke, *J. Am. Chem. Soc.*, 1968, 90, 4901.
421. H. E. Ungnade and F. V. Morriss, *J. Am. Chem. Soc.*, 1948, 70, 1898.
422. K. Yamada, Y. Kyotani, S. Manabe and M. Suzuki, *Tetrahedron*, 1979, 35, 293.
423. J. W. Huffman, *J. Org. Chem.*, 1959, 24, 1759.
424. C. J. Sih, D. Massuda, P. Corey, R. D. Gleim and F. Suzuki, *Tetrahedron Lett.*, 1979, 1285.
425. G. Stork, D. F. Taber and M. Marx, *Tetrahedron Lett.*, 1978, 2445.
426. R. A. Barnes and M. Sedlak, *J. Org. Chem.*, 1962, 27, 4562.
427. A. F. Ferris, *J. Org. Chem.*, 1955, 20, 780.
428. J. M. McEuen, R. P. Nelson and R. G. Lawton, *J. Org. Chem.*, 1970, 35, 690.
429. P. H. Boyle, W. Cocker, D. M. Grayson and P. V. R. Shannon, *J. Chem. Soc., Chem. Commun.*, 1971, 395.
430. T. V. Lee, S. M. Roberts, M. J. Dimsdale, R. F. Newton, D. K. Rainey and C. F. Webb, *J. Chem. Soc., Perkin Trans. 1*, 1978, 1176.

431. D. Seebach and M. Pohmakotr, *Helv. Chim. Acta*, 1979, 62, 843.
432. R. M. Scarborough, B. H. Toder and A. B. Smith, *J. Am. Chem. Soc.*, 1980, 102, 3904.
433. K. Mori and H. Ueda, *Tetrahedron Lett.*, 1981, 461.
434. G. Mehta and A. V. Reddy, *J. Chem. Soc., Chem. Commun.*, 1981, 756.
435. G. Stork and E. W. Logusch, *Tetrahedron Lett.*, 1979, 3361.
436. P. A. Grieco, S. Ferrino and G. Vidari, *J. Am. Chem. Soc.*, 1980, 102, 7586.
437. P. A. Grieco, G. Vidari, S. Ferrino and R. C. Haltowanger, *Tetrahedron Lett.*, 1980, 1619.
438. P. A. Rossy, W. Hoffman and N. Müller, *J. Org. Chem.*, 1980, 45, 617.
439. A. G. Schultz, W. K. Hagmann and M. Shen, *Tetrahedron Lett.*, 1979, 2965.
440. G. W. Kenner, K. M. Smith and J. F. Unsworth, *J. Chem. Soc., Chem. Commun.*, 1973, 43.
441. J. R. Carson and S. Wong, *J. Med. Chem.*, 1973, 16, 172.
442. R. K. Brown in 'Indoles', ed. W. J. Houlihan, Wiley, New York 1972, Part 1, p. 227; see p. 236.
443. U. Hörlein, *Chem. Ber.*, 1954, 87, 463.
444. D. D. Weller and H. Rapaport, *J. Am. Chem. Soc.*, 1976, 98, 6650.
445. G. D. Pandey and K. P. Tiwari, *Synth. Commun.*, 1980, 10, 43.
446. R. N. Lacey, *Brit. Pat.*, 745,853 (1956); *Chem. Abstr.*, 1957, 51, 491c.
447. W. S. Bowers, T. Ohta, J. S. Cleere and P. A. Marsella, *Science*, 1976, 193, 542.
448. *Pesticides*, p. 476.
449. G. W. K. Cavill and C. Houghton, *Aust. J. Chem.*, 1974, 27, 879.
450. J. D. Bower and G. R. Ramage, *J. Chem. Soc.*, 1955, 2834.
451. D. W. Dunwell, D. Evans, T. A. Hicks, C. H. Cashin and A. Kitchen, *J. Med. Chem.*, 1975, 18, 53.
452. E. J. Corey and A. K. Long, *J. Org. Chem.*, 1978, 43, 2208.
453. A. W. Johnson, *Chem. Ber.*, 1980, 82.
454. E. Seeger and A. Kottler, *U.S. Pat.*, 3,133,967 (1964); *Chem. Abstr.*, 1964, 61, 9436g.
455. C. H. Heathcock in Ap Simon, Vol. 2, p. 207 - 222.
456. W. G. Dauben, G. Ahlgren, T. J. Leitereg, W. C. Schwarzel and M. Yoshioko, *J. Am. Chem. Soc.*, 1972, 94, 8593.
457. J. C. Aumiller and J. A. Whittle, *J. Org. Chem.*, 1976, 41, 2955.

458. G. Stork and D. J. Morgans, *J. Am. Chem. Soc.*, 1979, **101**, 7110.
459. P. Fritzsch, *Liebigs Ann. Chem.*, 1897, **296**, 344.
460. F. H. Howell and D. A. H. Taylor, *J. Chem. Soc.*, 1956, 4252.
461. J. G. Srivastava and D. N. Chaudhury, *J. Org. Chem.*, 1962, **27**, 4337.
462. A. Kamal, A. Robertson and E. Tittensor, *J. Chem. Soc.*, 1950, 3375.
463. M. L. Greenlee, *J. Am. Chem. Soc.*, 1981, **103**, 2425.
464. H. Tucker, *J. Org. Chem.*, 1979, **44**, 2943.
465. J. H. Biel, H. L. Friedman, H. A. Leiser and E. P. Sprengeler, *J. Am. Chem. Soc.*, 1952, **74**, 1485.
466. *Drug Synthesis*, p. 336.
467. E. J. Soboczenski, *U.S. Pat.*, 3,235,360 (1966); *Chem. Abstr.*, 1966, **64**, 14196f.
468. W. C. Still, *J. Am. Chem. Soc.*, 1979, **101**, 2493.
469. E. E. Van Tamelen and G. T. Hildahl, *J. Am. Chem. Soc.*, 1956, **78**, 4405.
470. T. Sakai, H. Naoki, K. Takaki, and H. Kameoka, *Chem. Lett.*, 1981, 1257.
471. D. W. Clutton and M. B. Evans, *J. Chromatogr.*, 1978, **167**, 409.

FORMULA INDEX

Target molecules from both the text (T) and workbook (W) appear in order of molecular formula. Entries show main functional group in the first column and reason for synthesis or structural feature in the second column. The last column gives structure number, e.g. (23), and page number, e.g. T146.

Abbreviations

Ar	Aromatic	Pharm.	Pharmaceutical
CNS	Central nervous system	Prot.	Protection
		Stereo.	Stereochemistry
Ident.	Identification	Struct.	Structural
Int.	Intermediate	Syn.	Synthetic
Manuf.	Manufacturing	Unsat.	Unsaturated
Mech.	Mechanistic		

$C_3H_4ClIO_2$	Chloroformate	Prot. reagent	(9)	W82
$C_4HO_3^-$	Cyclobutenedione	Microbial toxin	(10)	W391
$C_4H_4N_2O_2$	Cyclic amide	Uracil	(47)	T250
$C_4H_6O_2$	Acid	Cyclopropane	(6)	W354
	α-Hydroxy ketone	Cyclobutanone	(37)	W384
C_4H_6S	Cyclic sulphide		(20)	T244
$C_4H_7NO_4$	α-Amino acid	Aspartic acid		W300
$C_4H_8O_2$	*cis*-Alkene	*cis*-Butenediol	(10)	T128
$C_4H_9ClO_2$	α-Chloro ether	Prot. reagent	(15)	T44
$C_4H_{10}N_2O$	γ-Amino amide	Psychotropic	(20)	W78
$C_4H_{10}OS$	Hydroxysulphide	Insecticide int.	(18)	W51
$C_5H_2NNaO_3S_2$	Amide, thiophene	Sweetener	(1)	W452
$C_5H_5N_4O_4$	Nitro furan	Antibacterial	(23)	T330
$C_5H_6O_2$	Alkyne, acid		(9)	W163
C_5H_7Br	Alkyne		(13)	T128
$C_5H_8N_2O$	Cyclic urea	Enzyme inhibitor int.	(76)	W351

Formula	Type	Application	(Ref)	Code
C_5H_8O	Enone		(15)	W193
$C_5H_8OS_2$	Cyclic sulphide		(4)	W63
$C_5H_9BrO_2$	Acetal		(41)	T49
$C_5H_{10}N_2O_3$	Amide	Easy-care fabrics	(60)	W348
$C_5H_{10}O_2$	α-Hydroxy-ketone		(4)	T185
	γ-Hydroxy-ketone		(3)	T66
$C_5H_{11}N$	Unsat. amine		(5)	W72
$C_5H_{11}NO$	Cyclic aminoketone		(14)	T147
$C_5H_{11}NO_2S$	α-Amino acid	Methionine	(44)	W60
$C_5H_{12}O_2$	Diol	Prot. reagent	(20)	W205
$C_5H_{12}OS$	Thiol, alcohol		(21)	T74
$C_6H_5ClN_2O_2$	Nitro aniline	Moluscicide int.	(24)	W27
C_6H_7ClOS	Ar thiol	Tetracycline int.	(14)	W22
$C_6H_7O_3$	Enone, acid		(33)	T157
C_6H_8	Diene	Cyclobutene	(6)	T120
$C_6H_8N_2$	Dinitrile	Nylon Manuf.	(22)	W274
$C_6H_8O_2$	α-Diketone	Perfume	(24)	T282
	Cyclic enone	Synth. int.	(48)	T195
$C_6H_8O_3$	Unsat. ketoester		(16)	W303
	Keto-lactone	Antibiotic int.	(7)	W324
$C_6H_8O_3S$	Ketoester	Cyclic sulphide	(27)	T246
$C_6H_9BrO_2$	Lactone		(31)	W341
$C_6H_{10}O$	Enone	Mesityl oxide	(28)	T150
	Ketone	Cyclopropane	(32)	T257
	Ketone	Cyclopropane	(11)	W355
	Alkyne, alcohol	Tranquiliser	(6)	T127
	Enone		(16)	W193
	Bicyclic alcohol	Mech. study	(45)	W364
$C_6H_{10}O_2$	Unsat. ester		(49)	T161
	Acid	Cyclopentane	(11)	W92
$C_6H_{10}O_3$	β-ketoester	Acetoacetate	(8)	T145
	Lactone	Mevalonolactone	(74)	T165
		and	(29)	W395
$C_6H_{10}O_6$	Diol diacid		(28)	W275
$C_6H_{11}NO_2$	Lactam		(20)	T183
$C_6H_{11}NO_2$	Amine, ketone, acid		(31)	T215
	Cyclic amino acid		(25)	W317
$C_6H_{12}O$	Epoxide		(29)	T243
$C_6H_{12}O_2$	Carboxylic acid	(28) T82, (16) T109,	(15)	W93
$C_6H_{12}O_4$	Hydroxy-acid	Mevalonic acid	(69)	T165
$C_6H_{14}O_2$	Diol	Synth. int.	(20)	T148
	Diol	Pinacol	(24)	T201
$C_6H_{16}N_2$	Diamine	Nylon manuf.	(21)	W274

Formula	Type	Application		Ref
$C_7H_4BrNO_4$	Nitro Ar acid	Synth. int.	(9)	W19
$C_7H_4ClF_3$	Halogenated toluene		(7)	T19
$C_7H_5BrO_2$	Ar diol	Salbutamol int.	(4)	T67
$C_7H_5ClO_3$	Hydroxy Ar acid	Moluscicide int.	(23)	W27
$C_7H_5NO_3S$	Imide	Saccharine	(19)	T14
$C_7H_6Cl_2O$	Bicyclic ketone	Cyclobutanone	(8)	T275
$C_7H_6N_2$	Heterocycle	Struct. study	(49)	W466
$C_7H_6O_4$	Lactone	Synth. int.	(35)	T158
C_7H_7BrO	Phenol		(13)	T12
$C_7H_7NO_3$	Hydroxy amino Ar acid	Anaesthetic int.	(9)	T20
$C_7H_7N_2O_4$	Nitro Ar	Polyurethane int.	(21)	W25
C_7H_8ClNO	Amino Ar ether	Tetracycline int.	(15)	W22
$C_7H_8O_2$	Bicyclic lactone	Prostaglandin int.	(16)	T33
C_7H_{10}	Diene	Cyclopentene	(22)	T130
$C_7H_{10}O$	Enone	Cyclopentenone	(28)	T213
	Enone	Cyclopentenone	(3)	T279
	Enone	Cyclopentenone	(1)	W397
	β,γ-Enone	Cyclohexenone	(24)	T184
$C_7H_{10}O_2$	Lactone	Synth. int.	(25)	W169
	Bicyclic lactone		(42)	W309
C_7H_{12}	Alkene	Cyclohexane	(9)	T121
	Diene	Diels-Alder reagent	(26)	T124
$C_7H_{12}O$	γ,δ-Enal		(42)	T289
	Aldehyde	Synth. int.	(8)	W91
	γ,δ-Unsat. ketone	Synth. int.	(40)	W415
	Bicyclic ether	Four-membered ring	(22)	W423
$C_7H_{12}O_2$	Alkene, acetal	Synth. reagent	(29)	T353
	Ketone, ether	Synth. int.	(29)	W55
$C_7H_{12}O_3$	Aldehyde, ester	Biotin int.	(21)	T227
	Epoxy-ester		(5)	T252
$C_7H_{13}BrO$	Aldehyde	Alkaloid int.	(18)	W232
$C_7H_{13}ClO_2$	Protected ketone	Synth. int.	(4)	T42
$C_7H_{13}NO_2$	Amide, aldehyde		(1)	W299
C_7H_{14}	Alkene		(7)	W146
$C_7H_{14}O$	Ketone	Perfume (24) T130,	(12)	W127
	Ketone		(12)	W137
$C_7H_{14}O_2$	Acid	Pheromone int.	(16)	W129
	Hydroxy-aldehyde	Pheromone int.	(14)	W444
$C_7H_{15}N$	Unsat. amine	Mech. int.	(20)	W129
$C_7H_{15}NO$	Hydroxy-amine	Int. for ring expansion	(8)	T43
C_7H_{16}	Alkane	Petrol research	(45)	T205
$C_7H_{16}O$	Alcohols	Mech. study (4)	(5)	T76
	Alcohol	Synth. int.	(14)	W100
$C_7H_{17}N$	Amine		(11)	T61

$C_8H_6Br_2O$	Ketone	Reagent for characterisation	(8)	T52
$C_8H_6Cl_2O_3$	Ar Ether Acid	Herbicide 2,4-D	(33)	T47
$C_8H_6O_2$	Ar aldehyde	Palanil int.	(11)	W27
$C_8H_6O_3$	Ar aldehyde acetal	Piperonal, perfume	(12)	T9
C_8H_7BrO	Ar acid bromide		(23)	T81
$C_8H_7ClN_2O_3$	Nitro Ar amide	Tetrazole int.	(13)	T54
$C_8H_7ClO_2$	Ar ester	Mechanism int.	(9)	W11
	Chloroformate	Protection reagent	(5)	T36
C_8H_7N	Ar nitrile		(15)	T13
$C_8H_8N_4$	Heterocyclic hydrazine	Antihypotensive	(79)	T24
$C_8H_8O_3$	Ar acid		(12)	W92
	Cyclic anhydride	Alkaloid int.	(13)	W17
$C_8H_9NO_2$	Hydroxy amide	Paracetamol	(4)	T35
	Imide	Fungicide int.	(1)	W29
$C_8H_9NO_3S$	Ar hydroxy nitro sulphide	Protection reagent	(1)	W80
$C_8H_{10}N_2O$	Amide nitrile	Barbiturate int.	(52)	T16
$C_8H_{10}O$	Ar ether	Perfume	(17)	T31
	Ar alcohol		(8)	T88
$C_8H_{10}O_3$	Unsat. lactone		(29)	T15
$C_8H_{10}O_4$	Diester	Cyclobutene	(3)	T26
$C_8H_{11}ClO_3$	Bicyclic ketone	Cyclobutanone	(24)	W39
$C_8H_{11}N$	Amine		(18)	T63
$C_8H_{12}N_2O$	Heterocyclic amide	Insecticide int.	(43)	W46
$C_8H_{12}O$	Dienone		(21)	W37
	Cyclohexenone		(16)	W37
	Bicyclic ketone	Cyclobutanone	(1)	W38
	- " -	Cyclopentanone	(32)	W40
$C_8H_{12}O_2$	β-Diketone	Dimedone	(39)	T17
	β,γ-Unsat Lactone	Pheromone int.	(19)	W39
$C_8H_{12}O_3$	β-Keto ester	Cyclopentanone	(11)	T14
	δ-Keto acid	Cyclohexanone	(4)	T17
	Acetal ketone	Synth. int.	(6)	W40
	Hydroxy ester	Prostaglandin int.	(31)	W11
$C_8H_{12}O_5$	β-Diketone, ester, ether	Synth. int.	(36)	T15
$C_8H_{13}NO_2$	α-Keto lactam	Synth. int.	(12)	W20
C_8H_{14}	Diene	Insecticide int.	(21)	T20
$C_8H_{14}N_2O$	Cyclic hydrazide	Pheromone int.	(24)	T24
$C_8H_{14}O$	γ,δ-Unsat. ketone	Manuf. int.(1) T1,	(18)	T11
			(52)	T29
	Bicyclic ether		(53)	W34
	β,γ-Unsat. aldehyde	Insecticide int.	(35)	W39
	Cyclohexanone	Pheromone int.	(27)	W42
$C_8H_{14}O_2$	Acetal	Pheromone	(18)	W19
	Keto ether		(12)	W25
	Unsat. acid	Synth int.	(4)	W39

$C_8H_{14}O_3$	γ-Hydroxy acid	Conformation Study	(15)	T211
	Aldehyde ester	Synth. int.	(5)	T218
	Hydroxy α,β Unsat ether	Mechanism study	(55)	T361
	Aldehyde ester	Alkaloid int.	(13)	W232
	ε-Hydroxy acid		(34)	W319
$C_8H_{15}Br$	Homoallylic bromide	Pheromone int.	(23)	W167
$C_8H_{15}NO$	Heterocycle	Synth. reagent	(21)	T148
	Amino ketone	Mechanism Study	(31)	W342
$C_8H_{15}NO_2$	Epoxy amide	Tranquiliser	(31)	T151
C_8H_{16}	Alkene		(10)	T121
$C_8H_{16}O$	Unsat. alcohol	Pheromone	(13)	W111
	Ketone	Perfume	(6)	W125
$C_8H_{16}O_2$	Acid	Stereochem. Study	(19)	W94
$C_8H_{16}O_3$	Aldehyde acetal	Synth. int.	(12)	W42
$C_8H_{17}NO$	Amino ether	Cyclohexane	(28)	T298
$C_8H_{18}O$	Alcohol	Petrol research	(36)	T85
	Alcohol	Pheromone	(1)	T86
	Alcohol	Synth. int.	(7)	W91
$C_8H_{20}N_2$	Diamine	Polymer manuf.	(13)	T181
C_9H_8OS	Ketone	Synth. int.	(19)	W52
$C_9H_8O_2$	α,β-Unsat acid	Cinnamic acid	(55)	T162
	Unsat. Ar acid	Polystyrene manuf.	(14)	W272
$C_9H_8O_4$	Phenol, ketone, acid	Salbutamol int.	(12)	T22
$C_9H_9Cl_2NO$	Ar amide	Weedkiller	(4)	T27
$C_9H_9Cl_3O_2$	Ar ester	Perfume	(1)	W89
C_9H_9N	Indole		(40)	T332
$C_9H_9O_4$	Tricyclic lactone, acid		(39)	W308
$C_9H_{10}ClNO_3$	Nitro Ar ether	Anti-malarial int.	(15)	T23
$C_9H_{10}O$	Ketone, Ar		(15)	W372
	Aldehyde, Ar	Perfume	(32)	T264
	Furan, diene	Synth. int.	(35)	W216
$C_9H_{10}O_2$	Ether, Ar ketone	Perfume	(2)	T7
$C_9H_{10}O_4$	Epoxy keto-lactone	Antibiotic int.	(7)	W441
$C_9H_{11}NO_2$	Ar amino ester	Local anaesthetic	(1)	T6
C_9H_{12}	Triene	Photochem. study	(1)	W417
$C_9H_{12}O$	Bicyclic ketone	Synth. int.	(52)	T311
	Bicyclic furan	Synth. int.	(11)	T327
	Tricyclic epoxide	Synth. int.	(36)	T299
	Bicyclic enone	Synth. int.	(19)	W304
$C_9H_{12}O_2$	Bicyclic β-diketone	Synth. int.	(33)	T237
	Bicyclic enone	Synth. int.	(26)	T221

$C_9H_{12}O_4$	Cyclic anhydride	Synth. int.	(49)	W264
$C_9H_{13}NO$	Bicyclic enone	Enamine	(37)	T248
$C_9H_{13}NO_2$	Alkyne, amide, ester	Sedative	(6)	W161
$C_9H_{14}Br_2O_2$	Acetal, cyclopropane	Synth. int.	(35)	T258
$C_9H_{14}N_2O$	Heterocycle, ether	Flavour/odour	(85)	T341
$C_9H_{14}O$	Cyclic γ,δ enone		(6)	T294
	Cyclic α,β-enone		(34)	T176
	Bicyclic alcohol	Cyclopentene	(22)	W412
	Bicyclic ether		(33)	T138
	γ,δ-Unsat. aldehyde		(5)	T132
	γ,δ-Unsat. ketone	Cyclohexanone	(17)	T118
$C_9H_{14}O_2$	Ketone, enol ether	Hormone int.	(8)	T186
	α,β-Unsat ester	Cyclohexane	(13)	T122
	β-Diketone	Cyclohexane dione	(32)	W239
	β-Diketone	Cyclohexanone	(71)	W228
	Diene acetal	Mech. Study	(29)	W270
	Bicyclic acetal	Folk medicine	(68)	W490
$C_9H_{14}O_3$	γ-keto-ester	Antibiotic int.	(5)	T210
	Ether, ester	Perfume	(1)	W440
	β-keto-ester	Synth. int.	(26)	T150
$C_9H_{15}BrO_2$	α-Hydroxy-ketone	Pharm. int.	(36)	W173
$C_9H_{15}NO$	Spiro lactam	Struct. ident.	(17)	W241
$C_9H_{15}NO_2S$	Nitro-sulphide	Biotin int.	(14)	T198
$C_9H_{15}NO_5$	Amide, esters	Amino acid int.	(7)	W301
$C_9H_{16}N_2O_2$	Nitriles, ethers	Crown ether int.	(19)	W114
$C_9H_{16}O_2$	Cyclohexanone	Stereochem	(20)	W140
$C_9H_{16}O_2$	γ,δ-Unsat. ester	Synth. int.	(45)	T289
	Acetal	Pheromone (lineatin)	(21)	W38?
	Acetal	Olive fly pheromone	(22)	W32?
	Acetal	Pheromone (chalcogran)	(57)	W29?
	Acetal	Pheromone (brevicomin)	(22)	W25?
	1,4-Diketone		(33)	W290
$C_9H_{16}O_3$	Ketone, alcohols	Pharm. research	(35)	W173
$C_9H_{17}Cl$	Unsat halide	Synth. int.	(49)	T296
$C_9H_{17}NO$	β-Amino-ketone	Heterocycle	(41)	T249
$C_9H_{18}O$	Ketone	Pheromone	(3)	T10
	Ketone	Synth. int.	(28)	W132
			(16)	W139
	Ketone	Stereochem. Study	(2)	W123
$C_9H_{18}O_2$	Ester		(38)	T26?
	Cyclic ether, alcohol		(25)	W132
	Ene diol	Cyclisation Study	(23)	W13

Formula	Type	Application	(n)	Ref
$C_9H_{18}O_3$	β-Hydroxy-ester	Synth. int.	(68)	T164
$C_9H_{20}N_2$	Diamine	Synth. int.	(13)	W246
$C_9H_{20}O$	Unsat. alcohol	Perfume	(1)	W96
$C_9H_{20}O_2$	1,2-Diol	Stereochem. Study	(6)	W253
$C_{10}H_8O$	Ar Ketone	Cyclopropane	(28)	T256
	Ar Epoxide	Structure	(26)	W150
$C_{10}H_8O_3$	Cyclic anhydride	Spiro compound	(37)	W426
$C_{10}H_8O_4$	Acetal, lactone	Alkaloid int.	(38)	W481
$C_{10}H_9N$	Quinoline	Perfume	(61)	T336
$C_{10}H_9NO_2$	Imide	Anti-convulsant	(23)	T212
$C_{10}H_{10}$	Polycyclic triene	Mech. Study	(57)	T312
	Ar cyclobutane	Structure	(9)	T133
$C_{10}H_{10}O$	Benzofuran	Pharm. int.	(29)	T330
$C_{10}H_{10}O_2$	Acetal, alkene	Perfume	(27)	W150
	β-Diketone	Perfume	(5)	T144
$C_{10}H_{10}O_3$	Acetal, ether	Alkaloid int.	(46)	W482
$C_{10}H_{12}ClNO_2$	Ar nitro	Mech. Study	(29)	T83
$C_{10}H_{12}NO_3P$	Ar nitrile, phosphonate	Manuf. int.	(10)	W271
$C_{10}H_{12}O$	Ketone, Ar	Perfume	(13)	W127
$C_{10}H_{12}O_2$	Enedione	Synth. int.	(39)	T307
	Ketone, phenol	Flavouring	(51)	T207
	Spiro α-diketone	Synth. int.	(18)	W401
	Spiro β-diketone	"	(28)	T228
$C_{10}H_{12}O_6$	Acid, lactones	B_{12} int.	(14)	T225
$C_{10}H_{13}Cl_3O_3$	β-Keto-ester	Insecticide int.	(30)	W395
$C_{10}H_{13}NO$	Heterocycle	Morpholine	(14)	W67
$C_{10}H_{13}NO_2$	Amine, Ar acetal	Amphetamine analogue	(5)	W242
$C_{10}H_{13}NO_3$	Amine, ether, ester	Perfume	(12)	W33
	Ar nitro ether	Industrial research	(15)	W13
$C_{10}H_{13}NO_4$	Phenol amino-acid	Pharm. Study	(12)	T43
$C_{10}H_{14}ClN$	Amine, Ar	Appetite suppressant	(7)	T180
$C_{10}H_{14}Cl_2O_2$	Cyclopropane, ester	Insecticide int.	(3)	T252
$C_{10}H_{14}O$	Bicyclic enone	Synth. int.	(47)	T300
	Bicyclic enone		(18)	T285
	Diene, aldehyde		(21)	T281
	Diene, ketone		(19)	T281
	Bicyclic ketone	Cyclobutanone	(33)	T273
	γ,δ-Enone	Cyclopropane	(13)	W356
	Alcohol, Ar	Perfume	(27)	T282
$C_{10}H_{14}O_2$	Bicyclic β-diketone		(13)	T262
	Epoxy-ketone	Synth. int.	(25)	W339
	Acetal	Perfume	(3)	T41

$C_{10}H_{14}O_3$	Keto-ester	Antibiotic int.	(9)	T210
$C_{10}H_{15}N$	Amine, Ar	Anti-obesity drug	(11)	W74
$C_{10}H_{15}NO$	Amino enone	Alkaloid int.	(25)	W236
	Amino ester	Synth. int.	(9)	W409
$C_{10}H_{16}$	Propellane	Cyclobutanes	(4)	W377
	Diene	Terpene	(32)	T138
$C_{10}H_{16}N_6S$	Guanidine, hetero-cycle	Ulcer drug	(101)	T344
$C_{10}H_{16}O$	Aldehyde, Spiro	Cyclopropane	(11)	T275
	Bicyclic aldehyde	Cyclobutane	(6)	T269
	Bicyclic ketone	Cyclopentanone	(40)	W406
	Enone	Cyclopentanone	(77)	T166
	γ,δ-Unsat. aldehyde	Perfumery research	(33)	W184
$C_{10}H_{16}O_2$	Ketone, ether	Cyclobutane	(12)	T378
	Diene, ester	Synth. pheromone	(93)	T168
$C_{10}H_{16}O_3$	Unsat. keto-acid	Insect hormone	(9)	W313
			(55)	W158
	Unsat. ketone, acetal	Manuf. int.	(64)	T363
	Ketone, acetal	Prostaglandin int.	(16)	T233
	δ-keto-acid	Cyclohexanone	(19)	T173
	Epoxy-ketone	Spiro, cyclohexane	(22)	W359
$C_{10}H_{16}O_4$	Acid, lactone	Struct. synth. int.	(1)	T315
$C_{10}H_{17}NO_3$	Ester	Heterocycle	(32)	T247
$C_{10}H_{17}O_3$	Keto-ester	Cyclopentanone	(39)	W320
	Cyclic unsat. alcohol	Juvabione int.	(43)	T324
$C_{10}H_{18}O$	Cyclohexanone	Stereochem. Study	(26)	T297
	Cyclic ketone	10-Membered ring	(33)	W277
$C_{10}H_{18}O_2$	Acetal	Multistriatin	(3)	T2,
		(20) T99, (28) T352,		
			(8)	W2
	Unsat. cyclic diol	Steroid int.	(25)	T112
$C_{10}H_{18}O_3$	γ-keto-ester	Synth. int.	(10)	W191
$C_{10}H_{18}O_4$	Diacid	Synth. int.	(25)	W274
$C_{10}H_{19}NO_2$	Diene, amine, ester	Synth. int.	(10)	W74
$C_{10}H_{20}O_2$	β-Hydroxy-ketone		(19)	T148
	Cyclic, 1,2-Diol	Synth. int.	(1)	W252
$C_{10}H_{21}N$	Amine, cyclohexane	Decongestant	(14)	W421
$C_{10}H_{21}NO$	Amino-ether	Cycloheptane	(39)	T49
$C_{10}H_{22}$	Alkane	Struct. int.	(34)	W277
$C_{10}H_{23}NO$	δ-Amino-ether	Anti-malarial	(3)	W124
	ϵ-Amino-ether	Anti-malarial	(21)	W103
$C_{11}H_9N_3O_2$	Nitro-imidazole	Anti-cancer drug	(96)	T343
$C_{11}H_{10}O$	Ar ketone	Cyclopropane	(34)	T306
$C_{11}H_{10}O_4S$	Phenol, ketoacid	Tetracycline int.	(26)	W196

$C_{11}H_{11}ClO_3$	Ar ether, acid	Analgesic	(14)	W42
	Ar keto acids	Fungicide ints.	(26)	T39,
			(1)	W38
$C_{11}H_{12}ClNO_2$	Ar amino-ester	Anti-tumour drug	(26)	T46
$C_{11}H_{12}Cl_2N_2O_5$	Nitro Ar diol, amide	Antibiotic	(8)	W243
$C_{11}H_{12}N_2O$	Heterocyclic amide	Analgesic	(56)	W347
$C_{11}H_{12}O$	Ar ketone	Cycloheptenone	(53)	T208
	Ketone	Cyclopropane	(30)	W361
	Enone		(1)	T152
$C_{11}H_{12}O_2$	Ar ketone, ether	Synth. int.	(47)	T206
	Ar ketone, ether	Tetracycline int.	(1)	W429
	Ar keto-aldehyde		(80)	T167
$C_{11}H_{12}O_3$	Tricyclic anhydride	Cyclohexene	(23)	W182
$C_{11}H_{13}ClO_2$	Ar ester	Stereochem. Studies	(43)	T50
$C_{11}H_{13}NO_4$	Diester	Pyridine	(55)	T335
$C_{11}H_{14}O$	Spiro enone	Synth. int.	(8)	W285
	Ar ketone	Perfume	(3)	T17
$C_{11}H_{14}O_2$	Bicyclic enedione	Anti-tumour int.	(9)	W399
	Phenol, aldehyde	Mech. Study	(10)	W19
$C_{11}H_{14}O_3$	Tricyclic keto-ester	Cyclobutanes	(21)	T271
$C_{11}H_{14}O_6$	Diene, triester	Mech. Study	(38)	T193
$C_{11}H_{15}ClO_2$	Ar 1,2-diol	Epilepsy drug	(17)	T188
$C_{11}H_{15}N$	Ar cyclopentylamine	Potential analgesic	(49)	T105
	Ar amine	Piperidine	(7)	W190
$C_{11}H_{15}NO_2$	Ar ester	Pyridine	(58)	T336
$C_{11}H_{16}O$	Tricyclic ketone	Cyclobutane	(22)	T305
	Tricyclic ketone	Cyclobutane	(12)	T303
	Bicyclic enone	Cyclohexenone	(24)	W235
$C_{11}H_{16}O_2$	Bicyclic ester	Enzyme model study	(24)	W402
	Bicyclic hydroxy-enone	Anti-tumour int.	(27)	W237
$C_{11}H_{16}O_3$	Tricyclic ketone, acetal	Cyclopropane	(8)	T303
	Bicyclic keto-ester	Conformation	(22)	W436
$C_{11}H_{16}O_4$	Unsat. diester	Alkaloid int.	(1)	T303
$C_{11}H_{16}O_6$	α-keto-aldehyde, diester	Synth. int.	(15)	T220
$C_{11}H_{17}NO_3$	Phenol, hydroxy-amine	Bronchodilator	(34)	T192
$C_{11}H_{18}N_4O_2$	Heterocyclic amide	Insecticide	(89)	T342
$C_{11}H_{18}O$	Bicyclic ketone	Cyclobutane	(1)	T268
	Enone	Synth. int.	(5)	T224

$C_{11}H_{18}O_2$	γ,δ-Unsat. ester	Cyclopentene	(8)	T260
	Diene, ester	Flavouring	(43)	W155
	Diene, ester	Insecticide int.	(59)	W224
	Spiroacetal	Cyclohexane	(1)	W45
$C_{11}H_{18}O_3$	Unsat. aldehyde, ester	Alkaloid int.	(17)	W232
$C_{11}H_{18}O_4$	Diketo-ester	Synth. int.	(1)	W323
$C_{11}H_{20}O_2$	Lactones (2)	Perfumes (19)	(20)	W287
	Enediol	Cyclopentane	(30)	W171
$C_{11}H_{21}O_2Br$	Acetal	Synth. int.	(11)	T70
$C_{11}H_{22}O_2$	Acid	Straight chain fatty acid	(16)	W326
$C_{11}H_{22}N_2$	Heterocycle	Pheromone	(44)	W464
$C_{12}H_{10}O_2$	Phenol	Struct. study	(5)	W18
$C_{12}H_{12}O$	Cyclic enone	Synth. int.	(20)	T200
	Tricyclic Ar ketone	Cyclopentenone	(8)	W409
	Ar dienal	Drug research	(39)	W217
$C_{12}H_{12}O_2$	Furan, alcohol	Insecticide int.	(29)	W330
	Cyclic enone, ether	Bullatenone	(54)	W223
	Cyclic enone, ether	Alleged bullatenone	(60)	T163
	Polycyclic diketone	Synth. int.	(19)	W446
$C_{12}H_{12}O_5$	Cyclic acetal anhydride	Synth. int.	(13)	T295
$C_{12}H_{13}NO_3$	Indole, acid	Pharm. int.	(48)	T334
$C_{12}H_{13}NO_3S$	Aldehyde, acid	Biotin int.	(10)	T110
$C_{12}H_{14}O$	Ketone, Ar		(20)	W373
$C_{12}H_{14}O_2$	Hydroxy-enone		(41)	W262
$C_{12}H_{14}O_3$	Ar epoxy-ester	Perfume	(21)	W358
	Ar ether lactone	Synth. int.	(47)	W294
$C_{12}H_{15}NO$	Amino-ketone	Pharm. int.	(43)	T249
$C_{12}H_{15}NO_2$	Unsat. nitrile, ether		(68)	T226
$C_{12}H_{15}NO_3$	Amide, ether	Insecticide	(23)	W412
$C_{12}H_{16}ClNO_2$	Amide, ether	Herbicide	(12)	W66
$C_{12}H_{16}N_2O_5$	Nitro Ar ether	Perfume	(4)	T17
$C_{12}H_{16}O$	Tricyclic ketone	Cyclopropane	(22)	T255
	Bicyclic ketone	Insecticide int.	(33)	W414
	Bicyclic dienone	Steroid int.	(33)	W215
	Unsat. Ar ether	Mech. Study	(19)	T32
	Ar ketone	Mech. Study	(14)	W12
$C_{12}H_{16}O_2$	Cyclic acetal		(14)	T244
	Ar diol	Rearrangement int.	(19)	W373
	Ether, ketone		(31)	W143
	Hydroxy-aldehyde	Synth. int.	(13)	T90
	Enone, ester	Cyclohexenone	(21)	W234

$C_{12}H_{16}O_3$	Diketoaldehyde	Synth. int.	(19)	W233
	Enone, ester	Synth. int.	(21)	W234
$C_{12}H_{16}O_4$	Acetals	Perfume	(8)	W47
$C_{12}H_{17}F_3N$	Ar amine	CNS drug	(21)	T63
$C_{12}H_{18}N_2O_2S$	Heterocyclic Amide	Barbiturate	(12)	T280
$C_{12}H_{18}O$	Ar ether	Perfume	(18)	T31
$C_{12}H_{18}O_2$	Tricyclic hydroxy-ketone	Cyclobutanone	(48)	W483
$C_{12}H_{18}O_4$	Ar ether, diol	Antibiotic int.	(19)	T110
$C_{12}H_{20}$	Alkyne	Eight-membered ring	(40)	T203
$C_{12}H_{20}O_2$	Ester	Synth. insect attractant	(11)	T134
	β-Hydroxy-ketone	Diene	(21)	W213
$C_{12}H_{20}O_3$	Keto-ester	Cyclohexanone	(40)	W426
$C_{12}H_{20}O_4$	Acetal, ester	Synth. int.	(53)	W295
$C_{12}H_{22}O$	Diene, alcohol	Pheromone	(38)	W154
	Enal	Enzyme inhibitor int.	(26)	W207
$C_{12}H_{22}O_2$	γ,δ-Unsat. ester	Synth. int.	(31)	W413
$C_{12}H_{22}O_2$	Lactone	Pheromone	(1)	W107
$C_{12}H_{22}O_6$	Diester, diether	Synth. int.	(11)	T225
$C_{12}H_{24}O$	Unsat. alcohol		(9)	T89
$C_{12}H_{24}O_2$	Ester		(23)	T112
$C_{12}H_{25}N$	Amine	Piperidine	(12)	T62,
			(5)	W189
$C_{12}H_{26}O$	Alcohol		(4)	T87
$C_{13}H_9ClO_2$	Ar acid	Liquid crystal study	(8)	T19
$C_{13}H_{10}Cl_2S$	Sulphide	Acaricide	(21)	T32
$C_{13}H_{10}O_4$	Enone, lactone	Synth. int.	(87)	T168
$C_{13}H_{12}$	Ar alkane	Perfume	(13)	W12
$C_{13}H_{12}O_5$	Lactones	Germination stimulant	(4)	W472
$C_{13}H_{13}NO_3$	β-Diketone, lactam	Synth. int.	(24)	T277
$C_{13}H_{14}O$	Tricyclic Ar ketone	Cyclopentenone	(6)	W409
	β,γ-Unsat. ketone	Synth. int.	(27)	T286
$C_{13}H_{14}O_2$	Cyclic enone, Ar	Synth. int.	(26)	W250
$C_{13}H_{15}ClO_2$	Cyclic ketone, Ar ether	Synth. int.	(17)	W84
$C_{13}H_{16}F_3N_2O_4$	Nitro Ar amine	Herbicide	(17)	T13
$C_{13}H_{16}O$	Ketone, Ar	Cycloheptanone	(10)	W371
	Unsat. Ar ether	Synth. int.	(1)	W145
$C_{13}H_{16}O_2$	Ketone, ether, Ar	Synth. int.	(29)	T236
$C_{13}H_{16}O_3$	Ketone, ether, Ar	Steroid int.	(13)	T285
	Ar ethers	Insect hormone	(39)	W463

$C_{13}H_{16}O_4$	Ar ketone, acetal	Synth. int.		(45)	T160
	Ar ε-keto-acid	Synth. int.		(1)	W311
	Diacid	Synth. int.		(16)	W315
$C_{13}H_{16}O_5$	Enal, acetals	Prostaglandin int.		(21)	W316
$C_{13}H_{17}NO_2$	Amino-lactone	Anaesthetic int.		(56)	W265
	Ar amine	Alkaloid int.		(33)	T287
$C_{13}H_{18}N_2O_2$	Heterocyclic amide	Herbicide		(58)	W488
$C_{13}H_{18}O$	Ketone, Ar			(9)	T116
	Ketone, Ar			(7)	T115
$C_{13}H_{18}O$	Hydroxy-ether	Rearrangement Study		(29)	T47
$C_{13}H_{18}O_4S$	Diol mono-tosylate	Mech. Study	(35)	(36)	T102
$C_{13}H_{19}NO_2$	Hydroxy-amine	Pharm. int.		(2)	T87
$C_{13}H_{19}NO_3$	Amine, ethers	Anti-depressant		(47)	T357
$C_{13}H_{20}Cl_2O$	Spiro cyclobutan-enones	α-Haloketones		(14)	W392
$C_{13}H_{20}O$	Bicyclic enone	Cyclopentene		(28)	W478
$C_{13}H_{21}N$	Amine	Stereochem. Study		(5)	T94
$C_{13}H_{21}NO_2$	α-Amino-acid	Analogue		(4)	W108
$C_{13}H_{21}NO_3$	Triol, amine	Salbutamol(19)T57		(14)	T254
$C_{13}H_{22}O$	Dienone	Terpene		(57)	T291
$C_{13}H_{22}O_2$	Alkyne, ester	Perfume		(1)	W188
	Acid	Pharm. int.		(41)	T266
$C_{13}H_{24}O$	Ketone	Synth. Study		(16)	T79
$C_{13}H_{24}O_3$	β-keto-ester	Enzyme inhibitor int.		(27)	W207
$C_{13}H_{25}NO_4$	Amino-diester	Synth. int.		(42)	W59
$C_{13}H_{29}NO_4$	Amino-diacetal	Synth. int.		(10)	T171
$C_{14}H_{12}O_2$	Ar α-hydroxy-ketone	Benzoin		(21)	T189
	Ester	Insect repellant		(3)	T26
$C_{14}H_{12}O_3$	Unsat. lactone	Natural product		(8)	T253
	Triketone	Rat poison		(9)	T146
	Unsat. keto-ester	Mech. Study		(15)	W43
$C_{14}H_{18}N_2O_5$	Dipeptide	Sweetening agent		(16)	T72
$C_{14}H_{18}O$	Ar ketone	Cyclopentenone		(5)	W408
	Ar enal	Perfume		(30)	T151
$C_{14}H_{18}O_2$	Ar ketone	Allomone int.		(2)	T301
	Acid, Ar	Cyclohexane		(8)	W126
$C_{14}H_{20}N_2O_5$	Nitrophenol	Fungicide int.		(17)	T24
$C_{14}H_{20}O$	Ether, Ar	Cyclohexane		(10)	W32
$C_{14}H_{20}O_5$	Keto-diester	Anti-tumour int.		(1)	T229
$C_{14}H_{21}NO$	Hydroxy-amine	Analgesic int.		(41)	T103
$C_{14}H_{22}$	Ar alkane			(12)	T30
	Diene	Mech. Study		(1)	W284
$C_{14}H_{22}O$	Tricyclic ketone	Terpene int.		(35)	W362
$C_{14}H_{22}O_2$	Enone, ether	Synth. int.		(32)	W450

$C_{14}H_{22}O_3$	Bicyclic hydroxy-ester	Alkaloid int.	(4)	W418
	Tricyclic hydroxy-acetal	Anti-tumour int.	(53)	W366
$C_{14}H_{22}O_5$	Keto-diester	Synth. int.	(26)	T174
$C_{14}H_{23}NO$	Amine, Ar ether	Drug analogue	(12)	W74
$C_{14}H_{24}O_2$	Hydroxy-ketone	Juvabione int.	(31)	W425
$C_{14}H_{26}O$	Ketone	Cyclohexane	(21)	W140
$C_{14}H_{26}O_2$	Alkyne, diol	Surfactant	(7)	T127
	Unsat. ester	Pheromone	(16)	T129
$C_{14}H_{26}O_3$	Anhydride		(24)	T81
$C_{15}H_{13}ClNNaO_3$	Pyrrole, acid salt	Analgesic	(12)	W455
$C_{15}H_{14}O$	Ketone, Ar	Cyclobutanone	(50)	W387
	Tricyclic alcohol	Synth. int.	(8)	W99
$C_{15}H_{14}O_2$	Acid, Ar		(24)	T112
$C_{15}H_{16}O_2$	1,2-Diol, Ar		(43)	T194
$C_{15}H_{17}N$	Amine, Ar	Stereochem. Study	(1)	W70
$C_{15}H_{17}NO_2$	Heterocyclic ester	Alkaloid int.	(23)	W459
$C_{15}H_{18}O_6$	Ester, lactone, ethers	Synth. int.	(11)	T232
$C_{15}H_{19}NO$	Heterocycle	Synth. int.	(63)	W348
$C_{15}H_{19}NO_2$	Cyclic β-amino-acid	Analgesic	(29)	T137
	Amine, acetal	Alkaloid	(34)	T354
$C_{15}H_{19}ClO$	Ketone, cyclohexane	Photochem. Study	(32)	W198
$C_{15}H_{20}O$	β,γ-Unsat. ketone	Photochem. Study	(22)	T149
$C_{15}H_{20}O_3$	Ketone, epoxides	Pheromone	(63)	W488
$C_{15}H_{20}SO_3$	Bicyclic tosylate	Mech. Study	(33)	W383
$C_{15}H_{21}N$	Bicyclic amine	Stimulant	(18)	T182
$C_{15}H_{21}NO_5$	Pyrrole keto-diester	Synth. int.	(8)	W454
$C_{15}H_{22}$	Triene	Vit D Model	(42)	W219
$C_{15}H_{22}O$	Tetraene aldehydes	Sinensals	(56) (57)	and T362
$C_{15}H_{22}O_2$	Bicyclic enedial	Anti-feedant	(39)	W185
$C_{15}H_{22}O_3$	Ketofuran ether	Stress metabolite	(39)	W333
$C_{15}H_{23}ClO_2$	Diol, Ar	Indust. reagent	(6)	W10
$C_{15}H_{23}NO$	Amide, Ar	Herbicide	(3)	W71
$C_{15}H_{23}NO_4$	Pyrrole diester	Porphyrin int.	(15)	T328
$C_{15}H_{23}NO_4S$	Amino tosylate	Mech. Study	(43)	W262
$C_{15}H_{24}$	Triene	Perfume	(10)	T348
	Tricyclic alkene	Flavour	(28)	T320
	Tricyclic alkene	Pheromone	(34)	W362
	Bicyclic diene	Flavour	(69)	W490
$C_{15}H_{24}O$	Phenol	Anti-oxidant	(5)	T8
$C_{15}H_{25}O_3N$	Hydroxy-amine	β-Blocker	(34)	W260

$C_{16}H_{11}N_3O_6S$	Nitro, azo, sulphonate	Soluble dye	(17)	T38
$C_{16}H_{12}Br_2O_2$	α-Diketone	Mech. Study	(17)	T199
$C_{16}H_{12}ClNO_3$	Heterocyclic acid	Anti-inflammatory	(53)	W467
$C_{16}H_{13}ClN_2O$	Heterocyclic amide	Pharm. int.	(48)	T250
$C_{16}H_{14}O$	Ar unsat. epoxide	Synth. int.	(31)	W361
$C_{16}H_{15}Cl_3O_2$	Ar ether	Biodegradeable DDT	(2)	W8
$C_{16}H_{16}O$	Cyclic ether, Ar		(52)	W346
$C_{16}H_{16}O_2$	Lactone	Synth. int.	(30)	T222
			(13)	W286
$C_{16}H_{18}$	Ar alkane		(41)	W280
$C_{16}H_{18}O_2$	Diol, Ar		(4)	W98
$C_{16}H_{20}O$	Bicyclic ketone, Ar	Stereochem. Study	(19)	W140
$C_{16}H_{21}NO_2$	Hydroxy-amine	Heart drug	(51)	W485
$C_{16}H_{22}O_4$	Tricyclic di-ester	Synth. int.	(6)	T133
$C_{16}H_{24}O_2$	β-Diketone	Cyclobutandione	(18)	T277
$C_{16}H_{26}N_2O_3$	Diamino-ester	Anaesthetic	(22)	T45
$C_{16}H_{30}O$	Diene, alcohol	Pheromone	(50)	W156
$C_{16}H_{30}O_2$	Unsat. acid	Synth. int.	(6)	W231
$C_{16}H_{32}O_2$	Hydroxy-ketone	Pheromone int.	(22)	T227
$C_{17}H_{12}I_2O_3$	Phenol, Ar ketone	Vasodilator	(27)	T330
$C_{17}H_{15}NO_3$	Amide, acid	Analgesic	(42)	W344
$C_{17}H_{16}O$	Ar ketone	Cyclopropane	(4)	W369
	Cyclobutane, ketone	Photochem. Study	(38)	W385
	Tricyclic ketone	Cyclopentenone	(1)	W407
$C_{17}H_{18}O$	Ar ketone	Pharm. int.	(11)	W127
	Ar ketone	Stereochem. Study	(6)	W136
	Ketone, Ar		(1)	W134
$C_{17}H_{20}S_2$	Thiol, thioether	Pharm. int.	(12)	T37
$C_{17}H_{22}N_2$	Amino-pyrrole	Cyclopentene	(5)	W453
$C_{17}H_{22}O_2$	Tricyclic ester	Cyclopentene	(18)	W434
$C_{17}H_{23}NO_2$	Amine, ester	Pharm. Study	(69)	W350
$C_{17}H_{25}NO_2$	Amine, ester	Pharm. Study	(5)	W30
			(6)	T77
	Amine, ester	Pharm. Study	(38)	W363
$C_{17}H_{28}N_2O_3$	Amines, ester, ether	Anaesthetic	(17)	W35
$C_{17}H_{28}O_4$	Bicyclic diester	Terpene int.	(12)	T270
$C_{17}H_{30}O$	Dienal	Pheromone	(5)	W441
$C_{18}H_{16}O$	Cyclic enone	Synth. int.	(39)	W426
$C_{18}H_{18}O_3$	Ketone, Ar ethers	Veterinary research	(38)	W291

Formula	Type	Application	(Ref)	Code
$C_{18}H_{18}O_4$	Ester, acid	Resolution	(6)	T28
$C_{18}H_{18}O_2$	Ether, ketone	Pharm. Study	(45)	T309
	Spiro enone	Synth. int.	(22)	T235
$C_{18}H_{20}O_3$	Ester	Pharm. research	(8)	W370
$C_{18}H_{22}ClNO$	Amino ether	Anti-histamine	(32)	T84
$C_{18}H_{22}O_2$	Polycyclic ketone	Terpene int.	(27)	T320
	Diol	Rearrangement Study	(25)	W257
$C_{18}H_{23}NO$	Amino ether	Parkinson drug	(23)	W53
$C_{18}H_{24}N_2O_6$	Dinitro Ar ester	Fungicide	(18)	T24
$C_{18}H_{24}O$	Bicyclic ketone	Mech. Study	(1)	W376
$C_{18}H_{24}O_4$	Diketo acid	Fungal metabolite	(15)	T350
$C_{18}H_{26}O$	Tricyclic Ar ether	Natural product	(2)	T347
$C_{18}H_{30}O$	Enone	Cyclopentanone	(28)	W214
$C_{18}H_{30}O_2$	Triene ester	Hormone int.	(14)	W474
$C_{18}H_{32}O_2$	Diene, ester	Pheromone	(21)	W167
$C_{19}H_{16}O_4$	Keto lactone, Ar	Rat poison	(36)	W462
$C_{19}H_{17}NO_4$	Amine, acetals	Alkaloid	(72)	T338
			(27)	W459
$C_{19}H_{20}O$	Cyclohexanone	CNS drug int.	(22)	W140
$C_{19}H_{20}O_2$	Alkyne diol	Mech. Study	(30)	W259
$C_{19}H_{21}NO_6$	Amide, acetal	Alkaloid int.	(1)	W241
$C_{19}H_{24}N_2$	Diamine	Anti-histamine	(14)	W76
$C_{19}H_{26}O_2$	Keto enone	Steroid	(30)	T175
$C_{19}H_{26}O_5$	Keto diester	Synth. int.	(1)	W229
$C_{19}H_{28}N_2$	Amine, indole	Anti-depressant	(43)	T333
$C_{19}H_{35}NO_2$	Amino ester	Anti-spasmodic	(20)	T296
$C_{19}H_{38}O$	Epoxide	Pheromone	(19)	T123
	Unsat. alcohol	Perfumery research	(17)	W102
$C_{20}H_{18}O_2$	Ar hydroxy-ether		(9)	W99
$C_{20}H_{21}NO_5$	Nitrophenyl ether	Enzyme Mech Study	(32)	W152
$C_{20}H_{25}NO$	Amino alcohol	Muscle relaxant	(13)	T78
	Amino alcohol	Spasmolytic	(28)	T191
$C_{20}H_{26}O$	Diphenol	Oestrogen	(45)	W281
$C_{20}H_{26}O_4$	Diester	Cyclohexene	(56)	T162
$C_{20}H_{27}NO_2$	Amine, Ar ethers	Uterus relaxant	(9)	W472
$C_{21}H_{20}O_3$	β-keto-ester	Cyclohexenone	(28)	T175
$C_{21}H_{22}O_7$	Phenol, keto-ester	Tetracycline int.	(3)	W431
$C_{21}H_{25}NO_2$	Amino ester	Anti-spasmodic	(53)	W486
$C_{21}H_{27}NO$	γ-Amino-ketone	Analgesic	(23)	W305
$C_{21}H_{28}O_2$	Polycyclic ether	Anti-oestrogen	(33)	W341
$C_{21}H_{29}NS_2$	Amino sulphide	Tranquiliser	(11)	T37
$C_{22}H_{19}NO$	α-Amino ketone	Synth. int.	(16)	W68
$C_{22}H_{22}O_4$	Diene, ester	Oestrogen	(26)	T201
$C_{22}H_{29}NO_2$	Amino-ester	Analgesic	(47)	W220
$C_{22}H_{44}O_2$	Acid		(17)	T109
$C_{23}H_{16}O_{11}$	Lactone, acid	Asthma drug	(57)	W486
$C_{23}H_{21}O_2P$	Ester, ylid	Synth. int.	(9)	W270

$C_{23}H_{46}$	Alkene	Synth. int.	(36)	W153
$C_{24}H_{16}N_2$	Ar unsat. nitriles	Whitener	(16)	T123
$C_{24}H_{22}O_5$	Acetal, ketone, ether	Synth. int.	(21)	W250
$C_{24}H_{38}O_6$	Acetal, enone, ester	Synth. int.	(19)	W476
$C_{25}H_{25}NO_3$	Ar amide, ethers	Alkaloid int.	(7)	W31
$C_{27}H_{42}ClNO_2$	Ether, ammonium salt	Antiseptic	(11)	W49
$C_{28}H_{24}O$	Cyclic ether	Five-membered ring	(5)	T88
$C_{30}H_{26}$	Ar diene	Lubrication research	(6)	W312

INDEX

An entry without qualification usually means that the synthesis of that compound is referred to. Entries are mostly the same as those in the text.

Acetals, 45 ff., 194, 206, 297.
Acid chlorides, ketones from, 123, 138, 191, 289.
Acorone intermediate, 285.
Acyl anion equivalents, 252, 256, 290.
Acylation, of amines, 27, 241.
    See also Friedel-Crafts reaction.
Acyloin reaction, silicon modification, 384, 402, 404.
Alanine, 263.
Alclofenac, 42.
Alcohols, 89 ff.
Aldehydes, 19, 42, 184, 232, 271, 396, 444.
Alkenes,
    1,2-difunctionalised compounds from, 257, 373.
    exocyclic, 284, 391.
    from alkynes, 165, 219, 314.
    from elimination reactions, 145 ff., 280.
    Wittig route to, 150 ff.
Alkylation,
    of amines, 74, 76, 103.
    of cyanides, 126.
    of enols, 126 ff., 134 ff., 164, 194.
    See also Friedel-Crafts reaction, Malonates.
Alkynes, 163, 188.
    anions from, 160 ff., 188, 254.
    hydration of, 172 ff.. 252, 255.
    as reagents for vinyl anion synthon, 165 ff.
Allylic bromination, 267 ff.
Allylic halides,
    alkylation by, 43, 73, 131, 306, 310, 357, 385, 413.
    as masked α-carbonyl cations, 304.
Allylic inversion, 414.
Allylic oxidation, 273.
Aluminium i-propoxide, 366.
Ambucaine, 35.
Amides,
    as protecting group, 25, 26, 28, 81, 245.
    from amines, 27, 241.
    from sulphonyl halides, 28.
    reduction of, 70 ff., 189.

Amines,
 from amides and imides, 70 ff., 189.
 from imines, 70 ff., 124.
 from nitriles, 73.
 from nitro compounds, 241, 421.
 from oximes, 130.
 from phthalimides, 73.
α-Amino acids, 60, 301.
Amphetamine analogues, 242.
Ampicillin, 31.
Anaesthetics and Analgesics,
 Alclofenac, 42.
 Ambucaine, 35.
 Cyclomethycaine, 57.
 Darvon, 220.
 Indoprofene, 344.
 Methadone, 305.
Anhydrides, 179, 182, 264, 426.
 cyclic, 29, 289, 368, 443.
 in Friedel-Crafts reaction, 39, 44.
Antihistamine drugs, 76, 90, 363.
 intermediate, 458.
Antimalarial drugs, 104, 124.
Antipyrine, 347.
Antitumour drugs, 366, 399.
Arndt-Eistert synthesis, 370.
Aromatic substitution,
 electrophilic, 8 ff. - direction and activation,
   15 ff., 33, 36, 197.
 nucleophilic, 22.
Aspartic acid, 301.

Baeyer-Villiger reaction, 317 ff., 379, 393-4, 445.
Bayluscide, 26-7.
Benomyl, 31.
Benzethonium chloride, 49.
Benzofurans, 457 ff.
Benzoxazole, 467-9.
Benzylic halides, 51.
 reactivity of, 32, 49, 55, 271.
Benzyne, 336.
BHT, 17.
Biotin intermediate, 110.
Birch reduction, 424, 427.
Bombykol, 156.
Boron trifluoride, 56.
Bowl conformation, 442.
Brevicomin, 257.
Bromination,
 allylic, 267 ff.
 aromatic, 20.
 of alkenes, 65.

α-Bromoesters, 65, 225, 296, 317, 425.
N-Bromosuccinimide, 69, 267 ff.
Bs, *p*-bromobenzene sulphonate, 423.
Bulnesol intermediate, 84.
Butam, 71.
cis-Butenediol, 351.
Butoxamine, 260.

Cadmium compounds, 123, 138, 191, 289, 356, 370.
Cannizarro reaction, 205, 206.
Captan, 29.
Carbanions, 89.
Carbenes, halo, 365, 382.
Carbofuran, 412.
Carbon acids, pKa of, 162.
Carbon dioxide, 93, 94.
Carbonium ions, 50, 74, 307, 372, 374.
α-Carbonyl cation, synthon and reagent for, 284, 304.
Carbonyl compounds,
    by alkylation of enols, 126 ff., 134 ff., 164, 194.
    by 1,1 C-C disconnection, 123 ff., 138, 191, 199.
    by Michael addition, 132 ff., 139 ff.
Carbonyl derivatives RCOX, 29 ff.
Carroll rearrangement, 415.
Carboxylic acids, 19, 27, 92 ff., 163, 390.
    by carboxylation of Grignard reagents, 92 ff., 105.
    from nitriles, 92.
Carvone, 339.
Catechol, 48.
Chain extension, 321, 368, 382.
    see also Arndt-Eistert synthesis.
Chalcogran, 296.
Chemoselectivity, 38 ff., 62, 82, 87, 112, 255, 304,
        339, 367, 456.
    guidelines for, 39.
    in hydrogenations, 78, 87, 187.
Chloracetic acid, alkylation of phenols by, 100.
    ester, 60.
Chloracetyl chloride, 63, 67.
Chloral, 9.
Chloramphenicol, 243.
Chlorination, aromatic, 10, 12.
Chloromethylation, 241.
Chromium trioxide, 172, 193, 259, 296, 310, 322, 432.
    CrO_3 pyridine (Collins), 111.
    CrO_3 acetone H_2SO_4 (Jones), 41.
Chrysanthemic acid, ester, 224.
    intermediate, 414.
Citronellol, 96.
Claisen ester condensation, 200 ff.
Claisen rearrangement, 413.
Clemmensen reduction, 293, 342.

Collins reagent, 111.
Common atom approach to polycyclic compounds, 435 ff.
Computer design, 470.
Conformational control in 6-membered rings, 119 ff.,
    142, 198, 440, 447, 450.
Convergence, 470.
Copper compounds, 2.
    Cu(I) in aromatic nucleophilic substitution, 21, 25.
        in Michael reactions, 139, 141, 142, 160-1,
        424.
    Cu(II) in oxidation of alcohols, 132,
Cresols, 424,
Cross condensations, 216 ff.
Cubebene, 362.
Cyanides - (nitriles), 274.
    alkylation of, 135, 137.
    ketones from, 124, 125.
    reduction of, 78, 79, 408.
Cyanoacetate, 227, 404, 408.
Cyanohydrins, 140.
Cyclisation reactions, 67, 115, 190, 196, 209, 211, 215,
    255, 293, 335 ff., 353, 429.
Cyclisation to control selectivity, 209 ff., 307, 310.
Cycloaddition,
    1,3-dipolar, 348 ff.
    2+2, 376 ff., 389.
Cyclobutanones, 384, 387, 389 ff., 438, 483.
Cycloheptanone, 158, 174, 235, 371, 479.
Cyclohexenones, 209, 379.
Cyclomethycaine, 57.
Cyclopentannelation, 304.
Cyclopentadiene, 310.
Cyclopentenones, 397 ff.
Cyclopropane acids, 363.
Cyclopropanes, 355, 364.
Cyclopropanones, 375.
Cyclopropyl ketones, 353 ff., 359, 369, 433, 437.

Darvon, 220.
Darzens reaction, 340, 358.
DBN, diazabicyclononene, 150, 448-9.
DBU, diazabicycloundecene, 150.
DCC, dicyclohexylcarbodiimide, 109, 226.
DDQ, dichlorodicyanoquinone, 417-8.
Dealkylation, 19.
Deamination, 19, 26.
Decarboxylation, 457.
    of β-Keto acids, 128.
    of malonic acids, 129 ff.
Demeton, 51.
Diazinone, 464.

Diazoacetates, 363.
Diazoalkanes and diazoketones, 362, 368, 370, 371, 415, 432.
Diazotisations, 12, 21, 25, 26, 37, 88.
DIBAL, 446.
Dibenzoyl peroxide, 267 ff., 355.
Diborane, 75, 111.
Dicarbonyl compounds,
    1,2- 264, 401.
    1,3- 200, 208, 223.
    1,4- 284 ff., 397 ff., 453.
    1,5- 232, 359.
    1,6- 158, 311 ff., 390.
Diels-Alder reactions, 175 ff., 198-9, 206, 216, 227, 251, 308, 310, 314, 316, 324, 326, 346, 368, 426, 432, 449-50.
    endo selectivity of, 180.
    regio selectivity of, 183, 310.
    stereochemistry of, 178, 418.
Dienes, 276, 284, 312, 490.
    by Wittig reaction, 154 ff., 220.
    from alkynes, 165.
    See also Diels-Alder reaction.
Diethyl carbonate, 209, 227, 357, 400 (dimethyl).
Difunctionalised compounds, 58 ff.
    1,1- 45 ff.
    1,2- 49 ff., 76, 80 ff., 90, 252 ff., 363.
    1,3- 55 ff., 77, 78, 104, 115, 131, 196, 200 ff., 205, 218, 265.
    1,4- 284 ff., 299 ff., 397.
    1,5- 229 ff.
    1,6- 311 ff.
Dihydroxycitronellal, 83.
Diketocoriolin, 399.
Diketones, 290.
    1,2- 264, 275, 399, 401.
    1,3- 223, 227, 238.
    1,5- 233.
Dimedone, 403.
Dimerisations, 274, 275.
    see also Pinacols.
Dimethyl sulphate, 34, 254, 295.
Diols, 10, 98, 131, 185, 205, 243, 252, 254, 257, 275, 373.
1,3-Dipolar cycloadditions, 348-9.
Disconnections,
    guidelines for, 5-7, 15, 96.
    1,1 C-C, 89 ff., 282.
    1,2 C-C, 91 ff.
    1,3 C-C, 132, 135, 140, 193, 282, 424.
    two-group C-X, 45 ff.
Double bond cleavage, 299 ff.

Elasnin, 207.
Electrochemical dimerisation, 274.
Electrocyclic reactions, 407 ff.
Electrophiles, one carbon, in aromatic syntheses, 10, 11, 21.
Elimination reactions,
    alkenes from, 145 ff.
    ketenes from, 389 ff.
    regioselectivity of, 147 ff.
Enamines, 228, 232-6, 286, 327, 380, 436, 454.
Enals, 207, 316.
Enol ether, 330.
Enolisation, regioselectivity in, 210 ff.
Enones, 193, 202, 209, 212, 217, 229, 235 ff., 239, 254, 262, 285, 304, 397.
Epichlorhydrin, 485-7.
Epoxides, 62, 113-5, 117-8, 170, 257, 339, 359, 374.
    chemoselectivity of formation, 62, 339.
    as reagents in synthesis, 65, 67, 286 ff.
    rearrangements of, 5, 372, 382, 387-8, 427.
    regioselectivity of reaction, 143, 263, 286.
    stereochemistry of reaction, 143-4, 442.
    see also ethylene oxide, MCPBA
Equol analogue, 291.
Ethylene oxide, 52, 77, 90, 92, 94, 97, 104, 168, 364.
Ethyl formate, 303, 332.
Ethyl orthoformate, 48, 84, 206.

Favorskii rearrangement, 405.
Fenchone, 437.
FGA, 277 ff., 295 ff., 326, 377, 383, 408, 425, 429, 478.
FGI on Diels-Alder adducts, 185.
Fischer indole synthesis, 457.
Fluorenone, 99.
Fragmentations, 116, 379, 446.
Friedel-Crafts reaction,
    acylation, 10, 14, 54, 127, 221, 261, 408.
    with cyclic anhydrides, 39, 43, 295, 430.
    aliphatic, 396, 399, 409.
    alkylation, 14, 21, 51, 91, 144.
    on aromatic heterocycles, 456.
Frontalin, 45, 193-5.
Functionalisation,
    allylic and benzylic, 267 ff.
    α-, of carbonyl compounds, 259.
    1,2- and 1,4- by C=C cleavage, 299 ff.
    1,4- without reconnection, 304.
Furans, 216, 330, 333.

Gabamide, 78.
Gascardic acid, 42.
Gelsemium alkaloid intermediate, 179.
Gin, flavour of, 490.
Glutamic acid, 297.
Gossyplure, 167.

Hagemann's ester, 434.
α-Halo carbonyl compounds, 65, 120.
Haloform reaction, 403.
Halogenation,
    of alkenes, 65.
    of esters, 65, 270.
    aromatic, 10, 12, 20, 27, 39.
    benzylic, 270.
Halo-lactonisation, 117, 119, 341, 448.
Hantszch synthesis, 454 ff.
Herbicides and weedkillers, 66, 71, 227.
Heterocycles,
    aliphatic, 4-ring, 337, 340.
            5-ring, 76, 335 ff., 342, 347.
            6-ring, 349.
            7-ring, 351.
    aromatic, 5-ring, 452.
            electrophilic substitution in, 456.
            benzofurans and indoles, 457.
            furans, 216, 330, 333.
            pyrroles, 453.
            thiophenes, 452.
            6-ring, 459.
            iso-quinolines, 459, 479.
            pyridines, 459, 466.
            with two hetero-atoms, 464.
            imidazoles, 466.
            oxazoles, 467.
            pyridazines, 464.
            pyrimidines, 464.
Histapyrrodine, 76.
Hydrazine, 73, 87.
Hydrocarbons, 277 ff.
Hydrogen peroxide, 81, 114, 275.
Hydroxylation, 258, 373.
Hydroxymethylation, 10.
Hypochlorous acid, 314.

Imidazoles, 466.
Imides, 29.
Imines, 410.
    cycloadditions of, 393.
    reduction of, 70 ff., 124.

Indoles, 457 ff.
Indoprofene, 344.
Insecticides, 51, 52, 224, 330, 412, 464.
Insertion reactions, 3-membered rings by, 358.
Intramolecular reactions, 215.
Iodolactonisation, 117, 119, 341, 448.
Ipomeamarone, 332-4.
Ipsenol, 426.
Isoniazide, 217.
Isoquinolines, 459, 479.
Iteration, 387.
Ivangulin, 366.

Jones reagent, 41.
Juvabione intermediate, 425.
Juvenile hormone, 62.

α-Kainic acid, 72.
Ketene dimers, 394.
Ketenes, 389 ff., 406, 438.
Ketones, 12, 52, 55, 125, 127, 132-137, 140, 173, 198, 213, 291, 342, 355, 362, 372, 373, 376, 408, 414-5, 429.
    aromatic, see Friedel-Crafts acylation.
    cyclic, reduction of, 367.
    from acids, 123.
    from alkynes, 172.
    from β-keto esters, 128, 139.
    from nitriles, 125, 135, 194.
    from nitro compounds, 248.
    from organo-cadmium compounds, 123, 138, 191, 289, 370.
    from pinacols, etc., 372.
Keto-acids, 158, 311, 313.
    1,4- 38.
    1,5- 380.
Keto-aldehydes, 233.
Keto-esters, 191, 207, 229, 320, 323, 395, 426.
Keto-lactam, 203.
Keto-lactones, 324, 441, 462.
Key reaction strategy, 477.
Knoevenagel reaction, 212, 270, 405.

Lactams, 337-8, 393.
Lactones, 170, 182, 231, 265, 286-7, 294, 307-10, 317-8, 341, 394-5, 444.
    by Baeyer-Villiger reaction, 379, 394, 445, 472.
    optically active, 297.
    use in synthesis, 330, 354.

Laevulinic acid, 295.
LDA, 399.
Lead tetra-acetate, 63, 328.
Lewis acids, 2, 12, 56, 310.
Lindlaar catalyst, 3, 167-9.
Lineatin, 381-2.
Lithium aluminium hydride reductions,
    of acids, 137, 195, 263, 315, 321, 346, 443, 446.
    of amides, 191.
    of esters, 42, 58, 83, 85, 112, 131, 167,
          226, 233, 237, 296, 326, 329,
          332, 342, 367, 385, 387,
          391, 423.
    of imines, 71.
    of ketones, 463.
    of oximes, 130.
    of unsaturated ketones, 404.
    of unsaturated nitro compounds, 242.
Lithium borohydride, 39.

Magnesium enolates, acylation on carbon, 224.
Maleic anhydride, 43-4, 308.
Malonates, 129, 131, 163, 195, 230, 239, 288, 302, 310,
        321, 359, 385-7, 391, 436.
Malonic acid, 270.
Mannich reaction, 58, 218 ff., 222, 237, 245, 293.
Marasmic acid, 176.
MCPBA (*meta*-chloroperbenzoic acid), 62, 115, 195, 257,
        317 ff., 446, 448.
Meparfynol, 161-2.
Mercuric salts in hydration of alkynes, 174, 252, 254-5.
Mesityl oxide, 290.
Mesyl chloride, 85, 233, 411.
Methadone, 305.
Methionine, 60.
Methyl damascanine, 33.
Methylenomycin, 441.
Mevalonolactone, 394.
Michael reactions, 58-9, 78, 139, 163, 205, 230, 234,
        239, 248, 265, 321, 329, 340, 343, 351,
        400, 424, 459.
Moniliformin, 391.
Morpholines, 67.
Multicholanic acid, 264.
Multistriatin, 2-5.
Muurolene, 490.

NBS - see N-Bromosuccinimide
Nef reaction, 251, 290.

Nitration, of acetanilides, 25-6.
          of aromatic ethers, 14, 19, 25, 34.
          of aromatic esters, 37.
          of chlorobenzenes, 28.
          of hydroxy benzoic acid, 37.
          of toluene, 12, 20.
Nitriles – see cyanides.
Nitro-compounds, aliphatic, 241 ff., 290.
     from oximes, 246.
     hydrolysis of, 249.
Nitrones, 349.
Nitrosation, 455, 458.

Oestrone ether, 341.
Opren, 467.
Optically active compounds, 107 ff., 129, 135, 221, 279, 441, 445.
Orphenadrine, 53.
Osmium tetroxide,
     cleavage of diol by, 316.
     hydroxylation by, 258, 373.
Oxalic acid, 425.
Oxazoles, 467.
Oxidation, allylic, 273.
Oximes, 130, 246, 300.
Ozone, 54,
     cleavage of C=C by, 159, 300 ff., 315, 324, 326, 390, 411, 432, 443.

Palanil, 270.
PCC, pyridinium chlorochromate, 42, 92, 231.
Pederamide, 169.
Peptide synthesis, 86-7.
Perfumes, 188.
     carnation, 127.
     civet, 440.
     gardenia, 127.
     geranium, 12.
     jasmine, 46.
     nutmeg, 33.
     peach, 287.
     pear, 154.
     strawberry 'aldehyde', 358.
Peristylane, 322.
Phentamine, 74.
Phentermine, 126.
Pheromones (and intermediates), 45-6, 107, 111, 128, 153, 167, 193-5, 257, 296, 327, 381, 424, 464, 488.
Phosgene, 82, 161, 345.
Phosphorus pentoxide, 53.

Photochemical 2+2 cycloadditions, 376.
    regioselectivity, 378.
Phthalimide, 73, 411.
Pinacol rearrangements, 372.
Pinacol reduction, 275, 373, 374.
β-Pinene, 339.
Piperonal, 151.
pKa of carbon acids, 162.
Polycyclic compounds, 435-9.
Polygodial, 185.
PPA, polyphosphoric acid, 293, 407, 409, 430, 435, 463.
Potassium permanganate,
    oxidation of alkyl groups, 12
    hydroxylations, 83.
Precocenes, 463.
Propargyl halides, 163-4, 172, 305.
Propellane, 377.
Propylhexedrine, 421.
Prostaglandin intermediates, 118-9, 316, 438.
Protecting groups, 80 ff., see also THP.
    for acids, 80.
    for alcohols, 112, 169.
    for aldehydes and ketones, 2, 41, 42.
    for amines, 25-8, 81, 245, 301.
    for thiols, 53.
    new, 80.
    use of, 83.
Pyrazines, 465.
Pyridines, 459 ff.
Pyrroles, 453 ff.
Pyruvic acid, 457.

Quassin intermediate, 449.
Queen bee substance, 158-9, 313.
Quinones, 414, 417-8.

Radicals, 267 ff., 335, 384.
Raney Nickel, 246, 248, 365.
Rearrangements, 368 ff.
    Carroll, 415.
    Claisen, 413.
    Cope, 413.
    Dienone-phenol, 17.
    Epoxide, 5, 372, 382, 387-8, 427.
    Favorskii, 405.
    Pinacol, 372.
    Sigmatropic, 407.
Reconnections, 158, 299 ff., 311 ff, 230, 381, 411, 442.
Redal, 243.
Reducing agent, hindered, 154.

Reduction,
 of acids, 111.
 of aldehydes, 104, 106.
 of alkenes, 339, 346, 405, 426.
 of alkynes, 3, 165 ff., 168-9, 171, 220, 298, 314.
 of amides, 75, 190-1.
 of aromatic rings, 421 ff., 427.
 of benzyl groups, 87, 109.
 of cyanides, 78-9, 408.
 of enones, 435.
 of halides, 197, 430.
 of ketones, 10.
 of imines, 71-2, 77, 124.
 of nitro groups, 12, 19, 25, 28, 34, 37, 246, 248.
 of oximes, 300.
 chemoselectivity of, 78.
 Lindlaar, 3, 167-9.
 partial, of dinitro compounds, 21, 39.
Reformatsky reaction, 225, 342, 425.
Regioselectivity, 134 ff.
 in alkylation of ketones, 134.
 in Diels-Alder reactions, 183, 310.
 in epoxide reactions, 143, 263, 286.
 in Michael reactions, 139.
Reimer-Tiemann reaction, 21.
Resolutions,
 of alcohols, 107.
 of acids, 137, 315.
Reversal of polarity, 65 ff.
Ring contraction, 373, 403, 405.
Ring expansion, 371, 373, 382, 388, 406, 410, 411.
Ring formation, factors affecting, 307, 310, 337, 341.
Rings, 3-membered, 353 ff., 433, 437.
  4-membered, 376 ff., 389 ff., 438.
  5-membered, 397 ff., 407 ff., 435, 437.
   from 1,4-diCO compounds, 397.
   from 1,5-diCO compounds, 401.
   from 1,6-diCO compounds, 399, 404.
  6-membered, 426.
   by carbonyl condensations, 417, 433.
   by Diels-Alder reactions, 418.
   by reduction of aromatic compounds, 421.
   by Birch reduction, 424.
Ritter reaction, 74-5.
Robinson annelation, 235-7, 380, 417, 427, 433.
Ruthenium tetroxide, 444.

iso-Saffrole, 150.
Salicylic ester, 423.
Santonin, 114.
Sarkomycin, 324-5.

Selenium dioxide, 273, 417-8.
Self-condensations, 212.
Semburin, 489.
Sigmatropic rearrangements, 409 ff., 412.
    -See also Claisen. Cope, and Carroll rearrangements.
Simmons-Smith reaction, 364, 367.
Sodium borohydride, 58, 77, 97, 106, 248, 250, 261, 289, 295, 335, 367, 411.
Sodium periodate, 444.
Specific enol equivalents, 223-5, 233, 239, 284.
Spiro compounds, 247, 285, 327, 359, 386, 392, 401, 426, 451.
Statistical effect, 40, 58.
Stereoselective reactions, 117, 364, 366, 386, 392, 423, 440, 448.
Stereospecific reactions, 111 ff., 114, 117.
Strategy, see Strategy Chapters I - XVII and especially Chapters 11, 28, 37, 40.
    branchpoint, 100 ff., 128, 135, 146.
    chain lengthening, 321, 368, 382.
    common atom, 311, 435 ff.
    convergence, 470.
    cyclisation, 67, 131, 190, 196, 211, 215, 429.
    FGA, 296, 326, 377, 383, 408, 425, 429, 478.
    key reaction, 477.
    reconnection, 299 ff., 311 ff., 330, 411, 442.
    simplification, 349, 356, 358, 371, 430.
    starting materials, 18, 27, 54, 261, 294, 485.
    symmetry, 13, 51, 63, 91, 99, 146, 177, 200, 274 ff., 316, 365, 376, 442, 445, 465.
Strawberry 'aldehyde', 358.
Strecker synthesis, 60-1, 108.
Sulphides, 51, 63, 80, 196, 453.
Sulphur ylids, 359, 360, 361.
Symmetry, strategy based on, 13, 51, 63, 91, 99, 146, 177, 200, 274 ff., 316, 365, 376, 442, 445, 465.
Synthesis to establish stereochemistry, 445, 488.
Synthons,
    acyl anion, 252, 256.
    amide anion, 73.
    α-carbonyl cation, 284, 304.

Tetracyclines, 22, 196-7, 429-431.
Thianaphthalene, 52.
Thiols, 22, 53.

Thionyl chloride, 37, 54, 77, 90, 104, 106, 123, 408-9, 415, 462.
Thiophene saccharine, 452.
THP, 112, 169, 298.
Thujaketone, 102.
Titanium trichloride, 248, 250, 398.
Toluene sulphonic acid, 107, 147, 206, 400.
Torreyol intermediate, 180.
Trienes, 219, 348, 417.
Trifluralin B, 17,
Trifluoroperacetic acid, 246.
Trimethylsilyl chloride, 256, 384, 449.
Trisporic acid intermediate, 295.

Unnatural (illogical) synthons,
    electrophilic, 259, 284.
    nucleophilic, 172, 252, 289, 320.
α,β-Unsaturated carbonyl compounds, 202 ff., 209, 212, 217, 224, 226. See also enones.

Vanadium perchlorate, 275.
Vernolepin analogues, 236.
β-Vetivone, 294.
Vinyl cyclopropane - cyclopentene rearrangement, 409.
Vinyl ketones, 3, 193.

Warfarin, 465.
Wittig reaction, 150 ff., 165, 217, 220, 222, 231, 270, 272, 285, 296, 304, 334, 390, 420.
    stereochemistry of, 151 ff., 296.
Wolf-Kishner reduction, 326, 378.

Zinc, 83.